Materials, Technology and Reliability for Advanced Interconnects and Low-k Dielectrics

MATERIALS RESEARCH SOCIETY
SYMPOSIUM PROCEEDINGS VOLUME 612

Materials, Technology and Reliability for Advanced Interconnects and Low-k Dielectrics

Symposium held April 23–27, 2000, San Francisco, California, U.S.A.

EDITORS:

G.S. Oehrlein
University of Maryland
College Park, Maryland, U.S.A.

K. Maex
IMEC
Leuven, Belgium

Y.-C. Joo
Seoul National University
Seoul, Korea

S. Ogawa
Matsushita Electronics Corporation
Kyoto, Japan

J.T. Wetzel
International SEMATECH
Austin, Texas, U.S.A.

Materials Research Society
Warrendale, Pennsylvania

CAMBRIDGE UNIVERSITY PRESS
Cambridge, New York, Melbourne, Madrid, Cape Town,
Singapore, São Paulo, Delhi, Mexico City

Cambridge University Press
32 Avenue of the Americas, New York NY 10013-2473, USA

Published in the United States of America by Cambridge University Press, New York

www.cambridge.org
Information on this title: www.cambridge.org/9781107413153

Materials Research Society
506 Keystone Drive, Warrendale, PA 15086
http://www.mrs.org

First published 2001
First paperback edition 2013

Single article reprints from this publication are available through
University Microfilms Inc., 300 North Zeeb Road, Ann Arbor, MI 48106

CODEN: MRSPDH

ISBN 978-1-107-41315-3 Paperback

CONTENTS

*Invited Paper

*Invited Paper

LOW-k DIELECTRICS—POROUS MATERIALS

*Invited Paper

POSTER SESSION: LOW-k DIELECTRICS

*Invited Paper

*Invited Paper

INTERCONNECTS

POSTER SESSION: INTERCONNECTS

*Invited Paper

*Invited Paper

JOINT SESSION: GRAIN EVOLUTION OF METALS

JOINT SESSION: PROCESS INTEGRATION
AND MANUFACTURABILITY

*Invited Paper

PREFACE

These proceedings are a record of Symposium D, "Materials, Technology and Reliability for Advanced Interconnects and Low-k Dielectrics," held April 23–27 at the 2000 MRS Spring Meeting in San Francisco, California, and highlight important achievements and challenges in advanced interconnects and low-k dielectrics as employed in the microelectronics industry. The replacement of Al alloys with Cu along with the introduction of new barrier materials to protect Cu from chemical attack, and the utilization of new dielectric materials with a lower relative dielectric constant k than SiO_2 in multi-level metallization structures of increasing complexity, are the major themes of evolution in this field. Invited reviews illustrate the significant progress that has been achieved, and they pointed out the challenges. Contributed papers presented by researchers from different countries demonstrated progress on current topics using a truly multidisciplinary approach. A few brief comments on each topic that will be dealt with in more detail in reviews and papers are presented here as an introduction to these Proceedings.

Interconnects must be reliable from mechanical, chemical and electrical points of view. Mechanical properties of multilevel metallization structures are addressed by examining the mechanical strength of the primary materials themselves and the properties of composite structures. Of particular interest are the mechanical properties of a variety of novel low-k materials and the adhesion between metal/dielectric films, fracture and delamination mechanisms. Liner or barrier layers on metal and dielectric surfaces or interfaces are employed to protect the metals and dielectrics from the chemical attack of impurities and prevent chemical interactions. New diagnostic approaches are required to characterize locally the variation of the mechanical properties at dielectric/barrier and barrier/metal interfaces. Characterization and control of the electromigration and stress migration characteristics of multilevel structures is basic to producing reliable interconnects.

The introduction of a true low-k dielectric material for use as the inter- and intra-line insulator promises the most significant reduction of RC-delays since the introduction of Cu. The widespread use of such a material has been slow because of the significant changes in process flows that is required, materials reliability concerns, and a marginal reduction of the dielectric constant for many of the alternative dielectric materials that have been considered. Process integration and compatibility of current low-k materials with conventional fabrication sequences is a significant issue. Nano-porous low-k materials are possibly the most important candidates that will satisfy the spectrum of technological requirements in the long-term.

The deposition of barrier and/or seed layers with satisfactory properties is a prerequisite of successful metallization. This demands the investigation of novel deposition techniques that provide sufficient control of deposited film properties, and the verification of the properties of the deposited films by a set of complementary measurement techniques. The characterization of deposited metal film properties, the evolution of surface texture and grains, either during synthesis or in post-processing, and the performance and reliability of the interconnects is also of central interest, especially for Cu. Satisfactory post-processing of deposited Cu films and barrier layers, in particular chemical mechanical planarization to define the Cu lines, is an important challenge that has to be overcome for manufacturable advanced multilevel metallization structures.

G.S. Oehrlein
K. Maex
Y.-C. Joo
S. Ogawa
J.T. Wetzel

November 2000

ACKNOWLEDGMENTS

Many people contributed time and energy to making this symposium a success. First, the speakers and authors, who presented their work at the Meeting, and wrote the papers that are contained in these proceedings; second, the session chairs who ran the symposium in a smooth fashion; third, the reviewers of the manuscripts who put in their efforts to guarantee that the papers contained in these proceedings are of high quality; fourth, the staff of the Materials Research Society, who provided the organizational framework of the symposium; and finally, the following sponsors of the Symposium, who provided financial support:

ASM International
ATMI
Dow Chemical Company
Dow Corning Company
Novellus Systems
International SEMATECH

We extend our sincere "Thank You" to all of those who contributed to this undertaking.

MATERIALS RESEARCH SOCIETY SYMPOSIUM PROCEEDINGS

MATERIALS RESEARCH SOCIETY SYMPOSIUM PROCEEDINGS

Prior Materials Research Society Symposium Proceedings available by contacting Materials Research Society

Mechanical Properties

Mat. Res. Soc. Symp. Proc. Vol. 612 © 2000 Materials Research Society

ON THE MECHANICAL INTEGRITY OF ULTRA LOW DIELECTRIC CONSTANT MATERIALS FOR USE IN ULSI BEOL STRUCTURES

E. O. SHAFFER II, K. E. HOWARD, M. E. MILLS AND P.H. TOWNSEND III
The Dow Chemical Company, Advanced Electronic Materials, Midland, MI 48674

ABSTRACT

Adherence to the prescript of Moore's law continues to drive materials development for new and lower dielectric constant materials for use as back-end-of-line (BEOL) interlayer dielectric in advanced logic IC's. As is the case for the current generation of low-K materials (<3.0), these ultra-low K materials (<2.2) will need to meet the variety of integration and reliability requirements for successful product development. Excluding the incorporation of fluorine to lower the material polarity, further reductions of dielectric constant can only be achieved by reduced density. Based upon the industry's experience with the current class of full density dielectrics, process integration may be challenging for ultra-low K materials. This anticipated difficulty derives from the profound differences in material properties, e.g. mechanical integrity, as one lowers the material density, which in turn confounds existing manufacturing processes that have evolved over 35 years based on silicon dioxide.

Minimizing these material and processing differences by extending leveraged learning from previous technology nodes is essential for timely and cost-efficient development process. As a result, material selection of a full density low-K is somewhat influenced by the ability of that material to be extended into future generations. Understanding how the material properties will change as its density is lowered is vital to this selection process. In this paper, we present a summary of models for calculating effective properties as a function of density and apply these to current low-K materials with emphasis on mechanical integrity. We will also review experimental methods for measuring the mechanical integrity of ultra-low K materials and compare the results to the various models described herein.

INTRODUCTION

The development of manufacturing process for advanced logic integrated circuits is strongly influenced by the material properties of the interlayer dielectric material (ILD). As a result, IC manufacturers want to minimize changes in the ILD as they proceed from one technology to the next. However, as already described in the SIA roadmap [1], ILD materials will have to change to meet the lower dielectric requirements of successive technology cycles. One efficient way to minimize these changes is to extend the utility of a dielectric material by introducing porosity. In this way many of the chemical and deposition processes developed for the bulk material are preserved while meeting the lower dielectric constant requirement. However, the introduction of porosity will change many of its properties most notably mechanical. The introduction of pores will cause a reduction in the mechanical integrity. Understanding how much change in mechanical integrity per change in dielectric constant is the focus of this paper.

For the purposes of this paper, mechanical integrity will be described by four principal properties: Young's modulus, ultimate strength, hardness and fracture toughness. The Young's modulus of the material is its resistance to deformation during tensile loading. More specifically, it is the slope of the uniaxial tension stress-strain curve in the limit of small strain. The strength of the material describes a continuum mechanics approach where failure occurs when the stress exceeds this strength. Most often, people compare the maximum principal stress to the strength in order to predict reliability. Hardness is a combination of strength and modulus and is measured using indentation experiments. The maximum indentation pressure divided by the contact area of the indentation defines the hardness [2]. Lastly, fracture toughness is the resistance of a material to cracking. This can be by resisting either crack initiation or propagation. For linear elastic materials with a sharp crack, a far field stress is intensified by the crack with a 1/2 singularity. This stress intensity, K, is proportional to the stress times the square root of distance in front of the crack in the limit of distance going to zero [3]. When the applied stress intensity, K, exceeds the fracture toughness of the film, K_c, the crack will proceed.

The challenge for reliability engineers is the selection of the appropriate mechanical integrity metric. For the past 35 years, the BEOL dielectric has been a ceramic material. As a rule, these materials are extremely stiff and brittle. Hence, hardness is a very effective predictor for mechanical integrity because hardness is related to crack initiation processes. However, many of the new ILD materials being evaluated are polymers that maintain mechanical integrity by resisting crack propagation, i.e. tough. For these materials hardness is not a useful measure of mechanical integrity and comparing hardness between the two classes of materials is not appropriate. For example, compare silica glass to Plexiglas™ plastic and ask yourself which one would you rather have protecting your eyes from flying debris. The bottom line is that plastic materials are not worse, simply different. Hence, many of the testing and fabrication methods used will have to be optimized.

In this paper, we will review some common scaling models that predict effective foam properties. We will then plot these predictions for two different dielectric materials: SiLK* semiconductor dielectric and PE-CVD silicon dioxide. SiLK* dielectric resin is a polymeric, thermoset resin that is spin-coated onto a silicon wafer rather than chemically vapor deposited. We will also compare the predictions for fracture toughness to experimental results. The implications of these results will be discussed.

EFFECTIVE PROPERTY MODELING OF FOAMED

A set of effective properties for foamed material has been derived by Gibson and Ashby and is described in reference [4]. Foam geometries can range from open cell foam to closed cell foams. The properties of the foam are derived from assuming a beam geometry that relates the amount of material in the cell edge versus cell face and the relative density of the foam.

Equations 1-4 describe the changes in the foam's Young's modulus, E'; the crush strength, σ_{cr}'; hardness, H'; and toughness, K_c' for both the open and closed cell case, respectively, as a function of density, ρ. The term ϕ relates the amount of material in the cell strut to the cell face. For a purely open cell f is unity. One challenge the construction before is that it assumes ϕ is independent of ρ, however, this is often not the case.

$$\frac{E^*}{E_o} = \left(\frac{\rho^*}{\rho_o}\right)^2 \text{Open cell}; \quad \frac{E^*}{E_o} = \phi^2\left(\frac{\rho^*}{\rho_o}\right)^2 + (1-\phi)\left(\frac{\rho^*}{\rho_o}\right) \text{Closed Cell} \qquad (1)$$

$$\frac{\sigma_{cr}^*}{\sigma_{cr}} = \left(\frac{\rho^*}{\rho_o}\right)^{3/2} \text{Open cell}; \quad \frac{\sigma_{cr}^*}{\sigma_{cr}} = \phi\left(\frac{\rho^*}{\rho_o}\right)^{3/2} + (1-\phi)\left(\frac{\rho^*}{\rho_o}\right) \text{Closed cell} \qquad (2)$$

$$\frac{H^*}{H} = 0.23\left(\frac{\rho^*}{\rho_o}\right)^{3/2}\left(1+\left(\frac{\rho^*}{\rho_o}\right)^{1/2}\right) \text{ Open; } \frac{H^*}{H} = 0.3\left(\phi\frac{\rho^*}{\rho_o}\right)^{3/2} + 0.4(1-\phi)\left(\frac{\rho^*}{\rho_o}\right) \text{ Closed} \qquad (3)$$

$$\frac{K_{lc}^*}{K_{lc}} = 0.65\left(\frac{\rho^*}{\rho_o}\right)^{3/2} \text{ Open cell; } \frac{K_{lc}^*}{K_{lc}} = \left(\frac{\rho^*}{\rho_o}\right)^{3/2} \text{ Closed cell} \qquad (4)$$

Other approaches for effective property calculations are available. For example, a composite spheres approach [5] can be used to calculate the effective modulus. Here the reduced modulus is related to the concentration of voids, the bulk modulus, K_o and the shear modulus of the matrix, G_o, shown in Equation 5. Another approach for calculating effective properties is to use percolation theory [6]. This is particularly useful for strength of material arguments. For a system of uniformly distributed pores with a distribution of pore sizes using an off-lattice model the strength can be approximated with an exponential decay model shown in Equation 6 where A is a constant. However, most systems do not pack with 100% efficiency. That is the system will percolate at a lower threshold. At this critical percolation limit the strength will go to zero.

$$\frac{E^*}{E_o} = 1 - \frac{c}{1-[(1-c)K_o/(K_o+1.33G_o)]} \qquad (5)$$

$$\frac{\sigma_{cr}^*}{\sigma_{cr}} = A\exp\left(-\left(\rho^*/\rho_o\right)\right) \qquad (6)$$

One other note, the coefficient of thermal expansion is typically not a function of the density over the ranges discussed here. This is because the amount of expansion does not depend on the cross-sectional area of the specimen. Imagine two pillars of steel holding a horizontal beam. As the temperature is cycled the beam will displace vertically due to the expansion of the pillars, independent of their cross-sectional area.

Equations 1-6 are used to calculate the changes in the properties of two dielectric materials: SiLK* resin and PE-CVD SiO$_2$. The matrix properties for SiLK* resin and SiO$_2$ are listed in Table 1. The properties are plotted against the calculated dielectric constant, which is related to the density of the matrix using the series model. The results for Young's modulus, crush strength and hardness are shown in Figures 1-3, respectively.

Table 1: Matrix Properties of SiLK* Resin and PE-CVD SiO$_2$

Property	SiLK* [7]	SiO$_2$ [8,9]
Young's Modulus, GPa	2.45	72
Strength, MPa	90	100
Hardness, GPa	0.38	7.5
Toughness, MPa-m$^{1/2}$	0.62	0.75
Dielectric Constant	2.65	4.1

As shown, the models predict that all three properties decrease with increasing porosity. From Figure 1, we see that the modulus of the SiLK* resin remains lower than silica at all dielectric constants. This is due to the high modulus of silica initially. On the other hand, the strength of SiLK* resin at equivalent dielectric constant is predicted to be greater than porous silica. This result is arises from the lower porosity required to achieve equivalent dielectric constant of silica. of more concern is the extremely low strength predicted by the open cell model. Many of the strategies considered for making porous silica result in open celled structures. In Figure 3, the hardness of both SiLK* resin and silica are plotted. Again, the silica has higher hardness over the dielectric range due to the much higher stiffness of the silica initially.

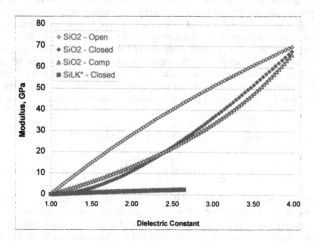

Figure 1: Predicted Young's Modulus versus change in dielectric constant as porosity is introduced into either SiLK* dielectric resin or silica. The models include the open and closed cell calculations from Equation 1 or the composite model from Equation 5.

EXPERIMENTAL PROCEDURES

Sample Fabrication

SiLK* resin is used as the matrix dielectric. To a SiLK* solution, an organic poragen is added at different weight percents. The poragen is selected based on its thermal degradation temperature. The porous SiLK* resins are coated onto silicon wafers using standard spin-coat procedures developed for non-porous SiLK* resin. AP4000 adhesion promoter is applied and baked prior to the SiLK* resin deposition to prevent adhesive failure.

Fracture Testing

Adhesion is measured using the modified-Edge Liftoff Test, m-ELT [10]. It consists of applying a thick backing material (e.g. epoxy) to the test structure which is on a rigid substrate.

The backing layer material must have a known stress-temperature profile and higher fracture toughness than the test structures. It also must have excellent adhesion to the test structure. The wafer is diced so that 90° edges to the substrate are formed. The sample is then cooled until debonding is observed. If the backing layer material is much thicker than the test resin, the applied fracture intensity, K_{app}, is given by Equation 7.

$$K_{app} = \sigma_o \sqrt{h/2} \qquad (7)$$

where, h and σ_O are the backing layer thickness and residual stress, respectively

Surface Analysis

After testing, the interface is examined to determine the locus of failure. X-ray photoelectron spectroscopy (XPS) is used on both sides of the interface to ensure that the failure was cohesive. Also, the fracture surface is scanned using tapping mode atomic force microscopy (TPAFM).

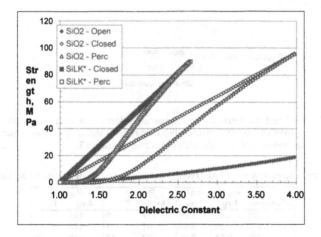

Figure 2: Predicted crush strength versus change in dielectric constant as porosity is introduced into either SiLK* dielectric resin or silica. The models include the open and closed cell calculations from Equation 2 and the percolation model from Equation 6.

RESULTS AND DISCUSSION

Samples of porous SiLK* resin were formulated with 20%, 25%, 30% and 35% poragen by weight of solids. The fracture toughness of the samples is listed in Table 2. Also listed in Table 2 are the RMS roughness, R_q, and average roughness, R_a, of the substrate-side failed surface. Comparing the results, no strong correlation is observed between surface roughness and toughness. The failed surfaces are extremely rough with peak heights on the order of the coating film thickness, 1 um. This is an indicator that the crack was propagating through the film. To ensure that the measured toughness by mELT is the cohesive toughness of the coating, atomic surveys are conducted using XPS of the failed surfaces. The results are listed in Table 3. The results show that the samples conclusively failed cohesively.

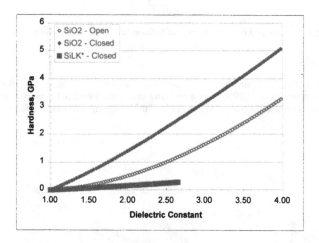

Figure 3: Predicted hardness versus change in dielectric constant as porosity is introduced into either SiLK* dielectric resin or silica. The models include the open and closed cell calculations from Equation 3.

Table 2: Fracture toughness and TPAFM roughness measurements of Porous SiLK* samples

%Poragen	Kc, MPa-m1/2		Ra, nm		Rq, nm	
	Avg	Dev	Avg	Dev	Avg	Dev
20	0.40	0.03	137	9	166	2
25	0.38	0.03	97	14	128	25
30	0.35	0.02	123	18	149	22
35	0.30	0.03	77	5	93	6

The fracture toughness results are plotted against dielectric constant and compared to the model predictions from Equation 4 in Figure 4. For the porous SiLK* resin samples we were not able to have the dielectric constants measured in time for this paper. So, we use the calculated dielectric constant based on the porosity and the series model. Also plotted in Figure 4 are results for different silica based dielectric materials [11, 12]. The predicted models are in good agreement with the data. The SiLK* resin remains tough for dielectric constants required for next generation ILD applications. This is due again to the low porosity required to achieve low dielectric constants. On the other hand, silica materials require 75% porosity to achieve a dielectric constant of 2.0. As a result, these materials are open cell and very brittle. This may present challenges during fabrication and packaging of IC devices.

Table 3: Atomic composition of the failed fracture surfaces.

Sample	Test Side	O(1s)	C(1s)	Si(2p)	Pi-pi*/C(1s)
20% Poragen	Substrate	2.4	97.6	n.d.	0.108
	Film	2.3	97.7	n.d.	0.109
25% Poragen	Substrate	2.5	97.5	n.d.	0.089
	Film	2.7	97.3	n.d.	0.089
30% Poragen	Substrate	3.3	96.7	n.d.	0.106
	Film	3.4	96.7	n.d.	0.101
35% Poragen	Substrate	3.2	96.9	n.d.	0.104
	Film	3.3	96.7	n.d.	0.102

CONCLUSIONS

In this paper, models were presented that predict the mechanical properties of interlayer dielectrics as a function of porosity. These properties include the Young's modulus, crush strength, hardness and toughness. Experiments were conducted on porous SiLK* resin films to compare toughness to the model predictions. Good agreement was found between model predictions and experimental results. The results for SiLK* semiconductor dielectric show that porous SiLK* resin maintains good mechanical integrity at a dielectric constant of 2.0 which should facilitate integration in advanced IC devices.

ACKNOWLEDGEMENTS

The authors wish to thank Greg Meyers for the AFM analysis and Dave Hawn for the XPS analysis. We also wish to thank Richelle Minda assistance with the mELT data.

REFERENCES

(1) SIA Roadmap, 1998.
(2) S. P. Timoshenko and J. N. Goodier, Theory of Elasticity, 3rd ed. (1987).
(3) M. Kanninen and C. Popelar, Advanced Fracture Mechanics, (1985).
(4) L. Gibson and M. Ashby, Cellular Solids, 2nd ed. (1997).
(5) R. M. Christensen, Mechanics of Composite Materials (1979).
(6) D. Stauffer, Introduction to Percolation Theory, (1985).
(7) P. Townsend . S. J. Martin, J. Godschalx, D. R. Romer, D. W. Smith, D. Castillo, R. DeVries, G. Buske, N. Rondan, S. Froelicher, J. Marshall, E. O. Shaffer and J. H. Im, Mat. Res. Soc. Symp. Proc., **476**, 9 (1997).
(8) B. Tapley, Eshbach's Handbook of Engineering Fundamentals, 4th ed. (1990).
(9) R. Cook, E. Liniger, D. Klaus, E. Simonyl and S. Cohen, Mat. Res. Soc. Symp. Proc., **511**, 33, (1998).
(10) E. O. Shaffer II, F. J. McGarry, L. Hoang, Poly. Sci & Eng., **36**, (18), 2381 (1996).
(11) A. Jain, S. Rogojevic, S. Nitta, V. Pisupatti, W. Gill, P. Wayner, J. Plawsky, Mat. Res. Soc. Symp. Proc., **565**, 29 (1999).

(12)B. Zhu, F., McGarry., et. al., presented at the Fall MRS, Boston, MA (1999).

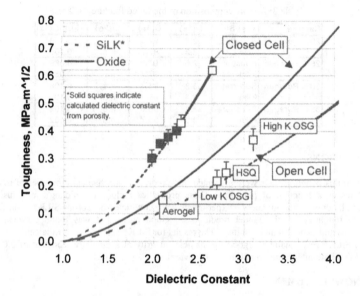

Figure 4: Predicted toughness versus change in dielectric constant as porosity is introduced into either SiLK* dielectric resin or silica. The models include the open and closed cell calculations from Equation 4. Also, plotted are results for various silica based ILD materials and SiLK* resin.

Mat. Res. Soc. Symp. Proc. Vol. 612 © 2000 Materials Research Society

THE USE OF THE FOUR-POINT BENDING TECHNIQUE FOR DETERMINING THE STRENGTH OF LOW K DIELECTRIC/BARRIER INTERFACE

Ting Tsui*, Cindy Goldberg**, Greg Braeckelman**, Stan Filipiak**, Bradley M. Ekstrom**, J.J. Lee**, Eric Jackson**, Matthew Herrick**, John Iacoponi*, Jeremy Martin*, and David Sieloff**.
* Advanced Micro Devices, Technology Development Group, AMD/Motorola Alliance, Austin, TX
** Advanced Product Research and Development Laboratory, Motorola Inc., Austin, TX

ABSTRACT

One of the important reliability challenges in integrating copper/Low-K dielectric technology has been adhesion between the Low-K dielectric and barrier metal. This investigation explored the applicability of the four-point bend technique for determining the adhesion strength of a fluorine doped low dielectric constant oxide in contact with tantalum barrier layer. Time of flight secondary ion mass spectroscopy (ToFSIMS) was used for surface chemical analyses of the delaminated surfaces to identify the fractured interface and its chemical compositions. The effect of annealing on mechanical strength was coupled with chemical analysis to discern the adhesion properties. Experimental results suggested that fluorine rich interfacial layer formation was associated with degraded adhesion characteristics between Low-K dielectric and tantalum barrier metal.

INTRODUCTION

To reduce capacitance delay and electrical resistance, copper and low dielectric constant materials were integrated into the back-end-of-line (BEOL) technology [1-8] at Advanced Micro Devices and Motorola. One of the common BEOL integrations in the semi-conductor industry uses a fluorine (F) doped oxide insulator, tantalum (Ta) metal barrier layer and electroplated copper (Cu) metallization [1-3, 5, 7-8]. Fluorinated oxide was chosen for the low-K dielectric as incorporation of fluorine into the insulator reduces the dielectric constant from ~4.2 to ~3.7. One of the critical integration issues of this stack structure has been delamination between the low-K dielectric and barrier metal layers [9]. The film buckling failure mode suggested that the average film stress in the stack is compressive in nature. In order to quantify the adhesion strength of the Low-K dielectric and Ta interface, four-point bending technique [5] was used. The delaminated surfaces by the four point bending test were analyzed by the time of flight secondary ion mass spectroscopy (ToFSIMS). Results indicated that the adhesion between the dielectric and barrier metal reduces with the formation of the fluorine rich interlayer. Actual fracture interface is between the interlayer and the fluorinated low-K dielectric.

EXPERIMENTAL PROCEDURE

Low-K Dielectric and Barrier Metal Deposition

Blanket fluorinated oxide films were deposited on oxidized silicon wafers at Motorola Advanced Product Research and Development Laboratory (APRDL) in Austin, Texas. The dielectric films were deposited using plasma enhanced chemical vapor deposition (PECVD) techniques. The thickness of the dielectric layer is approximately half micron. To study the effects of the fluorine ion concentration on the Low-K dielectric and Ta adhesion, blanket dielectric wafers were deposited at three different fluorine precursor flow rates. The normalized flow rates are 1, 3.75, and 6. Tantalum barrier metallization was deposited using ion metal plasma (IMP) technique with final thickness near 500A. The specimens were annealed at different temperatures to understand the thermal effects on the interface mechanical and chemical integrity.

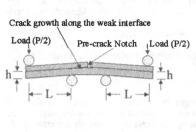

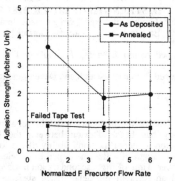

Figure 1. Schematic drawing of the 4-point bend test setup.

Figure 2. Adhesion strength of the fluorinated oxide/Ta barrier interface.

Four Point Bend Test

Four-point bend adhesion test [5] was used to measure interfacial strength in this study. Figure 1 shows the specimen orientation and the testing geometry. Specimens were prepared into strips with dimensions about 1 x 4 cm. The strips are then glued together using an epoxy with the film stacks facing each other. A pre-crack notch is created in the top silicon strip to produce a localized stress intensity region where delamination initiation is desired. During the adhesion experiment, instruments record the applying load (P) and the displacement of the loading pins (D). The strain energy release rate, G, which is the delamination driving force generated by the 4-point bending, can be expressed as:

$$G = \frac{21(1-\upsilon^2)M^2}{4Eb^2h^3} \quad [1]$$

where M=PL/2, E is the silicon elastic modulus, υ is the Poisson's ratio, and b and h are the width and thickness of the strip, respectively. L is the distance between the inner and outer pins as illustrated in Figure 1. P is the load that generates a stable crack growth at the weakest interface of the sample stack. By carefully cleaving the bottom silicon strips, the delaminated surfaces were characterized for unambiguous identification of the failing interface and its chemistry.

Time of Flight Secondary Ion Mass Spectroscopy (ToFSIMS)

Depth profile analyses were performed in a Cameca/ION-TOF TOF-SIMS IV instrument at APRDL using a pulsed Ar^+ ion beam interleaved with a low energy Cs^+ ion beam for sputter removal during depth profiling. The typical analytical raster area was 15 μm x 15 μm with the Ar^+ ion beam at 1 pA, coinciding with a 150 μm x 150 μm raster by the 0.6 keV Cs^+ sputter ion beam of 3 nA. Due to the lack of proper standard materials for quantification, the comparison of ToFSIMS depth profiles was based on ion intensity arising from the same matrix materials. With the assumption that the oxidation reactions at the

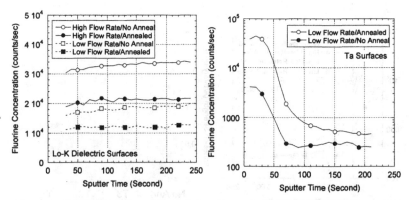

Figure 3. ToFSIMS depth profiles of the delaminated Lo-K dielectric surfaces.

Figure 4. ToFSIMS depth profiles of the delaminated Ta barrier metal surfaces.

fractured surfaces do not alter the surface chemistry significantly, ToFSIMS results can provide useful chemical information of the interface before the delamination failure.

RESULTS and DISCUSSIONS

The adhesion results of the as-deposited and annealed specimens are plotted as a function of fluorine precursor flow rate shown in Figure 2. The adhesion values are normalized and have arbitrary units. The figure shows that the as deposited Low-K dielectric/barrier adhesion strength is sensitive to the fluorine precursor flow rate, i.e. the fluorine concentration in the dielectric. The adhesion is the greatest for the sample with the lowest flow rate. It has an adhesion value of 3.6. The interfacial strength reduces rapidly, by a factor of 2, as the fluorine precursor flow rate increases and appears to be stabilized at 1.8 for greater gas flow rate. This may indicate fluorine ion saturation effects at the dielectric/Ta interface.

Results of the annealed samples are also plotted in Figure 2. The figure reveals the adhesion strength of all annealed samples are in the range of the 0.8 and 0.9 regardless of the fluorine precursor flow rate during the Low-K dielectric deposition process, i.e. initial fluorine concentration in the dielectric. From our prior experimental results, samples that failed tape tests have adhesion values less than 1. It is interesting to note that all of the annealed specimens have very similar adhesion values even though the initial fluorine concentrations in the dielectric are not identical. As will be discussed later, this may be due to the formation of the fluorine rich compound at the dielectric/metal barrier interface.

To understand the adhesion strength reduction after the thermal treatments, fracture surfaces of the four delaminated samples, circled in Figure 2, were examined by using ToFSIMS. They are as deposited and annealed samples deposited at the lowest and highest fluorine precursor flow rates. ToFSIMS depth profiling results of the Low-K dielectric fracture surfaces of the four samples are shown in Figure 3. The figure reveals the fluorine ion concentration plotted as a function of the sputtering time, i.e. sputtered depths. As expected, specimens with higher fluorine precursor flow rate contain more fluorine ions in the dielectric.

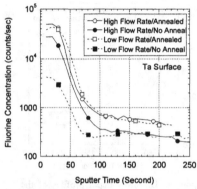

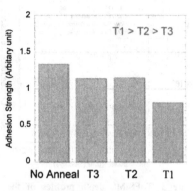

Figure 5. TOFSIMS depth profiles of the delaminated Ta barrier surfaces.

Figure 6. Adhesion strength of specimens under different thermal annealing process.

However, the amount of fluorine ions reduces with the thermal treatment. This is puzzling since the dielectric was capped with Ta and Cu films. The dielectric constant of the fluorine-doped oxide may increase due to the reduction of the fluorine ion concentration.

The ToFSIMS depth profiling results of the Ta fractured surfaces are shown in Figure 4. The data was obtained from specimens deposited with fluorine precursor operated at the slowest flow before and after anneal. The figure shows that the fluorine ion is highly concentrated at the fracture surface, which was the dielectric/barrier metal interface before delamination failure. Figure 4 also shows the extent of fluorine concentration increase, as much as ten times at the interface after annealing. This may suggested that a thermally driven diffusion reaction occurs at the Low-K dielectric/barrier metal interface during anneal.

The increase of the fluorine ion concentration at the dielectric/barrier metal interface was also observed for the specimen deposited with the highest fluorine precursor flow rate. The results are summarized in Figure 5. Data from figure 4 are also included for comparison. Figure 5 shows the amount of fluorine increase after annealing of the film deposited at high flow rate is smaller than the low flow rate sample. The interface fluorine ion count of the high flow rate specimen only doubled while the low flow rate sample increased by ten times. It is interesting to note that the final fluorine concentrations near the dielectric/barrier interfaces of the two annealed specimens are very close regardless of the initial fluorine ion concentration prior to the thermal anneal. One of the possible explanations is the formation of a stable compound consists of fluorine, dielectric material, and Ta at the interface. Once the compound is formed, the fluorine concentration, which depends on the stoichiometry of the compound, will be stabilized. Further increase in the fluorine ions of the dielectric only increase the thickness of the interlayer. Detailed examinations of other ToFSIMS spectrums confirmed that the locus of failure is locate at the interface of the interlayer and the fluorinated dielectric film.

In addition to the increase of the fluorine concentration at the barrier/Low-K dielectric interface, Figure 5 also reveals higher fluorine content in the bulk tantalum barrier metal. The amount of the fluorine concentration at sputter time longer than 150 seconds, which represents bulk Ta metal composition, increases after the thermal anneal for both specimens. This indicates tantalum metal does not completely terminate the flow of fluorine originated from the fluorinated oxide. This may partly explains the lost of fluorine shown in Figure 3.

To investigate the threshold of the dielectric/barrier metal interfacial reaction, experiments were repeated with two additional annealing conditions T2 and T3, where T1 > T2 > T3, and both T2 and T3 have shorter duration than T1. The specimen stack is similar to the previous experiment but with additional ~1μm thick copper film. The detailed results are plotted in Figure 6. The figure revealed that the T2 and T3 treated specimens have adhesion values between the as deposited and the T1 annealed specimens. This indicates the Ta-F diffusion reaction begins lower temperatures than T1.

CONCLUSIONS

1. The four-point bend adhesion test and ToFSIMS techniques can provide quantitative information on the mechanical and chemical integrity of the fluorine doped oxide/Ta barrier metal interface.
2. The adhesion strength degradation of the fluorinated oxide and Ta barrier metal can be attributed to the chemical reaction between fluorine and Ta during annealing. The reaction forms an interlayer between the dielectric and the tantalum metal.
3. Fluorine ions can diffuse into tantalum metal by thermal anneal.

REFERENCES

1. Y. Uchida, K. Taguchi, S. Sugahara, and M. Matsumura, "A fluorinated organic-silica film with extremely low dielectric constant," Japanese Journal of Applied Physics, Part 1, vol.38, no.4B p.2368-72.
2. T.E.F.M. Standaert, P.J. Matsuo, S.D. Allen, G.S. Oehrlein, and T.J. Dalton, "Patterning of fluorine-, hydrogen-, and carbon-containing SiO2 like low dielectric constant materials in high-density fluorocarbon plasmas: Comparison with SiO2," Journal of Vacuum Science & Technology A (Vacuum, Surfaces, and Films) vol.17, no.3 p.741-8.
3. Seoghyeong Lee and Jong-Wan Park, "Effect of oxygen post plasma treatment on characteristics of electron cyclotron resonance CVD fluorine-doped silicon dioxide films using SiF4 and O2 gas sources," Journal of the Electrochemical Society vol.146, no.2 p.697-701
4. B. Cruden, K. Chu, K. Gleason, and H. Sawin, "Thermal decomposition of low dielectric constant pulsed plasma fluorocarbon films. II. Effect of postdeposition annealing and ambients," Journal of the Electrochemical Society vol.146, no.12 p.4597-604.
5. Michael Lane, Nety Krishna, Imran Hasim, and Reinhold H. Dauskardt, "Adhesion and Reliability of Copper Interconnects with Ta and TaN Barrier Layers," Submitted to Journal of Materials Research, August 1999.
6. I. Banerjee, M. Harker, L. Wong, P.A. Coon, and K.K. Gleason, "Characterization of chemical vapor deposited amorphous fluorocarbons for low dielectric constant interlayer dielectrics," Journal of the Electrochemical Society vol.146, no.6 p.2219-24.
7. Chiang Chien, A.S. Mack, Pan Chuanbin, and D.B. Fraser, "Challenges and issues of low-K dielectrics," 1997 International Symposium on VLSI Technology, Systems, and Applications. Proceedings of Technical Papers, p.37-9
8. M.K. Jain, K.J. Taylor, G.A. Dixit, W.W. Lee, L.M. Ting, G.B. Shinn, S. Nag, R.H. Havemann, J.D. Luttmer, and M. Chang, "A novel high performance integration scheme using fluorinated-SiO2 and hydrogen silsesquioxane for capacitance reduction," 1996 Proceedings Thirteenth International VLSI Multilevel Interconnection Conference (VMIC) p.23-7
9. Richard J. Huang, Guarionex Morales, and Simon Chan, **Surface treatment of low-k SiOF to prevent metal interaction**, U.S. Patent Number US5994778, 1999.

Mat. Res. Soc. Symp. Proc. Vol. 612 © 2000 Materials Research Society

The Effect of Fatigue on the Adhesion and Subcritical Debonding of Benzocyclobutene/Silicon Dioxide Interfaces

Jeffrey M. Snodgrass and Reinhold H. Dauskardt
Department of Materials Science and Engineering, Stanford University
Stanford, CA 94305-2205, USA

ABSTRACT

The effect of fatigue loading on microelectronic thin film interfaces has until now been difficult to quantify. Most industrial fatigue testing uses HAST (Highly Accelerated Stress Testing) protocols, which inherently convolutes the effects of mechanical fatigue and the test environment. Our work focuses on isolating the deleterious effects of mechanical fatigue on interfaces, which we have found to be substantial. In this study, the integrity of a low-k polymer interface involving benzocyclobutene (BCB) and silica was examined under a variety of loading conditions. Critical (fast fracture) adhesion values were measured using standard interface fracture-mechanics geometries. Experiments were then conducted to measure the debond growth rate as a function of the applied strain energy release rate under both static and cyclic loading conditions. Our results show that even under room temperature conditions, debond growth rates measured under cyclic fatigue are considerably faster than those observed under static loading. Results are presented detailing the effects of interface chemistry (adhesion promoters), environmental moisture, and test temperature on the resistance of the interfaces to subcritical debonding. Strategies for increasing resistance of dielectric interfaces to fatigue debonding are outlined.

INTRODUCTION

In order to achieve the dielectric performance needed for next generation devices, silica is increasingly being replaced by polymer or polymer-like materials. Unfortunately these materials face a number of integration challenges related to their low elastic moduli and the large difference between their thermal expansion coefficients those of adjacent materials. These factors can lead to high film stresses and problems with adhesion. Indeed, adhesion and film integrity are often the most important factors in determining whether a new dielectric material can be integrated into a successful process.

The techniques outlined in this study describe a methodology that allows for the quantitative measurement of the parameters that influence adhesion and subcritical debonding. Much work has been completed in this area and numerous studies have been published detailing the effectiveness of these techniques in measuring interfacial adhesion [1, 2, 3]. The delaminating beam methods used are analyzed with well-founded fracture mechanics principles. Results are presented in terms of the critical strain energy release rate, G_C, measured in J/m^2, necessary to cause steady-state debonding. Time-dependent subcritical debonding is obtained by measuring the velocity of debond growth, da/dt, as a function of the applied strain energy release rate, G.

Recent work has begun to examine the effects of mechanical fatigue loading on polymer interfaces [4, 5]. It is believed that the resistance of interfaces to fatigue loading will be most important in determining long-term in-service reliability. Fatigue debond data is typically presented in terms of the debond growth per cycle, da/dN, as a function of the applied strain energy release rate range, ΔG.

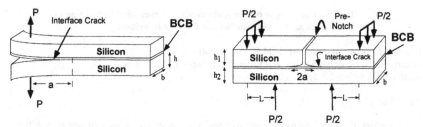

Figure 1. *The double cantilever beam (left) and mixed-mode delaminating beam (right) test geometries.*

EXPERIMENT

Sample Preparation

Interface test samples were prepared which contained a 2 μm thick benzocyclobutene resin layer (Cyclotene 3022-35, Dow Chemical Co., Midland Michigan) sandwiched between two silicon substrates. To fabricate the samples, four-inch silicon wafers were first cleaned for 20 minutes in 9:1 H_2SO_4:H_2O_2 at 120°C. Some wafers were first prepared with silane-based adhesion promoters, which have a silanol end-group designed to facilitate bonding with silica surfaces and a functional end-group designed to bond with the polymer overlayer. The two adhesion promoters used in this study were AP8000, which has an amino-based functional group, and AP3000, which has a vinyl functional group (both Dow Chemical Co.) After applying the adhesion promoter, BCB was spin-coated onto the wafer. The wafers were then diced into 40 mm x 40 mm squares, two squares were placed face to face, and the BCB on the two squares was cured together under pressure (<10 psi). The standard cure, which includes a final 60 minute soak at 250°C, was performed under an inert nitrogen atmosphere.

For the mixed mode samples, the bonded squares were diced into bend samples ≈ 3-5 mm wide. A notch was then cut into the center of the top beam in order to bifurcate the center crack along the interface. The normal mode samples were diced 5 mm wide and loading tabs were attached to the ends of the samples.

Critical Adhesion Measurement

Critical adhesion measurements were conducted under monotonic loading using either double cantilever beam (DCB) samples, where tractions are generally normal to the interface or with mixed mode delaminating beam (MMDB) samples, where the interface tractions contain both normal and shear components. Schematic drawings of the sample types are shown in Figure 1. The samples were tested either in a custom-built micromechanical test system with a piezoelectric actuator or on a servohydraulic test system under closed-loop load control.

For the DCB specimens, critical adhesion energies were measured in displacement control at a constant crosshead speed of 3 μm/s, which resulted in a debond growth rate of ≈30·10^{-6} m/s. From measurement of the load, P_C, at the start of debonding and the length of the debond, a, immediately prior to debond extension, the fracture energy of the interface, G_C, was calculated from the critical value of the strain energy release rate, G [6]:

$$G = \frac{12P^2a^2}{b^2h^3E'} \left(1 + 1.28\frac{h}{a} + 0.41\frac{h^2}{a^2}\right) \tag{1}$$

where b is the sample width, h is the half-height and E' is the biaxial elastic modulus. Debond lengths were measured *in situ* using both optical microscopy and beam compliance relationships. The phase angle of loading for the DCB geometry is $\approx 0°$.

Critical adhesion energy was also measured using the MMDB configuration. In this geometry, the strain energy release rate, G, is independent of crack length. P_C values were determined from the plateau region on a plot of load vs. displacement, using a test procedure previously described [7], and G_C is given by:

$$G_C = \frac{21P_C^2l^2}{16b^2h^3E'}, \tag{2}$$

where all of the variables are the same as those in Eq. 1, and l is the distance between the inner and outer loading pins. The MMDB adhesion tests were performed with a constant crosshead speed of 0.3 μm/s, which resulted in a crack velocity of approximately $25 \cdot 10^{-6}$ m/s. The mode mixity for the MMDB test is approximately 43°.

Subcritical Adhesion Measurement

Fatigue debond growth rates were measured using a test setup based on ASTM Standard E647-95 for Measurement of Fatigue Crack Growth Rates [8]. Fatigue debond tests were conducted using a decreasing ΔG, load-shedding scheme. The interface specimens were loaded with a sinusoidal frequency of 20 Hz and a load ratio, R, of 0.1. The crack length was continuously measured using sample compliance. Growth rates were typically measured over a range of da/dN from $\sim 10^{-5}$ to 10^{-9} m/cycle.

Static debond growth was measured through load-relaxation testing. In these tests, the specimen was loaded to a value of G below G_C, and the displacement was then fixed. The load was then measured as a function of time. Load relaxation under fixed displacement indicates debond extension has occurred. The initial load and crack length at the time when displacement is held fixed are designated as P_i, and a_i, and thereafter the instantaneous load is recorded as, P, as a function of time, t. Using compliance relationships, the velocity of the crack, da/dt, is determined using [9]:

$$\frac{da}{dt} = -\frac{dP}{dt} \cdot \frac{P_ia_i}{P^2}. \tag{3}$$

Actual debond length was verified optically before and after testing and through fractography after testing was completed. All load relaxation testing was performed in an environmental chamber [Cincinnati Sub Zero ZH-16, Cincinnati, OH] which precisely controlled the temperature and relative humidity at 25°C and 30%, respectively.

Table I. *Critical adhesion energies for different interface chemistries as a function on phase angle.*

Adhesion Promoter	DCB ($\Psi \sim 0°$)		MMDB ($\Psi \sim 43°$)	
	G_C (J/m²)	std. dev.	G_C (J/m²)	std. dev.
none	10.8	0.4	25.7	1.6
amino	18.9	0.2	44.7	3.1
vinyl	22.1	0.3	51.1	2.9

RESULTS

The results of critical adhesion testing for the three different interface chemistries are summarized in Table I. Both adhesion promoters were seen to provide a substantial increase in BCB/SiO₂ interface adhesion energy compared to interfaces with no adhesion promoter. The enhanced bonding results from the formation of strong covalent bonds across the BCB/silica interface. Of the two different types of adhesion promoters tested, interfaces with the vinyl-based functional group were about 10% stronger than interfaces with the amino functional group. This may result from more effective binding of the double-bond of the vinyl functional group into the BCB polymer network.

Figure 2 shows static debond growth rate curves for the three interface chemistries. It is evident from these curves that static subcritical debond behavior scales with the critical adhesion energies of the interfaces. Note that the lowest debond driving force at which measurable debonding occurs can be as low as 40% of the critical adhesion energy. Static subcritical debonding is believed to be caused largely by the corrosive action of environmental moisture on strained bonds at the debond tip. This conclusion is supported by the data in Figure 3, which shows the effect of humidity on the rate of static debond growth. Increasing the amount of humidity present during static subcritical debonding allows debonding to propagate at much lower applied driving forces. In addition, the slope of the debond growth curve is much steeper at low humidities, indicating a reduced susceptibility of the interface to subcritical debonding.

The effect of temperature can be difficult to study because of the convoluted effect of temperature on the activity of moisture present in the test environment and on the rates of thermally activated processes. Figure 4 shows the effect of temperature in an experiment where the partial pressure of water in the test environment was held constant (944 Pa). Clearly, there is relatively little effect of temperature, once the mitigating influence of differences in the activity of environmental water is removed. Indeed, even at low temperatures, the high humidity (10°C/77%R.H.) environment remained the most accelerating atmosphere for static debond growth.

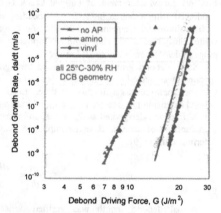

Figure 2. *The effect of interface chemistry functional group on subcritical debond growth curves.*

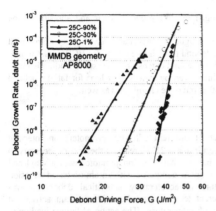

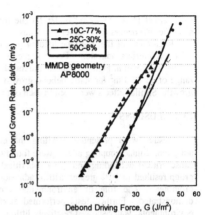

Figure 3. *The effect of humidity on static debond growth curves. Increasing humidity accelerates debond growth.*

Figure 4. *Effect of temperature on static debond growth, tested with a constant partial pressure of water*

Generally it has been seen that increasing the phase angle of loading in a given experiment results in increased interface adhesion energy [10]. This is true for the interfaces examined in the present study as well. The increase in adhesion energy has been attributed to the effects of interface roughness and asperity contact behind the debond tip. The critical data in Table I, as well as the static subcritical debond growth curve in Figure 5, show that mixed mode loading generally results in approximately a 2 fold increase in the critical adhesion energy and the strain energy release rate associated with a specific growth velocity under static loading. However, Figure 6 shows that the

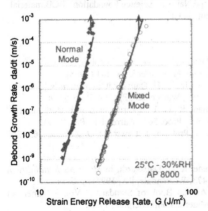

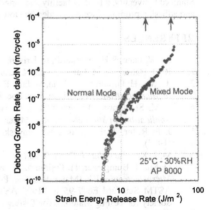

Figure 5. *Plot showing static subcritical debond growth in mixed mode and normal mode sample configurations.*

Figure 6. *Plot showing fatigue debond growth in mixed mode and normal mode sample configurations.*

same effect is not observed under fatigue loading, most likely due to wear and abrasion of the asperities the occurs after numerous cycles.

Regardless of the interface loading geometry, fatigue is observed to have a very detrimental effect on interface adhesion. For example, in the debond growth rate curves for $BCB/AP8000/SiO_2$ interfaces shown in Figures 5 and 6, the threshold for debond extension under static loading is in the range of 13-23 J/m^2 for static debond growth, while it can be as low as 6 J/m^2 for fatigue debond growth. Similar effects were seen in each of the other interface chemistries as well.

CONCLUSIONS

Results were presented detailing the effect of different adhesion promoters on interface adhesion. Silane coupling agents were found to result in a $\approx 2X$ increase in the energy necessary to cause critical debonding and grow subcritical debonds. An adhesion promoter with a vinyl end group resulted in 10% greater critical adhesion energy as compared with an amino-based adhesion promoter. The role of environmental moisture in accelerating subcritical debonding was demonstrated. Further, tests performed at different temperatures, but with constant activity of environmental water showed relatively little effect of temperature. The action of fatigue loading in accelerating debond growth rates and decreasing thresholds for subcritical debonding was described. Also, the effect of phase angle of loading was demonstrated for both static and fatigue debond growth. The protective effects of increasing phase angle were shown to be diminished under fatigue loading.

ACKNOWLEDGEMENTS

This work was supported by the Semiconductor Research Corporation, Packaging & Interconnect Systems. JMS was supported by the SRC Graduate Fellowship Program and the Intel Foundation Fellowship. The authors would like to thank Maura Jenkins and Dimitris Pantelidis for assistance in sample preparation. Samples were fabricated at the Center for Integrated Systems, Stanford University, a NUNN facility supported by the National Science Foundation. BCB materials and adhesion promoters were provided by Dow Chemical Co.

REFERENCES

1. M. W. Lane, R.H. Dauskardt, N. Krishna, I. Hashim, *J. Mater. Res.*, in publication.
2. R. H. Dauskardt, M. Lane, Q. Ma, N. Krishna, *Eng. Fract. Mech.* **61**, 141 (1998).
3. Q. Ma, H. Fujimoto, P. Flynn, V. Jain, F. Adibi-Rizi, R. H. Dauskardt, in *Materials Reliability in Microelectronics V*, (Mater. Res. Soc. Proc. **391**, Warrendale, PA, 1995) pp. 91-96.
4. J. M. Snodgrass, D. Pantelidis, J. C. Bravman, R. H. Dauskardt, in *Low-k Materials and Applications in Microelectronics*, (Mater. Res. Soc. Proc. **565**, Warrendale, PA, 1999).
5. J. E. Ritter, T. J. Lardner, W. Grayeski, G. C. Prakash, J. Lawrence, *J. Adhesion* **63**, 265 (1997).
6. M. F. Kanninen, *Int. J. Fract.* **9**, 83 (1973).
7. Q. Ma, J. Bumgarner, H. Fujimoto, M. W. Lane, R. H. Dauskardt, in *Materials Reliability in Microelectronics VII*, (Mater. Res. Soc. Proc. **473**, Warrendale, PA, 1997) pp. 3-14.
8. ASTM Standard E647-95 in "1995 ASTM Annual Book of Standards," Vol. 3 (American Society for Testing Materials for Testing and Materials, Philadelphia, 1995).
9. A.G. Evans, *Int. J. Fract.* **9**, 267 (1973).
10. K. M. Lechti, Y.-S. Chai, *J. of Appl. Mech.* **59**, 295 (1992).

Mat. Res. Soc. Symp. Proc. Vol. 612 © 2000 Materials Research Society

A QUANTITATIVE STUDY OF THE ADHESION BETWEEN COPPER, BARRIER AND ORGANIC LOW-K MATERIALS

F. Lanckmans*, S. H. Brongersma, I. Varga, H. Bender, E. Beyne, K. Maex*, IMEC Kapeldreef 75, B-3001 Leuven, BELGIUM,*also at E.E. Dept., K.U.-Leuven, BELGIUM

Abstract

The adhesion between several materials implemented in Cu/low-k integration is studied. Adhesion issues at different interfaces are important with regard to the reliability of back-end processing. Layered test structures are processed to study different interfaces. A tangential shear tester allows quantifying the adhesion force at the interface and provides a relative measurement to compare various materials. Failed interfaces are analyzed using auger electron spectroscopy (AES) and scanning electron microscopy (SEM). Among all studied structures, the strongest interface is seen between a barrier (Ti(N), Ta(N), W_xN) and Cu. A weaker interface proves to be between a low-k dielectric and Cu. However, the presence of a barrier increases the adhesion. The weakest interface occurs between an oxide cap and the low-k material, with a lower adhesion when the low-k material is fluorinated. The low-k/cap oxide interface forms a critical issue with regard to Cu/low-k integration processing such as chemical mechanical polishing (CMP). All test structures show no significant degradation of the adhesion after a thermal cycle up to 400°C.

Introduction

The integration of Cu and low dielectric constant (k) materials in back-end processing will replace aluminium and silicon dioxide by offering reduced signal propagation delay, crosstalk, and power dissipation [1]. The structures used in a Damascene approach consist of a multitude of layers (fig.1.a). The adhesion between these layers in multilevel metallization is crucial for the reliability of integrated circuits. Adhesion refers to the ability of the interface between two materials to resist mechanical separation [2]. Interfaces to be investigated are Cu/barrier, metal/low-k, and metal/cap oxide/low-k (fig.1.a). Other topics of concern are: the role of ambient gases, adhesion degradation during manufacturing and stress effects due to different mechanical properties of the materials [3]. In this work, layered test samples, which reflect interfaces in Cu Damascene technology, are quantitatively evaluated by shear testing to determine the adhesion force of the weakest interface. This provides a relative measurement to compare various materials. The failed interfaces are analyzed using AES or SEM.

Experimental

The test structure, on which the adhesion measurements are performed, was chosen taking into consideration the materials and interfaces occurring in the Cu Damascene technology (fig.1.a) and the size limits imposed by the shear tester. Both the test structure and the tip of the shear tester are shown in figure 1.b. The shear tester is able to apply a shearing force to a sample situated on a flat horizontal surface. The force is transferred to the sample through a metallic tip. During the measurement, the tip approaches the surface of the sample, moves back 0.5 μm upon touching and finally moves horizontally at constant height. After failure of the test structure at the weakest interface, the measurement is stopped and the maximum applied force is registered.

The basic test structure consists of a layered stack with Cu dots on top (figure 1.b). Dependent on the type of interface that is investigated, the stack is different. The most general structure has a PECVD (plasma enhanced chemical vapor deposition) oxide which is deposited on a Si wafer. Followed by a layer of a spin-on low-k material optionally embedded between two refractory metals (Ta(N), Ti(N)and WxN). Two different types of not fluorinated organic dielectrics are used: an aromatic hydrocarbon (SiLK^{TM}, Dow Chemical) and a polyarylene ether-based polymer (Flare 2.0^{TM}, Allied Signal). Fluorinated organic materials are obtained by exposing a not fluorinated dielectric to a NF$_3$ plasma (100W/10s) or by using a fluorocarbon based polymer. Some of the low-k materials also receive a N$_2$/O$_2$ plasma treatment to study the influence of etch processing on the adhesion. A thin sputtered Cu seed layer (150nm) is the next step followed by a resist layer coating ($\approx$15 μm). A lithographic process patterns the resist. The Cu dots are formed in the cavities of the resist by electroplating yielding circular test structures with a diameter around 100 μm. Finally the resist and Cu seed layer are removed using acetone and diluted HNO$_3$, respectively.

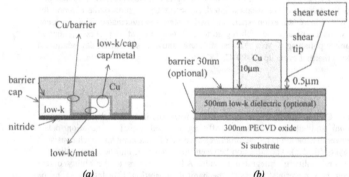

Figure 1: (a) Typical layer structure in single Damascene, (b) Schematic drawing of the tangential shear tester and test structure

Wafers with fewer layers are also prepared depending on the type of interface that is studied. To study the influence of a thermal cycle during processing, some samples receive heat treatments up to 700°C for 5 minutes in a nitrogen environment. All measurements are performed at room temperature. For every test structure and condition at least 16 dots are tested to obtain statistically relevant data. The uniformity of measured adhesion over the surface of a wafer is performed to investigate incidental differences in the starting conditions of the adhesion tests. Measured values of the maximum adhesion force are within a range of ±4 g (excluding data of some dots close to the edge of the wafer), indicating an acceptable uniformity.

Results and discussion

The basic adhesion between a substrate and a film refers to the sum of all interatomic interactions at the film substrate interface and can be quantified as the work necessary to separate the film from the substrate along the interface completely [2]. However, the measured adhesion is referred to as the experimental adhesion. The experimental adhesion depends on the basic adhesion but also on external factors (internal stress within the film, grain size, defects,...) and the influence of measuring technique i.e. the shear testing. The quantity, which is used to describe the experimental adhesion, is the maximum adhesion force F (g). F expresses the maximum force applied by the shear tester and this is when failure at the film/substrate interface occurs. This provides a relative measurement to compare the adhesion between various materials.

Barrier/Cu adhesion

Si/PECVD oxide/barrier/Cu layered structures are processed to investigate the adhesion between Cu and different barrier materials (Ti(N), Ta(N), W_xN). Figure 2.a shows the maximum average adhesion force F_{av} (g) as a function of different heat treatment (HT) conditions for different barriers. Cu directly on the oxide is used as a reference. The value of F_{av} for Cu on oxide is only of limited importance because a barrier layer must be used in order to avoid diffusion of Cu ions in the oxide.

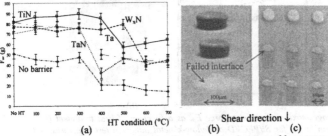

Figure 2 : a) Average maximum adhesion force F_{av} as a function of heat treatment (HT) temperature for 5min in N_2 for a Si/PECVD oxide/barrier/Cu adhesion test structure, b) Failure at the Cu/barrier interface, c) Failure inside the Cu dot.

AES and SEM reveal that the failure always occurs at the barrier/Cu interface (fig. 2.b) or oxide/Cu interface for the reference sample. The general trend is that the presence of a barrier increases the F_{av} comparing to samples without any barrier. This is because some form of chemical interaction at the metal/oxide interface mainly determines the adhesive strength. The adhesive strength is found to be dependent on the oxide free energy of the metal which is higher for refractory metals (Ti(N), Ta(N), W_xN) than for Cu [4]. Only small differences between the values of F_{av} for the different types of barriers are seen, indicating a similar bonding mechanism. An important finding was that the Ti/Cu interface is very strong, causing the structure to fail due to the cohesive failure of Cu (fig. 2.c). For this reason, the results are not included in figure 3.a ($F_{av} \approx 120g$). With regard to typical back-end processing temperatures (<400°C), no adhesion degradation is seen for thermal cycles up to 400°C.

Oxide/low-k/metal adhesion

The next structure investigates interfaces between two materials including a low-k dielectric: Si/PECVD oxide/low-k/(Ti(N))/Cu. This structure allows studying two types of interfaces: cap oxide/low-k and low-k/metal. The failure will occur at the weakest interface. The influence of thermal cycles up to 500°C on the adhesion is evaluated.

The average maximum adhesion force has no dependence on the type of not fluorinated dielectric for the same HT condition (fig. 3.a). The not fluorinated material is either a polyarylene ether-based material or an aromatic hydrocarbon. When treating the low-k dielectric with an O_2/N_2 plasma for 10s (100W) prior to Cu deposition, no change of the F_{av} is measured. F_{av} is also independent of the presence of a 30nm Ti or TiN barrier between the low-k and Cu. All these samples show no decrease of the maximum adhesion force for heat treatments of 5min up to 400°C. The decrease of F_{av} for heat treatments above this temperature is probably related with the thermal degradation of these dielectric materials that have a thermal stability temperature around 450°C. AES measurements and SEM pictures (fig. 3.b) reveal that in all cases the fracture occurred at the PECVD oxide/low-k interface. This explains the negligible influence of a barrier between the low-k and Cu on the measured adhesion force.

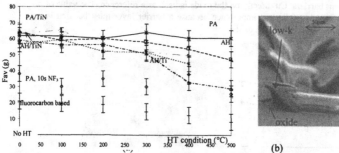

Figure 3 : a) Average maximum adhesion force F_{av} as a function of heat treatment (HT) temperature for 5min in N_2, for Si/PECVD oxide/low-k/metal adhesion test structures. AH and PA refer to an aromatic hydrocarbon and a polyarylene ether-based dielectric, respectively, b) Failure at the PECVD oxide/low-k interface.

Two types of fluorine containing organic dielectrics are also evaluated and shown in figure 3.a. The first is obtained by exposing the polyarylene ether-based polymer to a NF_3 plasma for 10s, causing a decrease of the dielectric constant of 4%. The second material is a fluorocarbon based dielectric and intrinsically contains fluorine. The first material shows no dependence of F_{av} for different heat treatments up to 400°C, whereas the F_{av} for the fluorocarbon slightly decreases for heat treatments at increasing temperatures. AES and SEM both indicate that the failure for both materials also occurred at the oxide/low-k interface. Increasing the amount of fluorine in the dielectric material decreases the maximum adhesion force. This could be related by the fact that an increase of fluorine in the organic dielectric causes less reactivity with the PECVD oxide [5].

This experiment shows that the PECVD oxide/low-k interface has a weaker adhesive strength than the low-k/metal interface. The oxide/low-k interface is important in Cu/low-k integration because oxide is often used as a cap material. The cap oxide serves a dual purpose as a hard mask for etch and as a CMP stop.

Metal/low-k adhesion

In order to study the low-k/metal interface, the low-k material is embedded between two metal layers. The following structures are processed and investigated: Si/PECVD oxide/Ti(N)/low-k/(Ta(N))/Cu. Figure 4 shows the results for the different structures. Two main types of structures can be distinguished depending on whether the Cu is in direct contact with the dielectric or a barrier is separating them. Both the not fluorinated polymers are used as the low-k dielectric.

First, the structure having Cu in direct contact with the low-k material is discussed: Ti(N)/low-k/Cu. For thermal cycles up to 450°C, no significant change of F_{av} is measured (fig. 4). Inspection of the failed surface indicates that the failure occurred at the low-k/Cu interface. This shows that the Ti(N)/low-k interface is stronger than the low-k/Cu interface. In literature, it is found that the bonding between a metal and a polymer is determined by the unfilled d-orbitals of the metal [5]. If a measurable property such as the adhesion is used to examine the polymer/metal interface, it is expected that refractory metals such as Ti or TiN have a higher bonding energy than those of noble metals (Cu), which have few or no unfilled d-orbitals available for bonding interactions [5]. This is experimentally confirmed. Indeed, the test structure failed at the low-k/Cu interface indicating the better adhesion of the polymer with Ti(N) as compared to Cu. A decrease of F_{av} occurs for heat treatments above 450°C. A mixed failure mode is seen using SEM. The failure occurred either at the low-k/Cu or Ti(N)/low-k interface. The behavior is most likely explained by the thermal degradation of the low-k material causing outgassing or breaking of chemical bonds [3].

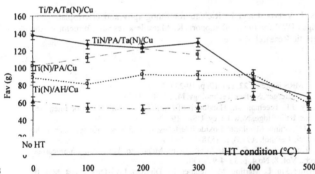

Fig 0   100   200   300   400   500 *tment (HT) temperature for 5min in N_2, for Si/PECVD oxide/Ti(N)/low-k/metal adhesion test structures. AH and PA refer to an aromatic hydrocarbon and a polyarylene ether-based dielectrics, respectively.*

Second, a structure where the low-k dielectric is embedded between two refractory metals is evaluated: Ti(N)/low-k/Ta(N)/Cu. (fig.4). A significant decrease of F_{av} is seen for thermal cycles at temperatures higher then 400°C. As mentioned before, the adhesion degradation is possibly caused by the thermal degradation of the low-k dielectric. The failure occurs at the Ti(N)/low-k interface and no difference is noticed between Ti and TiN. Although the dielectric is embedded between two different refractory metals (Ti(N) and Ta(N)), the failure mechanism is probably not related to a different interaction mechanism of the barriers with the low-k material but to the sequence of processing the test structure. It is shown that the metal/polymer interface is less strong than the polymer/metal interface if the metal is deposited using sputtering [6] [3]. In the last case the driving force for reactions is much higher because of the kinetic energy of the depositing metal atoms, resulting in a better adhesion.

Both sample structures indicate that the presence of a barrier between low-k and Cu increases the adhesion.

Conclusion

The adhesion between several contacting materials, which are used as well in typical Damascene structures, are investigated. Shear testing quantifies the adhesion between two materials. This method allows making a relative comparison between different materials. The strongest evaluated interface is the barrier/Cu interface. Several barriers are evaluated: Ti(N), Ta(N), W_xN. The Ti/Cu interface proves to be very strong because cohesive failure of the Cu occurs. A less strong interface is seen between a dielectric and Cu. In this case the PECVD oxide/Cu interface is weaker than the organic low-k/Cu interface. The presence of a barrier increases the interface strength. The PECVD oxide/low-k interface proves to be the weakest interface. The interface becomes even weaker if a fluorinated dielectric is used. PECVD oxide is often used as a cap material in Cu/low-k integration. Therefore the low-k/cap oxide interface becomes a critical issue in back-end processing, e.g. during the Cu CMP. No significant degradation of the adhesion is found after a thermal cycle up to 400°C.

Acknowledgements

We thank F. Nolmans, I. Vervoort and E. Richard for lithography and electroplating. F. Lanckmans acknowledges IWT for financial support. K. Maex is a research director of the National Fund for Scientific Research Flanders.

References

[1] W.W. Lee and P.S. Ho, MRS Bulletin, 22, No. 10, p. 19-23 (1997).
[2] H.K. Pulker, A.J. Perry, R. Berger, Surface Technology,14, 25-30 (1991)
[3] S.R. Wilson, C.J. Tracy, J.L. Freeman, Jr., Handbook of Multilevel Metallization for Integrated Circuits, Noyes Publications, Park Ridge, New Jersey, USA (1993)
[4] C. Peden, K.B. Kidd, N.D. Shinn, Metal/metal-oxide interfaces: A surface science approach to the study of adhesion, J. Vac. Sci. Technol. A9 (3), p. 1518-25 (1991)
[5] C.-A. Chang, Y.-K. Kim and A.G. Schrott, Adhesion of Metals to Thin-Film Fluorocarbon Polymers, J. Vac. Sci. Tech., Vol. 8, No.4, p. 3304-9 (1990)
[6] A. Alptekin, G. Czeremuszkin, L. Martinu, M. Meunier, M, Direnzo, Low-Diel. Const. Mat. II. Symp. Mater. Res. Soc., Pittsburgh, PA, USA, ix+208, p. 79-84 (1997)

Mat. Res. Soc. Symp. Proc. Vol. 612 © 2000 Materials Research Society

Nanoscale Elastic Imaging of Aluminum/Low-k Dielectric Interconnect Structures

G. S. Shekhawat[‡], O.V. Kolosov[*], G.A.D. Briggs[*], E. O. Shaffer[§], S. Martin[§], and R. E. Geer[‡]

[‡]Center for Advanced Thin Film Technology, University at Albany, SUNY, Albany, NY, 12222
[*]Department of Materials, Oxford University, OX1 3PH UK
[§]The Dow Chemical Company, Midland, MI, 48674

Abstract

A new characterization tool based on ultrasonic force microscopy (UFM) has been developed to image the nanometer scale mechanical properties of aluminum/low-k polymer damascence integrated circuit (IC) test structures. Aluminum and polymer regions are differentiated on the basis of elastic modulus with a spatial resolution ≤ 10 nm. This technique reveals a reactive-ion etch (RIE)-induced hardening of the low-k polymer that is manifested in the final IC test structure by a region of increased hardness at the aluminum/polymer interface. The ability to characterize nanometer scale mechanical properties of materials used for IC back-end-of-line (BEOL) manufacture offers new opportunities for metrological reliability evaluation of low-k integration processes.

Introduction

To maintain the continuation of the progressive decrease in integrated circuit (IC) device feature size and the concomitant increase in device performance requires the accelerated integration of low dielectric constant (low-k) materials to replace the SiO_2 dielectric. The introduction of such low-k materials reduces the inter- and intra-line capacitance in the IC wiring metallization layers used for IC device interconnections, hence resulting in reduced signal delay. For example, the 1999 International Technology Roadmap for Semiconductors calls for a dielectric constant for interlevel dielectrics (ILDs) between 1.6 and 2.2 at the 100nm device node, i.e. when the critical dimension (CD) of the device structure is equal to 100nm [1]. Currently, the only known solutions for sub-1.9 ILDs are porous materials with substantially reduced rigidity due to pore inclusion. This degraded elastic performance increases the potential for mechanical failure during chemical mechanical planarization (CMP) and thermal cycling of the interconnect levels due to the lower overall internal stress budget manageable by the low-k dielectric. To effectively optimize low-k integration within these constraints, it is necessary to develop a new metrology based on the accurate, nanometer scale measurement of mechanical properties of appropriate low-k interconnect test structures.

Pursuant to these goals, findings are presented regarding development of elastic imaging-based metrology for nanometer-scale mapping of mechanical rigidity of damascene processed low-k interconnect structures. Single level damascene aluminum/divinylsiloxane-bis-benzocyclobutene (BCB) test structures were fabricated and planarized. Nanoscale elastic imaging was undertaken via UFM on Al trench and contact pad regions. The acquired image contrast scaled with component rigidity, clearly differentiating metal and dielectric regions. Measured elastic features were independent of surface topography. Detailed mechanical imaging of metal/dielectric trench regions (0.32µm feature size) revealed an increase in the BCB rigidity at the Al interface. On spectroscopic investigation, the BCB polymer exhibited an enhanced oxygen concentration at the metal/dielectric interface attributable to the RIE process. We conclude this increase promotes local silicon oxide formation, enhancing the elastic modulus of the polymer. This imaging capability demonstrates the novel utility of nanoscale elastic imaging in providing process-dependent mechanical modification of low-k materials and underscores its potential as a metrology tool for low-k integration.

UFM: Operating Principles

Imaging scans were acquired using ultrasonic force microscopes configured from commercial AFM platforms (Model CP, Park Scientific Instruments, and SPM4200, JEOL Inc.) employing SiN cantilever contact tips. In elastic imaging mode a super-resonant ultrasonic vibration (f = 2.2MHz) is applied to the sample. In this dynamic regime the cantilever oscillation amplitude is inertially damped resulting in a periodic deformation of the sample surface by the SiN tip. Appropriate modulation of the sample ultrasonic waveform exploits the nonlinear response of the cantilever deflection signal that is subsequently measured via lock-in amplification [2]. The amplitude of the phase-locked deflection signal is proportional to the elastic/adhesive response of the sample surface to the immobile SiN tip. A standard raster utilizing conventional scanning probe software provides a sample contrast image based on the tip/sample interaction. Neglecting adhesive interactions, the resultant image contrast is proportional to the local sample contact stiffness.

Aluminum/BCB Damascene Test Structures

Single-level damascene test structures consisted of thermal chemical vapor deposited (CVD) aluminum in a RIE-patterned BCB matrix. Etching utilized a LAM 4520XL etcher (CF$_4$/O$_2$/Ar etch gas). Diffusion barriers/liners consist of a bilayer stack of physical vapor deposited (PVD)/CVD titanium nitride (TiN). Following Al CVD, CMP was undertaken with an HNO$_3$-based slurry [3]. CMP yielded a highly planarized single level Al/TiN/BCB interconnect structure. This was chosen for baseline qualification studies of nanoscale elastic imaging due to the elastic modulus contrast between Al (70 GPa) and BCB (4 GPa). Emerging metal/low-k material sets will offer greater mechanical contrast due to the increased modulus of Cu (110 GPa) and reduced modulus of porous dielectrics (< 2 GPa).

Results and Discussion

To explicitly demonstrate the elastic imaging capability of the UFM Fig. 1 displays AFM error-signal and UFM image scans obtained simultaneously in contact imaging mode. The former scan confirms the conventional force-feedback acquisition mode indicating, on average, a planar surface of the IC test structure. The UFM scan, in contrast, reveals a striking difference in the elastic properties of the aluminum and polymer regions of the test structure. Larger length-scale topography and elasticity images of the Al/BCB test structure are shown in Fig. 2. The image consists of a 0.32μm pitch Al trench field adjacent to a chessboard contact pad. Note the 10μm Al lead atop the trench field.

The surface topography of the metal/low-k surface shown in Fig. 2a is indicative of expected planarization nonuniformities encountered in damascene processing of Al/BCB test structures [4]. Planarization rates increase with Al fill factor resulting in topographical depressions of

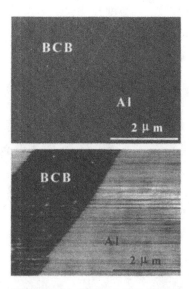

Figure 1. Error signal (top) and UFM
signal (bottom) of an aluminum and BCB
areas on the IC damascene test structure.
Note the elastic sensitivity of the UFM is
independent of the feedback mode of the
AFM platform.

contact leads and pads in comparison with Al trench or isolated fine-line features. The CMP
process utilized here also possessed a low BCB removal rate resulting in relative elevation of
BCB trench walls and fields compared to Al. Both variations are evident in Fig. 2a. In sharp
contrast, the UFM image of Fig. 2b displays no topographical sensitivity; the image contrast is
material specific, clearly delineating polymer and Al regions uniformly across the scan area.
This data, along with that of Fig. 1, clearly illustrates the elastic differentiation ability of UFM to
provide a baseline mechanical profile of a particular metal/low-k polymer integration scheme.

Figure 3 displays high-resolution topography and elastic images of the 0.32μm Al trench
region. Note the image contrast inversion between elasticity and topography data. Topographic
scans reveal depressed Al regions in Fig. 3a resulting from the BCB/Al CMP removal rate
nonuniformity. In contrast, Al trenches exhibit higher image contrast (enhanced rigidity
compared to the polymer) in Fig. 3b. Material identification of these features was confirmed by
comparing topographic and elastic signatures with the nearby BCB field. In all topographic
images studied, BCB trench wall elevation was observed, in addition to a reduced mechanical
response compared to Al lines. In these test structures it was not possible to distinguish the TiN
liner between the Al and the BCB regions, estimated <35 nm from focused ion beam scanning
electron microscopy (FIB-SEM) cross-sections. Aluminum oxidation may have obscured the

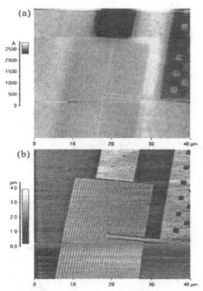

Figure 2. Topographic (a) and elastic (b) imaging scans of Al/BCB damascene interconnect test structure. In (b) image contrast is material (elastic) specific. Also, note the insensitivity of elastic image to topography seen in (a).

elastic contrast between the Al/TiN regions. Current investigations are underway to distinguish metal/liner interfaces using nanoscale elastic imaging.

A specific feature of interest in the elastic image of Fig. 3b is the contrast gradation at the aluminum/BCB interfaces. This is manifested in the UFM image as a linear region of elastic contrast intermediate between that of metal line and BCB running the length of the trench. These regions extend approximately 110nm into the polymer trench wall as determined from the adjacent AFM topograph. A separate confirmation of this spatial estimate was provided by FIB-SEM cross-sectioning of the trench area (not shown). These measurements confirm that the elastic gradation displayed between the Al lines in Fig. 3 occur within the BCB trench wall and outside the Al/TiN liner region.

If it is assumed the AFM/UFM tip/sample interaction is dominated by the surface elastic response, this gradation at the Al/BCB interface indicates a rapid increase in BCB rigidity as the vicinity of the interface is approached. Spectroscopic studies were undertaken on RIE-exposed BCB to ascertain possible process-induced compositional variations responsible for such mechanical variations. Fig. 4 compares XPS-determined composition from untreated and RIE-exposed BCB blanket films before and after a 30-second Ar sputter [5]. As expected, the surface

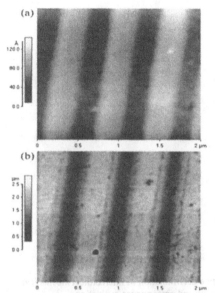

Figure 3. *High-resolution topographic (a) and elastic (b) scans of Al/BCB trench structure. Note elastic gradation at the Al/BCB interfaces.*

of the RIE-exposed BCB is oxygen rich and carbon deficient resulting in a silicon oxide rich surface. More significant is that this composition extends well into the BCB as shown by composition measurements made following a 30 second sputter which, from our best estimates, results in a removal of ~ 50 nm of BCB polymer.

A clearer manifestation of this effect is seen from XPS depth profiling of TiN/RIE-exposed BCB stacks (Fig. 5). Consistent with the data in Fig. 4, the oxygen concentration exhibits a pronounced peak at the TiN/RIE-exposed BCB interface. Compositional profiles from TiN/untreated-BCB stacks lack this RIE-induced oxygen peak.

Comparison of the aforementioned spectroscopic data with the elastic imaging of the Al/BCB test structures demonstrates concomitant mechanical and compositional variations of the low-k polymer at the trench wall interface, implying a direct relationship between RIE-induced oxidation of the BCB and increased polymer rigidity at the aforementioned interfaces. It is concluded from this data that the compositional modification of BCB by the RIE exposure results in the local formation of a rigid silicon-oxide-like polymeric layer that, in turn, produces a local increase in the mechanical rigidity as imaged by UFM. From a metrology perspective this demonstration illustrates the feasibility of imaging mechanical properties of metal/low-k interconnect structures with the same spatial resolution as conventional topography or scanning electron microscopy. Moreover, from a perspective of reliability, this imaging capability

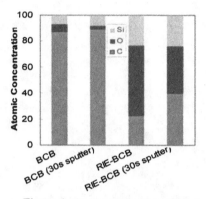

Figure 4. Atomic composition (%) of untreated and RIE-exposed BCB blanket films before and after 30 second Ar sputter. Note the RIE-induced composition variation and its penetration into the polymer.

presents an enabling technology for the quantitative determination of interconnect mechanical constants necessary for accurate predictive modeling and performance evaluation.

Conclusions

A new characterization tool based on UFM has been demonstrated with the capability of imaging the nanometer scale mechanical properties of an Al/low-k damascene test structure. It was shown the metal and dielectric regions of this test structure were differentiated on the basis of elastic modulus with a spatial resolution ≤ 10nm. Moreover, this technique has revealed

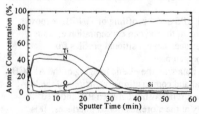

Figure 5. XPS depth profiling of blanket TiN/RIE-exposed BCB films. The oxygen peak at the interface is due to RIE-induced compositional changes of the polymer.

process-induced mechanical variations of the low-k dielectric polymer within the test structure itself, opening new opportunities for metrological evaluation of process dependent mechanical properties of low-k dielectrics and associated optimization of integration processes.

Acknowledgements

The authors gratefully acknowledge J. Hummel and J. Liu for BCB test structure patterning/CMP, and support from the Dow Chemical Company, DARPA, MARCO, and the New York State Center for Advanced Thin Film Technology.

References

1. The International Technology Roadmap for Semiconductors Semiconductor Industry Association, San Jose, CA, 1999.
2. Oleg Kolosov, Kazushi Yamanaka, Jpn. J. Appl. Phys. Vol. 32, 1993, pp. 1095-1098.
3. Heidi Gundlach, Robert Talevi, Zailong Bian, Guillermo Nuesca, Sujatha Sankaran, Kaushik Kumar, Alain E. Kaloyeros, Robert E. Geer, Joyce Liu, John Hummel, Edward O. Shaffer and Steven J. Martin, "Integration of CVD Al Interconnects in a Benzocyclobutene Low Dielectric Constant Polymer: A Feasibility Study", submitted to J. Vac. Sci. Tech. **B**.
4. Hayashi, T. Nakajima, K. Kikuta, Y. Tsuchiya, J. Kawahara, S. Takahashi, K. Ueno and S. Chicaki, 1996 Symposium on VLSI Technology: Digest of Technical Papers, 1996, pp. 88-89.
5. Robert Talevi, Heidi Gundlach, Zailong Bian, Andreas Knorr, Martin van Gestel, Alain E. Kaloyeros, and Robert E. Geer, "Material and Process Studies in the Integration of Plasma-Promoted Chemical Vapor Deposition (PPCVD) of Aluminum with Benzocyclobutene Low Dielectric Constant Polymer", J. Vac. Sci. Technol., **B18**, 252 (2000).

Mat. Res. Soc. Symp. Proc. Vol. 612 © 2000 Materials Research Society

CONCENTRATION AND STRESS EVOLUTION DURING ELECTROMIGRATION IN PASSIVATED Al(0.25 at. % Cu) CONDUCTOR LINES

H.-K. Kao*, G. S. Cargill III* and C.-K. Hu**

*Lehigh University, Bethlehem, PA 18015, gsc3@lehigh.edu
**IBM Research Division, T. J. Watson Research Center, Yorktown Heights, NY 10598.

ABSTRACT

We have used x-ray microbeam fluorescence and diffraction for in-situ measurements of electromigration-induced Cu diffusion and stress evolution in passivated, polycrystalline 10μm-wide, 200μm-long Al(0.25 at.% Cu) conductor lines. Cu migration is in the direction of the electron flow and is determined by the direction and magnitude of the current and by the temperature during electromigration. The effective charge and diffusivity of Cu in Al(Cu) have been obtained from analysis of the Cu concentration profiles. The evolution of electromigration-induced strains normal to the sample surface has been monitored by x-ray microbeam diffraction. A linear strain profile developed after about 9 hrs of electromigration with 1.5×10^5 A/cm^2 at 300°C, corresponding to 3MPa/μm equi-biaxial stress. From the Cu profile measured at the same time, the critical Cu concentration for significantly slowing down Al grain boundary diffusion is estimated to be ~0.15 at. %. These data also confirm that downstream Cu transport is accompanied by a counter flow of Al in the upstream direction.

INTRODUCTION

The role of copper in reducing the rate of electromigration (EM) damage in Al(Cu) conductor lines has been studied for many years [1-9]. Additions of a few percent of copper increase EM lifetimes by one or two orders of magnitude [2]. Cu atoms tend to segregate to the Al grain boundaries and apparently reduce the grain boundary mobility of Al atoms [4,9,10]. Most models for EM in Al(Cu) conductor lines [4,7,10] predict that an incubation time is needed, after the start of current flow, before the onset of significant Al migration. This delay in Al migration is attributed to the Cu concentration having to drop below some threshold level within a current-dependent, critical length of the cathode end of the conductor line. This threshold concentration has been estimated to be ~0.1 at.% [9,11], but no direct measurements have been reported. During EM in an Al(Cu) conductor line with blocking boundaries, a stress gradient is expected to develop in the part of the line where EM of Al occurs, but not in the part of the line where the Cu concentration remains above the threshold value [4,9,10] which drastically reduces Al migration. These issues are addressed in this paper, which describes synchrotron based real-time x-ray microbeam measurements of the dynamics of Cu motion and of stress development during EM in passivated Al(Cu) polycrystalline conductor lines.

EXPERIMENTAL

The samples and instrumentation used in the present experiments are described in ref. 15, and the experimental data shown in Fig. 2 of this paper are the same as those which were discussed in ref. 15. Measurements of Cu-K_α fluorescence show that the Cu concentration is uniform along the conductor lines before electromigration. These measurements were normalized to be consistent with the initial uniform concentration of 0.25 at.% Cu [16] determined from electron microprobe measurements. Cu concentration measurements were made by x-ray microbeam fluorescence during EM for several samples at different temperatures between 325°C and 275°C and for different current densities between 1.5×10^5 A/cm^2 and 3.0×10^5 A/cm^2. The electrical resistance of the line, the Cu fluorescence intensity at ten locations along the line, and the Al (111) plane spacing d_{111} at the same locations were monitored during 38 hrs of EM. Lattice spacings were measured for fiber axis oriented Al grains in near symmetrical scattering geometry with the scattering angle 2θ chosen in the range 25° to 27° for maximum scattered intensity at each measurement location. Stress values were calculated from the stress free lattice parameters [18,19] d_0 and from values of Young's modulus Y and Poisson's ratio ν for (111) fiber textured Al using the equation

$$\sigma_{//} = \frac{-Y}{2\nu}\varepsilon_{\perp} \tag{1}$$

appropriate for equi-biaxial stress, and this equation was also used to calculated changes in stress during EM $\Delta\sigma_{EM}$ from changes in d_{111} values.

RESULTS AND DISCUSSION

Changes of Cu concentration during EM

The Cu concentration distribution along a conductor line was monitored in real-time during a series of electromigration tests while controlling the direction and the magnitude of the electron flow. The results of the Cu concentration evolution at locations near each end of the line are shown in Fig. 1, together with the line's electrical resistance, as a function of time during EM. This experiment began with a forward electron current of 7.5 mA, which corresponds to a current density of 1.5×10^5 A/cm^2, flowing from the upstream end of the line where the reduced distance x/L=0 to the downstream end with x/L=1 at 300°C. After 11 hrs of electromigration in this direction, the electron flux was reversed, to flow from x/L=1 to x/L=0. Fig. 1(a) shows the Cu concentration at measurement locations near the ends of the line, x/L=0.10 and x/L=0.98, starting from the uniform initial reduced concentration of C/C$_0$=1. The initial concentration was C$_0$=0.25 at. %, or C/C$_0$=1.0. The Cu concentration at x/L=0.10 quickly droped to about C/C$_0$=0.5, while the Cu concentration at x/L=0.89 increased to about C/C$_0$=1.4. Cu was quickly depleted at the upstream location because Cu flux was blocked by the W diffusion barrier at that end of the line. The Cu concentration at x/L=0.89 increased during the first 11 hrs of EM to a value somewhat smaller than the equilibrium

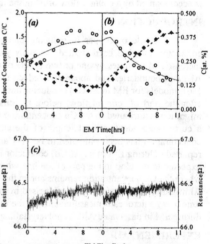

Figure 1(a),(b) *Evolution in the EM-induced Cu concentrations at locations 20μm (♦, x/L=0.10) and 178μm (○, x/L=0.89) of the line, for forward current (a) and for reverse current (b). Solid and dashed curves are model calculations with $Z^*_{Cu} = -8.6$ and $D^{eff}_{Cu} = 6.7 \times 10^{-10}$ cm^2/sec, as described in the text. (c),(d) Resistance of the Al(Cu) line during EM.*

solubility for Cu in Al at this temperature. This increase occurred because the Cu flux was blocked by the W diffusion barrier at the downstream end of the line, where an Al$_2$Cu precipitate forms and serves as a sink for Cu.

After 11 hrs of EM, the electron flux was reversed and EM was continued for another 11 hrs, as shown in Fig. 1(b) for the same two locations. For the reversed current, the Cu concentrations, in Fig. 1(b), for both locations change more rapidly than in Fig. 1(a) for forward current Cu electromigration. This effect can be understood as follows: The Cu concentration gradient along the conductor line at the end of the forward current period served as a chemical diffusion driving force in the same direction as the electromigration driving force when the electron flow was reversed. This caused the rate of decrease of Cu concentration at x/L=0.89 as well as the rate of increase of Cu concentration at x/L=0.1 to be faster during the first 4 hrs of reverse current electromigration than the rates of change in Fig. 1(a) for forward current EM from an initially uniform Cu concentration or for the reverse current EM in Fig. 1(b) after the Cu concentrations crossed about 4 hrs after reversing the current.

The resistance changes versus time for the Al(Cu) line are shown in Fig. 1(c),(d). The total line resistance changes for each period of the electromigration are less than 1%, which confirm that little EM of Al, leading to void growth, occurred during these tests. Therefore we can treat the Al as a stationary reference frame for the Cu migration, without considering interference or the coupling effects due to Al motion.

Model for Cu electromigration

The Al(Cu) conductor line is modeled as a confined, one dimensional continuum with perfectly blocking diffusion barriers at both ends. The Cu fluxes J_e due to the electrical current j and J_c due to the concentration gradient $\partial C_{Cu}/\partial x$ are expressed as

$$ J_e = \frac{D_{Cu}^{eff} C_{Cu}}{kT} \left(Z_{Cu}^* epj \right) \tag{2} $$

and

$$ J_c = \frac{D_{Cu}^{eff} C_{Cu}}{kT} \left(-kT \frac{\partial \ln C_{Cu}}{\partial x} \right). \tag{3} $$

Eq. (2) and Eq. (3) form the basis of a numerical model for Cu electromigration. The Cu concentration is calculated by tracking the atomic flux into and out of cells of incremental lengths Δx along the line axis, which span the full cross section of the line, during each time step Δt [17]. The boundary conditions are that there is no solute flux at either end of the conductor line. Once an Al$_2$Cu precipitate has formed at the downstream end of the line, the Cu concentration there is pinned at the Cu solubility value. When the current direction is reversed, the precipitate becomes a Cu source and eventually dissolves.

The change in Cu concentration in each cell is calculated using the continuity equation $\frac{\partial C}{\partial t} = -\frac{\partial J}{\partial x}$ with $J = J_e + J_c$ from the flux equations, Eq. (2)-(3), and the Cu concentration in each cell at each time step is obtained. From the Cu concentration the steady state value of $\frac{\partial \ln C_{Cu}}{\partial x}$ can be obtained, which is simply related to the value of effective charge Z_{Cu}^* by the equation [6]

$$\frac{\partial \ell n C_{Cu}}{\partial x} = -Z_{Cu}^* e\rho j .\tag{4}$$

With this value of Z_{Cu}^* from the experimentally obtained steady state composition profile, c.f. Fig. 2(a), the model calculations can be used to fit the experimental Cu concentration evolution profiles using D_{Cu}^{eff}, the Cu effective diffusivity, as an adjustable parameter.

Fig. 2(a) shows measurements of the Cu concentration profiles along a line at different times during electromigration with 15mA, or current density of 3×10^5 A/cm^2, at 310°C [15]. The initial Cu concentration was uniform along the line before electromigration. When current was passed through the line, the Cu concentration at the cathode end (x/L=0) depleted quickly. After about 15 hrs, the Cu concentration profile remained stationary. The dashed lines in Fig. 2(a) are the numerically calculated Cu concentration profiles at times before electromigration and at times 2.2 hrs and 27.5 hrs after the start of electromigration. Using the Cu concentration profile after 27.5 hrs of electromigration and Eq. (4), the effective charge Z_{Cu}^* was determined to be -8.6 ± 0.6. In Fig. 2(b), the Cu concentrations at location 52μm and 184μm from the cathode end are plotted versus time [15]. The solid curves in Fig. 2(b) are the best fit using $Z_{Cu}^* = -8.6$, as derived from Fig. 2(a),

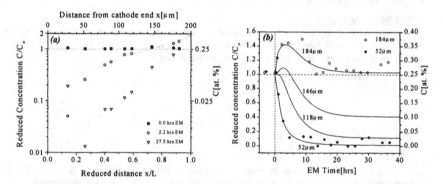

Figure 2(a) *Cu concentration profiles along the conductor line at different times. The numerically calculated profiles are shown as dashed lines.* **(b)** *Evolution of Cu concentration at locations 52μm and 184μm from the cathode end. The solid curves are numerically calculated. Calculations in (a) and (b) use effective charge $Z_{Cu}^* = -8.6$ and $D_{Cu}^{eff} = 1.5\times10^{-9}cm^2/sec$.*

and correspond to $D_{Cu}^{eff} = 1.5\times10^{-9}cm^2/sec$. The temperature dependence of Cu transport is also of interest. We are now analyzing results from experiments at different temperatures to obtain the activation energy for Cu electromigration [17]. An example is shown in Fig. 1(a).

Changes of strain during EM

Changes in d_{111} values were measured, as well as Cu concentrations, during 38 hrs of electromigration at 10 locations for one conductor line. Fig. 3(a) shows strain measurement results after 9 hrs of electromigration with 7.5mA, or current density of 1.5×10^5 A/cm^2, at 300°C. The decrease of the d_{111} values near the cathode end (x/L=0) of the line is due to the Al atoms removed from the grain boundaries by the electron flux, causing the in-plane tensile stress or reduction of in-plane compressive stress [14]. Farther down the conductor line, the accumulation of Al atoms within the grain boundaries creates compressive stress normal to the grain boundaries, causing d_{111} to increase due to the Poisson expansion along the film normal [14].

In Fig. 3(a), a nearly linear strain distribution is observed along about 60% of the total conductor line length from the cathode end. The error bars in Fig. 3(a) are the standard deviation from the average of three measurements at each location. The changes in in-plane compressive stress $\Delta\sigma_{EM}$ shown on the right axis of Fig. 3(a) were calculated with Eq. (1) from changes in the d_{111} values. As shown by the linear fit dashed line, a stress gradient $\Delta\sigma_{EM}/\Delta x$ of about 3MPa/μm extending over 60% of the line length developed after 9 hrs of electromigration.

The absence of stress changes in the remaining 40% of the line length nearest the anode is presumably due to the Cu concentration in this region being above the threshold value which greatly slows down Al diffusion [9,11]. This threshold Cu concentration can be estimated from the Cu concentration profile along the line shown in Fig. 3(b). If we choose the location at 118μm from the cathode end as the location where the Cu concentration is high enough to drastically reduce

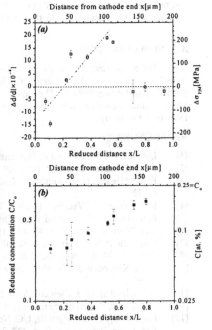

Figure 3(a) *EM-induced Al(111) plane spacing changes, (b) Cu concentration measured at different positions along the Al(Cu) conductor line after 9 hrs of EM.*

the Al grain boundary diffusion, the data in Fig. 3(b) indicate that the critical Cu concentration is ~0.15 at. %. The absence of significant changes in strain in the downstream 40% of the line length, although appreciable Cu enrichment has occurred in this region, also indicate that Cu electromigration by itself does not introduce significant compressive stresses. This observation supports the proposal by Shaw *et al.* [20,21] that Al and Cu movements are coupled during EM of Cu, so that downstream flow of Cu is balanced by upstream flow of Al.

CONCLUSIONS

Cu concentrations and strain distributions during electromigration of passivated Al(0.25 at. % Cu) conductor lines have been measured in real-time by x-ray microbeam fluorescence and diffraction. The Cu concentration distribution can be manipulated by controlling the current direction and magnitude. The evolution of the Cu concentration profile can be analyzed to obtain the effective charge Z_{Cu}^* and the effective grain boundary diffusivity D_{Cu}^{eff} for Cu in Al(Cu). Measurements show a linear strain distribution built-up among 60% of the line length from the cathode end during electromigration stressing with 1.5×10^5 A/cm^2 current density to produce a constant stress gradient of 3MPa/μm after 9 hrs. From these measurements, the critical Cu concentration for substantial reduction of Al diffusion [4,9,10] is estimated to be ~0.15 at. %.

ACKNOWLEDGMENTS

We acknowledge helpful discussions with R. Rosenberg and valuable advice and assistance from P.-C. Wang, K.J. Hwang, and A.C. Ho. This work has been supported by NSF grants DMR-9796284 and DMR-9896002. The experiments have been carried out at the National Synchrotron Light Source beamline X6C, Brookhaven National Laboratory, which is supported by the Department of Energy.

REFERENCES

[1]   I. Ames, F. d'Heurle, and R. Horstmann, IBM J. Res. Dev. **14**, 461 (1970).
[2]   F. M. d'Heurle, Proc. IEEE **59**, 1409 (1971).
[3]   F. M. d'Heurle, N. G. Ainslie, A. Gangulee, and M. C. Shine, J. Vac. Sci. Technol. **9**, 289 (1972).
[4]   R. Rosenberg, J. Vac. Sci. Technol. **9**, 263 (1972).
[5]   P. S. Ho and J. K. Howard, Appl. Phys. Lett. **27**, 261 (1975).
[6]   I. A. Blech, J. Appl. Phys. **48**, 472 (1977).
[7]   C.-K. Hu, P. S. Ho, and M. B. Small, J. Appl. Phys. **72**, 291 (1992).
[8]   C.-K. Hu, M. B. Small, and P. S. Ho, J. Appl. Phys. **74**, 969 (1993).
[9]   M. A. Korhonen, T. Liu, D. D. Brown, and C.Y. Li, Mat. Res. Soc. Symp. Proc. **391**, 411 (1995).
[10]   J. R. Lloyd and J. J. Clement, Appl. Phys. Lett. **69**, 2486 (1996).
[11]   J. R. Lloyd, Semicond. Sci. Technol. **12**, 1177 (1997).
[12]   P.-C. Wang, G. S. Cargill III, I. C. Noyan, E. G. Liniger, C.-K. Hu, and K. Y. Lee, Mat. Res. Soc. Symp. Proc. Symp. Proc. **427**, 35 (1996).
[13]   P.-C. Wang, G. S. Cargill III, I. C. Noyan, E. G. Liniger, C.-K. Hu, and K. Y. Lee, Mat. Res. Soc. Symp. Proc. Symp. Proc. **473**, 273 (1997).
[14]   P.-C. Wang, G. S. Cargill III, I. C. Noyan, and C.-K. Hu, Appl. Phys. Lett. **72**, 1296 (1998).
[15]   H.-K. Kao, G. S. Cargill III, K. J. Hwang, A. C. Ho, P.-C. Wang, and C.-K. Hu, Mat. Res. Soc. Symp. Proc. **563**, 163 (1999).
[16]   Note that the Cu concentration values given in ref. 15 are too large by a factor of two, and that the current and current density should be 15mA and $3 \times 10^5 A/cm^2$ instead of the larger values 25mA and $5 \times 10^5 A/cm^2$ given in ref. 15.
[17]   H.-K. Kao, G. S. Cargill III and C.-K. Hu, to be published.
[18]   W. B. Pearson, *A Handbook of Lattice Spacings and Structures of Metals and Alloys*, Pergamon, New York, 1958, p. 353.
[19]   *CRC Handbook of Materials Science*, C. T. Lynch, Ed., CRC Press, Cleveland, 1974, p. 341.
[20]   T. M. Shaw, C.-K. Hu, K. Y. Lee, and R. Rosenberg, Appl. Phys. Lett. **67**, 2296 (1995).
[21]   T. M. Shaw, C.-K. Hu, K. Y. Lee, and R. Rosenberg, Mat. Res. Soc. Symp. Proc. **428**, 187 (1996).

Interconnect Reliability

Mat. Res. Soc. Symp. Proc. Vol. 612 © 2000 Materials Research Society

Circuit-Level and Layout-Specific Interconnect Reliability Assessments

S.P. Hau-Riege[1], C.V. Thompson[1], C.S. Hau-Riege[1], V.K. Andleigh[1], Y. Chery[2], and D. Troxel[2]
[1]Department of Materials Science and Engineering, M.I.T., Cambridge, MA
[2]Department of Electrical Engineering and Computer Science, M.I.T., Cambridge, MA

ABSTRACT

We have developed a methodology and a prototype tool for making computationally efficient circuit-level assessments of interconnect reliability. A key component of this process has been the development of simple analytic models that relate the reliability of the complex structures in layouts to the simpler straight, junction-free lines of uniform width that are typically used in lifetime tests. We have considered interconnect trees as the fundamental reliability units, where trees can have multiple junctions and limbs, and can also have width variations. We have developed analytic methods for identifying trees which are immune to failure, and have demonstrated that computationally simple techniques lead to the identification of a large fraction of the trees in a circuit as immune to failure (i.e., that they are `immortal'). These trees therefore need not be considered in further analyses. Using simulations and analytic treatments we have also developed default models which allow estimation of the reliability of the remaining trees. These models have been tested and validated them through experiments on simple tree structures with junctions and line-width transitions. Our prototype circuit-level reliability analysis tool projects the reliability of circuits based on specific layouts, and provides a rank listing of the reliability of mortal trees. This allows the user to accept the assessment as is, to carry out more accurate but computationally-intensive analyses of the least reliable trees, or to modify the layout or process to address reliability concerns and reanalyze the reliability.

INTRODUCTION

In today's Si integrated circuit (IC) technology, several meters of metal interconnects are required to build a single high-performance circuit, so that in each IC many millions of metal segments exist. These metallic circuit elements are a great reliability concern owing mainly to electromigration. This concern increases with the level of integration, with each new generation of Si technology requiring the use of a larger number of narrower interconnects, stressed at ever-higher current densities.

Laid-out integrated circuits often have interconnects with junctions. In carrying out circuit-level reliability assessments, it is important to be able to assess the reliability of these more complex shapes, generally referred to as 'trees'. An interconnect tree consists of continuously connected high-conductivity metal within one layer of metallization. Trees terminate at diffusion barriers at vias and contacts, and, in the general case, can have more than one terminating branch when they include junctions. An example of an interconnect tree is shown in figure 1 (a). Most modeling and experimental characterization of interconnect reliability is focused on simple straight lines terminating at pads or vias, instead of interconnect trees. In the conventional reliability assessment methodology, the reliability of a tree is assessed by breaking it up into segments, as shown in figure 1 (b), and estimating the reliability of each segment separately using the results from straight, pad-to-pad or stud-to-stud lines [1]. This is generally inaccurate and often leads to overly optimistic reliability assessments, because the reliabilities of the different segments of the tree are *not* independent of each other. Rather, material can diffusive freely within a tree, and the stress evolution in different parts of the tree is coupled. Unlike lines on 'real' chips, pad-to-pad interconnects have large reservoirs of metal atoms at the line ends falsely indicating a high reliability. Similarly, in stud-to-stud interconnects the beneficial effects of back stress on reliability can be overestimated.

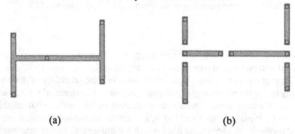

(a)   (b)

Figure 1: (a) Example of an interconnect tree, which is a continuously connected piece of metal within one layer of metallization and is bound by diffusion barriers. The stress evolution during electromigration stressing is a complex function of the current configuration, line lengths, and connectedness. (b) An interconnect tree broken up into segments. In conventional reliability assessment approaches, the reliability of the segments are independently estimated.

Present design strategies have generally been developed with the intent to overdesign interconnects to ensure that a circuit is immune to electromigration-induced failure. The current process of applying unnecessarily conservative and inflexible design rules is no longer feasible when pushing the limits in IC performance for a given technology. Interest has increased in developing new techniques for making realistic reliability assessments *during* the design and layout process (Reliability Computer Aided Design, RCAD), so that reliability analyses can be fed back into the design and layout process immediately. Reliability analyses applied in an RCAD framework can take into account the details of a layout and of circuit operation in order to achieve optimum performance while retaining high overall reliability. Because this requires that reliability assessments be made frequently for vast amounts of interconnect, a computationally efficient and flexible strategy for assessment of interconnect reliability is necessary.

OVERVIEW OF APPROACH

The goal of the research described in this paper is to develop capabilities for carrying out circuit-level and layout-specific reliability assessments. This requires identification and assessment of the reliability of the appropriate fundamental reliability units, which we feel are interconnect trees, and development of methods that minimize the computations required for full-circuit analyses. The latter goal has lead to the development of the hierarchical reliability analysis [2] schematically illustrated in figure 2. In this framework, trees are categorized into mortal and immortal units, using the notion that trees of certain shapes and under certain operating conditions are immortal to electromigration-induced failure. Immortal trees are identified using computationally simple algorithms [2-4], and 'filtered' from further analyses. The reliability of mortal trees is assessed through simulations of the kinetics of the failure processes. To reduce the number of trees that must be analyzed using computationally intensive simulation tools, the reliability of the mortal trees is estimated using a conservative, analytic default model that allows a reliability assessment that results in a ranking of the reliability of the mortal trees. Only the least reliable trees are analyzed in further detail, (using, for example, MIT/EmSim [5-8]). This hierarchical analysis has forms the basis of a prototype tool we have named ERNI, which works with the public domain layout tool MAGIC.

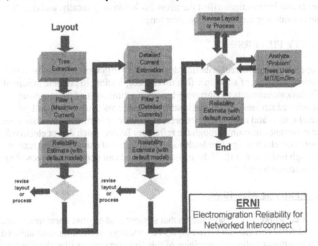

Figure 2: *Schematic illustration of the hierarchical reliability assessment methodology, as embodied in the tool ERNI.*

In the following sections, the Korhonen model for electromigration [9] is extended to interconnect trees, and the immortality filters are described. These filters are applied to all trees in the layout to identify potentially mortal units. The effectiveness of the first filtering step is demonstrated by applying the filters to several microprocessor layouts. Next, an analytic default model based on nodal reliability analyses is described, which is used to analyze the reliability of the potentially mortal units. The default model was validated through lifetime experiments on single nodes in trees. Finally, the electromigration-simulation tool is described.

EXTENSION OF KORHONEN MODEL TO TREES

The Korhonen model [9] for electromigration-induced stress evolution in interconnects was originally proposed for straight-line interconnects. The model has recently been extended to elbow-shaped interconnects as well as intersections of interconnects [2-3]. Elbow-shaped interconnects and interconnect intersections can be treated in a similar manner to straight-line interconnects, because the hydrostatic stress along the length of a diffusion path (averaged over the line cross section) is considered. Experiments on simple tree structures of different shapes, as described in reference [10], validate these assumptions. The effects of bends and intersections can be captured by a spatially-varying effective elastic modulus, B. B can be determined by calculating the hydrostatic stress as a function of sets of homogeneous dilational free strains [9, 11]. Thermal stresses are similar to stresses induced by free strains in the x-, y-, and z-direction. It has been shown through finite element modeling that the volume-averaged hydrostatic component of thermal stresses in passivated "L"- and "T"-shaped interconnects varies by less than 15% along the interconnect [12], so that B also varies less than 15% along the interconnect. Therefore, the effects of bends and intersections on electromigration is small. These results also show that bends and intersections affect the stress evolution only locally, and that these effects are negligible considering the total interconnect length.

IMMORTALITY FILTERS

During operation, atomic redistribution due to electromigration leads to the buildup of mechanical stresses at sites of an atomic flux divergence, leading to possible failure of the circuit. Reliability assessments are therefore based on estimations of the magnitude of the maximum developed stresses. High tensile stresses lead to void nucleation and growth. When refractory metal under- and over-layers are present, voids can form without causing immediate failure because current can shunt through the refractory layers which do not electromigrate. However, continued electromigration leads to void growth and can eventually lead to unacceptably high resistances [3]. If the resistance exceeds an upper limit imposed by the designer, the interconnect fails.

Description of Filtering Algorithms

Filippi and co-workers [13] have shown that straight stud-to-stud interconnect lines with refractory-metal ARC or diffusion barrier layers can undergo electromigration-induced voiding and void growth without failing. The origins of this "immortality" is illustrated in figure 3. Because the lines end in refractory metal diffusion barriers (W vias in Al technology or refractory metal liners in Cu technology), the vias are sites for electromigration-induced flux divergences. At the cathode (up-wind) end of the line, a tensile stress develops, and at the anode end (down-wind) of the line, a compressive stress develops. If the line is short enough, or the current density is low enough, a steady state stress gradient will develop, and the stress will remain too low to initiate failure through void nucleation or dielectric fracture (left side of figure 3). There is therefore a current-density-line-length product, jL, below which the line is immune to the *initiation* of damage.

The immortality concept can be extended to interconnect trees, such as the one shown in figure 1 (a), for both void-nucleation-limited failure [2] and resistance-saturation failure [3]. Further, it was shown that an effective jL product can be defined for an arbitrarily complex tree, and that this effective product can be compared with the jL products which are found to define immortality in stud-to-stud lines, to define immortality in trees. The effective jL product is found by summing the jL products of all paths through a tree, and choosing the maximum sum of the jL products as the effective jL product for the tree. This provides a means for developing a tree-based hierarchical approach to assessment of the reliability of integrated circuits.

Effectiveness of Filtering Algorithms

As illustrated in figure 4, trees can be extracted from layouts, and a worst-case effective jL product can be determined by assuming that all limbs of the trees are at the maximum current density allowed by the design rules. Those trees that are found to be immortal even under these worst-case conditions can then be eliminated from further consideration. With current density estimates for the remaining trees, other trees will be eliminated from further consideration, so that a much smaller population of *mortal* trees can be treated with simulations or default analytic models to make circuit-level reliability predictions.

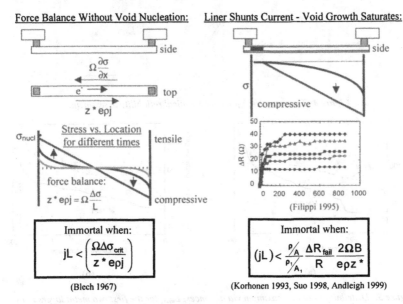

Force Balance Without Void Nucleation:

$$\Omega \frac{\partial \sigma}{\partial x}$$

$$z^* e\rho j$$

Stress vs. Location for different times

force balance:

$$z^* e\rho j = \Omega \frac{\Delta\sigma}{L}$$

Immortal when:

$$jL < \left(\frac{\Omega\Delta\sigma_{crit}}{z^* e\rho j} \right)$$

(Blech 1967)

Liner Shunts Current - Void Growth Saturates:

compressive

(Filippi 1995)

Immortal when:

$$(jL) < \frac{\rho/A}{\rho/A_l} \frac{\Delta R_{fail}}{R} \frac{2\Omega B}{e\rho z^*}$$

(Korhonen 1993, Suo 1998, Andleigh 1999)

Figure 3: *Illustrations of the origins of immortality in straight stud-to-stud lines, in which steady-state stress distributions develop without causing failures.*

We have carried out first-level analyses of the sort described above and outlined in figure 4 on microprocessor layouts available on the web. We have found, as expected, that at service conditions, the majority of interconnect trees are immortal, even when the worst-case assumption is made that all the limbs of all the trees are at the maximum current density (and the currents are in the same directions along the worst-case path). The results of this analysis are illustrated in figure 5. Assuming a design-limit current density of $j_{max} = 1\times10^5$ A/cm^2, a critical line-length-current-density product of 4000 A/cm [14] results in $L_{crit} = 200$ μm, so that 92% of metal-1 trees and 88% of metal-2 trees in case (a), and 95% of metal-1 trees and 71% of metal-2 trees in case (b) are identified as immortal to electromigration-induced failure. In this analysis, it is assumed that the maximum current density flows in every limb of the trees, which is very conservative. If more information about the currents such as the current direction or magnitude are known, the filters are expected to be even more effective.

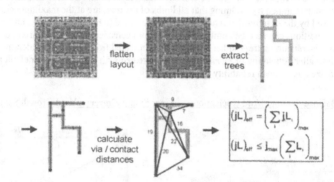

Figure 4: *Tree extraction for a hierarchical reliability assessment.*

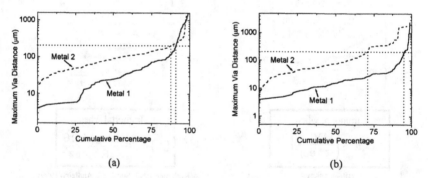

(a)   (b)

Figure 5: *Distributions of the maximum via distances, L_{max}, for the first two metal layers in two microprocessor layouts.*

ANALYTIC DEFAULT MODELS FOR MORTAL TREES

Failure in trees typically occurs at or near nodes, so that an analytic default model for the assessment of electromigration-limited reliability of interconnect trees can be based on nodal reliability analyses. In this section, the default model and its experimental validation are summarized. A detailed description is given in reference [15].

Description of Default Model

Failure can occur due to the resistance increase associated with voiding in the event of large tensile stresses, or due to cracks in the passivation and metallic extrusions in the event of large compressive stresses. Both failure modes are incorporated into our model. We will focus on voiding for now, and we will consider extrusions later.

When the tensile stress exceeds the critical stress necessary for void nucleation, σ_{nucl}, at time t_{nucl}, a void nucleates and starts growing. Eventually the size of the void leads to a resistance increase in one of the limbs exceeding an acceptable limit at time t_{growth}, at which time the node has failed. We will estimate t_{nucl} and t_{growth} separately, and we will take the longer time as the time to failure due to voiding. Similarly, we will estimate the time for extrusions to form, $t_{extrusion}$. We will conservatively take the time to failure of the node, t_{fail}, to be the minimum of the time for failure due to voiding and the time to failure due to extrusions. t_{fail} is estimated for each node in the tree, and the smallest t_{fail} is taken to be the lifetime of the tree.

The effect of each limb in the tree can be described by its diffusivity D_i, current density j_i, and length l_i. We will estimate t_{nucl} for void nucleation for a node connecting n subtrees. (If we imagine we cut a tree at a node, we obtained several smaller trees which we call *subtrees*.) Assuming a constant and time-independent diffusivity along the limb, the stress increase at a node draining into semi-infinite limbs, $\Delta\sigma$, is proportional to [15]

$$\frac{\sum_{i=1}^{n} D_i j_i}{\sum_{i=1}^{n} \sqrt{D_i}}. \tag{1}$$

We will treat the case of stress-dependent and therefore time-dependent diffusivities, as well as the case of near-bamboo microstructures, in which the diffusivity varies spatially along the Al interconnects, in more detail below. For now, the diffusivity is assumed to be constant. An electron flow into a node slows the build up of tensile stress, whereas an electron current out of a node can slow or accelerate the build up of tensile stress, depending on the magnitude of the current. In order to use expression (1) to conservatively estimate the stress evolution at a node, it must be considered that the stress evolution is determined by the whole subtree rather than the limbs immediately connected to the node. We replace each subtree with a semi-infinite limb with a diffusivity and current density chosen from that limb of the subtree that maximizes expression (1). The finite size of the subtree can lead to an overly optimistic reliability estimation only if the subtree slows the build up of tensile stress. If the subtree is too small, the back stress due to tensile stresses at the end of the subtree inhibits atoms from flowing into the node. To estimate the onset of these back stress effects, we associate an effective length with the subtree, which is the maximum path length within the subtree, and compare the distance the stresses are extending into the subtree with the effective length [15]. If back stress effects are present, the subtree is ignored, which is conservative. Once expression (1) is maximized, t_{nucl} can be calculated using $\sigma(t_{nucl}) = \sigma_{nucl}$.

In near-bamboo Al interconnects, the microstructure and therefore the diffusivity vary statistically along the interconnect. To obtain a truly conservative reliability estimate, the continuous range of possible diffusivities ranging from the diffusivity of fully-bamboo interconnects to the diffusivity of fully-polygranular interconnects has to be considered. Similarly, if the diffusivity is stress-dependent [6], the diffusivity can range from the minimum diffusivity at the maximum compressive stress to the maximum diffusivity at the maximum tensile stress. Given the continuous range of diffusivities for each limb, expression (1) has to be maximized using numerical methods [16].

Now we will estimate t_{growth}. Conservatively, we assume the void is present near the node from the beginning of the circuit operation. The local net atomic flow at the node will determine the size of the void and is given by

$$I = \frac{q \cdot \rho}{kT\Omega} \sum_i A_i D_i j_i , \qquad (2)$$

where A_i is the cross-sectional area of a limb. Similar to the void nucleation case, A_i, D_i and j_i have to be chosen from a limb within each sub-tree connected to the node, so that the expression $\sum_i A_i D_i j_i$ is maximized. In case all electron currents of a subtree are directed into the node, the most conservative assumption is zero-current in the subtree. For the calculation of I, it does not matter in which limb next to the node the void nucleates. However, the resistance increase is related to the line width, w, of the limbs, and therefore has to be evaluated for the limb which maximizes the resistance increase, ΔR, of the limb, given by

$$\Delta R \approx \rho_{shunt} \frac{I\Omega/(hw)}{wh_{shunt} + hw_{shunt}} . \qquad (3)$$

ρ_{shunt} is the electrical resistivity of the shunt layer, h, w_{shunt}, and w the line dimensions as sketched in figure 6, and $h_{shunt} = h_{shunt}^{top} + h_{shunt}^{bottom}$. For aluminum-based metallization schemes, typically $w_{shunt} = 0$, and for copper-based metallization schemes, typically $h_{shunt}^{top} = 0$. t_{growth} can be estimated with $\Delta R(t_{growth}) = \Delta R_{max}$, where ΔR_{max} is the maximum allowed resistance increase.

The second failure mode, i.e. failure due to too high compressive stresses, can be treated in a similar way as failure due to void nucleation, with the only difference being that the compressive rather than the tensile stresses at nodes are estimated.

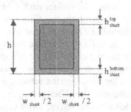

Figure 6: *The cross section of an interconnect.*

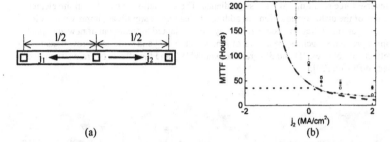

(a)   (b)

Figure 7: (b) *Comparison of experimental data with the default model for the structure shown in (a) with l = 500 μm, and w = 0.27 μm, T = 250°C, j₁=2x10⁶ A/cm².* *The dashed line shows the calculated times for void nucleation, t$_{nucl}$, the dotted line shows the times for void growth, t$_{growth}$, and the continuous line shows the estimated times to failure taken as the maximum of t$_{nucl}$ and t$_{growth}$. Overlaid are experimentally obtained median times to failure represented by solid square symbols and error bars indicating a 95% confidence interval. Also overlaid are times to failure obtained through simulations, represented by open circles.*

Experimental Validation of Default Model

We verified the default model through comparison to electromigration experiments on structures with the shape shown in figure 7 (a). For the different experiments, j_1 was kept constant, whereas j_2 was varied in magnitude and direction. The grain size of the as-deposited films is about 1μm, so after post-patterning annealing the 0.27μm-wide lines (annealing occurs during passivation, packaging, and pre-electromigration testing) are fully bamboo, so we do not consider the effects of the alloy additions or non-bamboo structures [6]. The median times to failure are plotted as a function of j_2 in figures 7 (b). Also shown in figure 7 (b) are the calculated t$_{nucl}$ with σ$_{nucl}$ = 500 MPa at 250°C. The nucleation stresses were obtained by matching experimental times to failure with simulation results. As in the experiment, t$_{growth}$ was evaluated for a 30% resistance increase. The comparison of experiments with the models shows that the default models predict the lifetime of the center node conservatively, and that the right functional dependence on j_2 is predicted. Similar experiments were performed on lines with widths w = 3.0 μm, so that the lines were fully polygranular. These results are described in reference [10]. Again, the comparison shows that the default model conservatively predicts the lifetime of the center node, and that the right functional dependence on j_2 is predicted.

ELECTROMIGRATION SIMULATIONS IN TREES

We have developed a tool, MIT/EmSim, for simulation of electromigration-induced stress evolution and failure of interconnects as a function of their length, width, thickness, and connectedness; as a function of materials selection and dimensions of conductor and clad/barrier layers; as a function of alloy additions and grain structures; and as a function of current density, temperature and thermal history [5, 7-8]. The calculations carried by MIT/EmSim are based on the Korhonen model [9] that accounts for the electron wind force on atomic diffusion as well as the reverse force resulting from stress gradients. The simulation is based on finite element solutions of the diffusion equation. In addition to leading to significant improvements in accuracy, our simulation method is relatively easily adapted to treatment of new and more complex problems, such as the effects of junctions [15, 17] and wide-to-narrow transitions [18], the effects of complex 2D and 3D grain networks [19], and the effects of grain structure on surface diffusion [20].

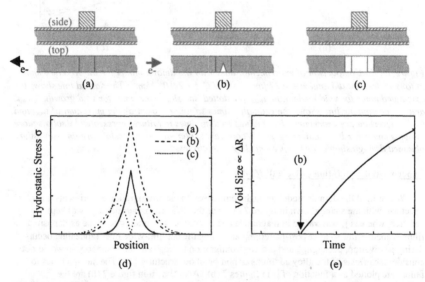

Figure 8: *Sequence of events leading to failure at a node with two active limbs. (a) A hydrostatic, tensile stress builds up in the line under the via. (b) If the tensile stress exceeds a threshold value, σ_{nucl}, a void nucleates and relieves the stress around the void. (c) The void grows and the resistance of the lines increases. (d) Evolution of the stress profiles near the via. (e) Resistance increase, ΔR, as a function of time. The rate of resistance increase eventually decreases due to back stress effects.*

We extended MIT/EmSim to allow the simulation of the effects of electromigration in interconnect trees [15, 17]. A typical failure scenario occurring at junctions in trees is shown in figure 8 (a) to (c). In this example, electrons are flowing from the top via into the interconnect to the left and to the right. A tensile stress peak develops at the junction, as shown in figure 8 (d). When the stress exceeds the critical stress necessary for void nucleation, a void forms, as shown in figure 8 (b), which relieves the stress around the void. The void grows, as shown in figure 8 (c), and the resistance starts to significantly increase, as shown in figure 8 (e). This happens because the current is forced to flow through the higher-resistivity shunt layer. Eventually, void growth slows down because of increasing back stress effects associated compressive stresses at the line ends.

The simulations has been used to predict the reliability of simple tree structures, such as 'I'-, 'U'-, 'L'-, and 'T'-shaped interconnects [10, 15] and for 'dotted-I' structures such as the one shown in figure 7 (a). Models, simulations and experimental results on the reliability of interconnect trees are shown to yield mutually consistent results. These experiments have been performed on Al(Cu) interconnects. However, the increase in ratio of wiring delay to the intrinsic transistor delay has provided the motivation for the IC industry to move from aluminum-based interconnects embedded in SiO_2 to copper-based metallization systems with inter-level dielectrics (ILD) having lower dielectric constants, k, than SiO_2 [21]. The Korhonen model for electromigration cannot be applied to Cu-based metallization systems without modifications. For example, electromigration experiments on Cu interconnects have shown that diffusion takes place primarily along the Cu/passivation interface, regardless of the grain boundary structure [22]. This has been incorporated into MIT/EmSim [8, 20]. In addition, low-k ILD's are often polymer-based, and are mechanically much softer than SiO_2 [23]. Typical examples of polymer-based ILD's are fluorinated poly(arylene ether) (FLARE) [24], SiLK [25], and polyimide [26], all of which have Young's moduli that are more than an order of magnitude smaller than that of SiO_2. The effects of changing the mechanical properties of the ILD along with changes of line aspect ratio and the presence of liner or barrier materials on the electromigration behavior of Cu interconnects can be studied through finite element modeling [10-11]. Similar experiment to those performed on Al(Cu) structures are being performed on Cu-based interconnects in order to characterize back-stress and immortality conditions.

DISCUSSION: PROTOTYPE RCAD TOOL 'ERNI'

Through completion of a full set of analytic models for hierarchical filtering of immortal trees, and for assessment of the reliability of mortal trees, the work described above has enabled the development of a prototype tool for carrying out circuit-level reliability assessments, as outlined in figure 2. This tool is called ERNI (electromigration reliability of networked interconnect) and can be used with the layout tool Magic (or the Java-based version, Majic) to provide a reliability estimate for a specific circuit, and to provide a list of mortal trees, ranked according to their reliability using the default models. The trees identified as least reliable can be shipped to MIT/EmSim for more accurate, but computationally intensive, reliability estimates. The layout of these trees might also be modified and the circuit reliability reassessed. Also, the effects of modifications in the processing of interconnects or in materials selection or dimensions (e.g. liner thickness) can be assessed.

ACKNOWLEDGMENTS

This research was supported by the SRC. S. P. Hau-Riege also holds an SRC graduate fellowship. We would like to thank Michael E. Thomas (formerly with National Semiconductor, now with Allied Signals) for providing processing of electromigration samples and access to equipment for electromigration testing.

REFERENCES

[1] H. A. Schafft, *IEEE Trans. ED* 34, 664 (1987).
[2] S. P. Riege, C. V. Thompson, and J. J. Clement, *IEEE Trans. ED* 45, 2254 (1998).

[3] J.J. Clement, S.P. Riege, R. Cvijetic, C.V. Thompson, *IEEE Trans. CAD* **18**, 576 (1999).

[4] C.V. Thompson, S.P. Riege, and V. Andleigh, *AIP Conference Proceedings* **491** of the 5th International Workshop on Stress-Induced Phenomena in Metallization, Stuttgart, Germany, pp.62 (1999).

[5] Y.-J. Park and C.V. Thompson, *J. Appl. Phys.* **82**, 4277 (1997).

[6] Y.-J. Park, V.K. Andleigh, and C.V. Thompson, *J. Appl. Phys.* **85**, 3546 (1999).

[7] A demonstration of MIT/EmSim is accessible on the World Wide Web at *http://nirvana.mit.edu/emsim.*

[8] V. K. Andleigh, V. T. Srikar, Y.-J. Park, and C. V. Thompson, *J. Appl. Phys.* **86**, 6737 (1999).

[9] M. A. Korhonen, P. Boergesen, K.N. Tu, and Che-Yu Li, *J. Appl. Phys.* **73**, 3790 (1993).

[10] S.P. Hau-Riege and C.V. Thompson, these proceedings.

[11] S.P. Hau-Riege and C. V. Thompson, "The Effects of the Mechanical Properties of the Confinement Material on Electromigration in Metallic Interconnects", submitted to J. Mat. Res.

[12] Y.-L. Shen, *Proc. 37th Int. Reliab. Phys. Symp.* 283 (1999).

[13] R.G. Filippi, G.A. Biery, and R.A. Wachnik, J. Appl. Phys. **78**, 3756 (1995).

[14] J.R. Kraayeveld, A.H. Verbruggen, W.-J. Willemsen, and S. Radelaar, *Appl. Phys. Lett.* **67**, 1226 (1995).

[15] S. P. Hau-Riege and C. V. Thompson, "*Modeling and Experimental Characterization of The Reliability of Interconnect Trees*", submitted to *J. Appl. Phys.*, submission number JR00-0361.

[16] J. Stoer and R. Bulirsch, Introduction to Numerical Analysis, Springer-Verlag, New York, NY (1980).

[17] S. P. Hau-Riege and C. V. Thompson, "*Electromigration-Saturation in a Simple Interconnect Tree*", submitted to *J. Appl. Phys.*

[18] C.S. Hau-Riege and C.V. Thomson, submitted to *J. Appl. Phys.*, submission number JR00-0197.

[19] W. Fayad and C.V. Thompson, An Analytic Model for the Development of Bamboo Microstructures in Thin Film Strips Undergoing Normal Grain Growth, submitted for publication to Phys. Rev. B, February 2000.

[20] W.Fayad and C.V. Thompson, Modeling texture Effects on Electromigration-Limited Reliability in Bamboo Interconnects, submitted for publication to J. Material Research February 2000.

[21] D. Edelstein, J. Heidenreich, R. Goldblatt, W. Cote, C. Uzoh, N. Lustig, P. Roper, T. McDevitt, W. Motsiff, A. Simon, J. Dukovic, R. Wachnik, H. Rathore, R. Schulz, L. Su, S. Luce, and J. Slattery, *IEEE Intl. Electron Devices Meeting Digest*, 773 (1997).

[22] C.-K. Hu, R. Rosenberg, and K.Y. Lee, *Applied Physics Letters* **74**, 2945 (1999).

[23] D.T. Price, R.J. Gutmann, and S.P. Murarka, *Thin Solid Films* **308-309**, 523 (1997).

[24] K.S.Y. Lau, J.S. Drage, N.P. Hacker, N.M. Rutherford, R.R. Katsanes, B.A. Korolev, T.A. Krajewski, S.P. Lefferts, H. Sayad, P.R. Sebahar, A.R. Smith, W.B. Wan, and E.C. White, *Proceedings Thirteenth International VLSI Multilevel Interconnection Conference (VMIC)*, pp. 92-7. Tampa, FL (1996).

[25] J. Waeterloos, M. Simmonds, A. Achen, and M. Meier, *Europ. Semicond.* **21**, 26 (1999).

[26] A.L.S. Loke, J.T. Wetzel, P.H. Townsend, T. Tanabe, R.N. Vrtis, M.P. Zussman, D. Kumar, C. Ryu, and S.S. Wong, *IEEE Trans. ED* **46**, 2178 (1999).

[27] Z. Suo. *Acta mater.* **46**, 3725 (1998).

Mat. Res. Soc. Symp. Proc. Vol. 612 © 2000 Materials Research Society

Electromigration Reliability of Dual-Damascene Cu/Oxide Interconnects

Ennis T. Ogawa,[1] Volker A. Blaschke,[2] Alex Bierwag,[1] Ki-Don Lee,[1] Hideki Matsuhashi,[1] David Griffiths,[2] Anup Ramamurthi,[1] Patrick R. Justison,[1] Robert H. Havemann,[2] and Paul S. Ho [1]
[1] Interconnect and Packaging Laboratory, Microelectronics Research Center,
The University of Texas at Austin, Austin, TX 78712-1100
[2] SEMATECH, 2706 Montopolis Dr.
Austin, TX 78741-6499

ABSTRACT

An electromigration study has determined the lifetime characteristics and failure mode of dual-damascene Cu/oxide interconnects at temperatures ranging between 200 and 325 °C at a current density of 1.0 MA/cm^2. A novel test structure design is used which incorporates a repeated chain of "Blech-type" line elements. The large interconnect ensemble permits a statistical approach to addressing interconnect reliability issues using typical failure analysis tools such as focused ion beam imaging. The larger sample size of the test structure thus enables efficient identification of "early failure" or extrinsic modes of interconnect failure associated with process development. The analysis so far indicates that two major damage modes are observable: (1) via-voiding and (2) voiding within the damascene trench.

INTRODUCTION

The arrival of dual-damascene Cu/oxide interconnects signifies an important change in the type of technologies necessary to achieve successful integration beyond 0.25 μm minimum feature dimension. Yet despite the formidable technological challenges ahead, detailed assessment of dual-damascene interconnect reliability has not been demonstrated publicly. The work here outlines a potentially useful methodology for electromigration (EM) reliability analysis. Two major issues have been targeted: (1) characterization of EM test structures using multiply-linked chains of interconnect elements and (2) determination of EM performance and damage formation mechanism - especially that of "early failures." [1]

EXPERIMENTAL DETAILS

Test Structures

The samples were prepared at Sematech using 200 mm wafers and consist of two-level interconnect structures based on a Ta/low temperature PVD seed Cu/ electroplated (EP) Cu stack. [2] Above the upper metal level (M2), a SiN$_x$ cap/ Al bond pad process is used for outside electrical connections. Heat treatment consists of a short excursion up to roughly 400 °C during nitride passivation, a 30 min. anneal at 325°C in forming gas after wafer processing, and a 35 min. cure at 330 °C for die attach. Consequently, the interconnects show "near bamboo" microstructure.

The test structure design (labeled LC) used here is a variation of the design used by I. A. Blech [3] where line elements of different lengths are serially aligned for critical length effect observation. In this version, however, an M2 dual-damascene integration scheme is used, where the line dimensions at metal 1 (M1) ensure EM failure above M1 and the serially arranged M2 interconnect elements each receive the same nominal current density. The M2 line elements vary in length from 10 to 300 μm using 14 distinct lengths. This chain is repeated six times such that 84 individual elements form the test structure. Extrusion monitor lines, interlaced between the existing interconnect chain, enable detection of interconnect anode shorting due to extrusion damage. The M2 line height x width is 0.8 μm x 0.6 μm. The via height is 0.7 μm with cross-section 0.5 μm x 0.5 μm, and the M1 cross-section is 0.8 μm x 1.0 μm with line length about 10 μm or less. Two types of test structures differing only by their interconnect line terminations have been examined. One type (labeled LP) utilizes a landing pad at the M2 line ends to ensure adequate overlay budget between via and M2 levels. The other design scheme uses no landing pad (NLP) at the M2 line ends for more faithful replication of industry-type interconnects.

EM Testing

EM testing is performed on a home-built system capable of housing 36, 16-pin, dual-in-line (DIP) packages. [4] The system consists of the high temperature oven, current monitor board to individually determine test structure current, current regulator system, and voltage measurement switching system to measure both test structure voltage and current monitor voltage. A schematic is shown in Figure 1. The system has been tested to about 400 °C (but is designed to reach higher temperatures) and exhibits temperature uniformity among all packages to within the resolution limits of the thermocouples located throughout the oven, about ±0.5 °C. The system is robust to different environments including high vacuum conditions, but usually a background pressure of about 30 Torr of house N_2 is used. Typically, two test structures are measured on a single DIP for a maximum capacity of 72 test structures per test. Joule heating analysis indicates that only test structures tested at 325 °C experience a significant temperature increase (< +4 °C).

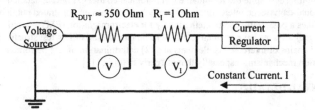

Figure 1. *Schematic of EM test setup under constant current conditions. R_I is a probe resistor of known resistance in series with R_{DUT} to enable accurate current measurement for each device under test (DUT). V and I then determine R_{DUT}.*

DISCUSSION

Resistometric Data and Interpretation

EM studies of these test structures prepared from a single wafer were done at 220, 275, 300, and 325 °C at a current density of 1×10^6 A/cm^2 (based on the M2 trench dimension). Well-behaved resistometric data is shown in Figure 2. Common to all resistance traces is the sudden appearance of a resistance increase some time after test initiation. This first occurrence is referred to as "onset" failure and is attributed to voiding of a region within an interconnect – suspected to be across the cathode (electron source) via bottom. Via-voiding requires very little material depletion at the via bottom to open the interconnect and is thus a possibly easy avenue to failure. After onset failure, recovery of the test structure resistance in the LC structure to a somewhat higher (within 20 Ω) baseline than before onset failure is observed. The recovery mechanism is not clear and definitely requires further study. Regardless, onset failure was chosen as the failure criterion for test structures since its rather short time interval would be significant when extrapolated to "at use" temperatures. This criterion is also useful in the sense that single interconnect failure within the chain causes structure failure – following a so-called weakest-link scenario. More recently processed samples during EM testing done at Sematech do not uniformly show onset behavior and recovery, indicating that this failure mechanism is probably not intrinsic in nature.

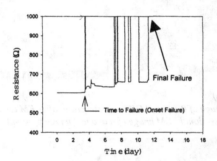

Figure 2. *A resistometric trace for LC/LP 0.5 μm test structure is shown for a sample tested at 220 °C and $1.0x10^6$ A/cm².*

Failure Analysis

After EM testing, examination of the test structures showed that some of the samples showed large amounts of surface damage, presumably from EM-driven mass transport; however, the rest of the samples showed little surface damage. Focused ion beam (FIB) imaging was then used to determine the failure site. A focused ion beam (FIB) image (shown in Figure 3a) using the ion beam induced contrast (IBIC) method clearly shows the location of final and catastrophic interconnect failure. [Note also the presence of extrusion damage - observable only on samples tested at 325 °C - at the anode ends of adjacent interconnects (see Figure 3b)]. The location of greatest IBIC contrast indicates that the final failure is presumably due to voiding at the cathode via bottom.

Figure 4 shows the scanning electron micrograph (SEM) in cross-section of the final failure site for a test similar structure tested at 220 °C and $1.0x10^6$ A/cm². The void was observed to form some distance from the Ta barrier located at the via bottom and may be the result of incomplete cleaning of the via bottom during damascene processing prior to Cu deposition. In the figure, the barrier at the via bottom has apparently remained continuous although the sidewall barrier has clearly been damaged. Via-failure is probably enhanced by the fact that the via cross-section is smaller than the M2 cross-section resulting in a nearly 2x higher current density in the via in comparison to the trench.

In contrast, samples later fabricated using improved processing steps and tested at Sematech have significantly longer median lifetime than this earlier sample. In these newer samples, interconnect failure is observed exclusively on M2 level (see Figures 5a and 5b) and shows that the via-voiding mechanism is attributable to process immaturity and is thus an early failure mechanism.

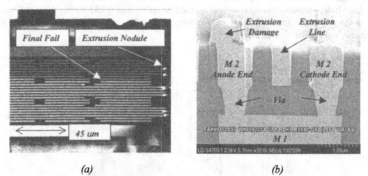

(a)   (b)

Figure 3. (a) FIB/IBIC image of 0.5 µm , LC/LP test structure, after EM testing at 325 °C and $1.0x10^6$ A/cm². (b) Cross-sectional SEM image of a anode damage in sample of Figure 4a.

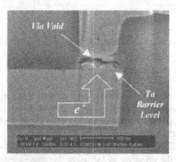

Figure 4. Cross-sectional SEM image of a via failure in a 0.5 µm , LC/NLP test structure, after EM testing at 220 °C and $1.0x10^6$ A/cm².

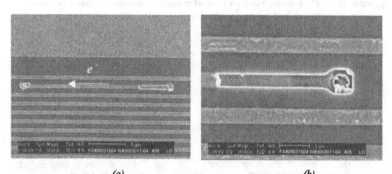

| (a) | (b) |

Figure 5. (a) Typical top SEM view of damage mode in more recently processed samples tested in the range 300 to 350 °C and 1.5×10^6 A/cm². (b) Closer view of EM damage at cathode end.

Failure Distribution

A cumulative failure distribution (CDF) plot is shown in Figure 6a of the set of non-landing pad (NLP) test structures examined at 220, 275, 300, and 325 °C. The CDF results for the landing pad (LP) test structures are similar to the NLP versions except that the failure times are typically more than twice as long as their NLP counterparts. This result probably arises from the larger overlay budget for the LP structures, the extra materials reservoir for mass transport at the M2 line ends, and a slightly wider via mouth from processing which lowers the effective current density within the via. An observed decrease of σ with increasing temperature is expected for test structures in multiple ensemble configurations failing under a single mode. [5] Finally, the failure data appears quite consistent with a single lognormal distribution (although it should not be exactly so because the lognormal is not a scaleable distribution [5]).

The curves for activation energy determination are shown in Figure 6b. The activation energy of 0.6 eV seems somewhat low [6] and likely results from a first generation process and/or the use of the onset failure criterion. Since the SiN_x/Cu interface is considered the faster diffusion interface compared to Ta/Cu, the SiN_x/Cu interface should be rate controlling. [6] Also, the Ta barrier used in the via should enable it to be longer lived relative to the M2 trench (except for the higher local current density). However, local process inhomogeneity can create statistically infrequent flaws in the vias made vulnerable to failure by the multiple ensemble nature of the test structure design. So, an early failure via with the higher local current density can win its race to fail relative to the trench and, hence, control test structure failure. This early failure is thus regarded as responsible for the low activation energy value.

The appearance of two damage modes but a more or less monomodal failure distribution appears somewhat confusing. The damage, however, was assessed after EM testing of all test samples such that failure analysis reveals damage in the test structures accrued well beyond that of onset failure. The choice of onset rather than a final fail requires further investigation. Choosing a final failure condition would alter the activation energy value; however, reliability is clearly limited by onset behavior. The more recently processed samples show an activation energy value of 1.1 eV, but also exhibit a somewhat bimodal distribution that is possibly a consequence of onset and final failure-type mixing.

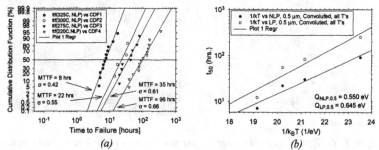

(a)   (b)

Figure 6. (a) Cumulative failure plots for NLP version of the examined test structures. (b) Determination of activation energy for both LP and NLP versions of the test structure.

CONCLUSION

The electromigration behavior of dual-damascene multiple ensemble interconnects permits simultaneous study of many interconnects under essentially identical conditions. The larger test ensemble permits isolation of infrequent and extrinsic failure modes associated with process immaturity. Additionally, failure analysis shows that two distinct aggregate damage modes are possible, dependent upon testing and local process conditions on the DUT, even if a single damage mechanism is expected to dominate. The activation energy reflects a system dominated by an early failure mechanism associated with via voiding. More detailed study to come should clarify this picture.

ACKNOWLEDGEMENTS

This study was supported by Sematech's Cu Integration group and with additional support from SRC/CAIST task #448.046. We greatly appreciate the work done by Laura Dennig, Nara Lieou, and Roger Reyes in obtaining the FIB and X-SEM images. We are also very grateful to Anne Nelsen and Mark Breene, respectively, for managerial and technical support at Sematech.

REFERENCES

1. M. Gall, Ph. D. dissertation, The University of Texas at Austin, May, 1999.
2. V. Blaschke, J. Mucha, B. Foran, Q. T. Jiang, K. Sidensol, A. Nelson, in *1998 Proceedings of the Advanced Metallization Conference*, pp.43-49.
3. I. A. Blech, J. Appl. Phys., Vol. 47, No. 4, 1203, 1976
4. A. J. Bierwag, M. S. thesis, The University of Texas at Austin, December, 1999.
5. J. R. Lloyd, J. Kitchin, J. Appl. Phys., Vol 69, No.4, 2117, 1991.
6. C.-K. Hu, R. Rosenberg, and K. Y. Lee, Appl. Phys. Lett., Vol. 74, No. 20, 2945, 1999.

Mat. Res. Soc. Symp. Proc. Vol. 612 © 2000 Materials Research Society

ELECTROMIGRATION CHARACTERIZATION VERSUS TEXTURE ANALYSIS IN DAMASCENE COPPER INTERCONNECTS.

Authors : T.BERGER*, L.ARNAUD, R.GONELLA*, I.TOUET, G.LORMAND**.

CEA, Direction des Technologies Avancées, 38054 Grenoble CEDEX, FRANCE.
*ST MICROELECTRONICS, 38926 Crolles CEDEX, FRANCE.
**GEMPPM UMR CNRS 5510, INSA, 69621 Villeurbanne CEDEX, FRANCE.

ABSTRACT

We have studied the effect of texture (X-ray diffraction pole figures) and grain morphology (Focus Ion Beam cross-sections) on the electromigration performances of copper damascene interconnects. Three different metallizations have been characterized : Chemical Vapor Deposition copper deposited on TiN (process A) and electroplated copper deposited either on Ta (process B) or TaN (process C). The reliability performance of these interconnects has been evaluated using both Wafer Level Reliability (WLR) and Package Level Reliability (PLR) tests on 4 and 0.6 µm wide lines using single metal level test structures. On the basis of the activation energy values and failure analysis observations, we concluded that interfacial diffusion plays a key role in the electromigration phenomenon for processes B and C whereas grain boundaries seem to be the active diffusion path for process A. The existence of several failure mechanisms during electromigration tests (interfacial or grain boundary diffusions), the impact of the damascene architecture on microstructure (sidewall textures and non columnar grain shapes) and the copper propensity for twinning seem to mask the impact of texture on the electromigration reliability of copper damascene interconnects.

INTRODUCTION

As a leading candidate for future interconnect material [1], Cu has been intensively studied during the last few years. The damascene process is nowadays the preferred process for Cu interconnects integration. Unlike the conventional substractive Reactive Ion Etching process (RIE), the damascene process begins with dielectric deposition. Photolithographic patterning and RIE are then used to define trenches in the dielectric. Diffusion barrier deposition and copper deposition follow the dielectric etching. The excess metal is finally removed through Chemical and Mechanical Polishing (CMP).
Texture and grain size distributions where found to be key factors controlling the electromigration behavior of Al-based interconnects. A controlled <111> texture and a large grain size provided longer electromigration (EMG) lifetimes and better electromigration performances [2,3].
In this study, we present a texture analysis (X-ray diffraction pole figures) of damascene Cu interconnects fabricated with three different processes. Several line widths ranging from 0.3µm to 3µm have been investigated for texture characterization. A qualitative grain morphology characterization has been carried out using Focus Ion Beam (FIB) cross-section imaging of the lines with a Scanning Electron Microscope (SEM). The electromigration reliability of these interconnects has been extensively studied for two line widths (w=4 and 0.6µm) using both highly and moderately accelerated tests.

SAMPLES PREPARATION

• Electromigration samples preparation :
For process A, Chemical Vapor Deposition (CVD) of copper and TiN were performed in a precision 5000 AMAT cluster tool. Two CVD chambers are available allowing successive deposition of the diffusion barrier and copper without vacuum break. A 10 nm thick layer of CVD TiN was deposited on patterned SiO_2 prior to copper deposition. CVD Cu deposition was performed at 190°C using an organic precursor (Cu(hfac)vtms).
For processes B and C, Ta and TaN barrier layers were both deposited using an Ionized Metal Plasma (IMP) assisted Physical Vapor Deposition (PVD) process. The aimed thickness of the diffusion barrier layers was 25nm. The barrier deposition was followed by the deposition of 120

nm of IMP-PVD Cu used as a seed-layer for subsequent electrolytic deposition of Cu. 1150 nm of electroplated (ECD) Cu were finally deposited.

All samples underwent an annealing at 400°C during 10 min prior to CMP to improve adhesion between Cu and the barrier layer. Samples were then polished to a final thickness of 0.4 μm. A subsequent 1μm SiO$_2$ passivation followed by an annealing at 425°C during 30 min was performed before pad opening. After pad opening, Al was deposited on pads to prevent Cu from oxidizing and to allow a reliable bonding for high temperature PLR electromigration tests.

- Texture samples :

Using similar processes and a different mask, control wafers were processed to allow subsequent texture and grain morphology characterization. Each die is composed of four lines networks with varying line widths : 3, 1, 0.5 and 0.3μm. The spacing between the lines is identical to the line width. We obtained a final copper thickness of 0.4μm for all metallizations.

TEXTURE CHARACTERIZATION : EXPERIMENTAL DETAILS AND RESULTS.

X-Ray Diffraction (XRD) pole figures were collected for <111> and <200> orientations using a standard four-circle goniometer. A 2.5°x5° grid was used with the polar angle χ ranging from 0° to 70° and the azimuthal angle ϕ from 0° to 355°. The counting time is 5 seconds per point. This procedure resulted in pole figures in which the 20° outer ring is missing because of severe defocusing. The pole figures have been corrected from background and normalized.

The results are presented in figure 1 for the wider lines (w=3μm) and in figure 2 for narrower lines (w=0.3μm). Wide lines of Process A exhibit a weak <200> fiber texture relative to the bottom of the damascene trench (see fig. 1a). This texture completely disappears for narrow lines (see fig. 2a). Wide lines of processes B and C exhibit a strong <111> bottom fiber texture (see fig. 2b and 2c). The texture is slightly stronger for process B. For 0.3μm wide lines, the <111> bottom texture intensity decreases and a sidewall fiber texture appears. This observation has already been reported by several authors [4,5] for ECD Cu damascene lines. Pole plots associated with narrow lines of processes B and C also highlight a <511> bottom fiber texture (see fig. 2b, 2c and 3) that is a direct consequence of the high degree of twinning in Cu.

GRAIN MORPHOLOGY CHARACTERIZATION

The grain morphology characterization of the lines has been performed using FIB cross-section imaging with a SEM on 3 and 0.3μm wide lines (which are close to the widths of the electrically

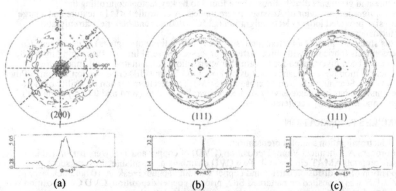

Figure 1. Pole figures (top) and associated pole plots (bottom) for wide lines (w=3μm) : (a) process A, (b) process B and (c) process C. The trench direction is from top to bottom. The pole plots represent the normalized intensity versus the polar angle for a given azimuthal angle (φ= 45° here).

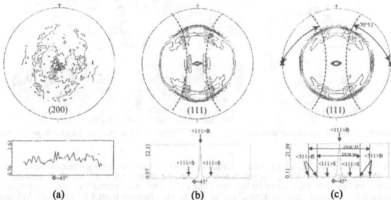

(a)	(b)	(c)

Figure 2. Pole figures (top) and associated pole plots (bottom) for narrow lines (w= 0.3μm) for (a) process A, (b) process B and (c) process C. The trench direction is from top to bottom. On the pole plots, B stands for "Bottom fiber texture", S stands for "Sidewall fiber texture".

tested lines). A mixing of backscattered and secondary electrons has been used to obtain an orientation contrast so as to underscore the microstructure of Cu. Observations of samples cross-sections (perpendicular to the electron flow) offers a clear insight on the possible impact of microstructure on the electromigration performance. These observations gave evidences of a high degree of twinning for Cu, whatever the deposition process may be, and a non-columnar morphology of the grains for both wide and narrow lines (see fig. 4). The achievement of FIB cross sections is nevertheless a time consuming task and did not allow us to extract statistically representative data for the grain size distributions. From a qualitative point of view, the average grain size was found to be larger for ECD copper metallizations. The wide lines are polycrystalline (see fig. 4b and 4d) whereas the narrow lines are quasi-bamboo (see fig. 4a and 4c) for all processes.

ELECTRICAL RESULTS

The electromigration performances of the three copper processes have been evaluated for two line widths (w=4 and 0.6μm) using both highly accelerated and moderately accelerated tests. Concerning the test structures design, the current leads of wide lines are twice as large as the lines. For narrow lines, "Babel tower" current leads [6] have been used to avoid, as far as possible, microstructural gradients. The tests carried out at high acceleration factors were performed at wafer level using the Constant Acceleration Factor method [7]. A 2% increase in

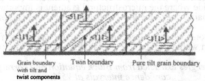

Grain boundary   Twin boundary   Pure tilt grain boundary
with tilt and
twist components

Figure 3. Cross-section of a copper damascene interconnect with a strong <111> bottom fiber texture (wide lines of process B and C). The presence of a twin boundary in the thickness of the trench induces a <511> bottom fiber texture highlighted in figures 2b and 2c. It should be noted that the high degree of twinning prevents the formation of pure tilt boundaries even for columnar grain shapes and strong bottom textures.

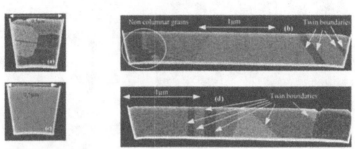

Figure 4. *SEM FIB cross-sections imaging : (a and c) process C, w=0.5μm, (b and d) process C, w=3μm.*

initial resistance was used as a failure criterion so as to maintain the mean temperature and current density constant during the test. The temperature ranged from 300°C to 400°C and the current densities were superior to 10 MA/cm². The moderately accelerated tests were performed at package level with T ranging from 210°C to 320°C and j ranging from 4 MA/cm² to 9 MA/cm². The activation energies were extracted using Black equation :

$$MTTF = A.(j/j_0)^{-n}.\exp[Ea/kT] \qquad (2)$$

where the MTTF is the median time to failure of a sample population tested at current density j and temperature T (in Kelvin), k is the Boltzmann's constant, $j_0=0.1$MA/cm², A is a normalization time, Ea is the apparent activation energy of the diffusion process and n is the current density exponent. At least 3 PLR tests and 3 WLR tests were performed for each metallization and both line widths. The electrical results will be presented in details elsewhere*. The activation energies extracted using a multi linear regression are listed in table 1. Confidence Intervals (CI) at 95% are also displayed. Concerning the extrapolated lifetimes at operating conditions (j=0.5 MA/cm² and T=140°C), they are one order of magnitude higher for process A with respect to processes B and C for both line widths. Processes B and C exhibit similar extrapolated lifetimes for wide and narrow lines.

FAILURE ANALYSIS

For wide lines of process A, our observations (see fig. 5) are consistent with the "grain boundary grooving model" developed by Glickman and Nathan [8] for drift velocity tests. This model explains the gap between reference values for grain boundary diffusion in copper (1.2eV[9], 0.92eV[10], 0.85eV[11]) and lower experimental values considering that surface diffusion along freshly created surfaces in the grooves can act as a healing mechanism. Therefore, we unambiguously concluded that grain boundaries are the main diffusion paths during the electromigration tests. Moreover, for narrow lines of process A, we demonstrated that grain boundary diffusion is still active in non bamboo sections : picture 5b was taken at the basis of the Babel tower at the anode side where the microstructural gradient is important but the occurrence of a failure remains improbable due to the low current density in this part of the test structure. We clearly observed in this area systematical Cu depletions (which were not responsible for lines failures). For bamboo sections, we also suspect diffusion at the upper interface to play a key role (see fig. 5c). This evolution would furthermore explain the slight increase of Ea with respect

Table 1. *Activation energies extracted from Black equation and corresponding statistical confidence intervals at 95%.*

	CVD Cu / CVD TiN	ECD Cu / IMP Ta	ECD Cu / IMP TaN
w=0.6μm	0.71eV [0.62; 0.80]	0.37eV [0.24; 0.50]	0.36eV [0.16; 0.55]
w=4μm	0.63eV [0.53; 0.73]	0.28eV [0.11; 0.45]	0.40eV [0.13; 0.67]

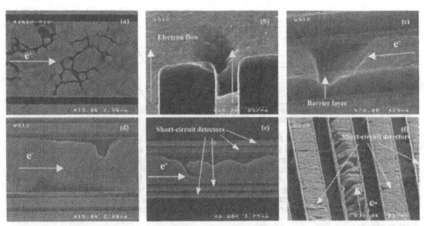

Figure 5. SEM pictures taken on tested structures (a) process A, w=4μm (b and c) process A, w=0.6μm (d) process B, w=4μm (e) process C, w=4μm, (f) process C, w=0.6μm.

to the wide lines although this increase is not statistically significant (see CI at 95% in table 1). For wide and narrow lines of ECD copper metallizations (see fig. 5d, 5e and 5f), we systematically observed large lateral depletions in contact with the barrier layers. Central depletion are much more unusual. Considering the low activation energy values obtained after the electrical tests and their relative stabilities with varying the line width (see CI at 95% in table1), diffusion at the interface "IMP-PVD copper/diffusion barrier" is suspected to be the active mechanism during electromigration tests for processes B and C.

DISCUSSION

The impact of texture on the electromigration performance has been demonstrated on polycrystalline Al-based interconnects. This impact of texture was explained based on the three following experimental observations :
- The grains had a columnar shape.
- The textures of aluminum interconnects were single bottom fiber texture.
- The main diffusion paths responsible for electromigration damages were grain boundaries.
It should be noticed that the two first observations are direct consequences of the substractive RIE process associated with Al-based interconnects. The result of these two first observations is that in highly bottom-textured interconnects (no sidewall texture), we find only pure tilt grain boundaries which can be describe by an array of edge dislocations oriented perpendicularly to the film plane and to the electron flow (see fig. 6). The higher the misorientation angle between the two adjacent grains, the smaller the spacing between the edge dislocations. Without any dislocation overlap in the grain boundaries planes (i.e. if the misorientation angle is not too important), these tilt boundaries don't provide long-range diffusion paths under an electromigration stress [12]. Moreover, the absence of a twist component for grain boundaries leads to a more homogeneous diffusion in the interconnect cross-section, with respect to the general case of a polycrystal, and minimizes atomic flux divergence.
In the case of the damascene architecture, the simultaneous growth of metal grains from the sidewalls and the bottom of the dielectric trench leads to a double fiber texture and grains with non-columnar shapes especially at lower corners of the trench (see fig. 4b). Furthermore, due to its low stack energy fault, Cu has a high propensity for twinning during the annealing steps of the processes. Even with a columnar grain shape and a strong single bottom fiber texture, the high degree of twinning would not lead to pure tilt boundaries (see fig. 3). Finally, in our experiments, failure analysis and experimental activation energy values provided evidences that interfacial

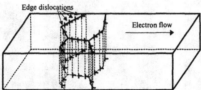

Figure 6. *Representation of pure tilt grain boundaries as edge dislocations networks in a strongly bottom-textured interconnect with columnar grains*

diffusion plays an essential role for both metallizations including ECD Cu. Thus, none of the three experimental observations valid for Al-based interconnects are verified in the case of our damascene copper lines. The previous arguments may explain why weakly or non-textured interconnects (i.e. wide and narrow lines of process A) can exhibit extrapolated lifetimes one order of magnitude higher than strongly textured lines (processes B and C). Hence, in the case of Cu damascene interconnects, the role of texture is not as clear as it used to be for conventional Al-based interconnects and an enhancement of electromigration lifetimes for Cu damascene interconnects through texture control will be a difficult problem to address.

CONCLUSION :

An analysis based on experimental activation energy values interpretation and failure analysis SEM observations allowed us to identify the active diffusion paths for three different Cu-based metallurgies. Interfacial diffusion is suspected to play a major role during electromigration tests for process B and C. We clearly demonstrated that the damascene architecture affects the microstructure of the tested lines : a sidewall texture is detected for high aspect ratio lines and grains exhibit a non-columnar shape at the lower corners of the damascene trenches. We also explained why the intrinsic copper propensity for twinning prevents a pure tilt boundaries network from existing in the Cu lines. Electromigration experiments clearly showed that weakly or non-textured interconnects can present longer lifetimes than strongly textured lines. Our conclusion is that the role of texture in copper damascene interconnects is much more difficult to apprehend than it used to be for conventionally etched Al-based interconnects and that highly textured damascene copper lines do not necessarily provide good electromigration performances.

ACKNOWLEDGEMENT

This work has been carried out in the framework of the CCMC agreement between ST Microelectronics, CEA-LETI and France Telecom CNET. We would like to thank R. Truche for the FIB cross-sections.

REFERENCES

[1] J.R. Lloyd and J.J. Clement, Thin Solid Films **262** (1995) 135-141
[2] S. Vaidya and A.K. Sinha, Thin Solid Films **75** (1981) 253-259
[3] D.B. Knorr and D.P. Tracy, Appl. Phys. Lett. **59** (25) (1991) 3241-3243
[4] L. Vanasupa, Electrochemical and Solid-State Letters, **2** (6) (1999) 275-277
[5] C. Lingk et al., Appl. Phys. Lett. **74** (5) (1999) 682-684
[6] L. Arnaud et al., Microelectronics Reliability **38** (1998) 1029-1034
[7] F. Giroux et al., Proc. IEEE Int. Conf. of Microelectronics Test Structures (1995) 229
[8] E. Glickman and M.Nathan, J. of Appl. Phys. **80** (7) (1996) 3782-3791
[9] B. Burton and G.W. Greenwood, Metal Science Journal **4** (1970) 215-218
[10] D. Gupta, Mat. Chem. Phys. **41** (1995) 199
[11] T. Surholt, Phys Rev. B **50** (1994) 3577
[12] R. R. Keller et al., Journal of Electronic Materials **26** (1997) 996-1001
* submitted to ESREF 2000

Mat. Res. Soc. Symp. Proc. Vol. 612 © 2000 Materials Research Society

Via Electromigration Lifetime Improvement of Aluminum Dual-Damascene Interconnects By Using Soft Low-k Organic SOG Interlayer Dielectrics

Hisashi Kaneko, Takamasa Usui, Sachiyo Ito and Masahiko Hasunuma
Process & Manufacturing Engineering Center, TOSHIBA Corp.
8, Shinsugita-cho, Isogo-ku, Yokohama-city, 235-8522, JAPAN
hisashi.kaneko@toshiba.co.jp

ABSTRACT

The via electromigration(EM) reliability of aluminum(Al) dual-damascene interconnects by using Niobium(Nb) new reflow liner is described. It has been found that the via EM lifetime was improved by introducing low-k organic spin on glass(SOG)-passivated structure than the conventional TEOS-SiO$_2$/SiN-passivated structure. Higher EM activation energy of 1.08 eV was obtained for the SOG-passivated structure than the conventional TEOS-passivated structure of 0.9 eV, even though no significant Al micro-crystal structure difference was found for both structures. It has been turned out that the low-k SOG material has the 1/7 Young's modulus (8 GPa) of TEOS-SiO$_2$ (57 GPa) or thermal SiO$_2$ (70 GPa). The small Young's modulus means that SOG is more elastically deformable and/or softer than TEOS or thermal SiO$_2$. This elastic deformation of the low-k SOG could retard the tensile stress evolution due to the Al atom migration near the cathode via, and elongated the time until the Al interconnect tensile stress exceeds the critical stress value for void nucleation. It has been concluded that the small-RC and reliable multi-level Al interconnect can be realized by the Nb-liner reflow-sputtered process with soft and low-k SOG dielectric materials.

INTRODUCTION

The dual-damascene multi-level interconnect process realizes the planarized metal/insulator surface at each level for keeping depth of focus margin for lithography, and has potential for the process cost down[1]. The key technology for the Al dual-damascene process is the superior hole and trench filling capability with least effective interconnect resistance increase. This filling capability strongly depends on the choice of the liner material for the sputter reflow[2]. By using the Niobium(Nb) liner, the filling capability of aspect ratio of 7.5 with φ0.17 μm via opening, and 30% decrease in the effective resistance than the conventional Ti liner could be achieved[3]. For high performance ULSIs, on the other hand, the reduction of the parasitic wiring capacitance or RC delay is highly necessary[4]. The low-k SOG dielectrics[5] and/or polyimide are the promising candidate for the inter layer dielectric (ILD) materials for smaller parasitic capacitance. However, the reliability knowledge about the Al damascene and/or dual damascene structure has been limited to the Al micro-crystal structure and texture studies[6-9], or the Al extrusion phenomena due to the poor mechanical properties of low-k materials[10,11].

In this paper, a reliability impact on the via EM phenomena of the Al dual damascene structure by using the low-k SOG material with dielectric constant of 2.5[5], and the conventional TEOS as the passivation has been studied[12]. The Al-0.5wt%Cu dual-damascene via EM test structure was fabricated by reflow-sputtering using Nb liner with the CVD-W lower-level interconnect. The observed EM lifetime difference between SOG- and

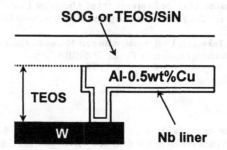

SOG or TEOS/SiN

Al-0.5wt%Cu

TEOS

W

Nb liner

Figure 1. Schematic illustration for the via EM test structure. The lower level interconnect was CVD-W line, and the 100 μm long upper level Al-0.5wt%Cu interconnect and the via were made simultaneously by reflow sputtering using Nb liner.

TEOS-passivated test structure has been discussed in terms of the difference of the Young's modulus of the passivation materials and the tensile stress evolution due to the EM induced Al atom migration.

EXPERIMENT

Figure 1 shows the schematic illustration of the two level via EM test structure. The 0.7 μm thick TEOS SiO_2 was deposited onto the 0.3 μm thick lower level CVD-W interconnect fabricated by chemical-mechanical polishing(CMP). The $\phi 0.25$ μm via hole and the 0.3 μm deep and 0.25 μm wide upper groove were simultaneously filled with Al-0.5wt%Cu by the 2-step reflow sputtering using 15 nm thick Nb liner at 450 °C. After the Al dual damascene structure was defined by the Al-CMP, the test structures were passivated by the 0.4 μm thick low-k SOG at 450 °C or by the 0.6 μm thick TEOS SiO_2 and 0.6 μm thick SiN. The length of the 0.25 μm wide Al damascene interconnect was 100 μm, and the interconnect was directly connected to the anode pad in order to avoid the back-flow effect[13]. The EM accelerated testing was carried out with the current density of 2 MA/cm^2 and at the temperatures of 200-250 °C. The critical length for 2 MA/cm^2 current density is around 30 μm, much less than the 100 μm of the present test structure. The electron flow was from the lower level W interconnect to the upper Al damascene interconnect, and therefore the EM induced void was expected to nucleate around the via. The failure criterion was a 10% resistance shift, and the current supply of each sample was stopped at a 20% shift.

RESULTS AND DISCUSSION

1. EM-induced void location

Figure 2 shows cross-sectional TEM photographs of Al dual damascene interconnects with TEOS SiO_2/SiN passivation (A and B), and with low-k SOG passivation (C), taken after EM testing. EM-induced voids were always accompanied by the horizontal grain boundary and/or

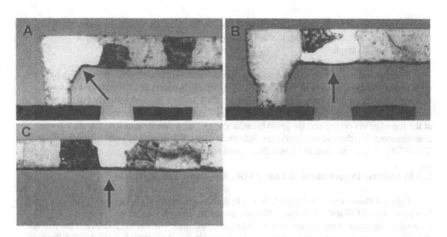

Figure 2. Cross-sectional TEM photographs after EM testing. Arrows indicate EM-induced voids.
Photo-A,B:TEOS-SiO₂/SiN-passivated samples. Photo-C:SOG-passivated sample.

grain boundary triple point. The voids observed in photo B and C were formed upstream of the bamboo grain boundary. This suggests that the dominant Al migrating path is the interface between bamboo Al grain and NbAl₃ reaction layer in the present metallization, and the top edge of the via corner (Al and TEOS and/or SOG interface in photo A) and the grain boundary triple point (in photo B and C) act as the void nucleation sites. No voids were observed at the bottom of the via and/or the interface between W and NbAl₃ reaction layer where the maximum Al atom flux divergence was generated. Many observations of the monolithic Al vias by TEM show that they are single-crystal. This microstructure would be due to the high enough reflow temperature (450 °C) to promote the grain growth in the small via. In addition to this favorable Al microstructure, adhesion of the Al to the NbAl₃ reaction layer, and NbAl₃ reaction layer to the W would be good. Therefore, the critical stress that enables the void nucleation at this interface is higher than the triple point in the Al damascene interconnect, resulting in the no void

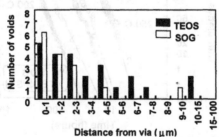

Figure 3. EM-induced void locations observed for both TEOS-SiO2 and SOG passivated Al Dual-Damascene interconnects.

nucleation at the via bottom. It is also noteworthy that there is not apparent grain microstructure difference between the TEOS SiO$_2$/SiN-passivated sample and low-k SOG-passivated sample.

The distribution of EM-induced void locations, measured as distances from the via by the optical microscopy is shown in Fig. 3. 24 TEOS SiO$_2$/SiN-passivated samples and 15 low-k SOG-passivated samples were used for analysis. Each sample had only one void in the test structure. Most of the voids were observed within 10 μm from the via. Voids would nucleate at the top edge via corner and the grain boundary triple point that is nearest to the via. It was also observed that there is no significant difference in the distribution of void locations between TEOS SiO$_2$/SiN-passivated and low-k SOG-passivated samples.

2. EM Lifetime Improvement in Low-k SOG Passivated Test Structure

Figure 4 shows the cumulative failure distribution for TEOS SiO$_2$/SiN-passivated and low-k SOG-passivated EM test structures. The observed EM lifetime was longer for the low-k SOG passivated structure at each temperature. Moreover, the difference in MTF between the two test structures becomes larger as the test temperature decreases. And, therefore the MTF difference will by significantly larger at a typical LSI operating temperature of 85 °C to 100 °C. From the cumulative failure distributions, the activation energy of the Al with low-k SOG passivation is 1.08 eV, while that of the TEOS SiO$_2$/SiN-passivation is 0.9 eV

As described in the former section, the Al microstructure was similar for both TEOS SiO$_2$/SiN-passivated and low-k SOG-passivated structures. It is reasonable to expect that the EM-induced Al and also the Cu solute migrating fluxes are the same for both passivated structure, and the Cu-sweep out model[14] that succeeds in explaining the existence of the incubation time by Cu doping to Al interconnect would not be a dominant mechanism in the

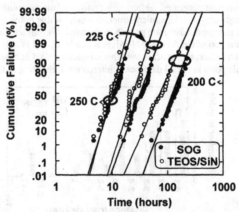

Figure 4. EM cumulative failure distribution for reflow sputtered Al-0.5wt%Cu/Nb-liner dual damascene interconnects with low-k SOG passivation and TEOS SiO$_2$/SiN-passivation

present metallization. Since the EM-induced void nucleation is also controlled by the tensile stress as the analogy of the stress-migration[15], the incubation time is also defined as the time that is necessary to create the critical tensile stress for void nucleation around the via by EM-induced Al atom migration[16]. Therefore, it is considered that there exists another mechanism that elongates the incubation time for the low-k SOG passivated structure. Figure 5 shows the thermal stresses in the 0.25 μm wide damascene interconnects with low-k SOG and TEOS-SiO_2/ SiN passivation as a function of temperature measured by the X-ray diffraction technique[17]. The initial tensile stress for the low-k SOG passivated sample was much lower than that of the TEOS-SiO_2/SiN passivated sample, even though the passivated temperature of the low-k SOG sample was higher. Moreover, the slopes of the low-k SOG passivated sample was smaller.

In order to clarify the difference in the EM lifetime and also the stress behavior during thermal cycle between the two systems further, the Young's moduli of low-k SOG and TEOS SiO_2 dielectrics were calculated by the micro-Vicker's hardness. Figure 6 shows indentation behavior of low-k SOG, TEOS SiO_2 and thermal SiO_2 measured by the nano-indentation technique. The slope for each material is proportional to the Young's modulus. The Young's moduli for low-k SOG and TEOS SiO_2 were deduced using known Young's modulus of thermal SiO_2 (70 GPa) The observed Young's modulus were 8 GPa for low-k SOG and 57 GPa for TEOS SiO_2, respectively. The Young's modulus of low-k SOG is about one seventh of that of TEOS SiO_2. This indicates that low-k SOG is much elastically deformable than TEOS SiO_2 and/or the low-k SOG is soft. From the results of the lower stress during thermal cycle and the smaller slope, and the smaller Young's modulus for low-k SOG material, it is concluded that the Al damascene interconnect stress has been suppressed by the easy elastic deformation of the low-k SOG passivation.

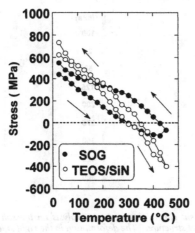

Figure 5. *Stress along the 0.25 μm wide damascene interconnect line with low-k SOG and TEOS-SiO2/SiN passivated structure as a function of temperature measured by the X-ray diffraction technique.*

To confirm the SOG deformation, the finite element elastic/plastic stress analysis was carried out on the SOG-passivated Al damascene structure, shown in Fig. 7. The 0.3 μm deep and 0.25 μm wide groove in the TEOS is filled with Al-Cu. This Al damascene interconnect was supposed to be passivated with the 0.4 μm thick low-k SOG at 450 °C. At this temperature, the stress of all material, i.e., the Al damascene interconnect, the TEOS and the SOG passivation, were stress free. Then this structure was cooled down to the room temperature, and was heated

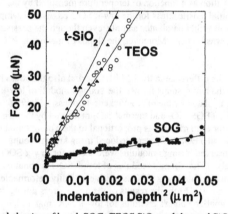

Figure 6. *Indentation behavior of low-k SOG, TEOS SiO₂ and thermal SiO₂ dielectrics.*

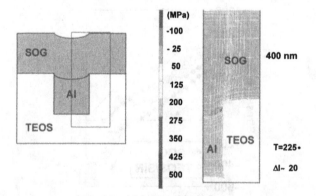

Figure 7. *Finite element elastic/plastic stress analysis on the soft low-k SOG passivated Al damascene interconnect structure. The deformation in the right side figure is magnified by the factor of 20. The stress shown in this figure is along the line.*

up to 225 °C, which was the EM test temperature. In this calculation, the plastic deformation of

Al damascene interconnect was also taken into account. The thermal expansion coefficients for

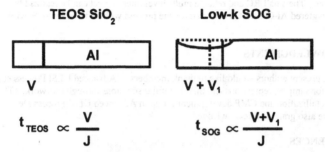

TEOS SiO$_2$

Low-k SOG

$$t_{TEOS} \propto \frac{V}{J} \qquad t_{SOG} \propto \frac{V+V_1}{J}$$

Figure 8. Schematic illustration to explain the EM-MTF difference for TEOS-SiO$_2$/SiN and low-k SOG passivation. V is the Al volume necessary to create the critical tensile stress for void nucleation around the via. V$_1$ is the additional Al volume required due to the ready elastic deformation of low-k SOG. J is the EM-induced Al atom flux, t$_{TEOS}$ and t$_{SOG}$ are the time to reach the critical stress for TEOS-SiO$_2$/SiN and low-k SOG passivation, respectively.

Al-Cu, TEOS and SOG were 23 ppm/K, 0.7 ppm/K and 3 ppm/K, respectively. The Young's modulus for Al-Cu was 80 GPa. The deformation shown in Fig. 7 is magnified by the factor of 20. The easy deformation of SOG is clearly observed, and the calculated Al interconnect stress along the line of 200-275 MPa well coincided with the measured stress value of 200 MPa by the X-ray diffraction technique shown in Fig. 5.

3. Model for the EM Lifetime Improvement by Soft Low-k SOG Elastic Deformation

Figure 8 shows schematic illustration to explain the lifetime difference with low-k SOG and TEOS-SiO$_2$/SiN passivated test structure. The V is the Al volume that is necessary to produce the critical tensile stress around the via for the TEOS-SiO$_2$/SiN passivation, and J is the EM-induced Al atom flux. Again, the J is the same for both low-k SOG and TEOS-SiO$_2$/SiN passivated sample, because the same Al crystal microstructure for both systems. t$_{TEOS}$, the time to reach the critical stress for void nucleation in the Al interconnect around the via for TEOS-SiO$_2$/SiN passivated sample is V/J. However, for the low-k SOG passivation, because of the ready elastic deformation, the additional Al volume of V$_1$ is required to create the same amount of the critical tensile stress. The EM lifetime difference, t$_{SOG}$- t$_{TEOS}$, is expressed as V$_1$/J. Therefore, the EM-MTF for low-k SOG passivation is longer than the TEOS-SiO$_2$/SiN passivation. Moreover, since J has the normal thermal activation component, e. g., J $\propto$exp(-Ea/kT), where Ea is real diffusional activation energy for the real migrating path, the lifetime difference becomes large at the lower accelerating temperature, as shown in Fig. 4.

CONCLUSION

The lifetime improvement for the via EM has been found for the low-k SOG passivated Al dual-damascene test structure. It is concluded that this improvement is due to the softness and/or ready elastic deformation of the low-k SOG material with Young's modulus of 8 GPa. This elastic deformation can suppress the Al damascene interconnect tensile stress around the via, and elongate the incubation time needed to create the critical stress for the EM-induced void nucleation. The small-RC and reliable multi-level interconnect can be realized by the Nb-liner reflow-sputtered Al-0.5wt%Cu dual-damascene process with soft and low-k SOG dielectric materials.

ACKNOWLEDGMENTS

The present authors would like to thank members of Advanced ULSI Process engineering Dept. V for sample preparations and their helpful discussions through this work. The members of the Metallization and CMP development group in Advanced ULSI Process Development Group are also gratefully acknowledged.

REFERENCES

1. C. W. Kaanta et al., "Dual Damascene : A ULSI Wiring Technology", in Proceedings of International VLSI Multilevel Interconnection Conference, 1991, p.144.
2. L. A. Clevenger et al., "A Novel Low Temperature CVD/PVD Al Filling Process for Producing Highly Reliable 0.175μm wiring/0.35μm Pitch Dual Damascene Interconnections in Gigabit Scale DRAMs", in Proceedings of International Interconnect Technology Conference, 1998, p. 137.
3. J. Wada et al., "Low Resistance Dual Damascene Process by New Al Reflow using Nb Liner", in Proceedings of IEEE 1998 VLSI Symp. on Technol., 1998, p.48.
4. D. C. Edelstein, G. A. Sai-Halasz and Y. J. Mii, "VLSI On-Chip Interconnection Performance Simulation and Measurements", IBM J. Res. Develop., 39, p. 383, 1995.
5. R. Nakata et al., "New Low-k Material "LKD^{TM}" for Al Damascene Process Application", Extended Abstracts of the 1999 International Conference on Solid State Devices and Materials, Tokyo, 1999, pp. 506-507.
6. J. M. E. Harper and K. P. Rodbell, "Microstructure Control in Semiconductor Metallization", J. Vac. Sci. Technol, B15, 1997, p. 763.
7. J. L. Hurd et al., "Linewidth and underlayer influence on texture in submicrometer-wide Al and AlCu lines", Appl. Phys. Lett., 72, 1998, p.326.
8. A. Furuya et al., "Electrical Characterization and Microstructure of 0.32μm-Pitch Aluminum-Damascene Interconnects", in Advanced Metallization Conference in 1998 Japan/Asia Session, 1998, p.33.
9. J. E. Sanchez, P. R. Besser and D. P. Field, "Microstructure of Damascene Processed Al-Cu Interconnects for Integrated Circuit Applications", in AIP Conference Proceedings of the Fourth International Workshop on Stress-Induced Phenomena in Metallization 418, 1997, p. 230.
10. S. Foley et al., "A Study of the Influence of Inter-Metal Dielectrics on Electromigration Performance", Microeletron. Reliab., 38, 1998, p. 107.
11. P. S. Ho, P. H. Wang and J. Kasthurigrangan, "Structure Integrity and Reliability of Low K Interconnects", in Advanced Metallization Conference in 1998 Japan/Asia Session, 1998, p.19.
12. T. Usui et al., "Significant Improvement in Electromigration of Reflow-Sputtered Al-0.5wt%Cu/Nb-liner Dual Damascene Interconnects with Low-k Organic SOG Dielectric", in Proceedings of 37th International Reliability Physics Symposium, 1999, p.221.
13. I. A. Blech, J. Appl. Phys., 47, 1976, p. 1203.
14. C. K. Hu, M. B. Small and P. S. Ho, "Electromigration in Al (Cu) two-level structures: effect

of Cu and kinetics of damage formation", *J. Appl. Phys.*, 74, 1993, p. 969.

15. H. Kaneko *et al.*, "A Newly Developed Model for the Stress-induced Slit-like Voiding", *in Proceedings of 28th International Reliability Physics Symposium*, 1990, p.194.

16. M. A. Korhonen *et al.*, "Stress evolution due to electromigration in confined metal lines", *J. Appl. Phys.*, 73, 1993, p. 3790.

17. A. Tezaki *et al.*, "Measurement of three Dimensional Stress and Modeling of Stress Induced Migration Failure in Aluminum Interconnects", *in Proceedings of 28th International Reliability Physics Symposium*, 1990, p. 221.

Mat. Res. Soc. Symp. Proc. Vol. 612 © 2000 Materials Research Society

STRESSMIGRATION BEHAVIOR OF MULTILEVEL ULSI AlCu-METALLIZATIONS

A.H. Fischer, A.E. Zitzelsberger, M. Hommel and A. von Glasow
Infineon Technologies AG, Reliability Methodology, Munich, Germany

ABSTRACT
With decreasing geometries of metal interconnects the demands on metallization reliability increase rapidly. In addition to electromigration, stress-induced voiding becomes a major problem, influencing lifetime and functionality of integrated circuits. This paper summarizes our studies on stressmigration behavior of various AlCu-multilevel metallizations. A model for an estimation of the median time to failure is presented.

INTRODUCTION
In reliability methodology stress-induced voiding is an issue of intensive discussion, due to the complexity of physical mechanisms acting in thin metal films and narrow interconnect lines during development and relaxation of mechanical stress. Many efforts were done in order to estimate the stressmigration-limited lifetime of ULSI AlCu-metallizations [1,2,3]. However, the assessment of the stressvoiding problem in process qualification often has a more qualitative character: Avoid the occurrence of stress-induced voids for high performance applications!
Three basic stressmigration-related risks can be distinguished. The *primary risk* is an increase in line resistance because of stress voids, affecting the circuit functionality. It can probably be controlled by providing a model which allows, similar to Black's equation for electromigration, the transformation of the failure distribution from highly accelerated to operation conditions. The *secondary risk* is the influence of stress voids on the electromigration (EM) performance. If the EM-failure is due to the growth of pre-existing voids, the current density exponent in Black's equation will be close to one [4] and as a consequence a lower EM-lifetime is obtained [5]. A *third risk* is conceivable due to the statistical nature of stress-void formation. Here, stressvoiding in narrow or short lines and at vias may cause higher resistances, early fails or weak links.
This paper will focus on the *primary risk* from the viewpoint of process qualification, where stressmigration monitoring is in general limited to resistance drift measurements after high temperature storage. Methods which otherwise allow the direct measurement of stress (e.g. X-ray diffraction), are unsuitable for this purpose.

Stress relaxation in interconnect lines during high temperature storage (HTS)
In the following a simplified scenario is considered to estimate the stressmigration mean time to failure. Due to the thermal mismatch between metal line, substrate and encapsulating oxide, a tensile stress is obtained at the interconnect during cooldown from high deposition temperatures. This thermally-induced stress can be expressed by:

$$\sigma_0 = E_{eff} \Delta\alpha (T_{dep} - T_{str}) \qquad (1)$$

where T_{dep} is the deposition temperature of the dielectric, T_{str} the storage temperature, E_{eff} the appropriate elastic modulus and $\Delta\alpha$ the difference in thermal coefficients of expansion between metal and surrounding medium. Several processes contribute to the stress relaxation, e.g. diffusional creep, dislocation climb or dislocation glide. The dominance of a particular mechanism is related to the storage temperature and the applied stress [6]. Since passivated interconnects are in a state of hydrostatic stress, diffusional creep is assumed to be the favoured relaxation process within the considered temperature and stress range [7]. In this case the plastic strain rate under the influence of an external stress σ at the temperature T_{str} is given by [8]:

$$\dot{\varepsilon}_{pl} = a \frac{D}{kT_{str}} \sigma \qquad (2)$$

where $D = D_0 \exp(-Q/kT)$ is the diffusivity with an activation energy Q of the respective relaxation process. The constant a contains the atomic volume as well as microstructural and geometrical parameters. It differs for both grain boundary and bulk diffusion. In a (quasi) steady state (T_{str}=const.) the total strain rate $\dot{\varepsilon} = \dot{\varepsilon}_{pl} + \dot{\varepsilon}_{elastic}$ is zero and from Hooke's law follows

$$\dot{\varepsilon}_{pl} = -\dot{\varepsilon}_{elastic} = -\dot{\sigma}/E_{eff}$$

hence with (2)

$$\dot{\sigma}(t) + aE \frac{D}{kT_{str}} \sigma(t) = 0$$

and therefrom

$$\sigma(t) = \sigma_o \exp(-t/\tau) \qquad (3)$$

with a time constant

$$\tau = kT_{str} / aE_{eff} D$$

It can be assumed that a certain amount of plastic strain, which correlates to a certain void volume, leads to an increase in line resistance. The absolute value of the resistance increase depends on line geometry, void size and void shape as well as on thickness and specific resistance of the redundant shunt layers. A device fail is defined by a certain relative resistance increase. The amount of plastic strain $\Delta \varepsilon_{pl}$ that leads to a fail can be expressed by:

$$\Delta \varepsilon_{pl} = \varepsilon_{pl}(t_{50str}) - \varepsilon_{pl}(t=0) = \int_0^{t_{50str}} \dot{\varepsilon}_{pl}(t)dt = \frac{aD}{kT_{str}} \sigma_o \tau(1 - \exp(-\frac{t_{50str}}{\tau}))$$

where t_{50str} is the median time to failure of a sample set which is stored at T_{str}. Assuming that t_{50str} is much smaller than the stress relaxation time constant, it can be approximated by:

$$t_{50str} \approx \frac{\Delta \varepsilon_{pl}}{a} \frac{kT_{str}}{D\sigma_0} = \frac{\Delta \varepsilon_{pl}}{a} \frac{kT_{str}}{D_0 \sigma_0} \exp \frac{Q}{kT_{str}} \qquad (4)$$

Since stress voids begin to develop right after the deposition process of the dielectric, some amount of the initial stress σ_0 is already relaxed before the resistance measurement. Therefore a reduced initial stress $\sigma^*_0 < \sigma_0$ has to be taken into account instead of σ_0 in (4). Thus with (1)

$$t_{50str} \approx C \frac{T_{str}}{T_{dep}^* - T_{str}} \exp \frac{Q}{kT_{str}} \qquad (5)$$

where $T^*_{dep} < T_{dep}$ is a temperature describing the reduced initial stress $\sigma_0^* = E_{eff} \Delta\alpha(T_{dep}^* - T_{str})$ that exists in the metal line at the start of the HTS-measurement. The constant C contains the constant parameters of (4). Equation (5) is similar to relations that have been proposed in [9,10].

EXPERIMENTAL

Stressmigration was investigated on Ti/AlCu(0.5%)/TiN metallizations of several technologies (e.g. Lot A: 4-level eDRAM, Lot B: 2-level DRAM). Resistance drifts, caused by stress voids, were observed after storage at various temperatures (T_{str}=125, 175, 225, 250, 275°C) in certain metal layers. The resistance of narrow meander-shaped line structures (35mm long) was measured on wafer level (60 to 80 structures per wafer) in readout intervals from several hours to several hundred hours. In all investigated samples voids were already observed before the start of the measurements. The occurance of these voids is connected with a high density plasma (HDP) process during the deposition of the SiO$_2$ intrametal dielectric. The voids could completely be suppressed by introducing an anneal process after the metal patterning together with a reduced HDP-deposition temperature.

RESULTS

For Lot A resistance drifts due to stress-induced voids (Fig.1) were found in metal layers M2 and M3. The drift behavior is related to the storage temperature (Fig.2). At 275°C the drift saturates already after ~1500h, whereas no saturation is observed at 175, 225 and 250°C even after 5000h. The highest resistance increases were found for 225 and 250°C. In order to define the time to failure of a single device a failure criterion of $\Delta R/R=5\%$ relative resistance drift is introduced. The resulting cumulative failure distributions of the samples show log-normal behavior at 175 and 225°C (Fig.3b). However, at higher temperatures deviations from log-normality were observed. The shape factors σ of the log-normal distributions vary between 0.2 and 0.9. In most cases σ was found to be slightly higher for wider lines (Fig.3a).

FIGURE1
SEM-images of stress-induced voids in metal lines of Lot A and B after 2000 hours storage at 225°C. Lot B contains mostly wedge-shaped voids.

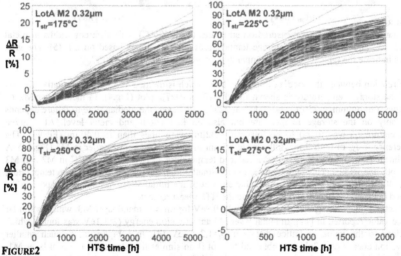

FIGURE2
Relative resistance drifts $\Delta R/R$ obtained on 0.32μm wide structures of Lot A (T_{dep}=400°C) in metal level M2 at different storage temperatures T_{str}.

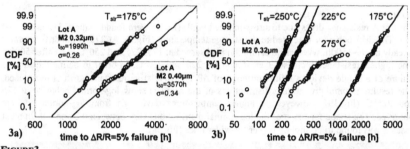

FIGURE3
Log-normal plots of cumulative failure distributions of M2-structures of Lot A (T_{dep}=400°C) with different line width **(3a)** and at various storage temperatures T_{str} for 0.32μm line width **(3b)**.

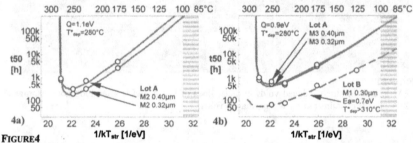

FIGURE4
Median time to failure t_{50} measured on structures (T_{dep}=400°C) with different widths in metal levels M2 and M3 at various storage temperatures T_{str}. The Fit is based on Eq. (5). The grey areas indicate the typical operation temperature range between 85 and 100°C.

The relation between the median time to failure, which is defined at 50% of cumulative failure, and the storage temperature is shown in a $\ln(t_{50})$-vs.-$1/kT_{str}$-plot (Fig.4). In metal layer M2 t_{50} reaches a minimum at 250°C and in M3 at 225°C. However, the median time to failure depends not only on the storage temperature but also on line width and metal level. At a given temperature t_{50} was found to be lower in 0.32μm wide lines than in 0.40μm (Fig.4a). The difference in t_{50} of both line widths is significant in M2 but becomes smaller in M3 (Fig.4b). A nonlinear fit based on eq. (5) gives a reduced temperature T^*_{dep} of about 280°C in M2 and M3 for both line widths. This temperature is much smaller than the real HDP-deposition temperature T_{dep}=400°C. It implies that a considerable amount of the initial stress has already relaxed in the period between the deposition process and the HTS measurement.

The activation energy Q was determined to be 0.9eV for lines in metal layer M3, which is similar to the value reported in [11]. In metal layer M2 an activation energy Q=1.1eV was found, which is close to the value of lattice diffusion Q_l=1.2-1.4eV [4]. In comparison, a much smaller activation energy Q=0.7eV was obtained for Lot B on similar line structures in metal layer M1. This activation energy implies grain boundary diffusion as the preferred mechanism of void formation, where values between 0.6 and 0.7eV are reported [4]. The mechanism is also supported by the occurance of a large number of wedge-shaped voids after high temperature storage (Fig.1), which have been associated with grain boundary diffusion as the path for void formation [11]. In contrast, a much smaller number of stress-induced voids is present in Lot A after HTS. Here, the average size of a single void in M2 or M3 was found to be considerable larger than in Lot B. The occurance of these voids can obviously be associated with lattice diffusion as an additional acting process, leading to higher activation energies.

Additional investigations were carried out to study the influence of the HDP-temperature T_{dep} on the failure distribution. It has been found, that a reduction of T_{dep} from 400 to 350°C increases the median time to failure by a factor 2...3 (Fig.5). This fact is probably due to the smaller thermally-induced stress at lower HDP-temperatures. As illustrated in the wafermaps (Fig.6), the resistance drift depends also on the chip position. High drifts were observed in the center and small or no drifts near the wafer edge. This is in contrast to Lot A (T_{dep}=400°C), where no center-to-edge effects were observed and all devices showed a resistance drift of at least 30% after 2000h HTS at 250°C (Fig.2). The influence of the type of cap oxide layer, which is deposited on top of the HDP-dielectric, on stressmigration was found to be rather small. Except the smaller shape factors obtained on samples with silane oxide-cap there were no significant differences in comparison to samples with a TEOS-cap (Fig.5). Additionally, large differences in the shape factors and only small variations in t_{50} were found on wafers of different lots.

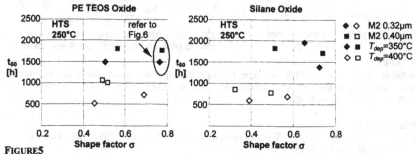

FIGURE5
Median time to failure and shape factor of the log-normal failure distributions obtained at various HDP-temperatures T_{dep} for samples with different cap layer types (TEOS, silane). The two data points measured for each structure and temperature belong to wafers of different lots.

CONCLUSION
The stressmigration behavior was investigated on several multilevel metallizations. The dependence of the median time to failure on the storage temperature was measured and found to be in good agreement with the proposed model. Based on this model, an activation energy of Q=0.7eV was determined for Lot B, implying grain boundary diffusion as the favoured process of void growth. Considerable larger values were obtained for Lot A, which associates lattice diffusion as an additional acting process. For storage temperatures up to 225°C the failure times were well described by a log-normal distribution. It was found that the median time to failure depends not only on the storage temperature but also on the line width and the metal level. A reduction of the HDP- temperature leads to significant higher t_{50}.

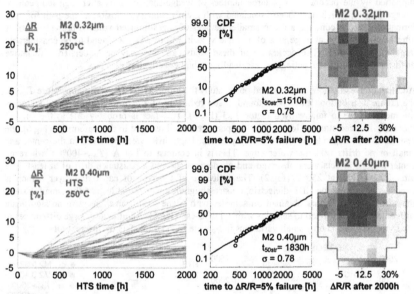

FIGURE 5
Relative resistance drifts $\Delta R/R$ over time, log-normal plots of the cumulative failure distribution and wafermaps of the drift after 2000h, measured on M2-structures of the same wafer with different widths (T_{dep}=350°C).

REFERENCES
[1]   C.-K. Hu, K.P. Rodbell, T.D. Sullivan, K.Y. Lee, D.P. Bouldin, IBM J. Res. Develop. Vol.38, No.4, July 1995
[2]   Okabayashi, Hidekazu, Mat. Sci. and Eng., R11, No.5, pp.191-241, Dec.1, 1993
[3]   S.E. Rauch, T.D. Sullivan, Proc. SPIE, vol. 1805, pp. 197-208, 1993
[4]   J.R. Lloyd, "Electromigration in integrated circuit conductors", J. Phys. D: Appl. Phys. 32 (1999), R109-118
[5]   A.E. Zitzelsberger and A.H. Fischer, "The influence of stress-induced voiding on the electromigration behavior of AlCu interconnects", to be published in this conference proceedings, MRS2000
[6]   H. J. Frost and M. F. Ashby, Deformation-Mech. Maps, Pergamon, Oxford, 1982, p.27
[7]   P. Besser, S. Brennan and J. C. Bravman, J. Mater. Res., 9 (1) (1994) 13-24
[8]   G. B. Gibbs: "Diffusion creep of a thin foil" Phil. Mag. 13 (1966) 589-593
[9]   T.D. Sullivan, proposed as JEDEC Standard, JC-14.2-98-189
[10] J.W. McPherson and C.F. Dunn, Jvst B5, 1321, (1987)
[11] C.J. Shute and J.B. Cohen, "Stress relaxation in Al-Cu thin films", Materials Science and Engineering, A149 (1992), 172-176

Mat. Res. Soc. Symp. Proc. Vol. 612 © 2000 Materials Research Society

A Percolative Approach to Electromigration Modelling

C. Pennetta[1,2], L. Reggiani[1,2], Gy. Trefán[2], F. Fantini[3], A. Scorzoni[4], I. DeMunari[5]
[1]Lecce University, Dept. of Innovation Engineering, Lecce, Italy
[2]National Institute for Material Science, INFM, Italy
[3]Modena University, Dept. of Engineering Sciences, Modena, Italy
[4]Perugia University, Dept. of Electronic and Information Engineering, Perugia, Italy
[5]Parma University, Centro MTI, Parma, Italy

ABSTRACT

We present a stochastic model which simulates electromigration damage in metallic interconnects by biased percolation of a random resistor network. The main features of experiments including Black's law and the log-normal distribution of the times to failure are well reproduced together with compositional effects showing up in early stage measurements made on Al-0.5%Cu and Al-1%Si lines.

INTRODUCTION AND MODEL

We present a stochastic approach which simulates electromigration (EM) damage in metallic interconnects in terms of percolation in a random resistor network [1]. The proposed approach has the inherent novelty of exhibiting specific stochastic features during degradation and healing processes associated with a current stress [2]. Here we present an extension of the previously proposed model [3] which allows us to reproduce most of the compositional effects (CE), often acting during the early stages of EM [4,5]. Typical examples of CE for Al-0.5%Cu lines under standard Median Time to Failure (MTF) measurements and for Al-1%Si lines under high resolution measurements made with the ratio of resistance (ROR) technique [6] will be reported in the next section and compared with the numerical results of the model.

We describe a metallic line as a two-dimensional square-lattice network of resistors (denoted as "regular" resistors, r_{reg}) laying on an insulating substrate at temperature T_0. The resistance of each n-th resistor depends on temperature as:

$$r_{reg,n}(T_n) = r_{ref}[1 + \alpha(T_n - T_{ref})] \qquad (1)$$

Here α is the temperature coefficient of resistance (TCR), T_n the local temperature, T_{ref} and r_{ref} the reference values for the TCR. When Joule heating is negligible, the resistors are all equal to $r_0 = r_{reg}(T_0)$. To save computational time, instead of using long rectangular networks we perform the calculations on a square $N \times N$ network, where N determines the linear sizes of the studied region, with the total number of resistors being $N_{tot} = 2 N^2$. Our network thus represents the region of dominant void growth of the film. As the line length is taken f times greater than the linear size of the studied region, the relative resistance variation of the whole line is obtained by multiplying the relative resistance variations of the network by the factor $1/(1+f)$. The network is contacted at the left and right hand sides to perfectly conducting bars, acting as electrical contacts, through which a constant stress current I is applied. The Joule heating induced by I is taken into account by defining T_n as:

$$T_n = T_0 + A\left[r_n i_n^2 + \frac{B}{N_{neigh}} \sum_{m=1}^{N_{neigh}} (r_{m,n} i_{m,n}^2 - r_n i_n^2) \right] \qquad (2)$$

where A is the thermal resistance of the single resistor and N_{neigh} the number of first neighbours of the n-*th* resistor. The value $B = 3/4$ is chosen to provide uniform heating of the perfect network (made by regular resistors only) [7].

The EM damage is simulated by allowing the breaking of regular resistors, i.e. by replacing $r_{reg} \rightarrow r_{OP}$ with $r_{OP} = 10^9 \cdot r_0$ representing an open circuit (OP) associated with the formation of voids inside the line. The complete failure of the line is thus associated with the existence of at least one continuous path of OP between the upper and lower sides of the network (percolation threshold) [8]. We express the probability W_{OP} for the n-*th* resistor to become OP as:

$$W_{OP} = exp(-E_{OP}/k_B T_n) \qquad (3)$$

where E_{OP} is an EM activation energy and k_B the Boltzmann constant [9]. A healing process, competing with the void formation is considered to occur with the same activation energy. A nonzero initial concentration p_{ini} of OP resistors randomly distributed can also be considered to account for different qualities of the lines.

The variation of composition due to Joule heating effects in Al-0.5%Cu lines is simulated by changing the resistance of the n-th resistor, from r_{reg} to a lower value r_{imp} (impurity resistors). These impurity resistors are associated with the formation of $CuAl_2$ precipitates. This resistance change is taken to occur with probability:

$$W_{r-i} = exp(-E_{r-i}/k_B T_n) \qquad (4)$$

with E_{r-i} a characterisctic activation energy. The mechanism of Cu dissolution into the Al matrix, antagonist to the precipitation, is simultaneously considered by taking a similar expression for the probability of the reverse process $r_{imp} \rightarrow r_{reg}$, which is determined by an activation energy E_{i-r}. As a consequence of the two competing processes, a steady state concentration of r_{imp}, dependent on substrate temperature and current induced heating is achieved after a certain amount of time. Similarly, the CE in Al-1%Si lines are simulated by changing the resistance of the n-th resistor, from r_{reg} to a higher value r_{imp}. Here the impurity resistors are associated with the heating induced dissolution of Si atoms into the Al matrix, and the antagonist process becomes the precipitation of Si clusters.

Monte Carlo simulations of the network evolution are carried out by using the following procedure. (i) Starting from the initial network, by solving Kirchhoff's loop equations we calculate i_n, the network resistance R, and by using Eq. (2) T_n. (ii) OP and r_{imp} are generated with the corresponding probabilities W_{OP} and W_{r-i}, while the remaining r_{reg} are changed according to

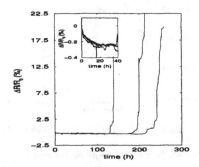
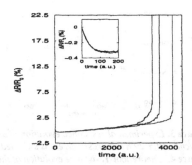

Figure. 1. *Experimental resistance evolutions of an Al-0.5%Cu line. Three typical curves are reported together with an enlargement of CE in the early stage of EM.*
Figure 2. *Theoretical resistance evolutions of the Al-0.5%Cu line in Fig. 1 performed with the parameters reported in the text.*

T_n. Then i_n and T_n are recalculated. (iii) OP and r_{imp} are recovered. (iv) i_n, T_n and R are recalculated. This procedure is iterated from (ii), thus the loop (ii)-(iv) corresponds to an iteration step, which is associated with a unit time step on an arbitrary time scale. The iteration proceeds until, depending on the parameter values, the following two possibilities are achieved: irreversible failure or steady-state evolution. More details about numerical simulations are reported in Ref. [10].

RESULTS

The results of experiments and simulations are reported in Figs. 1 to 6. Figure 1 shows a subset of experimental evolutions of the resistance measured with a standard MTF technique on 3000 μm long, 0.45 μm wide and 0.8 μm thick Al-0.5%Cu lines, stressed by a current of 3 MA/cm² at 219 °C. These data are obtained adopting a 2-metal level configuration with tungsten vias and using a 20 % relative resistance variation as failure criterion. The same measurements, performed using different stress conditions, on samples of the same population, indicate an activation energy $E_{OP} = 1.0$ eV and $n=1.95$ for the current exponent appearing in Black's law [11]. An initial non linear resistance decrease of about 10.000 ppm lasting about 10 h is followed by a nearly constant evolution of resistance and eventually by a series of violent bursts announcing the final failure. Focusing on the resistance decrease in the early stage of the evolution, this behavior could be mainly attributed to a thermal CE. The phenomenological interpretation of this behavior is as follows. The temperature of the stressed samples is the sum of the oven temperature and the Joule heating due to the applied stress current. Because of the relatively high temperature, a fraction of Cu dissolved into the Al matrix (after the last thermal treatment during fabrication) precipitates forming $CuAl_2$ clusters. The low resistivity of these clusters and the reduction of scattering centers of Cu in the solid solution, cause a resistance decrease of the strip which is naturally limited in time on the time scale of clusterization, here of about 10 h. Figure 2 reports a subset of evolutions of the relative resistance variations obtained using our model. By taking $N=70$ the following parameters are chosen according to experimental

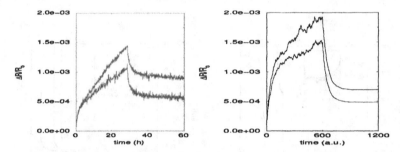

Figure 3. *Experimental resistance evolution of an Al-1%Si line stressed at T = 240 °C with j = 2 MA/cm². After 29 h the stress current is switched off.*

Figure 4.. *Theoretical resistance evolution of the Al-1%Si lines in Fig. 3 with the parameters reported in the text.*

conditions of Fig. 1: I=1.1x10⁻² A, T_0=219 °C, T_{ref}=0 °C, r_{ref}=296 Ω, α = 3.6·10⁻³ °C⁻¹. The following parameters have been adopted: A=1.5·10⁶ °C/W (to provide for an initial heating of the network of about 15 °C, comparable with that estimated for the lines in Fig. 1), p_{ini}=0.0987 (to account for an initial defectiveness of the line), f=30 (to reduce computational efforts), r_{imp}=350 Ω (to account for CE). Furthermore, to shorten the simulation time, we used for the activation energies the values: E_{OP}=0.43 eV, E_{r-i}=0.22 eV, E_{i-r}=0.17 eV. The initial decrease of the resistance due to CE is well reproduced. Moreover, strong bursts preceding the failure are evident. The overall experimental behaviour shown in Fig. 1 together with its statistical properties is satisfactorily reproduced by our simulations. A small discrepancy between experiments and simulations is detected in the intermediate resistance evolution, where the flat behavior found in the experiments is compared to a slow resistance increase in the simulations. This is actually what one could expect since presently our model does not take into account the well known electron-wind assisted precipitation of Cu [12], which most likely counteracts the slow resistance increase due to Al transport in Al.

In a second set of experiments, high resolution measurements (HRR), see Fig. 3, were performed using the ROR technique [5] and including an interruption of the stress current. In the HRR technique both the resistance variations of a stressed sample and the resistance variations on an unstressed stripe located close to the stressed one are acquired. A suitable mathematical algorithm cancels the test-line resistance variations due to the hot-plate temperature instability. Measurements on Al-1%Si, were found to exhibit the following main features. An initial non linear resistance increase of about 700 ppm lasting about 10 h is followed by a linear resistance increase until the stress current is switched off after 29 h. Hence the resistance drops to a nearly constant level of 600-900 ppm which is reached at the time of about 40 h, i.e. within a time scale which is nearly the same of the initial non-linear increase. Since the strips were subjected to a long pre-annealing at the stress temperature prior to being stressed, the phenomenological interpretation of this behavior attributes the transient nonlinear increase to a thermal CE associated with the increase in temperature (of a few degrees only) of the strip due to the stress current. This thermal increase originates the dissolution (with a decreasing size) of Si clusters (possibly Al-doped Si) originally embedded in the Al matrix. The resistivity of the strip thus increases because of the increased scattering efficiency due to the dissolved Si. This process is naturally limited in time by the kinetics of the dissolution process and is concluded after about 10 hours. Soon after, only EM effects remain as responsible of the systematic linear increase of resistance. Switching off the stress current causes an abrupt stop of both Joule heating and EM process, thus implying the reverse process of precipitation of Si clusters in Al and the consequent drop of resistance to the level due to EM damage only. Note that the net increase of resistance due to EM only is of 550-850 ppm on the time period of about 29 h, i.e. with a rate of about 19-29 ppm/h. Figure 4 presents the results of the corresponding simulations. In analogy with the previous case, the following parameters are changed as: $I=1.26 \times 10^{-2}$ A, $T_0=227$ °C, $p_{ini}=0$, $f=200$, $r_{imp}=2000$ Ω, $E_{r-i}=0.26$ eVΩ, $E_{r-i}=0.17$ eV. The two simulations differ in the values of the thermal resistance of the single resistor, taken to be $A = 1.3 \times 10^6$ °C/W for the lower curve and $A = 1.5 \times 10^6$ °C/W for the upper curve, respectively. At current switch-off the EM associated processes are switched off. The initial non-linear increase of the resistance as well as the recovery to the EM plateau after current switch-off are well reproduced by simulations. The overall agreement between simulations and experiments is thus considered to be satisfactory for the purpose of validating the stochastic approach proposed here. For completeness, two other significant results obtained from simulations are further reported.

Figure 5 shows a typical evolution of the damage pattern obtained for a network of 40×40 with the same parameters of Fig. 2. The simulated pattern just before complete failure with the typical vertical filamentation shape reproduces well the experimental pattern observed by SEM in a metallic line failed due to the formation of a void causing an open circuit [13].

In EM experiments it was also observed that the current exponent n appearing in Black's law exhibits values in the range $1 \leq n \leq 3$. The value of n is usually related to the importance of Joule heating effects [14] and thus depends on the stress current density. At high stress conditions used in accelerated tests ($j= 10$ MA/cm^2) values of $n \geq 2$ are usually reported [14]. We calculated therefore the MTF of a family of simulated lifetime tests. We considered 40 lifetime tests using networks with linear sizes $N=100$, stressed by currents in the range $0.1 \leq I \leq 2.5$ A. By neglecting CE and recovery of OP we chose $E_{OP} = 0.17$ eV, $T_0=27$ °C, $r_{ref}=1$ Ω, $\alpha = 10^{-3}$ °C^{-1} $A=5 \times 10^5$ °C/W. The results for the exponent n in Black's law are reported in Fig. 6. The figure shows that for large currents the value of n increases well over 2, while for small currents it tends to a value lower than 1. Present results correlate an increase of n to an increase of filamentation of the damage pattern. Similar interpretation of n can be found in Ref. [14].

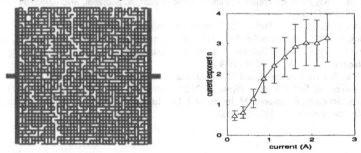

Figure 5.. *Simulated damage patterns just close to complete failure of a network with $N=40$ using the same parameters of Fig. 2.*
Figure. 6. *Dependence of the current exponent n on the stress obtained from simulations. The error bars denote numerical uncertainity.*

CONCLUSIONS

We have presented a percolative simulation of the degradation of metallic interconnects due to EM. The importance of recovery processes is here emphasized for the two relevant cases of defect healing during EM damage and transient compositional effects associated with Joule heating. The reproduction of the general features of experiments support the physical trustworthiness of the proposed model. Remaining discrepancies will likely be overcome by further refinements of the model.

ACKNOWLEDGEMENTS

Partial support was provided by the project "Physics of Nanostructures" of the Ministero dell'Università e della Ricerca Scientifica e Tecnologica (MURST), and by CNR through the MADESS II project.

REFERENCES

1   D. Stauffer and A. Aharony, *Introduction to Percolation Theory*, (Taylor and Francis, 1992).
2   Z. Li, C. L. Bauer, S. Mahajan, A. G. Milnes, *J. Appl. Phys.*, **72**, 1821 (1992).
3   Z. Gingl, C. Pennetta, L. B. Kiss, and L. Reggiani, *Semic. Sci. Technol.*, **11**, 1770 (1998).
4   F. Fantini, J. R. Lloyd, I. De Munari, and A. Scorzoni, *Microelectronic Engineering*, **40**, 207 (1998).
5   A. Scorzoni, I. De Munari, H. Stulens, V.D'Haeger, *Mat. Res. Soc. Symp. Proc.*, **391**, (1995) pp.513-519.
6   A. Scorzoni, S. Franceschini, R. Balboni, M. Impronta, I. De Munari, and F. Fantini, *Microelectron. Reliab.*, **37**, 1479 (1997).
7   C. Pennetta, L. Reggiani, L. B. Kiss, *Physica A*, **266**, 214 (1999).
8   C. Pennetta, L. Reggiani, G. Trefán, Proc. of MAM2000, in press.
9   C. Pennetta, L. Reggiani, G. Trefán, F. Fantini, I. De Munari, A. Scorzoni, *Microelectron. Reliab.*, **39**, 857 (1999).
10 C. Pennetta, L. Reggiani, G. Trefán, *IEEE Trans. on Electron. Devices*, in press.
11 J. R. Black, in IEEE International Reliability Physics Symposium (1967).
12 A. Scorzoni, I. De Munari, R. Balboni, F. Tamarri, A. Garulli and F. Fantini, *Microelectronics Reliab.*, **36**, 1691 (1996).
13 A. Scorzoni, B. Neri, C. Caprile, and F. Fantini, *Mat. Science Rep.*, **7**, 143 (1991).
14 M Tammaro and B. Setlik, *J. Appl. Phys.*, **85**, 7127 (1999).
15 S. Foley, A. Scorzoni, R. Balboni, M. Impronta, I. De Munari, A Mathewson, and F. Fantini, *Microelectron. Reliab.*, **38**, 1021 (1998).

Mat. Res. Soc. Symp. Proc. Vol. 612 © 2000 Materials Research Society

The Effects of Width Transitions on the Reliability of Interconnects

C. S. Hau-Riege*, C.V. Thompson*, and T.N. Marieb**,
*Massachusetts Institute of Technology, Cambridge, MA 02139
**Intel Corporation, Portland, OR 97124

Abstract
Experimental and modeling studies of Al-based interconnects with narrow-to-wide transitions show that the width transition is a site of atomic flux divergence due to the discontinuity in diffusivities between the narrow and wide segments, which have different microstructures. Lifetimes have been experimentally determined for populations of lines with transitions, which have varying line-width ratios and varying locations of the width transitions with respect to the line end. The electromigration failure rate is increased as the width-transition is moved closer to the electron-source via. Correlation of the effects width transitions on lifetimes allows determination of the critical stress range for void-nucleation failure, which was found to be 600 ± 108 MPa.

INTRODUCTION

Electromigration, which is the diffusion of atoms in an interconnect induced by an electric current, leads to changes in the local stress which originate at sites of atomic flux divergences, such as W- or Ti-filled vias. The development of tensile or compressive stresses can lead to interconnect failure due to voiding or extrusion, respectively. The atomic electromigration-induced flux (J) is described by:

$$J = \frac{DC}{kT}[\Omega\frac{\partial\sigma}{\partial x} + Z^*q\rho j],$$ (1)

where D is the atomic diffusivity, C is the atomic concentration, k is Boltzmann's constant, T is temperature, Ω is the atomic volume, σ is taken to be the hydrostatic stress, x is the distance along the interconnect length, Z^*q is the effective charge, ρ is the resistivity, and j is the current density. Failure by voiding will initiate once the stress somewhere in the line exceeds a critical stress, σ_{crit}, which has been previously estimated to fall somewhere in the range of 100 MPa to 1 GPa [1,2].

It has been well established that microstructure plays a strong role in governing the reliability of Al-based interconnects [3-5], where lines with "bamboo" microstructures, in which grain boundaries span the width and thickness of the line, have low diffusivities and electromigration rates and consequently have high median times to electromigration-induced failure (MTTF). Lines with polygranular microstructures, in which the median grain size is less than the width of the line, have continuous high-diffusivity paths along the line length and therefore have relatively high electromigration rates and a low median times to failure.

Electromigration has been well-studied for simple straight-line interconnects. Today's integrated circuits, however, have much more complex geometries. One common feature that exists in circuits but not test structures, is width transitions. These transitions are expected to affect reliability, especially for Al, due to associated microstructure transitions. In this study, we address the reliability issues resulting from the interaction (or lack thereof) of the stress evolution at a no-flux electron-source via and at a bamboo-to-polygranular microstructural transition for structures with a narrow-to-wide width transitions (and corresponding microstructural transition) at varying distances from the electron-source via as well as with varying width ratios (Figure 1).

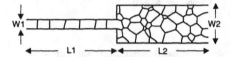

Figure 1. The width-transition structures used in this study included lines with varying length ratios (L1:L2) as well as varying width ratios (W1:W2).

EXPERIMENTAL

Two-level via-to-via Al 0.5 wt-% Cu interconnects with narrow-to-wide transitions were fabricated for this study. All structures were 300 μm long. In one round of experiments, width-transition structures had fixed narrow (W1) and wide (W2) widths at 0.6 and 2.4 μm, respectively, while the transition location (L1 in Figure 1) was varied. That is, L1 was 10, 20, 200 or 280 μm away from the electron-source via, and these samples will be referred to as L10, L20, L200 and L280, respectively. Transmission electron microscopy (TEM) studies revealed that the median grain size of 0.4 μm-thick annealed continuous films was 0.96 μm, so that the W1 sections, which were annealed after patterning, had bamboo microstructures and the microstructure in W2 sections was polygranular.

In a second round of experiments, L1 and W1 was fixed at 150 and 0.6 μm, respectively, while W2 was varied such that four different width ratios (i.e., 1:2, 1:4, 1:6, and 1:8) were tested. The median grain size of 0.75 μm-thick annealed continuous films was 1.4 μm. Therefore, all structures, except the ones with 1:2 width ratios have polygranular structures in the W2 segments.

Electromigration experiments on the width-transition structures were carried out at wafer level at 250 °C and at a current density of 2.5 MA/cm^2 in the narrow region. Joule heating in the narrow segment for all samples was measured to be below 3 degrees. All testing was conducted with electron flow from the narrow (bamboo) to wide (polygranular) segments. Failure was defined as a 30% increase in the measured resistance during a test. Failure site analysis was carried out by removing the top passivation layer using reactive ion etching (RIE), followed by scanning electron microscopy (SEM).

NUMERICAL METHODOLOGY

A finite element solver [7] based on the 1-D Korhonen model [8] was used to calculate the electromigration-induced stress evolution in the test structures. In this analysis, the diffusivity is set not only by the microstructure of the line (i.e. polygranular or bamboo) but also by the local stress state such that

$$D = D' \exp\left[\left(\frac{\Omega}{kT} + \frac{1}{B}\right)\sigma\right], \tag{2}$$

where B is the effective modulus of the metal-dielectric composite and D' has the usual Arhennius temperature dependence.

Key parameters which are input into the electromigration simulator include the diffusivity of Al atoms in different microstructures (i.e., bamboo and polygranular) and the initial thermal stress [9]. The bamboo diffusivity at test temperature was determined through experiments on single-crystal Al interconnects [10] while the cluster diffusivity was determined through electromigration experiments and simulations [9]. The bamboo to polygranular diffusivity ratio is 0.012 at the test temperature used in this study.

RESULTS AND DISCUSSION
Length Ratios

Figure 2 shows results from experiments. Structures with the width-transition close to the electron-source via (i.e., L10 and L20) have significantly lower median times to failure (MTTFs) compared to structures which have the transition far from the via (i.e. L200 and L280). Additionally, the MTTF of L10 lines was lower than that of L20 lines, while the MTTFs for L200 and L280 lines were nearly identical. It was also observed that all lines failed in an identical manner; an open failure at the electron-source via.

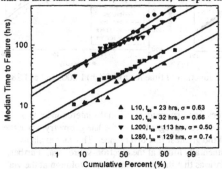

Figure 2. Compilation of electromigration data for width-transition structures with W1 = 0.6 μm, W2 = 2.4 μm, L1 + L2 = 300 μm, and transition located 10, 20, 200, and 280 μm away from the electron-source via.

Simulation of the stress evolution during electromigration was used to gain insight into these experimental results. It was found, as expected, that a flux divergence existed at the width transition as well as at the line-end vias. The tensile stress peak at the transitions is a consequence of the net depletion of atoms resulting from the higher flux of atoms through the polygranular region than from the bamboo region. The stress peaks at the electron-source and electron-sink vias result from the no-flux conditions at the via, where the stress build-up at the electron-source via is tensile and the stress build-

up at the electron-sink via is compressive. The magnitudes of all peak stresses increase
with time (Figure 3).

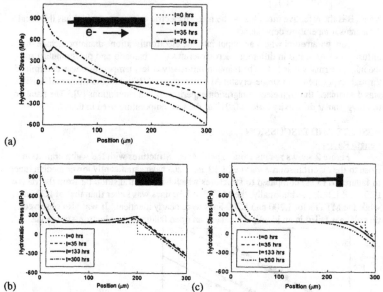

(a)

(b)

(c)

Figure 3. Calculated stress profiles as a function of time for a) L20, b) L200, and c) L280
structures.

Of the three flux-divergence sites, the maximum stress in the interconnect was
found to always occur at the electron-source via regardless of the line geometry, since the
via/interconnect interface is the site of the greatest flux divergence. This fact explains
why the failure site was always experimentally found at the electron-source via. Further,
it was found through a series of simulations that the maximum stress evolution at the via
was identical for structures whose width transitions were further than 75 μm from the
electron-source via, which includes L200 and L280 (Figure 4). Therefore, since the
maximum stress is independent of the width-transition location beyond a critical distance,
the reliabilities of L200 and L280 structures should be the same, as found in experiments.

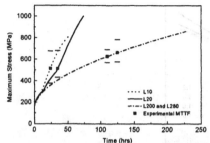

Figure 4. The maximum stress in L10, L20, L200 and L280 lines. The lines indicate results from simulations. Experimentally determined median times to failure (MTTFs) are superimposed on the lines to determine the critical stres s for void-nucleation.

The initial evolution of the maximum stress in L10 and L20 lines at early times was identical to that calculated for L200 and L280 lines. Only once there is interaction between the stress evolution at the via and a nearby width -transition, does the stress rise faster for L10 and L20 lines than for L200 and L280 lines (Figure 4). The close proximity of the two tensile stress sites allows for a more rapid stress rise because the width-transition serves as an atomic sink. In addition, the increase of the tensile stress in the narrow region enhances the diffusivity in this region (Equation 2).

Because all structures have the same mode of failure, a critical stress for void-nucleation can be determined by comparing experimental MTTFs with the evolution of the maximum stress for lines with different L1:L2 ratios. To do this, the stress value that corresponds to each MTTF was correlated using the appropriate maximum stress evolution for each of the four geometries (Figure 4). The critical stress is estimated to be 600 ± 108 MPa, which is well within the critical stress ranges experimentally determined by others for similar metallization schemes [1, 2].

Effects of Varying Width Ratios

Unlike the effects of varying length ratios, Figure 5 shows that varying width ratios for lines with L1 = L2 = 150 μm showed no effect on lifetime. However, the transition location in this set of structures is greater than the minimum distance (75 μm) required for interaction of the stress evolution at the via and the width transition in the experiments described above. The results shown in Figure 5 are therefore consistent with the results of Figure 4.

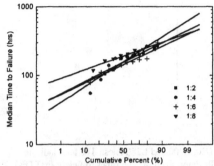

Figure 5. Failure times for lines with L1 = L2 = 150 μm and different width ratios (W1:W2). W1 was fixed at 0.6 μm.

The dependence of lifetimes on width ratios for lines with shorter transition distances (i.e., L20) were explored through simulations. In this case, it was found that a narrower W2 (or a greater width ratio) corresponded to a lower reliability for a given current magnitude and direction. This dependence can be explained by the fact that less mass flux is needed in narrower polygranular regions for the same stress increase. Therefore, with the same failure criterion, width-transition structures whose transition is within 75 μm of the electron-source via will fail faster as W2 is decreased as long as the structure of the W2 segment remains polygranular.

SUMMARY AND CONCLUSIONS

The effects of geometry and microstructure on AlCu interconnect reliability were studied through experimental characterization and simulation of electromigration-induced failures in interconnect lines with microstructural transitions. It was confirmed that microstructural transitions are sites for atomic flux divergences that can decrease the overall reliability of an interconnect. In our test structures, this occurs due to interaction between the electromigration-induced stress evolution at the electron-source via and the width transition, but this only occurs if the two flux-divergence sites are sufficiently close. In addition to the placement of the width transition along the length of the line, it was also found that width ratios can affect the reliability of width-transition structures, but, again, only when the transition is sufficiently close to the electron-source via.

BIBLIOGRAPHY
1. J. R. Lloyd and P. M. Smith, J. Vac. Sci. Technol. A1, 455 (1983).
2. R. J. Gleixner, B. M. Clemens, and W. D. Nix, J. Mater. Res. 12, 8 (1997).
3. C.V. Thompson, S.P. Riege, and V.A. Andleigh, AIP Conf. Proceed. 491, 62 (1999).
4. Vaidya, S., Fraser, D. B., Lindenberger, W.S., J. Appl. Phys. 51, 8, (1980).
5. J. Cho and C.V. Thompson, Appl. Phys. Lett. 54 , 25 (1989).
6. S.P. Riege, C.V. Thompson, and J.J. Clement, IEEE Trans. ED 45, 2254 (1998).
7. Y. J. Park, V.K. Andleigh, C.V. Thompson, J. Appl. Phys. 85, No. 7 (1999);
   http://nirvana.mit.edu/emsim/index.html

8. M.A. Korhonen, P. Børgensen, K.N Tu, C-Y Li, J. Appl. Phys. **73**, 3790 (1993).
9. C.S. Hau-Riege and C.V. Thompson, to be published in J. App. Phys. (2000).
10. V.T. Srikar and C.V. Thompson, Appl. Phys. Lett. **74**, 37 (1999)

Mat. Res. Soc. Symp. Proc. Vol. 612 © 2000 Materials Research Society

Novel Low-k Dual-phase Materials Prepared by PECVD

Alfred Grill and Vishnubhai Patel
IBM Research Division, T.J.Watson Research Center, Yorktown Heights, NY 10598

ABSTRACT

Dielectric materials based on Si, C, O, H (SiCOH) have been demonstrated previously with dielectric constants of about 2.8. This value could be potentially further reduced by increasing/introducing porosity in the SiCOH films. Depositing multiphase films containing at least one thermally unstable phase and annealing the films to remove this labile phase from the material could create the enhanced porosity. Dual-phase materials, SiCOH-CH, have been prepared in the present study by PECVD from mixtures of a SiCOH precursor with a hydrocarbon. The films have been characterized as-deposited and after thermal anneals of up to 4 hours at $400\,^{\circ}C$. The atomic composition of the films has been determined by RBS and FRES analysis and their optical properties have been determined by FTIR and n&k measurements. Metal-insulator-silicon structures have been used to measure the electrical properties of the dual-phase films. After an initial anneal at 400 °C, accompanied by a significant loss of CH and some SiH species and a thickness loss of up to 50%, the films stabilized. Depending on the deposition conditions and concentration of the CH precursor in the feed gas, the dielectric constant decreased by 10-15% during the stabilization anneal and reached values as low as 2.4. These initial results indicate the possibility to further reduce the dielectric constant of PECVD produced SiCOH films and the potential to incorporate such films in the interconnect structures of future ULSI chips.

INTRODUCTION

The electrical properties of the interconnect dielectric (ILD) become as critical as the those of the metal wiring for achieving high performance in the shrinking, high speed ULSI devices. Low dielectric constant materials are required for the ILD to reduce propagation delays, cross-talk noise between metal wires, and power dissipation from RC coupling. The research done on low-dielectric constant (low-k) materials is well reflected in the proceedings of this symposium of the last 5 years.[1] However, the search for an integratable low-k dielectric material is still a work-in progress which appears to elude the original roadmap for low-k interconnect dielectrics.[2] Some promising low-k materials can be deposited by spin-on techniques while other can be prepared by plasma enhanced chemical vapor deposition (PECVD) processes. The latter can produce films that are more crosslinked than spin-on polymeric films and be therefore mechanically tougher.

Low-k materials comprised of Si, C, O and H, (SiCOH films) and prepared by plasma enhanced chemical vapor deposition (PECVD) have been reported previously by the present authors [3] and are being offered by equipment vendors under different trade names.[2] The materials, often referred to as carbon-doped oxides, are characterized by dielectric constants of about 2.8, almost independent of the source.

Further lowering of the dielectric constant could potentially be achieved by increasing or introducing porosity in the SiCOH films. Depositing multiphase films containing at least one thermally unstable CH phase in addition to the SiCOH phase and annealing the films to remove the labile phase from the material could produce such enhanced porosity. (*The CH and SiCOH notations in this paper reflect the atomic compositions of the phases but not their stoichiometry*)

In the present work, we report a dual phase dielectric prepared by incorporating a CH phase in the SiCOH films and annealing the films to remove a significant fraction of the thermally less stable CH phase.

EXPERIMENT

The dual-phase SiCOH-CH films were prepared by the same PECVD method as described elsewhere [3] but adding a hydrocarbon (CH precursor) to the SiCOH precursor feed to the reactor. The substrates were placed on the powered electrode, thus acquiring a negative bias relative to the plasma. The RF power was kept at levels small enough to produce pure SiCOH films having low-k values of about 2.8, yet sufficiently high to dissociate the hydrocarbon and incorporate its CH fragments in the deposited films. After deposition, the films were annealed in helium for 4 hours at 400 °C to remove the less stable fraction of the films.

The dual-phase films were characterized by Rutherford backscattering (RBS) to determine the atomic composition with the exception of hydrogen and forward recoil elastic scattering (FRES) for the hydrogen content. Fourrier transform infrared analysis (FTIR) was used to characterize the different bonds in the films. The index of refraction, and optical gap were measured with a spectrometric reflectance tool, n&k Analyzer 1280 from n&k Technology, Inc. Dielectric constant (κ) measurements were performed on metal-insulator-silicon (MIS) structures using highly doped, electrically conductive Si substrates and Al dots structures. The backside of the Si wafer was coated with a blanket Al film to obtain good electrical contact. Breakdown and leakage currents were determined from I-V curves measured on the same structures.

RESULTS AND DISCUSSIONS

Composition and optical properties

Figure 1 presents the composition of as-deposited SiCOH-CH films as a function of the ratio of the precursors in the gas feed to the reactor. It can be seen that the C and H concentrations increase and the Si and O concentrations decrease continuously with increasing CH precursor fraction in the gas feed. The RBS/FRES analysis showed significant material loss during annealing, the loss increasing with increase concentration of the hydrocarbon in the gas feed. In addition, the annealing caused a decrease in the concentration of C and H and corresponding increase in the concentration of Si and O, indicating a preferential loss of CH specimens from the SiCOH-CH films.

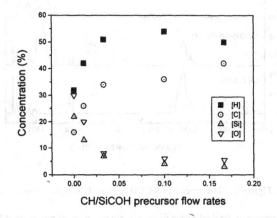

Figure 1. Composition of dual phase films vs precursor flow ratio

Figure 2 presents a comparison of FTIR spectra obtained from an as-deposited SiCOH film (a), a CH film deposited from the pure hydrocarbon (c), and a film obtained from a mixture of the SiCOH and CH precursors (b). The spectrum of the SiCOH film is characterized by the strong SiO peak at 1045 cm^{-1}, the Si-CH$_3$ peak at 1274 cm^{-1}, a doublet SiH$_x$ peak at 2233 and 2177 cm^{-1} and a relatively very small CH$_x$ peak at 2966 cm^{-1}. [3] In contrast to this spectrum, the spectrum of the CH film is characterized by a strong CHx band with several peaks around 2900 cm^{-1}.[4] The film deposited from a mixture of the SiCOH and CH precursors displays the peaks characteristic of both the SiCOH and the CH films, indicating at least the existence of a significant amount of CH bonds incorporated in the SiCOH structure (Figure 2, b). A closer look at the 1270 cm^{-1} peak indicates that it is in fact a doublet (See Figure 3, c), the secondary peak most probably corresponding to an epoxide type (COC) bond. The FTIR spectrum in Figure 2, b thus indicates the formation of a dual phase SiCOH-CH film, to the extent that the amorphous SiCOH and CH structures could be considered phases.

The effect of annealing on the FTIR spectrum of the dual phase film is illustrated in Figure 3. The comparison of the spectra in Figure 3, a and b, shows the strong decrease of the CH peak relative to the SiO peak as a result of the annealing, while the comparison of the expanded parts of the spectra in Figure 3, c and d, shows an associated decrease in the COC peak of the doublet at 1270 cm^{-1}. The results thus indicate a significant loss of the CH and probably also CO specimens from the dual phase films during annealing. The changes result in a decrease of the C concentration and an increase in the Si concentration in the annealed films as indicated by RBS/FRES.

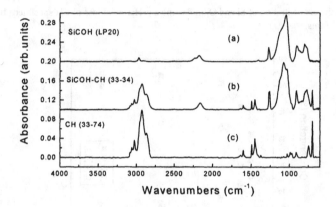

Figure 2. FTIR spectra of SiCOH, CH, and dual phase films

Two optical properties of the dual-phase films, namely the index of refraction at 633 nm and the optical gap are shown in Figure 4 as a function of the ratio of the precursors in the gas feed. The index of refraction increases and the optical gap decreases with increasing concentration of the CH precursor in the gas mixture. These changes are associated with corresponding increases in C concentration in films as indicated in Figure 1. Increasing the C concentration in the films causes an increase in the index of refraction, starting from values close to that of SiO_2, and corresponding decreases in the values of the optical gap. Removal of CH fractions by annealing caused only small changes in the optical gap and no significant changes in the index of refraction.

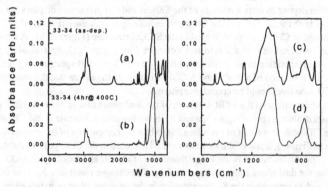

Figure 3. FTIR spectra of a SiCOH-CH film. (a), (c) – as deposited; (b) and (d) after annealing

Electrical properties

The electrical characteristics of the dual-phase films are presented in Figures 5 and 6. Figure 5 shows the dielectric constant of the annealed films as a function of the ratio of the precursors in the gas feed. The dielectric constant of a single phase SiCOH film prepared in similar conditions is 2.85 as reported previously.[3] With incorporation of the secondary CH phase in the films the dielectric constant of the annealed films decreases and reaches a minimum of 2.4 at a precursor ratio of about 0.06. The existence of a minimum in the dielectric values can be correlated to the thermal stability of the dual-phase films that were prepared with the intention to change during annealing. While FTIR indicated that the annealing causes a relatively larger loss of the CH specimens, RBS showed that the annealing is associated with loss of both CH and SiCOH fractions. These losses occurring during annealing could result in formation of nanoporosity in the films and lower density if the film would maintain its geometrical dimensions. The formation of porosity in the film will in turn reduce its the dielectric constant. The loss of material during annealing increases with increasing fraction of the CH phase, therefore the corresponding decrease of the dielectric constant observed in Figure 5. However, measurement of steps generated in the films showed that the film thickness also decreases during annealing. The thickness reduction increased with increasing fraction of CH precursor in the gas feed and reached values up to -50% for the investigated conditions. Thus, the potential porosity formation during annealing is competing with the reduction (collapsing) of the film thickness and for large fractions of CH phase this competition can result in a reversal in the changes of the film density during annealing. While this is to a certain extent a speculative assumption at this stage, it can nevertheless explain the minimum observed in the dielectric constant.

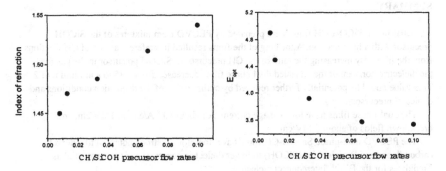

Figure 4. Index of refraction at 633 nm and optical gap of as deposited SiCOH-CH films.

A typical dependence of the leakage currents on the electric field is illustrated in Figure 6 for an as-deposited and annealed dual-phase film. The leakage current is essentially unaffected by the annealing and has a low value of 2.10^{-9} A/cm^2 at 1 MV/cm, typical for SiCOH films. The breakdown fields of the investigated dual phase SiCOH-CH films are in the range of 4.5 to 6 MV/cm. These values make the films suitable candidates for the back-end dielectric of ULSI interconnects.

The behavior observed in the investigated films is most probably related to the precursor used for the CH phase and the dissociation of both the CH and SiCOH precursors in the plasma. It may therefore be possible to further reduce the dielectric constant by optimizing the deposition conditions and the choice of precursors.

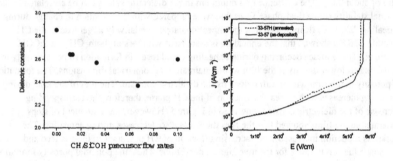

Figure 5. (left) Dielectric constant of annealed SiCOH-CH films.
Figure 6. (right) Leakage currents in as-deposited and annealed SiCOH-CH films.

SUMMARY

Dual-phase SiCOH-CH films were prepared by PECVD from mixtures of the SiCOH precursor with a hydrocarbon. Annealing of the films resulted in preferential loss of CH fractions from the films. By increasing the ratio of the CH precursor to SiCOH precursor in the gas feed the dielectric constant of the annealed dual phase films decreased from 2.85 to a minimum of 2.4. This value could be potentially further reduced by optimization of the deposition conditions and choice of precursors.

The dual-phase films have low leakage currents, of about 10^{-9} A/cm^2 at 1MV/cm, and breakdown fields of about 5 MV/cm.

The PECVD dual-phase SiCOH-CH films have, therefore, a strong potential to extend the carbon doped oxide dielectrics (SiCOH) to lower dielectric constants and their potential as candidates for the BEOL interconnect dielectric.

ACKNOWLEDGEMENTS

The authors are thankful to S.M.Gates for the initial discussions, D.Klaus for electrical characterizations, R.Carruthers for the preparation of the SIM structures, and A.Kellock from Almaden Research Center for RBS/FRES analysis.

REFERENCES

1. *Mat.Res.Soc.Symp.Proc.,* **381** (1995), **443** (1996), **476** (1997), **511** (1998), **565** (1999),
2. L. Peters, *Semiconductor International,* **23,** 52 (2000).
3. A. Grill, V. Patel, C. Jahnes, S.A. Cohen, and L. Perraud, *Mat.Res.Soc.Symp.Proc.* **565,** (1999) 107.
4. A. Grill and B. Meyerson, in *Synthetic Diamond: Emerging CVD Science and Technology,* K.E.Spear and J.P.Dismukes, editors, (John Wiley and Sons, New York, 1994), p.91.

Low-k Dielectrics

Mat. Res. Soc. Symp. Proc. Vol. 612 © 2000 Materials Research Society

Study of SiH4-based PECVD Low-k Carbon-doped Silicon Oxide

Hongning Yang, Douglas J. Tweet, Lisa H. Stecker, Wei Pan, David R. Evans, and S.-T. Hsu
Sharp Laboratories of America, Inc., 5700 NW Pacific Rim Blvd., Camas, WA 98607

ABSTRACT

In previous studies, low-k carbon-doped silicon oxide (SiOC) films were deposited using organosilicon precursor: $(CH_3)_x SiH_{4-x}$. In this paper, we present the properties of PECVD low-k SiOC films produced by using conventional SiH_4 based gas precursors. The SiH_4 based SiOC films have similar gross physical and electrical characteristics to those of $(CH_3)_x SiH_{4-x}$ based SiOC. Since the precursors are inexpensive, commercially available and convenient to operate for existing tools, the process should not require additional cost as compared with that of PECVD silicon dioxide. We demonstrate the feasibility of integrating Cu with SiOC on damascene interconnection. The evaluation on electrical performance of the Cu/ SiOC based damascene structure will be discussed.

INTRODUCTION

Recently, carbon-doped silicon oxide (SiOC) has been emerging as one of the best low-k candidates for delivering the required performance of interconnect dielectric (ILD) in future ULSI devices [1]. SiOC has a dielectric constant ranging between 2.5 and 3.2 with sufficient mechanical strength and adhesion properties. SiOC preserves some beneficial characteristics of SiO_2, so that the integration of SiOC with Cu can utilize those traditional well-developed processes and technology, such as PECVD, etch and CMP. Clearly, the replacement of SiO_2 by SiOC shows advantages in a simple and low cost transition from SiO_2 to low-k ILD.

The reduction of dielectric constant can be achieved by doping methyl (-CH_3) group into SiO_2 network to form SiOC, where some of Si-O bonds are replaced by Si-CH_3 bonds. Since the termination of O-Si-O cross-linking by -CH_3 creates volume, the higher carbon concentration in SiOC can thus lead to a lower dielectric constant. In previous studies, PECVD SiOC films were deposited using organosilicon precursor: $(CH_3)_x SiH_{4-x}$ [1 - 4], such as methylsilane (x = 1) [5], trimethylsilane (x = 3) [1 – 4] and tetramethylsilane (x = 4). In this paper, we show that SiOC films can be produced by using conventional SiH_4 based gas precursors, which are the mixture of SiH_4 and hydrocarbon gas. The SiH_4 based SiOC films have similar gross physical and electrical characteristics to those of $(CH_3)_x SiH_{4-x}$ based SiOC. The dielectric constant ranges from 2.8 to 3.2 with density from 1.4 to 1.7 g/cm^3 correspondingly. Leakage current is measured in an order of 10^{-9} A/cm^2 at a field of 10^6 V/cm. The films are stable against thermal anneal at > 400 °C. Post anneal for as-deposited film may not be necessary if appropriate process conditions are chosen. We have also demonstrated the feasibility for the integration of Cu and SiOC on damascene interconnection. Since the SiH_4 based precursors are relatively inexpensive, commercially available and convenient to operate for existing tools, the process may not require additional cost as compared with that of PECVD silicon dioxide. This might be a practical advantage for SiOC to be considered as a new low-k ILD.

EXPERIMENTS

SiOC films were prepared in a commercial PECVD system (OXFORD Plasmalab 100 system), equipped with a parallel plate reactor and dual frequency (high frequency: 13.56 MHz and low frequency: 100 – 900 kHz) RF power supplies. During deposition, a mixture of SiH₄, N₂O and one or two types of hydrocarbon gas is used for SiOC gas feeds in discharge. The films were deposited onto 6-inch Si substrates placed on a heated chuck (ground electrode) with temperatures ranging from 25 to 700 °C.

The thickness and refractive index of SiOC films were determined by a spectroscopic ellipsometer (SENTECH Instruments Gmbh). We employed an X-ray reflectivity (XRR) technique to determine film density using a Philips X'Pert MRD Diffractometer. The dielectric constant k is determined from C-V measurement (100 kHz) on Al-dot/SiOC/Si capacitors. FTIR, XPS and RBS were employed to determine the bonding and chemical concentration in SiOC films. The film stress was measured using a Tencor FLX 2300. The isothermal test of thermal stability for SiOC was performed in a vacuum chamber with a base pressure of 10^{-5} Torr.

RESULTS AND DISCUSSION

PECVD Deposition and Post Anneal

The flow ratio of the precursor gas is a major factor for controlling film properties. Figure 1 is a plot of dielectric constant and refractive index as a function of SiH₄/hydrocarbon ratio (arbitrary units), where the film is deposited at 250 °C. The dielectric constant of SiOC largely depends on the flow of hydrocarbon gas due to the fact that higher flow rate can increase the carbon concentration in SiOC film.

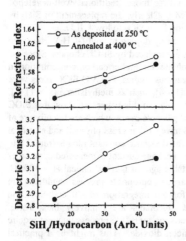

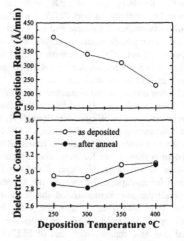

Fig. 1   Dielectric constant and refractive index vs. SiH₄/hydrocarbon ratio

Fig. 2   Dielectric constant and deposition rate as a function of temperature

Fig. 2 shows temperature effect on the deposition, where SiH_4/hydrocarbon ratio is fixed at 15. Although at higher deposition temperature, the deposition rate is reduced significantly, the dielectric constant only changes slightly. Clearly, carbon is more difficult to doped into SiO_2 at higher deposition temperature.

Post anneal is also an important step to reduce the dielectric constant, as indicated in Fig. 1 and Fig. 2, where film is annealed at 400 °C for 2 hours. After anneal the dielectric constant can be lowered as low as 2.8.

It is important to note that the deposition at 400 °C can allow us to save the anneal step without sacrificing much on the dielectric constant, where k is about 3.0 – 3.1 at 400 °C. This certainly can reduce the cost from the view of integration process.

Film Properties

Table I lists the measured density, stress, dielectric constant and atomic concentration of a SiOC film before and after anneal. The film is deposited at 300 °C and annealed at 400 °C for 2 hours. The density of SiOC is accurately determined from X-ray reflectivity measurements. Below a certain "critical angle" 100% of the x-rays are reflected from the film surface. Above that angle oscillations in reflected x-ray intensity are observed. The average electron density (and thus mass density) is found from the critical angle. The x-ray data was modeled using Philips' WinGixa program written by Leenaers and de Boer, which is based on Maxwell's equations [6, 7]. The atomic concentration is determined from Rutherford backscattering (Charles Evans & Associates).

Table I: SiOC Film properties

SiOC Film	Density (g/cm³)	Stress (MPa)	Dielectric Constant	Atomic Concentration (%)			
				Si (±2%)	O (±3%)	C (±5%)	H
As-deposited	1.47±0.02	-30	2.94	9.0	13.5	31.0	46.5
After anneal	1.46±0.02	+28	2.81	9.2	13.3	33.0	44.0

After anneal, the SiOC film changes slightly. Outgassing may cause the reduction of H-concentration by ~ 2%. Some of these reductions might result from the decrease of Si-OH bond, as evidenced from FTIR spectrum shown in Fig. 3. Since OH bond is mainly associated with orientation polarization, the anneal process can thus lower the dielectric constant, as indicated from Table I.

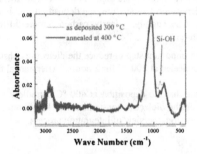

Fig. 3 FTIR spectrum of SiOC film before and after anneal.

Etching Properties

The high-density plasma (HDP) etching of SiOC shows a higher etching rate compared with SiO_2 and SiN. Either SiO2 or SiN can be used as etching stop layer for SiOC/SiOC dual damascene structure. It is also important to test how SiOC is capable of resisting O_2 plasma exposure [8]. Fig. 4 compares FTIR spectra of a SiOC film before and after HDP O_2 plasma etching at ~ 0 °C. No significant change can be observed in the FTIR characteristics. However, such a resistance to O_2 plasma can be significantly weakened if wafer temperature is higher. We have found that at 300 °C, O_2 plasma dramatically changes the FTIR characteristics and the film is more or less similar to that of SiO_2, indicating that the exposure degrades the film integrity and reduces significantly the carbon concentration. This reminds us to be extremely cautious when using photoresist ashing process, where the temperature condition has to be modified as compared with that of SiO_2 process.

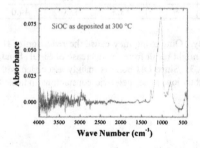

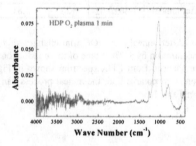

Fig. 4(a) FTIR spectrum of SiOC film (as deposited).

Fig. 4(b) FTIR spectrum after HDP O_2 plasma etching.

Cu/SiOC Damascene Interconnection

One type of Cu/SiOC dual damascene structure, Cu/SiOC/SiOC, is displayed schematically in Fig 5, where low-k SiOC is used as ILD for both top trench layer (metal wiring) and underneath via layer (via contact).

The patterning of SiOC stacking layers is by conventional plasma etching recipe similar to that for SiO_2. TiN is then sputtered before copper deposition as diffusion barrier. MOCVD copper was used to fill the trench and via using Cu-hfac-tmvs precursor. Excellent adhesion of MOCVD copper film to TiN diffusion barrier film is achieved through a combination of process and equipment design. Copper is then patterned by CMP method using commercially available system and slurry.

The CMP polishing is a real test for the adhesion and mechanical strength of the SiOC damascene stacking layer structure. Fig. 6 is a SEM cross section image of Cu damascene interconnection after CMP. It indicates that the integration approach for Cu/SiOC damascene is feasible.

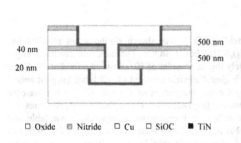

Fig. 5 Schematic dual damascene structure.

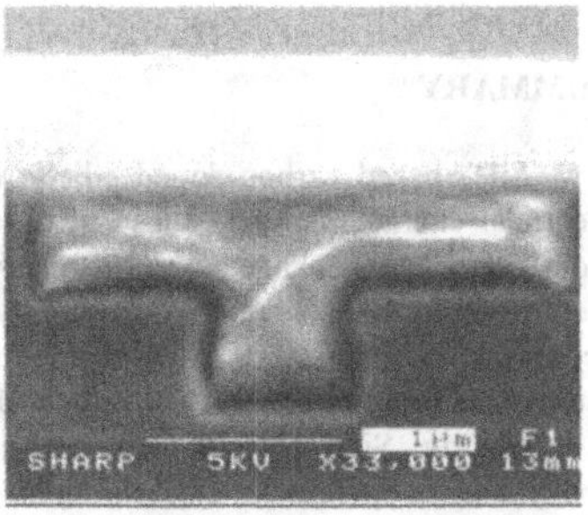

Fig. 6 SEM cross section of Cu/SiOC damascene interconnection

Electrical Properties

The leakage current of SiOC films were measured both from Al-dot/SiOC/Si capacitors and from comb circuit structure (line spacing ~ 0.95 μm). The results are plotted in Fig. 7(a) and (b), respectively. For comparison, we also plot the result from SiO_2 films. Shown in Fig. 6(a), at an electric field of 0.8×10^6 *V/cm*, the leakage current is about 4×10^{-9} *A/cm^2*, which is compared with

~ 5×10^{-10} A/cm^2 for SiO$_2$. For the line-line leakage shown in Fig. 7(b), SiOC shows a slightly lower leakage current that that of SiO$_2$. This, however, is still consistent with Fig. 7(a) because the current is measured at a field of ~ 0.25×10^6 V/cm, at which the leakage current of SiO$_2$ is indeed slightly higher.

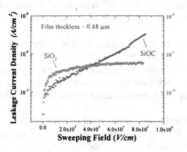

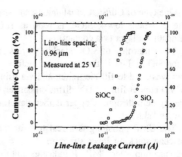

Fig. 7(a) Leakage current measured from Al-dot/ILD/Si capacitors.

Fig. 7(b) Leakage current measured from comb circuit structure.

SUMMARY

SiH$_4$ based carbon-doped silicon oxide (SiOC) have shown similar gross physical and electrical characteristics to those of (CH$_3$)$_x$SiH$_{4-x}$ based SiOC. The dielectric constant ranges from 2.8 to 3.2. Leakage current is relatively low in an order of 10^{-9} A/cm^2 at a field of 10^6 V/cm. Deposition at 400 °C results in a slightly higher dielectric constant of ~ 3.1 but the post anneal process is not necessary. The feasibility for the integration of Cu and SiOC on damascene interconnection has been demonstrated, showing good adhesion, etching and CMP properties.

Clearly, SiH$_4$ based precursors are relatively inexpensive, commercially available and convenient to operate for existing tools. No additional cost is needed as compared with that of PECVD silicon dioxide. Thus, SiH$_4$ based SiOC may be considered as a favorable candidate for a new low-k ILD.

REFERENCE

1. For a review, see M. J. Loboda in *Advanced Metalization Symposium,* Orlando, FL. 1999.
2. S. McClatchie, K. Beekman, A. Kiermasz, Proceedings of DUMIC Conference, Santa Clara, CA, 1998, pp. 311 (1998).
3. M. Naik and et. al., Proceedings 1999 IEEE Int'l Interconnect Technology Conference, pp. 181 (1999).
4. A. Grill and et. al., Mat. Res. Soc. Proc. **565**, 107 (1999).
5. Wai-Fan Yau and et. al., in *Advanced Metalization Symposium,* Orlando, FL. 1999.

6. L.G. Parratt, Phys. Rev. **95**, 359 (1954).
7. D.K.G. de Boer, Phys. Rev. **B44**, 498 (1991).
8. Chooo Kun Ryn and et. al., in *Advanced Metalization Symposium,* Orlando, FL. 1999.

Mat. Res. Soc. Symp. Proc. Vol. 612 © 2000 Materials Research Society

Integration and Characterization of Low Carbon Content $SiO_xC_yH_z$ Low κ Materials for < 0.18 µm Dual Damascene Application

Ju-Hyung Lee, Nasreen Chopra[*], Jim Ma, Yung-Cheng Lu, Tzu-Fang Huang, Ralf Willecke, Wai-Fan Yau, David Cheung, Ellie Yieh
Applied Materials Inc., 3320 Scott Blvd., Santa Clara, CA 95054
[*]Agilent Inc., 350 Sharon Park Drive #N209, Menlo Park, CA 94025

ABSTRACT

A CVD-based low κ film was evaluated for inter-metal dielectric in < 0.18 µm generation devices. The film was deposited by conventional rf PECVD method using organosilane compound and oxygen. The measured dielectric constant of the film was 2.7~2.75. The κ value of the film was stable over several weeks and the moisture absorption was minimal. The chemical composition was in the form of $SiO_xC_yH_z$, where the carbon content was less than 5 atomic %. Blanket film integration study was conducted to find out the manufacturing compatibility. The largest increase in κ value occurred during etching and ashing steps. However, SIMS compositional analysis revealed that the damage from these steps were limited to within top 300 Å, and the initial low κ value was recovered after the top damaged layer was removed by CMP. The final integrated dielectric constant was less than 3.0. The film density was measured as 1.4, compared to 2.3 g/cm^3 of conventional SiO_2. The low density of the film resulted from the termination of SiO_2 network structures by $Si-CH_3$ and $Si-H$.

INTRODUCTION

Silicon dioxide (SiO_2) has been dominantly applied as inter metal (IMD) and inter layer dielectric (ILD) throughout the history of microelectronics industry. The dominance is mainly attributed to its desirable electrical, thermomechanical characteristics as well as simplicity in integration with other materials. As the interconnection complexity required for high performance devices rapidly increases, however, several issues have arisen with traditional interconnection scheme, including enhanced RC delay effect, signal integrity due to cross talking and high power dissipation[1]. As part of the efforts to resolve these issues, new low dielectric constant material needs to be developed. Several approaches have been widely pursued in the past several years. Organic[2,3], inorganic[4,5] or hybrid[6,7,8] materials have been widely evaluated by either spin-on or chemical vapor deposition methods. Some of the issues with organic materials include thermal stability and reduced mechanical strength, which impose significant challenges on film integration. Inorganic low κ materials, in general, exhibit less integration issues, but κ stability still needs to be improved and there is difficulty in obtaining manufacturable materials with κ less than 3.0. Inorganic/organic hybrid material has received wide attention to obtain reliable low κ films while minimizing integration difficulties. In the development of hybrid low κ material, it is desirable to maintain low carbon content. High carbon content material has potential issues of via poisoning, etch and ash difficulties and thermomechanical instability. In this paper is presented the low carbon content low κ materials with carbon content less than 5%.

EXPERIEMENT

All film deposition was conducted in an Applied Materials DL κ^{TM} chamber, a parallel plate PECVD chamber with the substrate temperature controlled by heat exchanger. The substrate temperature was typically controlled to less than 50 °C. The film curing after the deposition was done in a typical CVD vacuum chamber at 400 °C for 30 minutes in N_2 ambient. The film thickness and refractive index was measured using spectroscopic ellipsometer with wavelength range of 250 to 750 nm. All the electrical measurements were done uisng Hg probe. For capacitance measurement, a thin thermal oxide layer was grown before the film deposition and the dielectric constant of the film was calculated using a simple expression for capacitance in series.

A series of integration steps leading to single damascene structures were conducted on the films in order to test the manufacturing compatibility of the film. A low pressure inductively coupled plasma reactor was used for both etching (C_xF_y chemistry) and ashing (O_2 chemistry). The film was immersed in ACT970^{TM} at 80 °C for 30 minutes and then in SR1^{TM} solution at room temperature for 10 minutes. The film was then baked out for 1 min at 400 °C and treated with plasma (He+H_2), which is to simulate the pre-clean step often used before barrier/seed layer deposition. The film was polished using Applied Materials Mirra^{TM} system and baked shortly after the CMP step at 400 °C in N_2. In this study, a simple procedure was applied using humidity chamber to accelerate the susceptibility to moisture absorption. The wafer was placed in the humidity chamber for about 17 hr, while the chamber condition is maintained at 85% relative humidity at 85°C. The ramping up the temperature and humidity from atmospheric environment or vice versa was set up in the way to ensure no moisture condensation on the film. Finally, the atomic film composition was determined by Hydrogen Forward Scattering (HFS) for hydrogen and Rutherford Back Scattering for silicon, oxygen and carbon. Si, O and C.

RESULTS AND DISCUSSION

From the process variables' effect on film characteristics, it was found that the dielectric constant increases with applied rf power, which indicates ion-induced bombardment adversely affect the film porosity. Therefore, the rf power was applied to the chamber in the manner that the ion bombardment, thereby film densification and k increase is minimized. In Fig.1 is shown the FTIR spectra of as-deposited and after-anneal films. It is shown that the overall FTIR spectra is very close to conventional SiO_2 except the presence of SiH (2200 cm^{-1}) and SiCH$_3$ (1250 cm^{-1}). The annealed film compared to the as-deposited film has slightly lower Si-H and Si-CH$_3$, but higher Si-O bond concentration. The small peaks near 1600 cm^{-1} and broad peak near 3600 cm^{-1} with as-deposited film indicate the presence of Si-OH or absorbed water in the film. In the anneal step, it is believed that the excess non-bonded or loosely-bonded species as well as absorbed water are driven out of the film, forming more open micro structures in the film. The porosity is further increased by termination of SiO_2 network structures by Si-CH$_3$ and Si-H.

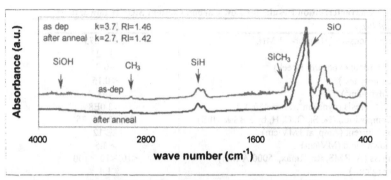

Fig.1 FTIR spectra of as-deposited and annealed films

In Fig.2 is illustrated the correlation of κ value with film bond concentrations determined by FTIR. It indicates that the κ value generally decreases with increase in Si-CH$_3$ content, but fairly independent of Si-H content. This could be explained by the molecular size of network terminating species. The effective porosity formation is much greater when the network structure is terminated by bulky –CH$_3$. In Table 1 is summarized the characterization results of the prepared film. The measured dielectric constant was 2.7~2.75 and was stable for several weeks. The κ increase in relatively high temperature (85 °C) and high humidity (85 %) environment was also minimal. The density of 1.4 g/cm^3 compared to PECVD SiO$_2$ (2.3 g/cm^3) indicates that the film is in open, porous micro structure. As shown by the hardness and elastic modulus data, the mechanical strength of the film is comparable to SiO$_2$ and order of magnitude better than typical organic low κ films. It should be also noted that the film maintains good dielectric characteristics, as evidenced by breakdown voltage and leakage current measurement. Film surface morphology is particularly important in film integration, since the planarization in each step is crucial in dual damascene scheme.

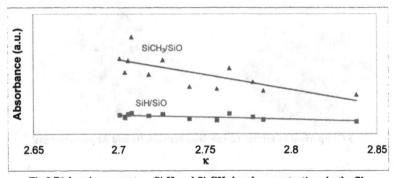

Fig.2 Dielectric constant vs. Si-H and Si-CH$_3$ bond concentrations in the film

Table 1.Characterization Results for Low Carbon Content SiOCH Low κ film

Material Properties	Results
Dielectric Constant (Hg probe, 1 MHz)	2.7 ~ 2.75
RI	1.42
Δκ (clean room exposure, 1week)	~0.0
Δκ (humidity test)	< 0.15
Stress (dyne/cm^2)	4E8
Stress Hysteresis (dyne/cm^2,RT to 450 °C)	< 2.0E8
Film Composition (%, Si, O, C, H, by RBS & HFS)	25, 35, <5, 35
Leakage Current (Amp, at 1MV/cm)	2E-12
Breakdown Field (MV/cm)	> 3.5
Roughness (Å, RMS, Ra, Rmax, 5000 Å film)	<10, <10, <100
Density (g/cm^3)	1.4
Hardness (Gpa)	1.8
Elastic Modulus (Gpa)	11.1

The measured roughness from AFM image was comparable to the conventional SiO$_2$ films deposited by PECVD method. The effect of the carrier gas on the surface morphology is compared for helium and argon. As shown in the AFM image(Fig.3), the film surface with Ar is composed of bigger nodules compared to the surface with He. Both κ value and moisture absorption performance, however, were comparable between the two carrier gases.

The evolution of dielectric constant with single damascene integration steps is illustrated in Fig.4. It was found that the dielectric constant increases by about 10% after the etch step, but it did not change after the following O$_2$ ashing step. In the separate test where only ashing was performed without etching, it was found that similar percentage of k increase was found. It is also shown in the figure that the k increase from the etching and ashing steps was partially recovered after a short bake out step at 400°C. This indicates that κ increase is partially due to Si-OH formation during the etching or ashing processes.

He   Ar

Fig.3 Comparison of surface morphology between He and Ar as carrier gas

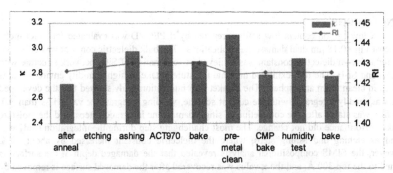

Fig.4 Changes of dielectric constant with single damascene integration steps

The remainder of the k increase from the etching and ashing processes is believed to originate from permanent structural damage to the film. To find out the damage extent, a compositional analysis was done on the film that was exposed to O_2 ashing. In Fig.5 is illustrated the compositional change before and after the ashing step. It was found that the compositional change in the film is limited to less than top 300 Å. Relatively little variation in refractive index throughout the integration step also indicates no structural change in bulk film. This <300 Å top damaged layer is confirmed by the following CMP steps, after which the initial κ value of the film was fully recovered. No scratch marks or peeling were observed after CMP, which resulted from relatively high hardness of the film. Small increase in κ value was found after ACT970 and SR1 chemical treatment. The increase is believed due to absorbed water in the film and the κ value was recovered by a following short bake step. From this test, it is demonstrated that the prepared film is compatible with the wet chemical solutions commonly applied in Cu damascene processes. It should be also noted that the moisture uptake characteristics has not changed after the completion of whole single damascene sequence, and stayed at $\Delta\kappa$ of <0.15. The final sample was monitored for κ value increase under clean room environment, but no change was observed.

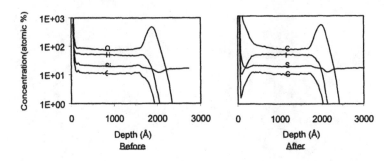

Fig.5 SIMS compositional profile for samples before and after ashing

CONCLUSIONS

A low carbon content low κ film prepared by rf PECVD was evaluated for inter metal dielectrics in <0.18 μm dual damascene applications. The bulk dielectric constant was 2.7~2.75. The reduction in dielectric constant was achieved by termination of SiO_2 network structure with $Si-CH_3$ and Si-H. The film exhibited minimal moisture uptake in high humidity environment as well as in clean room atmosphere. The blanket film integration study showed that the developed film was readily integrated with the current scheme, yielding integrated κ value less than 3.0. The film was stable after the completion of single damascene integration steps, and the moisture uptake performance did not change. The most challenging step during the integration study was found as etching and ashing steps, where the dielectric constant increased by about 10%. However, the SIMS compositional analysis revealed that the damaged depth at the surface is limited to the top 300 Å, and the κ value was recovered after the damaged surface is removed by CMP.

REFERENCES

[1] S-P. Jeng, et al, *VLSI Tech. Symp.Tech. Dig.*, 1994, p73
[2] D. Stoakley, et al, *Polymer Preprints*, **34**, 1993, p381
[3] G-R. Yang, et al, *J. Electron. Mater.*, **25**, 1996, p1778
[4] C. Jin, et al, *MRS Bulletin*, Oct., 1997, p39
[5] H. Igarashi, et al, *VLSI Tech. Symp.Tech. Dig.*, 1994, p73
[6] R.D. Miller, et al, *MRS Bulletin*, Oct., 1997, p44
[7] A. Nara, et al, Jpn. J. Apl. Phys., Part1, **36**, 1997, p1477
[8] A. Grill, et al, *J. of Appl. Phys.*, **85**, 1999, p3314

Mat. Res. Soc. Symp. Proc. Vol. 612 © 2000 Materials Research Society

SYNTHETIC CONTROL AND PROPERTIES OF PROCESSIBLE POLY(METHYLSILSESQUIOXANE)S

Jin–Kyu Lee,* Kookheon Char,& Hie-Joon Kim,* Hee-Woo Rhee,+ Hyun-Wook Ro,* Dae Young Yoo,* and Do Y. Yoon.*
*School of Chemistry and Molecular Engineering and &School of Chemical Engineering, Seoul National University, Seoul 151-742, Korea
+ Department of Chemical Engineering, Sogang University, Seoul 121-742, Korea

ABSTRACT

Processible poly(methylsilsesquioxane)s (PMSSQs) were prepared in THF solution under nitrogen atmosphere in the presence of HCl catalyst. It was found that various reaction parameters such as concentration, temperature, reaction time, the amount of water, and the amount of acid catalyst could affect the molecular weight and the amount of functional end groups of PMSSQ samples. Thin films prepared from our PMSSQ samples by spin -coating followed by curing to 420°C exhibited a much better crack resistance than those presented in the literature, while the dielectric constant remained practically the same, i.e., ca. 2.7.

INTRODUCTION

Silsesquioxanes with empirical formula of $(RSiO_{3/2})_x$, where R is hydrogen or an organic group, have been studied for many years since the first commercialization of silicone polymers as an electronic insulation materials at high temperature.[1] Poly(phenylsilsesquioxane), PPSSQ, was the most widely studied material and many well-controlled synthetic methods were patented.[2-4] Recently, silsesquioxanes have again attracted much attention as a promising candidate for a low dielectric insulator in semiconductor industry. Since semiconductor devices are becoming smaller and device packing densities are increasing rapidly, both signal delays due to the combined resistance and capacitance (RC) coupling and the crosstalk between metal interconnects have been found to be a serious problem.[5,6] According to the International Technology Roadmap for Semiconductors, when the dimension in integrated circuits shrinks to 0.13 μm around year 2002, this interconnect delay is believed to dominate the overall device cycle time.[7] In addition to switching the metal electrode from Al to more conductive Cu, therefore, the introduction of low-dielectric insulator materials to the integrated circuits becomes another key requirement.

Many researchers have tried to decrease the dielectric constant of an insulator by incorporating nanometer size pores in polymer matrices.[8] Among many candidates for the insulators, the inorganic/organic polymer hybrid system shows quite interesting and promising results, where silicon based polymers such as poly(methylsilsesquioxane) (PMSSQ) were used as a matrix and organic oligomers (known as a porogen) were introduced to generate pores within the matrix.[9] PMSSQ has been well known as an insulator matrix owing to its attractive low dielectric constant of ca. 2.7, low moisture absorption, excellent thermal stability up to 500°C, and reasonable mechanical hardness.[1,10] Synthetic methods for preparing soluble and stable PMSSQ by acid or base catalyzed hydrolytic polymerization of methyltrifunctionalsilane have been improved since the early patent publication by Japan Synthetic Rubber (JSR) in 1978 (as shown in Scheme 1).[11-14] However, they have not yet been understood well enough to control the molecular weight over a wide range as well as the amount and nature of unreacted functional end groups, which are believed to be quite important for the formation of nanometer size pores and the mechanical properties of final thin film insulators.

EXPERIMENT

Methyltrimetoxysilane (MTMS), purchased from Aldrich, was used without further purification. THF and HCl, purchased from Dae Jung Chemical Co., Korea, were used for the polymerization. All the polymerization reactions were performed under a nitrogen atmosphere by a standard Schlenk line technique unless otherwise specified. ^{1}H-NMR and ^{29}Si-NMR spectra were obtained at room temperature with TMS (tetramethylsilane) as an internal standard using Bruker DPX-300 MHz and Bruker DRX-500 MHz, respectively. NMR spectra were obtained in acetone-d_6 unless otherwise noted. Gel permeation chromatographic (GPC) analyses were carried out using house-made GPC equipped with a mixed bed Jordi column and Waters 2410 differential refractometer. THF was used as an eluent at a flow rate of 1.0 ml/min and the GPC column was calibrated with respect to polystyrene standards (Polymer Standard Service, USA, Inc.) which ranged from 500 – 1.0 x 10^6 in molecular weight.

RESULTS AND DISSCUSION

We have investigated the effects of several synthetic factors on generating processible PMSSQs with a high content of functional end groups, which strongly affect the miscibility of the PMSSQ matrix with porogens. The synthetic factors to be considered are solvent, temperature, stirring rate, humidity, molar ratio of catalyst to MTMS (R_1), and molar ratio of water to MTMS (R_2). When the known literature method[13] was employed to make PMSSQs in an open system, where the concentration of monomer and polymer increases during the reaction process due to the evaporation of solvent and water, it was quite difficult to reproduce polymers which exhibit the same molecular weight as mentioned in the literature and to control the amount of functional end groups. In order to generate reproducible solid PMSSQs with controlled molecular weight and amount of functional end group, therefore, MTMS was polymerized in refluxing THF solution (28%(w/w) solution) under nitrogen atmosphere in the presence of HCl catalyst. The amount of HCl catalyst affected the polymerization rate through the parameter R_1, and $R_1 = 0.03$ turned out to be the optimum condition to obtain processible solid PMSSQ before gelation occured. As shown in Scheme 1, the polymerization takes place through the two different reactions and the variation of R_2 value affects the reaction rate and the equilibrium state of each reaction. Consequently, one can expect the hydrolysis reaction to become more favorable with large R_2 values and the equilibrium of the condensation reaction to shift back to the reactants yielding low molecular weight polymers with a large amount of –OH end group. Therefore, R_2 was varied form 1.0 to 15 in order to check the important influence of the R_2 value on both molecular weight and amount of the functional end group, and the resulting polymer products were characterized by ^{1}H-, ^{29}Si-NMR and GPC. Some representative results are summarized in Table 1 and Table 2.

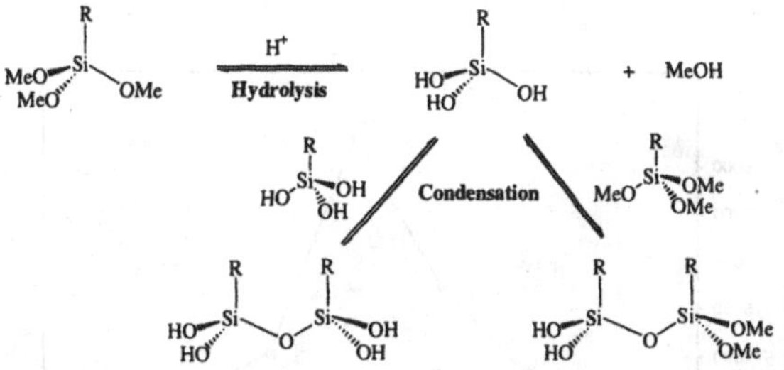

Scheme 1. Acid catalyzed polymerization mechanism of polysilsesquioxane.
$R = H$ or alkyl, $X = $ alkoxy (OR)

When R_2 was increased, the molecular weight of PMSSQ's increased until R_2 reached 2.2 (although the stoichiometric value of R_2 is 1.5) showing the maximum molecular weight, as shown in Table 1 and Figure 1. After the maximum point, the molecular weight of PMSSQ quickly decreased upon further increase of R_2. The amount of functional end groups is also affected by the R_2 value as expected. The change in the amount of functional end groups as a function of the R_2 value is shown in Table 2. High molecular weight polymers, which have undergone more condensation reactions, have a relatively small percent of functional end groups and the amount of –OH group increases with larger R_2 values. From ^{29}Si-NMR data, the number of Si-O-Si linkage on Si atom could be identified and it was confirmed that all the PMSSQ polymers we prepared contained only T_2 and T_3 structures.[13,14] PMSSQ polymers have a broad molecular weight distribution with an apparent polydispersity index (PDI) value ranging from 1.7 to 3.7 due to the polycondensation mechanism. When the monomer concentration of polymerization solution was increased from 28% to 43%, even higher molecular weight (up to 130,000 measured with GPC based on PS standards) polymer with a broader PDI value of 26.1 could be obtained. Those high molecular weight polymers were still soluble and stable in solid or solution state. In order to obtain a narrow molecular weight distribution of the PMSSQ sample, the original polymerized samples could then be fractionated by a combination of toluene and acetonitrile solvents to yield a relatively narrow distribution with PDI values around 1.7 ~ 2.0.

Table 1. Average molecular weights of PMSSQ samples prepared with different R_2s[a]

R_2	Mw[b]	Mn[b]	PDI
1.3	2221	1290	1.7
1.6	3631	1754	2.0
1.9	5469	2172	2.5
2.2	8635	2356	3.7
2.5	4062	1781	2.3
2.7	3493	1707	2.0
3.0	2538	1364	1.8

[a] R_2 = (mole of H_2O)/(mole of methyltrimethoxysilane)
[b] Determined from GPC with polystyrene standards as a reference

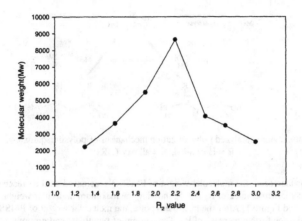

Figure 1. Weight-averaged molecular weight (M_w) of PMSSQ as a function of R_2 value

Table 2. NMR analysis results of PMSSQ samples prepared in this study [a]

R_2	Si-CH_3[b]	Si-OCH_3[b]	Si-OH[b]	Functional end group (%)[c]	Si-OH (%)[d]	Refractive index[e]
1.3	8.0	1.0	0.5	15.5	5.0	-
1.6	18.8	1.0	1.3	12.3	6.3	-
1.9	27.9	1.0	1.4	7.6	2.1	1.37
2.2	84.7	1.0	3.8	4.8	4.3	1.37
2.5	70.5	1.0	4.8	7.6	6.2	-
2.7	89.1	1.0	6.5	7.8	6.7	1.37
3.0	69.3	1.0	6.3	9.5	8.2	1.38

[a] NMR spectra were taken in aceton-d_6 and integration values were used to determine the amount of each functional groups.
[b] Normalized value to the integration value of Si-OCH_3
[c] Functional end group (%) = [(Si-OCH_3 + Si-OH) / (Si-CH_3 + Si-OCH_3 + Si-OH)] x 100
[d] Si-OH (%) = [Si-OH / (Si-CH_3 + Si-OCH_3 + Si-OH)] x 100
[e] Measured from variable angle ellipsometer

Thin films of PMSSQ polymers with a high content of functional end groups were fabricated from methyl isobutyl ketone (MIBK) solution by spin-coating on silicon wafers followed by curing to 430°C, and their mechanical and electrical properties were examined. Upon initiating a crack by loading weights on the film with microvickers, the crack propagation velocity was measured and compared with the values reported by Cook and Liniger[15], who measured the crack velocities as function of film thickness for various PMSSQ films prepared by different synthetic methods. Preliminary results show that PMSSQ films of ca. 1.0 μm thickness, prepared with our samples of Mn = 4,400 and Mw = 16,000 (according to GPC measurement on PS standards), exhibit an average crack velocity smaller than of 10^{10} m/sec in water. This value is found to be significantly smaller, by ca. 10^4, as compared with the smallest value reported by Cook and Liniger for their PMSSQ samples.[15] Further studies with various molecular weight fractions are currently in progress. As shown in Table 2, the molecular weight of PMSSQ and the amount of functional end groups do not seem to affect the refractive indices of the cured PMSSQ thin films. Dielectric constants of the PMSSQ thin films prepared were found to be ca. 2.7 in agreement with the literature values. Detailed studies of introducing porogen oligomers and generating nanometer size pores in the prepared PMSSQ matrices are currently under investigation.

CONCLUSIONS

We have developed a simple synthetic method to control the molecular weight of PMSSQ as well as the amount of functional end group. Thin solid PMSSQ films fully cured to 430°C showed quite promising mechanical and electrical properties. This novel simple synthetic method for PMSSQ will be used to synthesize the next-generation low-dielectric thin film insulators, and the detailed results on nanoporous PMSSQ films will be reported in a forthcoming paper.

ACKNOWLEDGENTS

This work is partially supported by the Korean Collaborative Project for Excellence in Basic System IC Technology (98-B4-CO-00-01-00-02), and the Brain Korea 21 Program by the Korean Ministry of Education.

REFERENCES

1. Baney, R. H.; Itoh, M.; Sakakibara, A.; Suzuki, T. *Chem. Rev.* **1995**, *95*, 1409.
2. Adachi, H.; Adachi, E.; Hayashi, O.; Okahashi, K. Japanese Patent Kokai-H-1-26639, 1989; *Chem. Abstr.* **1989**, *111*, 58566.
3. Adachi, H.; Adachi, E.; Hayashi, O.; Okahashi, K. Japanese Patent Kokai-H-1-92224, 1989; U.S. Patent 5,081,202, 1989; *Chem. Abstr.* **1989**, *111*, 154663.
4. Yamazaki, N.; Nakahama, S.; Goto, J.; Nagawa, T.; Hirao, A. *Contemp. Top. Polym. Sci.* **1984**, *4*, 105.
5. Peters, L. *Semicond. Int.* **1998**, *Sept*, 64.
6. Lee, W. W.; Ho, P., Eds. Low-Dielectric Constant Materials. *MRS Bull.* **1997**, *22(10)*.

7. International Technology Roadmap for Semiconductors; Semiconductor Industry Association; 1999.
8. (a) Tanev, P. T.; Pinnavaia, T. J. *Science* **1995**, *267*, 865. (b) Kresge, C. T.; Leonowics, M. E.; Roth, W. L.; Vartuli, J. C.; Beck, J. B. *Nature* **1992**, *359*, 710. (c) Huo, Q.; Margolese, D. I.; Ciesla, V.; Feng, P.; Gier, T. E.; Sieger, P.; Leon, R.; Petroff, P. M.; Schuth, F.; Stucky, G. D. *Nature* **1994**, *368*, 317. (d) Tamaki, R.; Chujo, Y. *J. Mater. Chem.* **1998**, *8*, 1113. (e) Yang, H.; Coombs, N.; Ozin, G. A. *J. Mater. Chem.* **1998**, *8*, 1205.
9. (a) Hedrick, J. L.; Miller, R. D.; Hawker, C. J.; Carter, K. R.; Volksen, W.; Yoon, D. Y.; Trollsas, M. *Adv. Mater.* **1998**, *10*, 1049. (b) Mikoshiba, S.; Hayase, S. *J. Mater. Chem.* **1999**, *9*, 591. (c) Nguyen, C. V.; Carter, K. R.; Hawker, C. J.; Hedrick, J. L.; Jaffe, R. L.; Miller, R. D.; Remenar, J. F.; Rhee, H. -W.; Rice, P. M.; Toney, M. F.; Trollsas, M.; Yoon, D. Y. *Chem. Mat.* **1999**, 11, 3080.
10. Loy, D. A.; Shea, K. J. *Chem. Rev.* **1995**, *95*, 1431.
11. Suminoe, T.; Matsumura, Y.; Tomomitsu, O. Japanese Patent Kokai-S-53-88099, 1978; *Chem. Abstr.* **1978**, *89*, 180824.
12. Abe, Y.; Hatano, H.; Gunji, T. *J. Polym. Sci. Part A: Polym. Chem.* **1995**, *33*, 751.
13. Takamura, N.; Gunji, T.; Hatano, H.; Abe, Y. *J. Polym. Sci. Part A: Polym. Chem.* **1999**, *37*, 1017.
14. Kudo, T.; Gordon, M. S. *J. Am. Chem. Soc.* **1998**, *120*, 11432.
15. Cook, R. F.; Liniger, E. *J. Electrochem. Soc.* **1999**, *146*, 4439.

Mat. Res. Soc. Symp. Proc. Vol. 612 © 2000 Materials Research Society

THEORETICAL AND EXPERIMENTAL ANALYSIS OF THE
LOW DIELECTRIC CONSTANT OF FLUORINATED SILICA

A. DEMKOV*, S. ZOLLNER*, R. LIU*, D. WERHO*, M. KOTTKE*, R.B. GREGORY*,
M. ANGYAL*, S. FILIPIAK*, G.B. ADAMS**
*Motorola Semiconductor Products Sector, Mesa, AZ
**Department of Physics, Arizona State University, Tempe, AZ

ABSTRACT

Fluorinated silica has a dielectric constant lower than that of F-free SiO_2 and is a potential interlayer dielectric. We investigate the F-doped SiO_2 with *ab-initio* modeling and various characterization techniques searching to explain the dielectric constant reduction. FTIR transmission and spectroscopic ellipsometry give us information about the ionic and electronic contributions to ε. Nuclear reaction analysis and Auger spectrometry measure F composition. XPS and FTIR provide information on the atomic structure of the film. We use several cells of cristobalite to model fluorinated silica using the electronic structure theory. The ground state geometry, vibrational density of states, electronic band structure, and Born effective charges are analyzed. The calculations suggest that it is the ionic component of the dielectric constant that is mostly effected by the F incorporation.

INTRODUCTION

Interest in insulating films with low static dielectric constants ε (or ε_{DC}, or k) increases, as transistor gate lengths shrink below 0.25 μm, [1]. The parasitic capacitance between metal lines causes signal delay, but could be reduced with these so-called low-k dielectrics. whereas The SEMATECH roadmap calls for materials with $\varepsilon<2.5$, significantly lower than that of currently used SiO_2-based interlevel dielectrics ($\varepsilon\approx4.2$). Very low values of ε could be achieved in carbon based materials ($\varepsilon=2.2$) [2], low-density silicas ($\varepsilon=1.3$) [3,4], or polymers ($\varepsilon=2.2$) [5]. While these novel materials are being developed and integrated, a more evolutionary approach is to reduce ε of the existing silica-based dielectrics. Fluoridation has been shown to reduce ε of silica films based on tetraethylorthosilicate $(C_2H_5O)_4Si$ (TEOS).

Variations in precursors and deposition conditions may result in materials showing different behavior. Han and Aydil reported SiO_xF_y films grown by plasma-enhanced chemical vapor deposition (PECVD) with SiF_4 and O_2 as feed gases and 100 W radio frequency (rf) power [6,7]. They observed a reduction of ε with increasing F concentration. On the other hand, Hasegawa *et al.* used a high-density plasma with an rf power of less than 30 W, with TEOS, O_2, and CF_4 gases [8] and found the opposite trend in ε as a function of F concentration. This difference may be due to the following reason: SiF_4 forms a molecular solid at temperatures below –90°C with a bcc lattice. It is not a framework structure. Oxidation of the SiF_4 gas introduces oxygen bridges between the Si atoms and allows formation of a framework structure. The oxidation is only partial, however, and results in a highly porous film with an appreciable concentration of silicon difluoride sites. On the other hand, TEOS contains the $(SiO_4)^{4+}$ tetrahedral unit which is the building block of the (4;2) silica framework. The introduction of fluorine disrupts the formation of this framework and creates defects, but fewer than in the first case (growth from SiF_4 and O_2). The topology of the two materials is therefore different. We will show that the framework topology plays an important role in the dielectric properties of the film.

EXPERIMENT

We used a PECVD system using SiF_4, TEOS, O_2, and other gases to deposit the films in this study. A varying SiF_4 flow rate was used to increase the F content of the films. A 20% reduction of ε, compared to undoped TEOS has been achieved, see Fig. 1(a). ε was seen to saturate at 3.4 for high SiF_4 flow rates. This suggests an upper limit to the amount of F participating in the reduction of ε that might be incorporated into the silica film. The film composition (F content) was analyzed using Rutherford backscattering spectrometry (RBS) and nuclear reaction analysis (NRA) techniques. Auger depth profiling (using NRA standards) found that the F concentration was uniform with depth. It is shown in Fig. 1(b) as a function of normalized SiF_4 flow rate.

The dielectric constant $\varepsilon(633$ nm) measured by spectroscopic ellipsometry (SE), see Fig. 2 (b), provides a good estimate for the electronic contribution ε_∞ to the reduction in ε_{DC}. We find $\varepsilon(633$ nm)=2.13 for 0% fluorine (undoped TEOS) and $\varepsilon(632$ nm)=2.05 at 11 at.% F. This accounts for only 15% of the total reduction of ε_{DC} (which is 0.8) at this concentration, see Fig. 1(a). This reduction of ε_∞ may be caused either by an increase in the effective energy gap or a decrease in the electron density per unit volume. In any case, our results indicate that the leading cause of the total dielectric constant reduction is not an electronic effect.

Infrared absorption spectra of F-doped TEOS (compared to undoped TEOS) reveal a new peak at 937 cm^{-1}, see Fig. 3, which we tentatively assign to a localized Si-F stretch mode derived from a four-fold coordinated Si with three bridging O atoms and one terminal F. Although the absorption coefficient of this mode increases linearly with F concentration, this straight line does not intercept the origin, suggesting that there is more than one bonding configuration. We also note a slight blueshift of the Si-O stretching mode with increasing F concentration.

XPS data were acquired on the surface of the F-doped TEOS film. Peak deconvolution is consistent with the presence of two peaks, see Fig. 2(a). The peak at the higher binding energy (686.6 eV) is consistent with terminal Si-F bonding. The peak appearing at lower energy (684.8 eV) could possibly be due to ionic fluoride (F^{-1}) or possibly a F bridging two Si atoms.

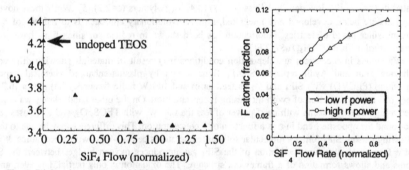

Fig. 1: Dielectric constant at 100 kHz (a) and fluorine content (b) versus normalized SiF_4 flow.

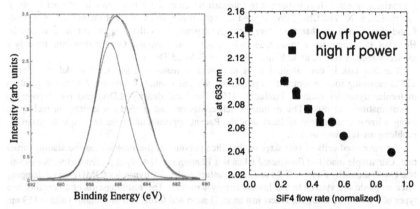

Fig. 2: (a) Deconvoluted XPS F1s core-level spectrum for highly F-doped TEOS film. (b) Reduction of the electronic contribution (measured at 633 nm) to ε measured by ellipsometry.

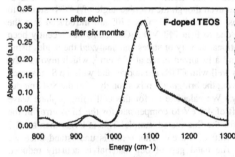

Fig. 3: FTIR absorbance of F-doped TEOS with high SiF_4 flow rate showing the Si-F stretch is at 937 cm^{-1} and the Si-O stretch near 1080 cm^{-1}. With aging, both peaks shift to slightly lower energies without changing their amplitudes very much.

To summarize, we found that ε as low as 3.42 can be achieved by F doping of TEOS in agreement with previous results [1,7]. SE shows a 0.1 reduction in ε_∞ when the F concentration is increased form 0 to 11 at.%, which cannot account for a change of 0.8 of the total ε. We have found that for our process conditions the maximum F concentration in TEOS is 11 at.%. The Auger analysis, calibrated by NRA experiments, indicates a uniform distribution of F in the film. The XPS and FTIR data suggest at least two types of F in the as-grown material. FTIR identifies a F-related mode at 937 cm^{-1} and finds a slight increase in frequency of the Si-O mode with increasing F concentration.

THEORY

Within the adiabatic approximation the crystal polarization and therefore the dielectric constant can be described by a sum of the electronic and lattice terms. Relative contributions of these two terms change from one material to another, e.g., in Si the dielectric response is predominantly electronic, while in high-ε perovskites, such as $BaTiO_3$, it is mainly due to the lattice polarizability. Silica is an intermediate case, where the two contributions are approximately equal. It is important to establish which of these two terms is affected by the F incorporation. Xu and Ching investigated the optical properties of silica theoretically [9]. They found that ε_∞ and the band gap are strongly dependent on the volume per SiO_2 molecule. However, ε_∞ is insensitive to changes in bond angles and lengths. Our SE data indicate only a moderate reduction of ε_∞ upon F incorporation in F-doped TEOS.

The first task in our calculations is to provide a reasonable structural model of F-doped TEOS capturing some essential film properties. We use a simplified density functional quantum molecular dynamics method, Fireball96 [10]. The local density (LDA) and pseudopotential approximations are used. The method is self-consistent and carried out entirely in real space using a linear combination of local orbitals. Recent applications of the technique to materials problems are reviewed in [11].

We have used cells of two sizes of the silica polymorph β-cristobalite as the starting structure. Our simple model of fluorinated silica is a 12 atom cell ($I\bar{4}2d$) of β-cristobalite. We substitute two O atoms by F (16 at.% F) and run quantum molecular dynamics (QMD) with a damping force to guide the system to the lowest energy structure. The resulting structure contains two types of F atoms: A **bridging** F acts just as an O atom and forms two Si-F bonds (with 0.159 nm and 0.162 nm bond length) and an Si-F-Si bond angle of 140°, well within the reported range of the bridging F^1 (115-180°). In addition, we find a **terminal** F atom that forms only one bond to Si (0.152 nm bond length). The F-F distance in this model is 0.4 nm. Analysis of the Si-O-Si bond angle distribution in the cell shows that even though the average angle of 147° is similar to that in F-free cristobalite, angles as low as 120° and as high as 170° are introduced into the system. The increase of the Si-O-Si angle due to F in PECVD SiO_xF_y films has recently been reported [12]. We have calculated the vibrational density of states and analyzed the modes using the inverse participation ratio (IPR). We find a localized mode at 935 cm^{-1}, which involves the terminal F. The calculated frequency agrees well with FTIR, see above, and with the Si-F stretch frequency reported in [1,13]. (Modes involving the bridging F mix strongly with the Si-O modes and could not be identified experimentally.) We calculate ε_∞ for this cell using a plane wave method (CASTEP) [14]. Using a rigid shift of 4.0 eV to compensate for the LDA error in the fundamental energy gap, we find $\varepsilon_\infty=2.69$ for the fluorinated sample! This is larger than $\varepsilon_\infty=2.2$ calculated for a F-free cell with the same shift of 4.0 eV. This result is understood once we examine the electronic density of states. The band gap of our material is actually reduced compared to that of undoped silica. When the terminal SiO_3F structure formed, it left a dangling bond on the adjacent Si, which results in formation of a midgap electronic state. At our F concentration of 16 at.%, the density of these defects is 5.6×10^{21} cm^{-3}. This gives rise to an occupied band in the middle of the SiO_2 gap. This band gap reduction results in the increase of ε_∞. The SE results show a small reduction of ε_∞ with increasing F concentration, from 2.13 in undoped TEOS to a minimum of 2.05 in F-doped TEOS. We conclude that the presence of dangling Si bonds in F-doped TEOS is inconsistent with experiment. The dangling bonds could also be saturated by H termination, but this is ruled out by the absence of Si-H vibrations in our FTIR spectra.

We now refine our structural model. Because of stoichiometric constraints, the local composition should be SiF_2O (oxidation states of Si, F, O are +4, -1, and -2). We consider two 46-atom models with the composition $Si_{15}F_2O_{29}$ (doubled C9 cell, F concentration is 4.6 at.%). After the QMD relaxation, the lowest energy structure (F-F distance 0.3 nm, Si-F distance 0.149 nm) has several dangling bonds producing levels in the band gap. The structure without dangling bonds (F-F distance 0.31 nm, Si-F distance 0.150 nm) is about 1 eV higher in energy, but has a clean gap. ε_∞ is slightly larger for the lower-energy structure. The energy difference can be understood in terms of strain. One Si and three O atoms have been substituted by two F atoms to satisfy the stoichiometry. The system is forced to close a void from a missing SiO_3, see Fig. 4. The first Si-Si neighbor distance in SiO_2 is about 0.31 nm, while the second Si-Si neighbor distance is 0.49 nm. The resulting Si-Si distance found for the dangling bond free cell is 0.34 nm or just 10% larger than in equilibrium. With our cell size of 0.716 x 0.716 x 1.432 nm^3, the missing units are not sufficiently separated to absorb the lattice distortion by an angle adjustment. This shows how a chemical requirement puts an upper limit on the amount of F that can be incorporated into the (4;2) silica network. The bond length along the chain connecting two sites with "missing" Si range from 0.162 nm to 0.176 nm, and the O-Si-O angles range from 147° to 170°. Note that the most destabilizing angels below 130° encountered in a 12-atom model have been eliminated.

The bridging F (often regarded as a Lewis base) may transform into the terminal one or leave the sample altogether, as seen in the aging studies reported earlier [15]. The vibrational density of states indicates only a minor hardening of the modes, in agreement with our FTIR data. Our preliminary calculations indicate that Born effective charges associated with the Si-O-Si bond bending are higher than those associated with the Si-O bond stretching. The bending force constants are significantly lower than the stretching ones [10]. It appears therefore, that by disrupting the SiO_2 framework with terminal F we eliminate the leading contribution to the lattice susceptibility. The susceptibility is proportional to the square of the Born effective charge of the vibrational mode, and inversely proportional to the square of its frequency. That suggests a plausible mechanism for the dielectric constant reduction in fluorinated films. In addition, the bending force constant increases with the increase if the Si-O bond length [10]. This may explain the observed hardening of the Si-O mode with increasing F concentration.

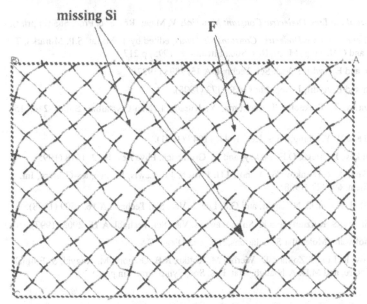

Fig. 4: Wire schematic of the F doped SiO_2. Note the lattice strain due to the missing SiO_3 unit.

CONCLUSIONS

We have examined the structural and dielectric properties of F-doped TEOS films prepared by PECVD experimentally and theoretically. We find a 0.8 reduction of ε at 11 at.% F. The inability of the film to incorporate more than 11 at.% F while maintaining good quality is attributed to strain effects caused by F on the (4;2) silica framework. Ellipsometry data show that the reduction of ε is not primarily an electronic effect. We believe that the main effect of F on ε is to reduce the lattice polarizability by removing oxygen bridges, which give large contributions to the lattice susceptibility of the silicon dioxide.

ACKNOWLEDGEMENTS

The authors wish to thank Anatoli Korkin for his insights into the chemistry of TEOS CVD, and Gerry Lucovsky for helpful discussions. The nuclear reaction analysis to determine F content was performed at the University of Arizona by L.C. MyIntyre and M.D. Ashbaugh.

REFERENCES

1. G. Lucovsky and H. Yang, J. Vac. Sci. Technol. A **15**, 1509 (1997).

2. A. Grill et al., in *Low-Dielectric Constant Materials II*, edited by A. Lagendijk, H. Treichel, K. Uram, and A. Jones (Mater. Res. Soc., Pittsburgh, 1996), p. 155.

3. D. Zhao, P. Yang, N. Melosh, J. Feng, B.F. Chmelka, and G.D. Stucky, Adv. Mater. **10**, 1380 (1998).

4. K.C. Yu et al., in *Low-Dielectric Constant Materials V*, Mater. Res. Soc. Proc. **565** (in print).

5. J. Wetzel et al., in *Low-Dielectric Constant Materials*, edited by T.-M. Lu, S.P. Murarka, T.-S. Kuan, and C.H. Ting (Mater. Res. Soc., Pittsburgh, 1995) p.217.

6. S.M. Han and E. Aydil, J. Vac. Sci. Technol. A **15**, 2893 (1997).

7. S.M. Han and E. Aydil, J. App. Phys. **83**, 2172 (1998).

8. S. Hasegawa, T. Tsukaoka, T. Inokuma, Y. Kurata, J. Non-Crystalline Solids **240**, 154 (1998).

9. Y.-N. Xu and W.Y. Ching, Phys. Rev. B **44**, 11048 (1991).

10. A.A. Demkov, J. Ortega, O.F. Sankey, and M. Grumbach, Phys Rev. B **52**, 1618 (1995).

11. O.F. Sankey, A.A. Demkov, W. Windl, J.H. Fritsch, J.P. Lewis, M. Fuentes-Cabrera, Int. J. Quant. Chem. **69**, 327 (1998).

12. K. Kim, D.H. Kwon, G. Nallapati, and G.S. Lee, J. Vac. Sci. Technol. A **16**, 1509 (1998).

13. M. Yoshimary, S. Koizumi, and K. Shimokawa, J. Vac. Sci. Technol. A **15**, 2908 (1997).

14. *Cerius2* software, Molecular Simulations, Inc., San Diego, CA.

15. A. Demkov, R. Liu, S. Zollner, D. Werho, M. Kottke, R.B. Gregory, M. Angyal, S. Filipiak, L.C. McIntyre, and M.D. Ashbaugh, Mat. Res. Soc. Symp. Proc. (in print).

Low-k Dielectrics—
Porous Materials

Mat. Res. Soc. Symp. Proc. Vol. 612 © 2000 Materials Research Society

Structure and Property Characterization of Porous Low-k Dielectric Constant Thin Films using X-ray Reflectivity and Small Angle Neutron Scattering

Eric K. Lin, Wen-li Wu, Changming Jin[1], and Jeffrey T. Wetzel[1]
Polymers Division, Materials Science and Engineering Laboratory,
National Institute of Standards and Technology
100 Bureau Drive, Stop 8541, Gaithersburg, MD 20899-8541 USA
[1]SEMATECH
2706 Montopolis Drive, Austin, TX 78741-6499 USA

ABSTRACT

High-resolution X-ray reflectivity and small angle neutron scattering measurements are used as complementary techniques to characterize the structure and properties of porous thin films for use as low-k interlevel dielectric (ILD) materials. With the addition of elemental composition information, the average pore size, porosity, pore connectivity, matrix density, average film density, film thickness, coefficient of thermal expansion, and moisture uptake of porous thin films are determined. Examples from different classes of materials and two analysis methods for small angle neutron scattering data are presented and discussed.

INTRODUCTION

With decreasing feature sizes, next generation integrated circuits require materials with lower dielectric constants. A lower dielectric constant is needed to increase signal propagation speed, reduce power consumption, and to reduce crosstalk between adjacent conducting lines. One strategy to lower the dielectric constant is the incorporation of nanometer scale pores into a solid dielectric material [1,2]. The voids in the film effectively reduce the overall dielectric constant of the material. Although the introduction of voids effectively lowers the dielectric constant, other properties required for successful integration into devices including thermal, chemical, adhesive, electrical, and mechanical properties may suffer with the introduction of voids into the film. Unlike conventional fully dense dielectric materials, structural information such as the film porosity, connectivity, and pore size distribution is needed to understand and improve other critical properties of the material. Structural information as a function of processing methods and processing parameters provides materials engineers information needed to optimize these properties.

The measurement of structural and material properties of porous low-k dielectric films is challenging because the films are typically 1 μm thick and must be characterized as prepared on a silicon substrate. To date, several experimental methodologies have been developed to characterize the structure of porous low-k dielectric thin films. Gidley et al. use positronium annihilation lifetime spectroscopy to measure the average pore size and pore size distribution of porous low-k thin films [3]. Dultsev and Baklanov use ellipsometric porosimetry to also determine the average pore size and pore size distribution [4]. Wu et al. at the National Institute of Standards and Technology (NIST) use a combination of high resolution specular X-ray reflectivity (SXR) and small angle neutron scattering (SANS) to measure the average pore size, porosity, pore connectivity, matrix density, coefficient of thermal expansion, and moisture uptake of porous low-k dielectric thin films [5].

In this paper, we summarize the NIST methodology and demonstrate its application to three qualitatively different porous thin films. The analysis methods and equations needed to extract the desired information are also presented. In addition to the two-phase model used previously, we introduce another analysis formalism based upon the Porod invariant to characterize samples that cannot be analyzed using the simple two-phase model. The goal of this work is to characterize the structure and properties of porous thin films and to provide information to correlate structural information with other measured physical properties.

EXPERIMENTAL METHODOLOGY

The NIST methodology combines SXR and SANS data to characterize silica-based porous thin films. Each individual measurement technique provides several important parameters describing the porous thin film. Other critical parameters such as the film porosity and matrix density require the use of information from more than one technique. A common input necessary for each technique is the elemental composition of the film. The elemental compositions are determined through Rutherford backscattering spectroscopy (RBS) (for silicon, oxygen, and carbon) and forward recoil elastic spectroscopy (FRES) (for hydrogen). The chemical composition is needed to convert electron density to mass density for the x-ray reflectivity data and to convert neutron scattering contrast to mass density for the small angle scattering neutron data.

In the following sections, each technique will be briefly described along with the appropriate analysis formalism to demonstrate what information can be obtained using these techniques. Three samples will be used as examples. The samples are all silica-based thin films but have qualitatively different structures. The first sample, A, is an aerogel material. The second, B, is a CVD porous material and the last sample, C, is also a porous silica film. Additional details about the samples are beyond the scope of this paper. Here, we focus on the characterization methodology and the information that can be obtained using our approach.

High Resolution Specular X-ray Reflectivity

High-resolution specular X-ray reflectivity (SXR) is a powerful experimental technique to accurately measure the structure of thin films in the direction normal to the film surface. In particular, the film thickness, film quality (roughness and uniformity) and average film density can be determined with a high degree of precision. The coefficient of thermal expansion (CTE) can also be determined from measurements of the film thickness at different temperatures.

High-resolution X-ray reflectivity at the specular condition with identical incident and detector angles, θ, was measured using a θ–2θ configuration with a fine focus copper X-ray tube as the radiation source. Typically, the reflected intensity is measured at grazing incidence angles ranging from 0.01° to 2°. The incident beam is conditioned with a four-bounce germanium [220] monochromater. The beam is further conditioned before the detector with a three-bounce germanium [220] crystal. The resulting beam has a wavelength, λ, of 1.54 Å, a wavelength spread, $\Delta\lambda/\lambda = 1.3\times10^{-4}$, and an angular divergence of 12 arcsec. With a goniometer having an angular reproducibility of 0.0001°, this instrument has the precision and resolution necessary to observe interference oscillations in the reflectivity data from films up to 1.5 μm thick.

In Figure 1, the X-ray reflectivity curve is shown for sample A plotting the logarithm of the reflected intensity (I_r/I_o) as a function of q (where q = $(4\pi/\lambda)\sin\theta$). At low q values, the X-ray

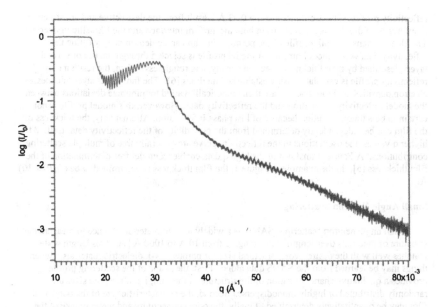

Figure 1. X-ray reflectivity data from silica-based porous thin film A presented as the logarithm of the reflected intensity vs. q where $q = (4\pi/\lambda)\sin\theta$.

beam is nearly completely reflected with a reflectivity of one. As q increases, the reflectivity dramatically drops at two separate critical angles, θ_c, the first at approximately q=0.017 Å^{-1} and the second at approximately q=0.03 Å^{-1}. Each value of θ_c is related to the electron density of different layers in the sample. The first critical angle arises from the electron density of the porous thin film and the second critical angle arises from the silicon substrate. The oscillations that appear in the reflectivity curve between these two critical angles arise from a waveguiding region that is very sensitive to both the thickness and electron density depth profile of the film. Given the elemental composition, the average electron density of the porous thin film can be converted into an average mass density of the film. For sample A, the average mass density of the film is 0.71 ± 0.01 g/cm^3. The average mass density of the film is also related to the porosity and matrix density of the film through the equation

$$\rho_{eff} = \rho_w(1 - P) \tag{1}$$

where ρ_w is the density of the matrix or wall material and P is the porosity of the film. At this point, an assumption of the matrix mass density can provide a numerical estimate of the film porosity. However, no information about the pore size can be obtained using SXR.

In addition to the average mass density of the film, the film thickness can be determined from more detailed analysis of the reflectivity data or the periodicity of the oscillations in the reflectivity profile. The oscillations at $q > 0.03$ $\mathring{A}^{-1}$ result from the destructive and constructive interference of the X-rays reflected from both the air/film interface and the film/silicon interface. The electron density depth profile of the porous thin film can be determined by fitting the reflectivity data with a model profile. A model profile is selected through the use of several layers described by their thickness, electron density, and roughness. Then, the resultant X-ray reflectivity profile is calculated using established methods [6]. The individual layer thicknesses, electron densities, and roughnesses are then numerically varied to minimize deviations between the model reflectivity calculation and the reflectivity data. However, this model profile is not certain to be a unique solution because of lost phase information. Alternatively, the thickness of the film can be independently determined from the periodicity of the reflectivity data itself. At higher q values, the oscillations in the reflectivity curve are generally free of multiple scattering contributions. A Fourier transform of the high q data enables a model free determination of the film thickness [5]. In the example in Figure 1, the film thickness is determined to be (7610 ± 10) $\mathring{A}$.

Small Angle Neutron Scattering

Small angle neutron scattering (SANS) is a widely used in materials science to measure the structure of materials over length scales ranging from 10 $\mathring{A}$ to 1000 $\mathring{A}$ [7]. This length scale matches well with the desired void structural size for porous low-k dielectric materials. Several things may be learned from the SANS data alone. First, the shape of the scattering profile provides a qualitative characterization of the structure of the film, i.e. if the pore sizes are randomly distributed or highly monodisperse. Second, the characteristic size of the pores in the film may be quantitatively determined. Finally, the pore connectivity and water uptake of the film may be determined from changes in the scattered intensity upon the immersion of the samples in the appropriate deuterated solvent [5].

Small angle neutron scattering measurements were performed on the NG1 SANS instrument at the NIST Center for Neutron Research. The porous thin film samples were placed with the beam parallel to the film surface normal. The films were stacked together (with up to 8 films) in order to enhance the scattering signal. The single crystal silicon substrates are essentially transparent to the neutron beam and the scattered intensity arises almost completely from the structure in the porous thin films. The neutron beam wavelength, λ, was 6 $\mathring{A}$ with $\Delta\lambda/\lambda = 0.12$. The sample to detector distance was 3.6 m and the detector was offset from the beam normal by $3.5°$ to increase the observable range of scattering angles. The scattered intensity was collected on a 2-D detector and the data were reduced using standard data reduction methods. The scattered intensity is presented as a function of q (where $q = (4\pi/\lambda)\sin(\theta/2)$ and θ is the scattering angle) [7]. The 2-D data are circularly averaged and placed on an absolute intensity scale using pure water as a secondary standard. Figure 2 shows the circularly averaged data all three porous thin films.

The neutron scattering data can be qualitatively interpreted from the shape of the scattering curves. In q space (reciprocal space), the data at a given q value can be thought of as an observation of the sample at a particular length scale. Low q values represent larger sizes in real space and higher q values move to smaller length scales. From Figure 2, it is clear that the shapes of the scattering curves and the relative intensities of the scattering from each sample are very different. Sample A exhibits the highest scattering intensity and a monotonically decreasing

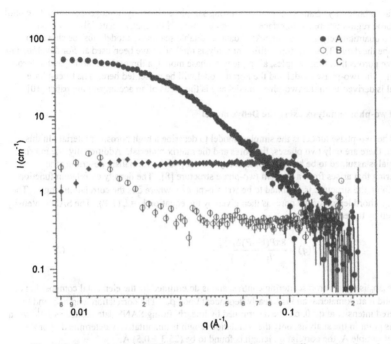

Figure 2. Representative SANS data from three qualitatively different silica-based porous thin films. The absolute intensity is shown as a function of q where $q = (4\pi/\lambda)\sin(\theta/2)$. Sample A is an aerogel material, sample B is a CVD prepared material, and sample C is a porous silica sample.

scattering profile with q. This suggests that sample A has structural features over all observed length scales and is indeed well characterized by the random two-phase model of Debye to be described later [8]. In contrast, sample B has a very low scattering intensity suggesting the presence of fewer scattering sites. There are features at low q values, indicating some heterogeneities at larger length scales, but the scattering intensity at $q > 0.03$ Å^{-1} arises primarily from inelastic or background scattering. Sample C is intermediate to A and B in terms of scattering power, but has an interesting scattering profile. The intensity is almost flat over q ranging from 0.02 Å^{-1} to 0.1 Å^{-1}. This indicates that the film appears homogeneous over the breadth of these length scales. At $q > 0.1$ Å^{-1}, the scattered intensity drops very sharply with a power law dependence very close to q^{-4}. This scaling power of the intensity with q is important and was first described by Porod [9]. Porod demonstrated that the scattered intensity of two-phase systems with infinitely sharp interfaces must decrease as q^{-4} at length scales (real space) that are both smaller than the average size and curvature of the domains. The power law generally reflects the requirement that at larger q values, the scattering arises from the total interfacial boundary. Sample C is an interesting sample with nearly homogeneous structure until the Porod region where the interface between the pore and the matrix material is very sharp.

To quantitatively analyze the SANS data, a suitable scattering model must be chosen to describe the data. Thus far, three different analysis methods have been used and/or developed to analyze porous thin film samples, a simple two-phase model, a three-phase model, and a Porod model. The two-phase model and the Porod model will be presented here. The three-phase model is derived from the two-phase model and is the focus of an accompanying paper [10].

Two-phase analysis using the Debye model

The two-phase model is the simplest model to describe a high porosity material. In this model, there are only two phases, the pores and the matrix material. Additionally, the matrix material is assumed to be homogenous. Debye developed the formalism describing the scattering that arises from a random two-phase structure [8]. The density correlation function describing the structure is assumed to be $\gamma(r) = \exp(-r/\xi)$, where ξ is the correlation length. The average chord length of the pores is then given by the equation $l_c = \xi/(1-P)$. The SANS intensity is given by the equation

$$I(q) = \frac{8\pi P(1-P)\Delta\rho_n^2\xi^3}{\left(1+q^2\xi^2\right)^2} \qquad (2)$$

where $\Delta\rho_n$ is the neutron scattering contrast and is determined by the elemental composition of the solid matrix material and is linearly dependent upon ρ_w. The correlation length, ξ, and the scattered intensity at $q = 0$, can be determined by linearly fitting SANS data plotted as $1/I^{1/2}$ vs. q. At this point in the analysis, only the correlation length is quantitatively determined. For the case of sample A, the correlation length is found to be (25.3 ± 0.5) Å.

To determine the film porosity, P, and the matrix mass density, we must use additional information from SXR. Given $I(0)$ and ξ, equation (2) becomes a function only of ρ_w and P. From the SXR formalism, equation (1) is also a function of ρ_w and P. With two equations and two unknowns, ρ_w and P, we can solve for these two quantities for the porous thin film. For sample A, we find that porosity is $(56 \pm 1)\%$ by volume, the matrix mass density is (1.63 ± 0.05) g/cm^3, and the average chord length of the pore is (57.5 ± 0.5) Å.

The pore connectivity and moisture uptake of the film can also be determined using the Debye formalism. The samples are placed into quartz cells and immersed in either deuterated toluene (d-toluene) or deuterated water (D2O). The d-toluene solvent is chosen because it readily wets the samples provided to us thus far. If either of the deuterated solvents penetrates open and interconnected pores, the absolute value of the scattered intensity changes because of the large contrast change in $\Delta\rho_n$ from air or vacuum in the pores to a deuterated material. If all the pores within a sample were filled, the entire scattered intensity would increase by roughly a factor of 18.3. If the increase in scattered intensity is less than 18.3, then only a fraction of the pores are filled with the solvent. In a similar manner, the moisture uptake of D2O may also be determined. In this methodology, pore connectivity represents the fraction of pores that are interconnected and accessible to a solvent at the outside surface. The details of this approach may be found in an earlier publication [5].

Porod invariant analysis

Although many of the samples are well described by the Debye two-phase model, several samples exhibit SANS profiles that cannot be analyzed in this framework. To address these samples, the Porod invariant analysis developed to describe small angle scattering from a two-phase system is applied to determine the average pore size of the porous low-k dielectric thin film. To apply this method, the sample must be a two-phase system with sharp interfacial boundaries. The scattering curve must decrease as q^{-4} as predicted using Porod's law. To extract the characteristic length scale of the pores, the average chord length in the pore structure, we use the Porod invariant of the scattered intensity given by

$$Q^{*}_{exp} = \int_0^\infty q^2 I(q) dq \qquad (3).$$

Conceptually, the Porod invariant relates the experimental intensity to the average mean square of the scattering length fluctuations [9]. The measured invariant can be used to directly determine the average chord length, l_c, through the relationship [9]

$$l_c = \frac{\pi}{Q^{*}_{exp}} \int_0^\infty q I(q) dq \qquad (4).$$

An important advantage of this procedure is that the average chord length of the materials can be determined without reference to an absolute intensity. This procedure is applied to sample C shown in Figure 2. For this scattering curve, the intensity drops off at $q > 0.1$ Å^{-1} values as q^{-4}. This power law indicates that the interface between the pores and the matrix material is sharp and the integrals in equation (3) and (4) converge. The average chord length for sample C is determined to be (20.8 ± 0.5) Å. It should be noted that since the Porod limit is reached, the data are extrapolated to high enough q values to ensure convergence of the integral in equation (3). This is necessary in this case because the scattered intensity cannot be measured at high enough q values because the observed length scale reaches the size of constituent atoms.

To determine the porosity of the film using this formalism, we use the theoretical definition of the invariant given by the following equation

$$Q^{*}_{th} = 2\pi^2 \Delta\rho_N^2 P(1-P) \qquad (5).$$

As in the Debye two-phase model, there are two equations, equation (2) and equation (5), which are functions of ρ_w and P (given the elemental composition). In this case, Q_{th}^{*} is replaced with Q_{exp}^{*}. We implicitly assume that the structure of the thin film is well described by only two homogeneous phases, the pores and the matrix material, and that the interface between the pores and the matrix is very sharp. Using equations (2) and (5), the porosity and matrix density of sample C are found to be $(64 \pm 1)\%$ by volume and (2.19 ± 0.05) g/cm^3, respectively.

SUMMARY

We have demonstrated that the complementary use of high-resolution specular x-ray reflectivity and small angle neutron scattering is a powerful method for the characterization of the structure and properties of porous thin films for use as low-k dielectric materials. These techniques are able to measure porous thin films up to 1 µm thick as prepared on silicon substrates. In addition to the measurement of film porosity, thickness, roughness, coefficient of thermal expansion, average film density, average pore size, and pore connectivity, this methodology uniquely provides the matrix wall density. Two different frameworks are applied to specific porous thin film SANS data to obtain important structural information, the average chord length, of the films. The data obtained from these methods and others can be used to correlate structural properties to processing method and conditions as well as other important material properties other than dielectric constant.

ACKNOWLEDGMENTS

The authors thank Gary W. Lynn for his help in making the measurements and Professor Russell Composto and Howard Wang for performing the ion beam measurements for the determination of the elemental composition of these porous thin films.

REFERENCES

1. L. W. Hrubesh, L. E. Keene, and V. R. Latorre, J. Mater. Res. **8**, 1736 (1993).
2. C. Jin, J. D. Luttmer, D. M. Smith, and T. A. Ramos, MRS Bull. **22**, 39 (1997).
3. D. W. Gidley, W. E. Frieze, T. L. Dull, A. F. Yee, C. V. Nguyen and D. Y. Yoon, Appl. Phys. Lett., **76 (10)**, 1282 (2000).
4. F. N. Dultsev and M. H. Baklanov, Elec. Solid State Lett., **2**, 192 (1999).
5. W. L. Wu, W. E. Wallace, E. K. Lin, G. W. Lynn, C. J. Glinka, E. T. Ryan and H. M. Ho, J. Appl. Phys., **87**, 1193 (2000).
6. J. Lekner, *Theory of Reflection* (Nijhoff, Dordrecht, 1987).
7. J. S. Higgins and H. C. Benoit, *Polymers and Neutron Scattering* (Oxford University Press, Oxford, 1994).
8. P. Debye, H. R. Anderson, and H. Brumberger, J. Appl. Phys. **28**, 679 (1957).
9. G. Porod, Kolloidn Zh. **124**, 83 (1951).
10. W. L. Wu, E. K. Lin, C. Jin, and J. T. Wetzel, In an accompanying paper in this session (2000).

Mat. Res. Soc. Symp. Proc. Vol. 612 © 2000 Materials Research Society

Characterisation of Low-K Dielectric Films by Ellipsometric Porosimetry

M.R.Baklanov and K.P.Mogilnikov[1]
IMEC, B-3001 Leuven, Belgium, baklanov@imec.be
[1]Institute of Semiconductor Physics, 630090 Novosibisrk, Russia

ABSTRACT

Ellipsometric porosimetry (EP) is a simple and effective method for the characterization of the porosity (volume of both open and close pores), average pore size, specific surface area and pore size distribution (PSD) in thin porous films deposited on top of any smooth solid substrate. Because a laser probe is used, small surface area can be analyzed. Therefore, EP can be used on patterned wafers and it is compatible with microelectronic technology. This method is a new version of adsorption (BET) porosimetry. *In situ* ellipsometry is used to determine the amount of adsorptive which adsorbed/condensed in the film. Change in refractive index is used to calculate of the quantity of adsorptive present in the film. EP also allows the study of thermal stability, adsorption and swelling properties of low-K dielectric films. Room temperature EP based on the adsorption of vapor of some suitable organic solvents and method of calculation of porosity and PSD is discussed. Examination of the validity of Gurvitsch rule for various organic adsorptives (toluene, heptane, carbon tetrachloride and isopropyl alcohol) has been carried out to assess the reliability of measurements of pore size distribution by the ellipsometric porosimetry.

INTRODUCTION

Molecules polarity (μ-permanent dipole moment) and polarisability (α) define the relative permittivity ε_r of a substance. The quantitative relation between the relative permittivity and the electrical properties of molecules is expressed by Debye equation

$$\frac{\varepsilon_r - 1}{\varepsilon_r + 2} = \frac{4\pi N}{3\varepsilon_o}\left(\alpha + \frac{\mu^2}{3kT}\right) \qquad (1)$$

where N is the number of molecules per unit volume. If there is no contribution from the permanent electric dipole moments to the polarization (the molecules are non-polar) the same expression is called the Clausius-Mossotti equation. No or minimal dipole moment is typical for organic polymers and choice of organic polymers to decrease the permittivity value of dielectric films for the advanced interconnect application is defined by this fact. Thus, the organic polymers allow decrease the permittivity of the interlayer dielectric until 2.5-2.8. Further decrease of the permittivity value can only be realized by using of porous dielectrics:

$$\frac{\varepsilon_r - 1}{\varepsilon_r + 2} = \sum_i \frac{4\pi N_i}{3\varepsilon_0}\alpha_i = V\frac{\varepsilon_a - 1}{\varepsilon_a + 2} + (1-V)\frac{\varepsilon_s - 1}{\varepsilon_s + 2} \qquad (2)$$

where $V = (pore\ volume / film\ volume)$ is the relative film porosity, ε_a and ε_s are the permittivity of air and the film skeleton respectively. Porous dielectric films are thus becoming extremely important for the future ULSI technology. Film porosity and pore size distribution (PSD) define dielectric, mechanical, thermal and chemical properties of the films and their feasibility to be used in the ULSI technology. Increasing the porosity drives the dielectric constant down, but it degrades the mechanical and chemical properties of the film. Therefore the characterization of the film porosity and pore size distribution is an important issue in advanced microelectronics.

There are different ways of measurement of the porosity and pore size distribution in bulk materials: (*stereology analysis* such as microscope techniques; *intrusive methods* such as gas adsorption, mercury porosimetry etc.; and/or *non-intrusive methods* such as radiation scattering, wave propagation etc.) [1]. However, most of these methods have different limitations to be applied for the thin film characterization.

Recently, small-angle X-ray and neutron scattering (SAXS and SANS) and Positron Annihilation Lifetime Spectroscopy (PALS) were used to measure the mean pore size and pore size distribution of thin films [2,3]. However these methods use quite expensive equipment and complicated analysis and are not easy to use them for the in-line monitoring of porous low-K dielectric films.

Adsorption porosimetry is traditionally considered as a reference to examine the feasibility of new techniques for the PSD measurement since it is a reliable and simple method to determine PSD. The adsorption characteristics and basic equations necessary for the PSD calculations have been analysed and are well-documented [4]. A common method to apply adsorption porosimetry is to monitor adsorption and desorption of nitrogen vapour near the boiling point by direct weighing of the adsorbate which adsorbed/condensed in the pores. However, the sensitivity of the traditional microbalance technique used for weighing allows analyse only large powder like samples; film from several silicon wafers must be scrubbed off in order to get enough material to be analysed by microbalance porosimetry.

Several papers on the measurement of PSD by special modifications of the adsorption porosimetry for thin films have already been published. In these papers the measurement of the adsorption/ desorption isotherms were carried out by quartz crystal microbalance (QCM) [5,6], surface wave sensor (SAW) [7] and by ellipsometry [8-10]. In the QCM and SAW methods the porous film must be deposited on top of a special sensor. In principle, these methods are closest to classical adsorption porosimetry because they use mass determination of the adsorbed vapors. However, the necessity to deposit the film on top of special sensors limits their application in the microelectronic industry.

The principal feature of EP is to utilize the change of optical characteristics of the porous films during the vapor adsorption and desorption to determine the mass of an adsorbate condensed/adsorbed in pores instead of direct weighing.

The most important advantages of EP are:

1. All measurements can be carried out in a porous film, deposited directly on top of a silicon wafer or any smooth solid substrate. Normally a 50 nm thick film is enough to do a reliable analysis. Technological layers between the silicon substrate and the dielectric film do not induce any problems for the measurements.

2. Because a laser probe is used, small surface areas can be analyzed. Therefore EP can be used on patterned wafers and is compatible with microelectronic technology.

3. Room temperature PSD measurements are carried out using EP with organic solvents [5,6, 8-10] and do not have the problems conjured with low temperature nitrogen porosimetry.

The adsorbate amount inside of pores is calculated from the measured change of optical characteristics of the porous film during the vapor adsorption/desorption. The optical characteristics of the skeleton of the porous film and of the liquid adsorptive are used in these calculations. The various models give similar result [11], however, the Lorentz-Lorenz equation is more widely used.

The correct choice of an adsorptive for the room temperature porosimetry is an important issue. The adsorptive should be a volatile liquid, because of the need to work near the equilibrium pressure P_o. It was found [5,9,10] that a number of organic solvents can be used for this purpose.

In this work, fundamentals and practical applications of EP are discussed. *In situ* ellipsometric measurements were performed using a custom-built high-vacuum tool. Different types of porous films were used in the experiments. Adsorption of toluene ($C_6H_5CH_3$), heptane (C_7H_{16}), carbon tetrachloride (CCl_4) and isopropyl alcohol (IPA: i-C_3H_7OH)) vapors was used for the comparative analysis.

FUNDAMENTALS OF ELLIPSOMETRIC POROSIMETRY

EP is a combination of a non-intrusive (wave propagation) and an intrusive (vapor adsorption) methods. Therefore, this method allows measuring both full film porosity from the value of refractive index and open porosity from the amount of adsorbed solvent.

Full porosity. Relation between the optical characteristics and the material composition in a multicomponent system is described by the Lorentz-Lorenz equation:

$$B = \Sigma N_i \alpha_i = \frac{3(n^2 - 1)}{4\pi(n^2 + 2)} \tag{3}$$

where B is the polarizability of a unit of volume, N_i and α_i are the number of molecules and the molecular polarizability of the material components, $n = \sqrt{\varepsilon_r}$ is the refractive index of the film. If n_s is the refractive index of the dense part of material (film skeleton) with the volume polarizability B_s, n_p is the measured refractive index of the porous film and B_p is the volume polarizability calculated from n_p, the relative film porosity (V) is equal to:

$$V = 1 - \frac{B_p}{B_s} = 1 - \left[\frac{(n_p^2 - 1)}{(n_p^2 + 2)}\right] \Bigg/ \left[\frac{(n_s^2 - 1)}{(n_s^2 + 2)}\right] \tag{4}$$

The equation (4) is a consequence of equation (2). Accuracy of the measurement of the film porosity (full pore volume) can be improved by using of multiangle or spectroscopic ellipsometric measurements.

Open pores and pore interconnectivity. It is obvious that the above mentioned approach and equation (2) are also valid if pores are filled by a liquid (condensed adsorbate) with the known refractive index. In this case the adsorbate amount in the pores are calculated using the refractive index and density of the liquid adsorbtive. Ellipsometry allows us to measure both the refractive index and the film thickness d (i.e. the effect of the film swelling). The adsorptive volume in pores is calculated as

$$V_{ads} = \frac{V_m}{\alpha_{ads} \cdot d_1} (B_1 d_1 - B_0 d_0)$$ (5)

where $V_{ads} = (adsorbate\ volume/film\ volume)$ is the relative volume of the liquid adsorbate in the pores (volume of open or active pores), B_0 and B_1 are the volume polarizability of the film before and after adsorption, d_0 and d_1 are the film thickness before and after adsorption, respectively. V_{mol} is the molecular volume of the adsorptive and α_{ads} is the polarizability of the adsorptive molecule.

The relative volume of open and close pores is calculated by comparison of the results obtained by equations 4 and 5. This analysis gives an information related to the pore interconnectivity.

Pore size distribution. Calculation of the pore size distribution uses the Kelvin equation (6) and the phenomenon of progressive emptying of a porous system initially filled at $P=Po$. The calculations are based on analysis of hysteresis loop that appear due to the processes of capillary condensation and desorption of a vapor out of porous adsorbent. The hysteresis loops appear because the effective radius of curvature of condensed liquid meniscus is different during the adsorption and desorption. The adsorptive vapor condenses in pores at the vapor pressure (P) less than the equilibrium pressure of a flat liquid surface (P_o). Dependence of the relative pressure (P/P_o) on the meniscus curvature is described by the Kelvin equation:

$$\frac{1}{r_1} + \frac{1}{r_2} = -\frac{RT}{\gamma V_L \cos\theta} \ln\left(\frac{P}{P_0}\right)$$ (6)

where γ and V_L are surface tension and molar volume of the liquid adsorptive, respectively. θ is the contact angle of the adsorptive. The principal curvature radii r_1 and r_2 define pore sizes. Adsorptives with contact angle close to 0 are preferable. In the case of cylindrical pores, $r_1=r_2$ and

$$\left(\frac{1}{r_1} + \frac{1}{r_2}\right) = \frac{2}{r_k}$$ (6a)

The radius r_k is dimension characteristic of the capillary and often termed the Kelvin radius. If the radius of a cylindrical pore is r_p, then $r_p = r_k + t$, where t is the thickness of the layer already adsorbed on the pore walls (before capillary condensation occurs). Values of t are obtained from the data for the adsorption of the same adsorptive on a non-porous sample having a chemically similar surface and is defined by the BET equation:

$$t = \frac{d_o C \cdot K \cdot (P/P_o)}{[1 - K(P/P_o)] \cdot [1 + K(C-1)(P/P_o)]}$$ (7)

where d_o is the thickness of one monolayer, C is the BET constant, K is a coefficient satisfying to the requirement that at $P=P_o$ $t \leq 5$-6 monolayers [4]. Therefore, development of the room temperature porosimetry needs in knowledge of molecular characteristics of the adsorptive (molecular volume V_L, surface tension γ), and adsoprtion characteristics (BET constant C). The calculated t values are used for the correction of the measured pore radius. This procedure is termed *t-correction* and especially important for pore with radii less than 5 nm.

The initial experimental data for the calculation of the adsorption isotherm and PSD are the ellipsometric characteristics Δ and Ψ. These characteristics are defined by complex ratio ρ of the reflection coefficients R_p and R_s of light polarised respectively parallel and perpendicular to the plane of incidence on the sample, as

$$\rho = R_p / R_s = \tan \Psi \exp(i\Delta) \qquad (8)$$

A special computer software, developed at the Institute of Semiconductor Physics, allows calculating the change of the film thickness and refractive index of the film during the adsorption and desorption, the pore size distribution and cumulative surface area. Molecular characteristics necessary for the PSD calculations have also been investigated and documented for a number of suitable adsorptives.

The change of the adsorbate volume is calculated from the change of the refractive index using equation (5). The dependence of the adsorbate volume on the relative pressure P/Po is an adsorption isotherm, which is used to calculate the PSD.

Specific surface area is an important characteristic of porous materials because it defines their chemical behavior. The BET method for calculation of the specific surface area A (pore wall density) is well known and is widely used. This method involves two steps: evaluation of the monolayer capacity n_m from the adsorption isotherm, and conversion of n_m into A by means of the molecular area a_m (surface area occupied by one adsorbate molecule). To obtain a reliable value of n_m from the isotherm it is necessary that the monolayer shall be virtually complete before the build-up of higher layers commences. This requirement is met if the BET parameter C is not too low.

EP can efficiently use this method by using of some special adsorptives; however, in the most cases it is difficult to satisfy the requirement related to the high BET constant. If the pore size is small (less than 10 nm) the adsorptives with small BET parameter are more preferable because they allow decreasing t-correction.

For this reason, our software calculates *cumulative surface area* (CSA) [4]. The specific surface area of each small group of pores δA_i are calculated from the corresponding pore volume and pore radius as $\delta A_i = \delta V_i / r_i$ *(for cylindrical pores)*. By summing the values of δA_i over the whole pore system a value of the cumulative surface area is obtained.

Gurvitsch test. A possible problem of the reliability of the PSD calculation from the adsorption data can be related to the range of validity of the Kelvin equation. The most important and difficult problems deal with the curvature effect (change of surface tension of the adsorptive in fine pores) and surface tensile effects inside of fine pores. Even the most popular and traditional adsorptive such as nitrogen sometimes meets these problems and can give wrong information about the pore size and distribution [4]. Therefore, when using a new adsorptive, it is necessary to find additional proves to exclude the calculation errors. One way is to compare the measurement with results obtained by different methods (for instance, nitrogen porosimetry, SANS, PALS etc.) if these data are available. However, the easiest method to examine the measurements reliability is the Gurvitsch test that can be carried out using the same experimental tool. The idea of the Gurvitsch rule, embodied many years ago [12], is that a volume of a liquid (by use of the normal liquid density) should be the same for all applicable adsorptives in a given porous solid. In pores with size equal to several molecular sizes, change of the surface tension and molecular volume of adsorbate can be significant. Normally, it is difficult to predict where is the practical limitation of each adsorptive from this point of view. If to use several adsorbtives with different values of surface tension and molecular volume, and if they give the same value of pore size, this is a reliable proof that the used adsorptives are still outside of these limitations.

Another advantage of the Gurvitsch test with different solvents is that the use of adsorptives with different values of $T/\gamma V_L$ allows to expand the scope of the Kelvin method beyond the practical limit set by nitrogen adsorption (25 nm) [4]. The use of a suitable adsorptive can shift the leap of the desorption curve to region more convenient for the measurements and also make possible to reduce the magnitude of the t-correction.

An important application of the Gurvitsch test is related to microporous films. Difference in 'effective' pore volume found with the different vapors adsorbed on the microporous film reflect the effects of packing of molecules which may differ from the bulk (density effect) [14]. The use of probe molecules with different size allows to get an idea about PSD in microporous film.

EXPERIMENTAL RESULTS AND DISCUSSION

Mesoporous films. Figure 1 shows the typical behavior of the ellipsometric angles Δ and Ψ observed during the toluene vapor adsorption and desorption in mesoporous films. The initial point corresponds to the zero relative pressure (P=0) and the final one to the relative pressure equal to unity (P=Po). It is demonstrated (Figure 1a) that the experimental points are far from the theoretical curve calculated for the change of the film thickness at a constant refractive index (n=1.183 is the refractive index of the xerogel film used in this experiment). All experimental data are well described by a model using the change of the refractive index without remarkable change of the film thickness. Some difference in the Δ and Ψ values observed during the toluene adsorption and desorption in the xerogel film deals with insignificant and reversible swelling and

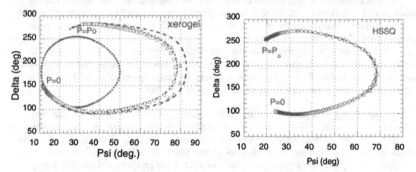

Figure 1. *Change of the ellipsometric angles Δ and Ψ during adsorption and desorption of the toluene vapor in mesoporous xerogel and HSSQ films. .... n=1.183, d-various; o - adsorption; Δ - desorption; ---- d=465 nm, n – various; - - - - d=445 nm, n - various.*

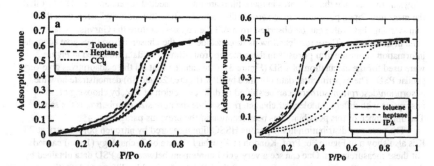

Figure 2. Adsorption/desorption isotherms of different adsorbates in mesoporous xerogel (a) and HSSQ (b) films.

shrinkage. One can see that this effect is much less in the mesoporous hydrogen silsesquioxane (HSSQ) film. This fact suggests that toughness of the HSSQ film is higher in comparison with the xerogel film and HSSQ film no not have any tendency to swell during the vapor adsorption.

Full porosity of the xerogel and HSSQ films were determined by multiangle ellipsometric measurements before adsorption (at P=0) and calculated by equation (4). These values were equal to 0.57 (57%) and 0.48 (48%), respectively. Concentration of open pores was determined by the solvents adsorption and equation (5).

Adsorption isotherms calculated from the experimental data presented in Figure 1 are shown in Figure 2. Experimental data obtained by adsorption and desorption of the heptane and IPA vapors are also shown in these figures. One can see that the saturation point for the different adsorptives are very close one to another and correspond to open porosity equal to 0.64 for the

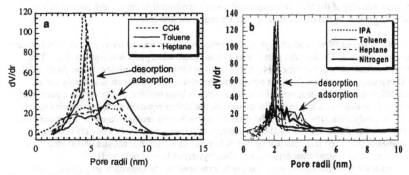

Figure 3. Pore size distribution in mesoporous xerogel (a) and HSSQ (b) films calculated from the adsorption and desorption isotherms of different solvents.

xerogel film and 0.48 for the HSSQ film. Comparison of these values with full film porosity allows us to conclude that all pores in these films are interconnected (open, active). The fact that the measured "open porosity" of the xerogel film is higher than "full porosity" deals with the film swelling that is also can be observed from Δ/Ψ behavior; therefore the Gurvitsch rule is valid in these cases. These results allow us to conclude that the chosen adsorptives give adequate information about the film porosity in these mesoporous films. The data presented in Figure 2 were used for the calculation of PSD (Figure 3). One can see that all three adsorptives give very similar PSD. The experimental data obtained by IPA (figures 2b and 3b) demonstrate how the adsorption/desorption isotherm can be shifted to different region of P/P_0 by choosing of adsorptive with different molecular characteristics while the pore size distribution is the same. The importance of this possibility has been mention in the previous paragraph.

The pore size distribution in the porous HSSQ film measured by nitrogen porosimetry at 78 K is also shown in Figure 3b. Prof. Kondoh at Kyushu Institute of Technology (Japan) carried out these measurements. One can see a very good agreement between the PSD data obtained by the low-temperature nitrogen porosimetry and the EP data obtained by different adsorptives at room temperature. This comparison is an additional proof of reliability of the EP analysis.

The average pore radius in these films were calculated from the desorption curves (a traditional approach in the adsorption porosimetry [4]). They are close to 4.5 nm and 2.3 nm. In the case of cylindrical meniscus that forms during the adsorption, r_2 in the equation (6) is equal to ∞ and the mean pore size calculated from the adsorption curve should be 2 times higher than the value calculated from the desorption curve (it is assumed that only spherical meniscus form during the desorption). The experimental data for the xerogel and the HSSQ films show that the model of cylindrical pores is not precisely correct, however it is still an acceptable approach for the PSD calculations.

The PSD widths obtained from the desorption and adsorption curves are quite different. This phenomenon is related to the branched distribution of pores inside the film. While the solvent desorption occurs through definite effective necks characterizing a surface region of the film, the adsorbate is distributed in the pore branches inside the film during the adsorption and a wide range of different types of meniscus can form at the same time. Therefore, the desorption curves provide a simplified image and are more straightforward for a standard characterization of the effective porosity of the film. However, analysis of the adsorption curves and corresponding PSD can provide additional ideas related to the real pore structure.

Microporous films. The adsorption/desorption isotherms for the microporous silica film are shown in Figure 4. The full film porosity measured before adsorption was equal to 0.25. The adsorption isotherms are of type 1 [4] that are typical for a microporous substance. Analysis of PSD in a microporous film from the adsorption/desorption measurements is generally not so straightforward as for mesoporous films. It has already been realized that in very fine pores with widths of the order of a few molecular diameters, the Kelvin equation no longer remains strictly valid. Not only would the values of the surface tension and the molar volume deviate from those of the bulk liquid adsorptive, but also the concept of a meniscus would eventually become meaningless [4]. Therefore, calculation of PSD by Kelvin equation is not straightforward because it can give significant errors (even if the hysteresis loop is observed).

Deviation from the Kelvin equation can be examined by the Gurvitsch test. Figure 4 shows that the adsorption/desorption isotherms for the different adsorptives are different for the

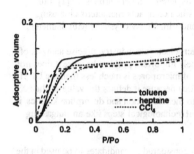

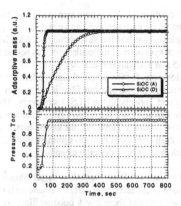

Figure 4. Adsorption/desorption isotherms in a microporous silica film.

Figure 5. Kinetics of the toluene adsorption in microporous SiOC films with different pore size [16].

microporous film while they are very similar for the mesoporous film (Figure 2b). A quite large difference between the full porosity and value calculated from the volume of the adsorbed vapors (0.25 and ≥ 0.12 respectively) is another feature of the microporous films. This difference means that concentration of open pores is less than 50%. The concentration of open pores calculated from the condensed adsorbate volume was equal to 0.15 for toluene, 0.13 for CCl_4 and 0.127 for heptane. As mentioned above the observed difference is related to the different adsorbate/surface interaction and also to problems with the liquid penetration into fine micropores.

More detailed analysis of microporous films can be carried out using *t-plots* (Lippens and de Boer), *α-plots* (Sing) or *DR plots* (Dubinin and Radushkevich) analysis [4]. These methods are also based on analysis of adsorption/desorption isotherms. In the first two cases (the *t*-plot and *α*-plot analysis) the pore shape model methods give a pore hydraulic radius which represents the ratio porous volume/surface. The *DR* model is based on thermodynamic description of micropore filling. The micropore filling of a system with a distribution of pore sizes can be described by the sum of the contribution from individual pore groups that are characterized by its own pore volume and characteristic energy. Application of these methods for the EP measurements and necessary calculation methods and software are in the stage of development.

Figure 5 shows another example of behaviour of the microporous films and gives an idea for the simple estimation of pore size in a microporous film. Two different microporous silicon oxycarbide (SiOC) films with full porosity 18-20% were examined by the toluene adsorption [16]. The adsorbate easily penetrates into the SiOC (D) film (the adsorbate amount increases simultaneously with vapour pressure). No hysteresis loop was observed in this case. In the case of SiOC (A) the increase of the adsorbate amount occurs much slowly than change of t he vapour pressure. The observed phenomena suggest that the pore diameter in SiOC (D) was higher than the kinetic diameter of the toluene molecule (≈0.6 nm) while the pore diameter of SiOC (A) is comparable to the kinetic diameter of toluene and the toluene diffusion is the limiting step of the micropores filling. This simple test by using of adsorptives with different diameters can provide an information related to pore size in microporous film. The approach based on the use of various probe molecules of known size and shape is considered as a very promising [1]. For example this method has successfully been applied for silica compacts characterisation using spherical symmetrical molecules, such as neopentane and trimethylsyloxysilane with diameters of 0.65 and 1.15 nm respectively [17].

However, for the practical issues related to the microelectronic industry, precise analysis of the pore size in microporous films is still not an extremely important problem because it does not have direct impacts on the ULSI technology. The size of micropores is much less than the ULSI feature size and only the relative pore volume is important because it defines the value of the dielectric constant. However, the information related to the adsorption and desorption kinetics is useful itself because it helps to optimise some important technological steps like an outgassing procedure after post-dry etch cleaning.

Polymer swelling. Organic polymer films are also considered as candidates to be used in the advanced ULSI technology when demand to the dielectric constant is around 2.5-2.8. Typical representatives of these low-k polymer films are, for instance, polyimide, SiLK from Dow Chemical Company and Flare from AlliedSignal. They are not porous, but do however consist of so called "free volume" with molecular size. Additional feature of the organic polymers is low toughness and, therefore, adsorptive molecules can penetrate into the film by activated diffusion. As a result, the polymer swelling can be observed. It has been pointed that in some cases the equilibrium swelling is sufficient to double the original volume or more, and materials which are originally hard and glassy become rubbers or gels [18].

Such type of swelling can occur, for instance, during the resist development and strip or post dry etch cleaning using organic solvents. Therefore, it is important to have an information related to the polymer permeability and swelling to examine their stability during different technological operations.

In situ ellipsometry is obviously a suitable tool to study the polymer swelling during adsorption and has already been used for this purpose. The EP tool can be used to obtain an extremely important information related to the solvent desorption and to optimise outgassing conditions after different technological steps using organic solvents. Because the diffusion process is activated, it is not easy to remove this solvent from the polymer. Normally it is necessary to use significant annealing to desorb all remaining solvent [15].

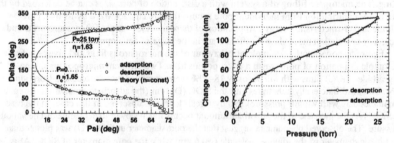

Figure 6. *Change of the ellipsometric characteristics Δ and Ψ and of the polymer film thickness during the toluene adsorption of organic low-K polymer film.*

Figure 6a shows a typical change of the ellipsometric characteristics Δ and Ψ during the toluene adsorption in a low-K polymer film. A principal feature of this curve in comparison with the above discussed inorganic low -K films is the change of the film thickness without remarkable change of the refractive index. Therefore the change of the ellipsometric characteristics are completely related to the polymer swelling by permeation of the toluene molecules into the film. Finally this film becomes a 2-component system because "swelled free volume" are filled by adsorptive. This phenomenon provide insignificant decrease of polymer refractive from 1.65 to 1.63 (refractive index of toluene is equal to 1.49). The shape of the adsorption/ desorption isotherm is typical for the polymer swelling (Figure 6b). Complete desorption occurs at very low pressure. Only pumping down until 10^{-5} torr or annealing until 350°C provides complete desorption of the solvent. Information we got from this observation allowed us to find conditions necessary for the polymer outgassing.

CONCLUSIONS

A new modification of the adsorption porosimetry (ellipsometric porosimetry) has been developed for thin film application. It is a reliable and simple method for non -destructive characterization of mesoporous dielectric films. This method allows the measurement of pore size distribution at room temperature in thin films directly deposited on Si or any smooth solid substrate. Intermediate layers between the silicon substrate and the porous film do not create any problem for the measurements if their optical characteristics are known. A small external surface area (<1 mm^2) is sufficient to carry out this analysis, making the method well suited for the microelectronic industry.

Ellipsometric porosimetry provides similar information as microbalance porosimetry. However, it is more informative than the traditional microbalance porosimetry: structural changes (swelling) during the adsorption/desorption processes can also be analyzed. This last information is extremely important in advanced microelectronic technology, for instance, for optimization of the wet post-etch cleaning conditions and new chemistries considered to be used in the organic low-K dielectric/Cu assemblies.

Various adsorptives for the room temperature porosimetry have been selected and examined. The molecular and optical characteristics of some organic solvents have been studied in relation with the PSD calculation. Verification of the Gurvitsch rule proves that these adsorptives are applicable for the room temperature porosimetry and that the corresponding data obtained by the room temperature ellipsometric porosimetry are reliable.

An additional advantage of the ellipsometric porosimetry is a possibility to obtain the information related to adsorption and desorption of different organic solvents at room temperature. This information can be directly used to predict behavior of organic solvents during the post dry etch cleaning and for optimization of these processes.

Finally, experimental tools and advanced software for the PSD calculations from the ellipsometric measurements of the adsorption of the organic solvents vapors in porous films have been developed. Round robin evaluation of various porous films and several different techniques such as EP, nitrogen porosimetry, SANS and SAXS have shown a good agreement between the results of the analysis. These results will be reported in our future publications.

ACKNOWLEDGMENTS

It is our pleasure to thank I. Leonov and D.Shamiryan for the help in experimental work, V.G.Polovinkin for the software development and E.Kondoh for the measurements of PSD by nitrogen porosimetry and useful discussions. We are thankful to K.Maex and S.Vanhaelemeersch for support of this work.

REFERENCES

1. A.Julbe and D.J.Ramsay, Methods for the characterization of porous structure in membrane materials, *Fundamentals of Inorganic membrane Science and Technology, Chapter 4*, ed. A.J.Burrgraaf and L.Cot (Elsevier, 1996) pp.67-118.
2. W.Wu, W.E.Wallace, E.Lin, G.W.Lynn, C.J.Glinka, R.T.Ryan and H.Ho, *J.Appl.Phys.*, **87**, 1 1193 (2000).
3. D.W.Gidley, W.E.Frieze, T.L.Dull, J.Sun, A.F.Yee, C.V.Nguyen and D.Y.Yoon, *Appl.Phys.Lett.*, **76**, 10, 1282 (2000).
4. S.J.Gregg and S.W.Sing, *Adsorption, Surface Area and Porosity, 2nd ed.*, (Academic Press, NY, 1982).
5. M. R. Baklanov, F. N. Dultsev and S. M. Repinsky, *Poverkhnost'*, **11**, 1445 (1988) (Russ.).
6. M. R. Baklanov, L. L. Vasilyeva, T. A. Gavrilova, F. N. Dultsev, K. P. Mogilnikov and L. A. Nenasheva, *Thin Solid Films*, **171**, 43 (1989).
7. Y.Spooner et. al. *Presented at 1999 SEMATECH Workshop*, Orlando, FL, (1999).
8. F.N.Dultsev and M.R.Baklanov, *Electrochem.Sol.St.Lett.* **2**(4), 192 (1999).
9. K.P.Mogilnikov, V.G.Polovinkin, F.N.Dultsev and M.R.Baklanov, *MRS Proceeding "Low-Dielectric Constant Materials Y*, ed. J.P.Hummel, K.Endo, W.W.Lee, M.E.Mills, S-Q.Wang, v.565 (2000).
10. M.R.Baklanov, K.P.Mogilnikov, V.G.Polovinkin and F.N.Dultsev, *J.Vac.Sci.Technol.B*, **18**, (2000).
11. G.A.Muranova and A.F.Perveev, *Sov.J.Opt.Technol.*, **60**, 2, 92 (1993).
12. L.Gurvitsch, *J.Phys.Chem.Russ.*, **47**, 805 (1915).
13. A.P.Karnaukhov, *Kinet.Kataliz,* **8**, 172 (1967); K.G.Ione, A.P.Karnaukhov and E.E.Kuon, *Kinet.Kataliz*, **12**, 457 (1971).
14. J.D.F.Ramsay, *private communication*.
15. M.R.Baklanov, M.Muroyama, M.Judelewicz, E.Kondoh, H.Li, J.J.Waeterloos, S.Vanhaelemeersch and K.Maex, *J.Vac.Sci.Technol.*, **B17 (5)**, 2136-2146, 1999.
16. D.G.Shamiryan, M.R.Baklanov et al., to be published
17. J.D.F.Ramsay and R.G.Avery, *Adsorption in silica compacts containing pores odf molecular size, in: Pore Structure and Properties of Materials, Part 1 , Proc.Int.Symp.IUPAC*, Academia Prague, 1973, pp.B37-B45.
18. C.E.Rogers, *Permeation of gases and vapours in Polymers, in: Polymer Permeability, ed. J.Comin, Chapter 2*, Elsevier, 19.., pp. 11-73.

Mat. Res. Soc. Symp. Proc. Vol. 612 © 2000 Materials Research Society

Probing Pore Characteristics in Low-K Thin Films Using Positronium Annihilation Lifetime Spectroscopy

D. W. Gidley, W. E. Frieze, T. L. Dull[1], J. N. Sun[1], and A. F. Yee[1]
Department of Physics, University of Michigan, Ann Arbor, MI 48109
[1]Department of Materials Science and Engineering, University of Michigan, Ann Arbor, MI 48109

ABSTRACT

Depth profiled positronium annihilation lifetime spectroscopy (PALS) has been used to probe the pore characteristics (size, distribution, and interconnectivity) in thin, porous films, including silica and organic-based films. The technique is sensitive to all pores (both interconnected and closed) in the size range from 0.3 nm to 300 nm, even in films buried under a diffusion barrier. PALS may be particularly useful in deducing the pore-size distribution in closed-pore systems where gas absorption methods are not available. In this technique a focussed beam of several keV positrons forms positronium (Ps, the electron-positron bound state) with a depth distribution that depends on the selected positron beam energy. Ps inherently localizes in the pores where its natural (vacuum) annihilation lifetime of 142 ns is reduced by collisions with the pore surfaces. The collisionally reduced Ps lifetime is correlated with pore size and is the key feature in transforming a Ps lifetime distribution into a pore size distribution. In thin silica films that have been made porous by a variety of methods the pores are found to be interconnected and an average pore size is determined. In a mesoporous methyl-silsesquioxane film with nominally closed pores a pore size distribution has been determined. The sensitivity of PALS to metal overlayer interdiffusion is demonstrated. PALS is a non-destructive, depth profiling technique with the only requirement that positrons can be implanted into the porous film where Ps can form.

INTRODUCTION

There is currently a great deal of interest in introducing and characterizing nanometer-sized voids into thin silica and polymer films. Such porous films are being intensely pursued by the microelectronics industry as a strategy for reducing the dielectric constant of interlayer insulators in microelectronic devices. Unfortunately, there are relatively few techniques capable of probing the pore characteristics (average size, size distribution, and interconnectivity) in sub-micron films on thick substrates. This is particularly true if the voids are closed (not interconnected) so that gas absorption techniques are not available. Transmission electron microscope (TEM) images, for example, are inherently challenging to interpret in such amorphous insulators. Neutron scattering [1] and beam-based positronium annihilation lifetime spectroscopy (PALS) [2] have recently been used to determine an average pore size in silica films and beam-based Doppler broadening positron annihilation spectroscopy [3] has been used to probe open-volume in silsesquioxane films. In closed-pore systems PALS may be uniquely capable of deducing a pore-size *distribution* [4], even in films buried under diffusion barriers. In this paper we will review the methodology and recent results of PALS.

PALS EXPERIMENTAL TECHNIQUE

In using PALS with thin films, an electrostatically focused beam of several keV positrons is generated in a high vacuum system using a radioactive beta-decay source of ^{22}Na (Figure 1). This beam is deflected onto a target sample as shown in the inset to this figure and forms positronium (Ps, the electron-positron bound state) throughout the film thickness. Positrons striking the sample generate secondary electrons that are detected in a channel plate (CEMA) and the Ps lifetime for each event is the time between this CEMA signal and the subsequent detection of annihilation gamma rays in a plastic scintillator. Ps inherently localizes in the pores where its natural annihilation lifetime of 142 ns is reduced by annihilation with molecular electrons during collisions with the pore surface. The collisionally reduced Ps lifetime is correlated with void size and forms the basis of the technique.

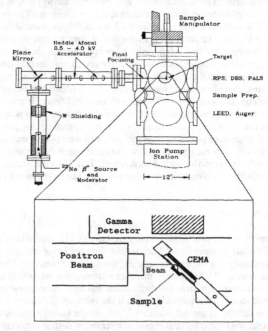

Figure 1. Schematic of the Depth-Profiled Positron Spectrometer.

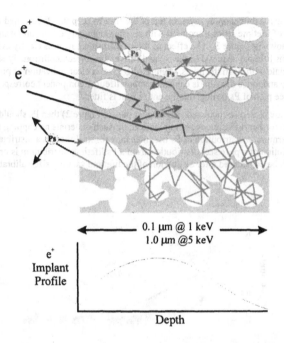

Figure 2. Positronium formation in porous materials. The shape of a typical positron implantation profile is depicted in the lower panel.

The formation of Ps in a porous insulator is depicted in Figure 2. The positron slows down through collisions in the material from its initial beam energy of several keV to several eV. It can either capture a bound molecular electron or recombine with free "spur" electrons generated by ionizing collisions to form the electron-positron bound state of positronium, Ps. This Ps, which initially has a few eV of kinetic energy, begins to diffuse and thermalize in the insulator. In porous films it localizes in the void volume where the Ps binding energy is not reduced by the dielectric constant of the surrounding material. However, even when it is thermalized in the pores, Ps is still colliding with the pore walls and the resulting Ps lifetime is shortened by positron annihilation with molecularly bound electrons in addition to the captured electron. Furthermore, Ps may diffuse over long distances that can be greater than the porous film thickness if the pores are interconnected. As a result Ps can easily diffuse out of the film and into the surrounding vacuum as depicted in Figure 2. The observable effect on the Ps lifetime is that most of the Ps annihilates with the vacuum lifetime of 142 ns, a telltale indicator that the pores in the film are interconnected. This is what we have found so far [2] for all porous silica films, regardless of manufacturing process. They have interconnected pores and Ps diffuses within the pore volume with most Ps finding its way into the vacuum. To extract information on the average pore size (technically, the mean free path for Ps in the interconnected pores) it is necessary to deposit a thin capping layer on top of the film to keep the Ps corralled in the porous film. Examples of lifetime spectra acquired with an aluminum-capped and an uncapped porous silica film are shown in Figure 3. The effect of the Ps diffusion barrier is clearly evident. In this example the 41 ns lifetime acquired in the capped film is the correct, collisionally-shortened lifetime to associate with Ps in the pores. Thus PALS gives a clear indication of pore interconnectivity and, once the film is capped, a single lifetime component corresponding to the average mean free path of Ps throughout the entire film is fitted.

If the pores are closed (as depicted in the upper part of Figure 2) then Ps should be trapped in a pore with no further diffusion occurring. Indeed, in such materials no capping layer is required. Furthermore, a distribution of Ps lifetimes may result if there is a distribution of pore sizes. Deconvolution of a pore size distribution from a Ps lifetime distribution is one goal of our research and will be considered after discussion of the lifetime vs. pore size calibration.

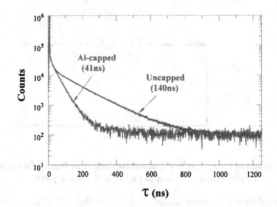

Figure 3. PALS spectra of uncapped and Al-capped porous silica films.

Calibration of Pore Size and Open Pore Systems

It is important to calibrate Ps lifetimes with pore sizes. In the very large pore (classical) regime (pores with mean free paths of order 100 nm), calibration was performed using high porosity (90-98%) silica powders [5]. In the other extreme (sub-nm pores), the quantum mechanical model first developed by Tao [6] and Eldrup [7] has been empirically used to calibrate Ps lifetimes of several nanoseconds with pore size, such as those in polymers [8]. In this simple model Ps is localized in a spherical infinitely deep potential well and only annihilates with molecular electrons when it is within a short distance of the pore surface. With only the ground state of Ps being considered in this Tao-Eldrup model, it is insufficient for characterizing larger voids where the pore diameter approaches the thermal De Broglie wavelength of positronium (about 6nm). Thermally excited states of Ps atoms in the pore must be included in the calculation. As a result, the effect of sample temperature should also appear in the calibration of lifetime versus pore size.

To fully extend the quantum mechanical model to the classical, large-pore limit, we have modified the Tao-Eldrup model in order to characterize both micro and mesopores. To summarize the results presented in Reference [2], a rectangular pore shape is assumed for calculational simplicity and it is assumed that there is no Ps-surface interaction. It is assumed that the Ps atoms randomly sample all of the states in the rectangular well with a probability governed by the Maxwell-Boltzman distribution. At a given temperature, a lifetime vs. pore dimension curve can then be calculated.

It is useful to convert such curves from rectangular pore dimension to a classical mean-free path, $l=4V/S$, where V/S is the volume-to-surface area ratio of the rectangular pore used in the calculation. The mean-free path is a linear measure of pore size that is independent of pore geometry. Furthermore, in large (classical) pores the Ps lifetime depends only on the mean-free path. As will be shown below, even well into the quantum mechanical regime, the Ps lifetime depends almost entirely on the mean-free path and is only modestly dependent on the detailed pore geometry. For these reasons, and for calculational simplicity when deriving pore-size distributions in closed pore systems, it is desirable to work with the mean-free path as opposed to physical pore dimension.

Figure 4 shows lifetime vs. mean-free path curves at several different temperatures. Cubic pores were used in the calculations. The model includes only one fitting parameter that is determined by existing experimental data for l below 2 nm. At room temperature, the model extrapolates perfectly through precision measurements [5] in large-pore silica powders as shown. Ps lifetimes measured in bulk silica gels with pores calibrated by gas absorption methods are in quite reasonable agreement [9] with Figure 4, but display a large scatter. Recently, identical silica films were studied by small angle neutron scattering (SANS)[1] and by beam-PALS [2] and the deduced average cord/mean free path was 6.5 ± 0.1 nm and 7.5 ± 0.3 nm, respectively. Further agreement between PALS and SANS at the 15 nm size scale in porous poly (arylene ether) films will soon be reported, as will a comparison of PALS with a gas absorption technique

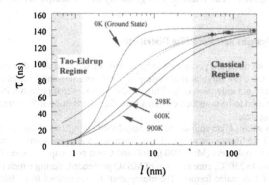

Figure 4. Pore-size calibration calculated at different temperatures using a cubical pore shape.

in a silica film with 3 nm-sized pores. These isolated comparisons suggest that our extension of the Tao-Eldrup model is valid and accurate for a range of materials. A systematic round robin of four identical sample films is presently underway to compare the results of PALS, SANS, ellipsometric porosimetry, gas absorption, and x-ray scattering. Results are to be presented during the conference.

The temperature dependence of the Ps lifetime, τ, was tested by PALS on several mesoporous samples and the very good agreement will be presented in the following sections (see also Figure 3 in Reference [2]). Note that all the calibrations of τ vs. pore size in this work are based on our theoretical extension of this model. To demonstrate the slight dependence on pore geometry, refer to Figure 5 in which Ps lifetime as a function of mean-free path has been plotted at two different temperatures using cubes and infinitely long square channels in the calculations.

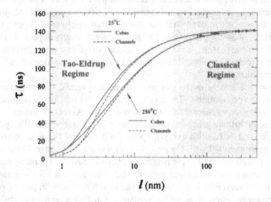

Figure 5. The effect of pore geometry at two sample temperatures.

Closed Pore Systems and Pore Size Distribution

The existence of closed pores with varying sizes should produce a distribution of Ps lifetimes since there is a singular relationship between pore size and τ indicated in the Tao-Eldrup model and our extension of it. This is the key feature that permits a PALS continuum lifetime distribution to be transformed into a pore-size distribution if Ps is trapped in isolated voids of varying sizes.

Our PALS study [4] on spin-on porous poly-MSSQ thin films clearly presents a distribution of lifetimes. These films were prepared by spin-casting a homogeneous mixture of methyl silsesquioxane prepolymers ($Mn \approx 1000$ g/mol) and 6-arm poly-caprolactone (PCL). Upon heating the spin-on films to 250°C, cure reaction of MSSQ proceeded, during which phase separation occurred and PCL domains formed. The mesoporous films received from IBM were developed by ramping the cured poly-MSSQ/PCL films to 430°C and holding for 2hrs before cooling down to room temperature in house nitrogen, during which process, the PCL phase thermally decomposed and left behind pores. Lifetime spectra acquired before and after decomposition of the PCL are compared in Figure 6. Initially we used the program POSFIT to fit discrete lifetimes to this spectrum. Prior to PCL decomposition the "unfoamed" film presents lifetimes consistent with those in bulk MSSQ (0.4 ns, 1.5 ns, and 6 ns), but no mesoporous lifetime components are observed. The fully decomposed porogen "foamed" sample shows long-lived events that cannot be adequately fitted with a single lifetime. At least two components (around 45 ns and 120 ns) are required in the fitting and the long lifetime component may be indicative of possible diffusion and escape of Ps into the vacuum. This suspicion was confirmed by spectra acquired with a 800 A sputter-deposited Al capping layer that confines Ps to the porous film. The 120 ns component disappeared as expected and lifetime components around 20 ns and 60 ns emerged, suggesting that only a portion of the pores are interconnected; a distribution of Ps lifetimes is still required for adequate fitting. Thus, these films may have a complicated, partially interconnected, pore system.

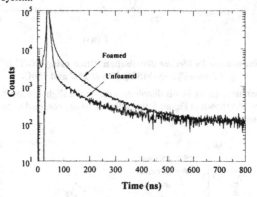

Figure 6. PALS spectra of foamed and unfoamed poly-MSSQ films.

The lifetime spectra shown in Figure 6 are given by dN_{Ps} / dt. To focus on the mesoporous part of the time spectrum we fit data for t beyond 60 ns, thus avoiding the bulk components with lifetimes less than 6 ns. We use a version of the continuum fitting program, CONTIN [10], specialized for exponential lifetime analysis. The result of this fitting program for the capped, fully-decomposed, film is shown in Figure 7 for two sample temperatures. (It should be noted the uniqueness of the fitted distributions presented in Figure 7 is still an open issue [4].) Given a statistically acceptable Ps lifetime distribution, $(1/N_{Ps})(dN_{Ps}/d\tau)$, determined from fitting the lifetime spectrum with CONTIN, the goal is to transform this distribution of Ps lifetimes into a distribution of void sizes, or more specifically, the specific void volume as a function of mean-free path. A key step in the transformation is a correction that must be made for Ps diffusing to and preferentially trapping in pores of large surface area. See Reference [4] for a detailed derivation. The fractional pore volume distribution as a function of mean-free path, l, is given by

$$\frac{1}{V}\frac{dV}{dl} = \frac{1}{N_{Ps}}\frac{dN_{Ps}}{d\tau}\frac{d\tau}{dl}l.$$

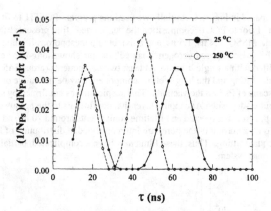

Figure 7. Normalized Ps lifetime distribution (fitted using CONTIN) of the capped "foamed" poly-MSSQ film at 25C and 250C.

There is an inherent uncertainty in this distribution due to the slight dependence of Ps lifetime on the pore geometry. As shown in Figure 5, this effect is mainly confined to pores with mean free paths less than 2 nm.

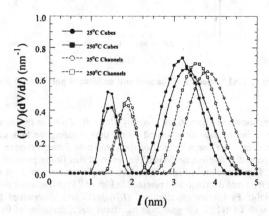

Figure 8. Volume fraction distribution in the mean free path, l, of a pore for spectra acquired at two different sample temperatures and analyzed using two different pore model dimensionalities.

Figure 8 shows the void volume distributions obtained from the lifetime distributions in Figure 7 using the calibrations of Figure 5. The solid symbols and curves correspond to the high and low temperature results derived using a closed, cubic pore model. The open symbols and dashed curves are the same spectra analyzed using an infinitely long, square-channel pore model. Although the lifetime distributions in Figure 7 were acquired at two different temperatures and are therefore quite different, the deduced pore size distributions in Figure 8 are quite similar. This is an important systematic test of the model that demonstrates that Ps is thermally distributed throughout the entire void volume and not adsorbed on the void surface. The systematically larger pore sizes deduced using the channels as compared with the cubic pores is indicative of the typical systematic error that would be assigned in the determination of these distributions. Presumably, the "correct" distribution falls somewhere between these limiting cases of pore dimensionality. Given the uniqueness issues surrounding CONTIN fitting (mentioned above) and the fact that an electron micrograph presented in reference [4] could not fully substantiate the distributions in Figure 8, we recommend further research on this difficult deconvolution of pore size distribution.

Metal Interdiffusion Into the Porous Film and Diffusion Barrier Integrity

In addition to determining the average pore size in open pore systems and extracting a pore size distribution in closed pore systems, PALS can be used to study the integrity of diffusion barriers and the diffusion of metal overlayers into the porous film. In the process of studying an open-porosity silica film, we attempted to cap the film by depositing 800 Å of thermally evaporated aluminum without the use of any sample cooling. Instead of the 140 ns component being shortened to a 70 ns component indicative of Ps diffusing within the interconnected pore structure of the film, the long-lived component totally disappeared and a short, 3 ns, component appeared in its place. We confirmed that the sample warmed up during the deposition, resulting in aluminum diffusion into the pore structure. A metal coating on the pores would explain the appearance of the 3 ns component since Ps annihilation would be enhanced by the high density of free electrons. We verified that the silica film did indeed have a large open pore structure by room temperature sputter-deposition of an Al cap and fitting a single 70 ns lifetime component. These results are shown in Figure 9.

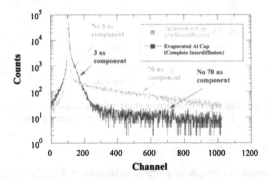

Figure 9. PALS spectra of identical silica films: one with a cold sputter-deposited Al cap and another that underwent uncontrolled heating during Al thermal evaporation.

To test the metal interdiffusion hypothesis we performed a systematically controlled study involving an Al sputter-capped silica film with an inherent lifetime component of 98 ns. We heat-treated the film at progressively higher temperatures and collected data at room temperature in between each heating cycle. We monitored the intensity of the 98 ns component and searched for the appearance of a short 3 ns lifetime component. The results are shown in Figure 10. At approximately 450 C we see the apparent onset of Al interdiffusion. This temperature is consistent with that of bulk annealing of Al and is probably significantly higher than that experienced by the sample during the thermal evaporation process. We tentatively conclude that the sample temperature during *vapor* deposition is much more critical than subsequent processing temperatures in preventing metal interdiffusion.

In a complementary study we are collaborating with SEMATECH to assess the integrity of candidate materials for use as thin diffusion barriers to prevent Cu and Al interdiffusion into the low-K films. Candidate materials are deposited with varying thicknesses as thin capping layers on silica films with interconnected pores. To qualify as a diffusion barrier *for Ps* such a capping layer must not allow any Ps to escape through the open silica pore structure into the vacuum (escape is readily manifested by a 140 ns component in the fitted PALS spectrum). Heat treatment of the film is also performed to test for thermal stability of the diffusion barrier. In this manner candidate materials and minimum critical barrier thickness can be identified. It is presumed that a diffusion barrier should at least be able to barrier Ps diffusion. However, other effects such as grain boundary diffusion of Cu atoms through a barrier will not be accounted for with Ps and would require separate testing.

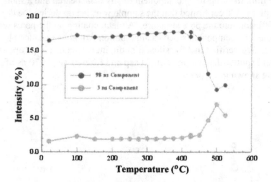

Figure 10. Controlled Al interdiffusion study using a silica film with a characteristic Ps lifetime of 98 ns within its interconnected pores.

CONCLUSION

PALS is a non-destructive, depth-profiling technique with the rather simple requirement that positrons can be implanted into the porous film where Ps can form. It is sensitive to *all* the void volume greater than a few angstroms in size, regardless of whether the pores are open or closed. The technique can readily distinguish open from closed porosity. In open (interconnected) pores a single Ps lifetime is fitted corresponding to the mean free path in the void volume throughout the film thickness. In this case capping of the film is required to keep Ps from escaping into the vacuum. This property can also be exploited to test the integrity of candidate materials for diffusion barriers and determine critical barrier thicknesses. Metal interdiffusion into the open pores can also be observed since metal coating of the pore surfaces drastically reduces the Ps lifetime. In closed pores with varying sizes the PALS spectrum contains a distribution of Ps lifetimes that should correspond to the pore size distribution. Such a distribution of Ps lifetimes has definitely been observed and a pore shape independent method to deconvolve a fractional void volume distribution in mean free path has been proposed.

ACKNOWLEDGEMENTS

We thank Simon Lin, Todd Ryan, and Huei-Min Ho at SEMATECH and Do Yoon at Seoul National University for helpful discussions. Cattien Nguyen, formerly at IBM, prepared the mesoporous MSSQ films. Sputter-coating by David Beglau at ECD Inc. is gratefully acknowledged. This research is supported by NSF grant ECS–9732804 and by the Low-K Dielectric Program at SEMATECH.

REFERENCES

1. W. -L Wu, W. E. Wallace, E. K. Lin, G. W. Lynn, C. J. Glinka, E. T. Ryan and H. -M. Ho, *Appl. Phys.* **87**, 1193 (2000).
2. 2. D. W. Gidley, W. E. Frieze, A. F. Yee, T. L. Dull, E. T. Ryan, and H. –M. Ho, *Phys. Rev. B.* **60** *(Rapid Comm.)*, R5157 (1999).
3. M. P. Petkov, M. H. Weber, K. G. Lynn, K. P. Rodbell, and S. A. Cohen, *J. Appl. Phys.* **86**, 3104 (1999).
4. D. W. Gidley, W. E. Frieze, T. L. Dull, J. Sun, A. F. Yee, C. V. Nguyen, and D. Y. Yoon, *Appl. Phys. Lett.* **76**, 1282 (2000).
5. D. W. Gidley, K. A. Marko, and A. Rich, *Phys. Rev. Lett.* **36**, 395 (1976).
6. S. J. Tao, *J. Chem. Phys.* **56**, 5499 (1972).
7. M. Eldrup, D. Lighbody, and J. N. Sherwood, *Chem. Phys.* **63**, 51 (1981).
8. Y. Y. Wang, Y. Nakanishi, Y. C. Jean, and T. C. Sandreczki, *J. Polym. Sci., Part B: Polym. Phys.* **28**, 1431 (1990).
9. K. Ito, H. Nakanishi, and Y. Ujihira, *J. Phys. Chem. B* **103**, 4555 (1999) and references therein.
10. S. W. Provencher, *Comput. Phys. Commun.* **27**, 213 (1982).

Mat. Res. Soc. Symp. Proc. Vol. 612 © 2000 Materials Research Society

Ultra Low k Mesoporous Silica Films: Synthesis, Film Properties and One-Level Copper Damascene Evaluation

Changming Jin,[#] J. Liu, X. Li, C. Coyle, J. Birnbaum, G. E. Fryxell, R. E. Williford and S. Baskaran*, Pacific Northwest National Laboratory, 902 Battelle Boulevard
Mail Stop K2-44, PO Box 999, Richland, WA 99352
[#]International SEMATECH, 2706 Montopolis Dr., Austin , TX 78741
*author for correspondence: suresh.baskaran@pnl.gov

ABSTRACT

Spin-on mesoporous silica films were prepared on eight-inch wafers at SEMATECH by condensation of a silicate network around surfactant micellar structures. Copper single-damascene one-level test structures were built using mesoporous silica as the intermetal dielectric. No major structural failures were observed after chemical mechanical planarization on both blanket films and patterned wafers, indicating relatively good mechanical integrity for a highly porous structure. A simple silane spin-coating step used at SEMATECH on the films appears to be insufficient for complete dehydroxylation and silylation. The electrical test results on the metal comb structures showed good capacitance and leakage current distributions. However, capacitance and leakage current changes were observed after each post-CMP process step, and these changes could be correlated to moisture desorption/outgassing, which was also noted during k measurement. With controlled film synthesis and dehydroxylation conditions, mesoporous silica films with $k \leq 2.0$ and elastic modulus of 4.0 GPa have been synthesized at PNNL. The results of the Cu one-level metal screening tests at SEMATECH combined with properties obtained at PNNL indicate that mesoporous molecularly-templated silicate films hold promise as ultra low k intermetal dielectrics.

INTRODUCTION

The semiconductor industry is currently targeting new intermetal dielectric films with $k < 2.5$ for interconnects in the sub-150 nm technology node. As the metal line width continues to decrease, and the packing density of metal lines on the semiconductors continues to increase, dielectric films with $k < 2.0$ will be soon required. Porous silica films with nanometer-scale porosity are potentially useful as ultra low dielectric constant films in advanced semiconductor interconnects. Mesoporous silica films with pore sizes in the range of 1 to 10 nm have been synthesized from solution precursors by a surfactant-templating process (1-3) in which the pores are formed in a spin coated (4-7) or dip coated film (8-9) upon removal of the surfactant. Although the term "mesoporous" refers generally to materials with pore sizes > 1 nm, it is now most commonly used to describe materials prepared by the surfactant-templating approach. Mesoporous silica films with low dielectric constants (≈ 1.8 to ≈ 2.5) have been demonstrated using cationic and non-ionic surfactants (10-12). This molecularly templated synthesis approach allows rational control of the porosity, pore size, pore shape, film texture and thickness, and can result in good mechanical properties in the film (12).

In this paper, we present an assessment of the deposition process and properties of a mesoporous silica film on 8-inch wafers, and the film performance through a one-level copper damascene module at SEMATECH. A summary of the dielectric and mechanical properties measured at PNNL for this class of materials is also provided.

FILM PREPARATION

The mesoporous silica films were deposited by a spin-on sol-gel process (4, 6-7, 10) as described previously. The deposition solution consisted of a silica precursor, deionized water, an acid catalyst, alcohol and a surfactant. The solution was spin coated on 200 mm diameter wafers using a production-scale spinner at SEMATECH. The deposition process did not require careful atmospheric control or gelation steps, and is potentially scalable to larger (e.g. 300 mm) wafers. Films 0.5 and 1.0 microns in thickness were deposited. The coated wafers were heated on a hot plate to 400°C to form the mesoporous structure in the films.

Porous silica films typically require careful exposure to silane compounds, e.g. hexamethyl-disilazane or trimethylchlorosilane (13) or other organic/organosilicon compounds to dehydroxylate and silylate the films, especially the pore surfaces. Special chambers on prototype spinner equipment are being considered for dehydroxylation of nanoporous silica films (14). In order to obtain a preliminary assessment of the surfactant-derived mesoporous films, the films were simply spin coated with a silane and then heated to 400°C in forming gas.

FILM PROPERTIES

The properties of the films on 200 mm wafers are shown in Table I.

Table I: Properties of Mesoporous Silica Films Prepared at SEMATECH

Refractive Index (spectroscopic ellipsometry)	1.16
Standard deviation on film thickness by ellipsometry (1.0 and 0.5 μm films)	< 2%
Dielectric constant (MIS)	k > 2.4*
Small-angle neutron scattering (SANS) Porosity Pore size Pore connectivity	64.4% ≈ 5 nm 100%
Adhesion (Tape pull test)	SiO2-pass Si-pass Ta-pass TaN-pass
Elastic modulus (Hysitron picoindenter at PNNL, and MTS nanoindenter)	3 GPa
Thermal conductivity (3□ method)	0.18 W/m.K

* dielectric constant varied above k of 2.4 with time, indicating that the silylation performed at SEMATECH was not an effective dehydroxylation step.

Based on the refractive index measured (1.16) immediately after surfactant pyrolysis, and models correlating refractive index to porosity, dielectric constants ≤2.25 can be achieved in this film if the film is effectively dehydroxylated. The dielectric constant of over 2.4 obtained at various times over a period of several months indicated that the silane spin wash was not an effective dehydroxylation procedure. The high porosity inferred from the refractive index was confirmed by SANS experiments. The highly porous film exhibited good adhesion to various substrate surfaces in a tape pull test. Elastic modulus was measured at PNNL using a Hysitron picoindenter on films prepared at SEMATECH. Modulus was measured as a function of indentation depth, and the representative value (3 GPa) measured at a depth of 5-10% of film thickness was in agreement with measurements obtained on a MTS nanoindenter.

Concurrent with synthesis, characterization and evaluation at SEMATECH, additional work to tailor process conditions and film properties was also conducted at PNNL. A summary of the important properties of mesoporous silica films prepared under controlled process conditions, including dehydroxylation, at PNNL on 100 mm wafers is shown in Table II.

Table II: Properties of Mesoporous Silica Films Prepared at PNNL

k (measured in ambient laboratory conditions)	Refractive index by ellipsometry (after removal of pore former, and before dehdroxylation)	Elastic modulus (Nano/picoindenter)
≤ 2.25	1.16	≈ 7 GPa
≤ 2.00	1.16	≈ 4 GPa
≤ 1.80	1.13	≈ 2.4 GPa

BLANKET FILM TESTS

A double layer stack of PNNL mesoporous films on 200 mm wafers with a SiO_2 layer in between, as illustrated in Figure 1, was constructed to evaluate the thermal stability of multilayer stacks. The film stack was annealed at 400 °C for 1 hour under N_2 ambient in an oven and inspected for cracks, delamination, and other visual film defects. The annealing was further repeated three times and inspected. No film defects were visible by optical microscopy.

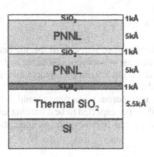

Figure 1 Illustration of double layer PNNL mesoporous silica film stack used for thermal annealing test

Blanket chemical-mechanical planarization (CMP) was used to test film adhesion and mechanical strength. The film stack used for the blanket CMP test is shown in Figure 2. 5 kÅ of PNNL's mesoporous silica film was deposited on a 1 kÅ SiN / 5.5 kÅ thermal SiO_2 / Si substrate. The mesoporous film was capped with a 1 kÅ CVD SiO$_2$ followed by the deposition of 250 Å TaN or Ta and 1 kÅ PVD Cu seed. No delamination is observed after the CMP process. In another CMP test, PNNL's mesoporous silica film was deposited on a substrate without any capping layer and was polished directly using a standard Cu CMP recipe. Again, the film did not show any optical defects, indicating relatively good mechanical integrity for a highly porous structure.

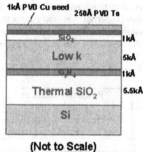

(Not to Scale)

Figure 2   Illustration of PNNL mesoporous silica film stack used for blanket CMP test

ONE-LEVEL METAL STRUCTURE EVALUATION

With the PNNL mesoporous silica films passing both the double stack annealing and blanket film CMP, evaluation of the material moved to the one level metal (1LM) module screening stage to determine any major issues with the material during integration. A 1 kÅ CVD oxide layer was deposited on every wafer to cap the PNNL mesoporous silica film. The wafers were baked at 400°C for 120 seconds before the oxide deposition. The wafers were then patterned using a DUV process.

Etch and Ash

The capped film was etched using a two step etch recipe, the first step to open the cap oxide layer and the second one to etch the mesoporous silica film and stop at the SiN underlayer. 0.35 um / 0.40 um (line/space) and 1.0 µm/1.6 µm (line/space) mesoporous silica trenches were studied. Inpection of the etched trenches by cross sectional SEM showed good etch profiles with sidewalls at ≈85°. Since the mesoporous silica films should etch like porous silica, etch selectivity with respect to SiN is not expected to be a limiting issue. After etch, the mesoporous silica was subjected to an ash process to remove the photoresist. No wet clean process was performed after ash.

Barrier/Cu Deposition and Cu Fill

A 250 Å TaN barrier layer and 1000 Å Cu seed were deposited with a standard physical vapor deposition tool. The wafers were baked for 120 seconds at 350 C prior to the PVD metal deposition. Cross sectional and longitudinal SEM images of 0.35 um / 0.40 um (line/space) and 1.0 um / 1.6 um (line/space) trenches after 250 Å TaN and 1 kÅ Cu seed deposition are shown in Figure 3. The longitudinal images were obtained by cleaving along the trenches. Effective TaN barrier and Cu seed sidewall coverage for trenches of both small and large dimensions was observed. The longitudinal SEM pictures display columnar and globular structures. The columnar structure is most likely caused by a microchanneling effect during etch. In the oxide cap open step of the etch process, some areas of the oxide cap may be etched through earlier than other areas. This non-uniform etch front can be magnified in the porous silica film, since the porous silica film etches much faster than the oxide cap. Microchanneling is a result of the non-uniform etch of the oxide cap layer and mesoporous silica film. The globular structure is likely a result of the TaN and Cu deposition.

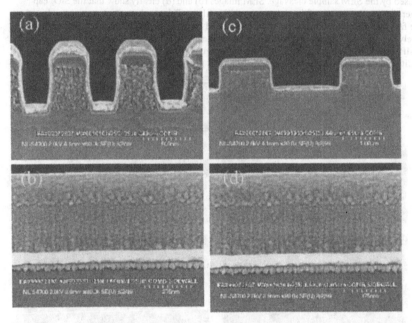

Figure 3. Cross sectional and longitudinal SEM images of (a) and (b): 0.35 um / 0.40 um (line/space) and (c) and (d): 1.0 um / 1.6 um (line/space) trenches.

After PVD TaN barrier and Cu seed deposition, electroplated Cu was deposited on the wafers using a Cu plating tool, with the final Cu thickness being ≈ 1 um.

CMP of Patterned Wafers and Electrical Tests

CMP is one of the most challenging process steps for porous materials because of the downward and shear stress applied onto wafers and the relatively low mechanical strength of porous materials. As mentioned earlier, the PNNL mesoporous silica film performed well in the blanket film CMP tests both with TaN and Cu on top of the film, and upon polishing the porous film directly. The patterned wafers were polished on a CMP tool using a standard SiO_2/Cu recipe.

Wafers were subjected to a final DI water buff step for 30 seconds. Special attention was given to closely monitor possible delamination during and after the CMP process.

No visual defects were observed on patterned wafers during and after the CMP process. Optical microscope and cross sectional SEM images of both patterned and field regions of PNNL patterned 1LM (one level metal) wafers after CMP are shown in Figure 4. No optical defects were observed in the patterned (a) or field (c) regions. The SEM samples were prepared by direct cleavage. The cracking in the mesoporous silica film shown in figures 4 (b) and 4 (d) is caused by the SEM sample cleavage. SEM images (b) and (d) clearly show that the SiO_2 cap layer was present on top of the mesoporous silica films. This result was not unexpected based on the fact that the mesoporous silica films had passed blanket film CMP tests. A patterned wafer is likely to pass the CMP process without failure if the material has passed tougher blanket film CMP tests. The Cu lines embedded in the mesoporous silica films on patterned wafers provide additional mechanical support for porous films and prevent adhesive and cohesive failures.

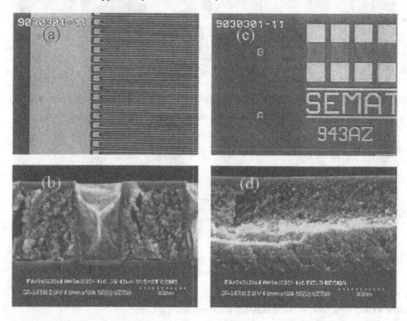

Figure 4   Optical microscope and cross sectional SEM images of patterned PNNL mesoporous silica wafers after CMP. (a) and (b): patterned region (c) and (d): field region. Neither delamination nor other defects were observed in both areas. The SiO_2 cap is clearly present in the SEM images (b) and (d). Cracks in the PNNL mesoporous silica films apparent in the SEM images occurred during SEM sample preparation.

In spite of concerns due to ineffective dehydroxylation on the films on eight-inch wafers, electrical tests were anyway performed after CMP. A 1 KA CVD SiN layer was then deposited to passivate the 1LM structures. The wafers were baked for 120 seconds at 400C before the SiN deposition. Probe bond pads were then patterned and etched. Following photoresist ash, the samples were electrically tested. Line to line capacitance distribution was measured on 0.35 um / 0.4 um (line/space) and 100 um long COMB structures. Each COMB had 84 fingers. Line to line leakage current distribution was also measured at 5 V on the same COMB structures. Electrical tests were performed after annealing repeatedly for 1 hour at 400 C under N_2 ambient.

The electrical test results (not displayed here) showed good capacitance and leakage current distributions. However, the capacitance values and the leakage currents decreased after each processing step after CMP. These results were attributed to moisture outgassing during each of these process steps. These observations further reinforced the need for using dehydroxylation procedures other than spin-coating of silanes on the mesoporous silica film.

CONCLUSIONS

Cu single damascene 1LM test structures have been built using PNNL mesoporous silica low-k material as the intermetal dielectric. No major structural failures were observed after CMP on both blanket films and pattered wafers, indicating relatively good mechanical integrity for a highly porous structure. The electrical test results showed good capacitance and leakage current distributions, and the capacitance and leakage current changes after each post CMP process step could be explained by moisture desorption/outgassing and SiN passivation. A simple silane spin-coating step was insufficient for dehydroxylation , and the film exhibited moisture adsorption both during material property characterization and in 1LM module screening. Although sidewall damage after the etch and ash was not characterized in detail, a robust photoresist ash process would be recommended to minimize damage to functional organic groups on pore surfaces.

It appears that mesoporous films with dielectric constants in the range 2.0-2.2 can be fabricated with adequate mechanical integrity for withstanding interconnect process and performance requirements. Mesoporous films with $k \leq 1.8$ and elastic modulus of 2.4 GPa can also be synthesized. Although mesoporous films with modulus in this range can be polished using conventional CMP tools with tailored CMP process parameters, a film modulus of ≥ 5 GPa is preferred. In summary, the screening tests at SEMATECH combined with basic properties measured by PNNL indicate that molecularly-templated mesoporous silicate films could be a viable approach to ultra low k intermetal dielectrics.

REFERENCES

1. H. Yang, A. Kuperman, N. Coombs,S. Mamiche-Afara, G. A. Ozin, Nature (1996) V.379, 703.
2. H. Yang, N. Coombs, I. Sokolov, G. A. Ozin, G.A., Nature (1996) V.381, 589.
3. I. A. Aksay, M. Trau, S. Manne. I. Honma, N. Yao, L. Zhou, P. Fenter, P. M. Eisenberger, S. M. Gruner, Science (1996) 273, 892.
4. J. Liu, J. R. Bontha, A. Y. Kim and S. Baskaran, MRS Symp. Proc. (1996) Vol. 431, pp245-50

5.   M. Ogawa, Chem. Commun. (1996) 1149-50
6.   P. J. Bruinsma, N. J. Hess, J. R. Bontha, J. Liu and S. Baskaran, MRS Symp. Proc. (1997) Vol. 445, pp105-110
7.   P. J. Bruinsma, J. R. Bontha, J. Liu and S. Baskaran, U. S. Patent No. 5,922, 299, July 13, 1999
8.   Y. Lu, R. Ganguli, C. A. Drewien, M. T. Anderson, C. J. Brinker, W. Gong, Y. Guo, H. Soyez, B. Dunn, M. H. Huang and J. I. Zink, Nature (1997) 389, pp364-68
9.   D. Zhao, P. Yang, N. Melosh, J. Feng, BF Chmelka, and GD Stucky, Advanced Materials (1998) Vol. 10 No. 16, pp 1380-1385
10. S. Baskaran, J. Liu, K. Domansky, N. Kohler, X. Li, C. Coyle, G. E. Fryxell, S. Thevuthasan and R. E. Williford, Proceedings of SEMATECH Ultra Low K Workshop, pp 55-80 (1999) SEMATECH, Austin, TX
11. C. J. Brinker, Y. Lu and H. Fan, Proceedings of SEMATECH Ultra Low K Workshop, pp 05-38 (1999) SEMATECH, Austin, TX
12. S. Baskaran, J. Liu, K. Domansky, N. Kohler, X. Li, C. Coyle, G. E. Fryxell, S. Thevuthasan and R. E. Williford, Advanced Materials (2000) Vol. 12 No. 4, pp 291-94
13. C. Jin, J. D. Luttmer, D. M. Smith, T. A. Ramos, Mat. Res. Soc. Bull. (1997) Vol 10, pp 39-43
14. D. Toma, M. Muramatsu, S. Nagashima, K. Takeshita, S. Kojima, H. Fujii, K. Nishimura, M. Slessor, R. Roberts and L. Forester, Proceedings of SEMATECH Ultra Low K Workshop, pp 469-84 (1999) SEMATECH, Austin, TX

Acknowledgements

Pacific Northwest National Laboratory is operated for the US Department of Energy (DoE) by Battelle Memorial Institute. This research was supported by the Laboratory Technology Research Program of DoE's Office of Science, with SEMATECH as the CRADA partner, and, in part, by Battelle Memorial Institute.   The Univeristy of Texas at Austin and the National Institute of Standards and Technology assisted SEMATECH in characterization of the films.

Mat. Res. Soc. Symp. Proc. Vol. 612 © 2000 Materials Research Society

A Multilevel Metal Interconnect Technology with Intra-Metal Air-Gap for Quarter-Micron-and-Beyond High-Performance Processes

Mark Lin [1,2], Chun-Yen Chang [1], Tiao-Yuan Huang [1,3], Mout-Lim Lin [2], and Horng-Chin Lin [3]

[1] Institute of Electronics, National Chiao Tung University, Hsinchu, Taiwan
[2] United Semiconductor Corp., Science-Based Industrial Park, Hsinchu, Taiwan
[3] Natiomal Nano Device Laboratories, Hsinchu, Taiwan

ABSTRACT

A multilevel metal interconnect with air-gap has been developed to reduce RC delay time for quick turn-around-time foundry manufacturing. The air-gap method has been successfully applied to 0.25 μm foundry technology. Measurements on ring oscillators confirm that the smallest delay time is indeed achieved with the air-gap method, compared with that using either conventional high-density-plasma (HDP) oxide or low-dielectric-constant spin-on-glass (SOG). In addition, we have also developed fitting equations for the delay time, thus provide a handy method for predicting the RC delay time. The oscillator delay time is also found to be critically dependent on not only the size, but also the position of the air-gap. Best delay time reduction is obtained when the air-gap is positioned in extended both above and below the metal lines to effectively reduce the fringing capacitance.

INTRODUCTION

As devices continue to scale down to deep sub-half-micron regime, the intrinsic transistor delay is no longer the limiting factor in the overall circuit delay. Instead, the RC delay due to the long metal interconnect now represents the most significant portion of the total circuit delay [1]. Therefore, how to minimize the interconnect RC delay becomes very important for achieving high-performance ULSI circuit. Many new materials with low dielectric constant (i.e., low k), including fluorinated SiO_2 (FSG), hydrogen silsesquioxane (HSQ) [2], and polymers, etc. have been proposed to minimize the intra-metal parasitic capacitance. However, these processes are often complicated, while the delay time improvement is usually not sufficient for ULSI circuit application. Recently, air-gap formed between metal lines during SiO_2 deposition has been proposed for reducing interconnect delay [3][4]. The effective relative dielectric constant, k_{eff}, is 1.75~2.5, which is much smaller than that achievable by conventional low-k dielectric materials. In this study, we report the successful development of a multilevel metal interconnect technology with air-gap between metal lines, which is well suited for quick turn-around-time foundry manufacturing of high-performance 0.25-μm-and-beyond ULSI circuits.

EXPERIMENTAL

8-inch wafers with 0.25 μm design rules were used in this study. After the metal lines were patterned, three different types of intra-metal dielectric materials were fabricated in this study. Specifically, the standard samples with SiO_2 dielectric (k~4.4) were deposited by a high-density plasma-enhanced chemical vapor deposition (i.e., HDP samples), while samples with a low-dielectric-constant material, i.e., low-k hydrogen silsesquioxane (HSQ, K~3.3), were coated by a conventional coater. Both splits were included to serve as controls, while samples with the new air-gap (i.e., void) structure between metal lines were formed by carefully tuning the SiO_2 deposition condition. Afterwards, a plasma enhanced chemical vapor deposition (PECVD) SiO_2 was deposited to serve as the inter-metal dielectric (IMD) layer. All metal lines consist of stack layers of TiN/AlCu/TiN, with a total thickness of 0.6 micron. A cross-sectional view of the resultant interconnect structure with the new air-gap structure is shown in Figure. 1. Figure 2 shows the configuration of 101-stage ring oscillators used in this study with (Figure. 2b) and without (Figure. 2a) deliberate metal wire load. The delay time was calculated by measuring the frequency of ring oscillators. Leakage current between metal lines was also checked to ensure the integrity of the interconnect system.

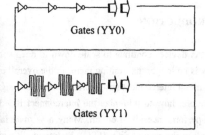

Figure.1 SEM photograph of air-gap structure in 0.4--μm lines and 0.3-μm spaces.

Figure.2 The configuration of ring an array of Oscillators

RESULTS AND DISSUSIONS

Figure 3 displays SEM photographs of interconnect structures with different gap-filling dielectric materials. The space between metal lines is fully filled by dielectric materials for the conventional HDP (Figure. 3b) and low-k HSQ (Figure. 3c) samples, except for the air-gap samples developed in this study (Figure. 3a). The gap-filling layers in the HSQ samples are composed of PECVD oxide sub-layer and an inorganic SOG top layer; while for conventional HDP samples, only a single HDP layer is used. SEM photographs of the air-gap structure fabricated using PECVD process with various spaces between metal lines show that the size of the air-gap reduces and the position of the air-gap shifts up when the spacing between metal lines increases (figure not shown).

Figure. 3 SEM photographs of different gap-filling dielectric materials.

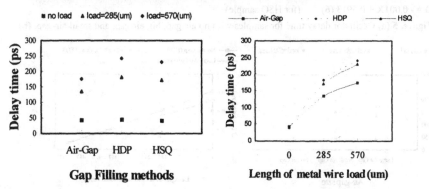

Figure. 4 (a) Ring oscillator delay time for samples with various gap-filling methods. (b) Delay time as a function of metal wire length for various gap-filling methods.

Figure 4 shows the delay time measured on various ring oscillators fabricated with three different gap-filling materials. Speed data were measured on ring oscillators with various metal wire load lengths, namely, 0 micron (denoted as YY0), 285 micron (YY1), and 570(YY2) µm, respectively. From Figure. 4(a), it can be seen that the oscillator delay time of YY0 (i.e., without deliberate metal wire loading) is almost the same for air-gap, HDP, and HSQ samples (i.e., 43.4, 44.5, and 42ps, respectively). However, by adding a 285-μm metal wire load (i.e., YY1) to the oscillator, samples with air-gap clearly display the smallest delay time (i.e., 133.6ps), compared to 183.2ps for the conventional HDP samples, and 172.5ps for low-k HSQ samples. The improvement is even more dramatic when the 570-μm metal wire load is attached to each stage of the oscillator. Specifically, the delay time is 174.8, 245, and 230.2ps, for air-gap, HDP, and HSQ samples, respectively.

In Figure. 4(b), the oscillator delay time is plotted as a function of the length of metal wire load for three different gap-filling materials. By defining the wire length = X μm and the oscillator delay time = Y ps, we could obtain the following fitting equations:

$Y = - 0.0003X^2 + 0.4156X + 43.4$ (1) (air-gap sample)

$Y = - 0.0005X^2 + 0.6165X + 44.5$ (2) (HDP sample)

$Y = - 0.0005X^2 + 0.5913X + 42$ (3) (HSQ sample)

Next, by defining dY/dX = the oscillator delay time per unit wire load (D), we could obtain the following equations:

$D = - 0.0006X + 0.4156$ (4)   (for air-gap sample)

$D = - 0.001X + 0.6165$ (5)   (for HDP sample)

$D = - 0.001X + 0.5913$ (6)   (for HSQ sample)

Figure. 5 (a) Oscillator delay time for samples with no air-gap, big air-gap, and small air-gap. (b)

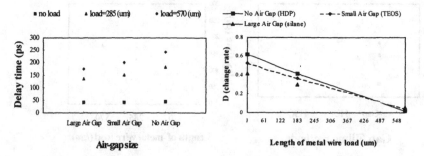

Oscillator delay time per unit wire load (D) versus length of metal wire load (X).

When X << 690 μm, D = 0.4156, 0.6165, and 0.5913 for air-gap, HDP, and HSQ samples, respectively. From these results, it is clear that the air-gap sample displays the slowest increase, while the conventional HDP sample the largest increase in delay time with increasing metal wire load. The fitting equations derived above thus allow us to predict the RC delay time of the resultant circuit, if the length of the metal wire load with a given process design rule is known. It is very helpful in estimating the effect of dielectric material on RC delay time by these equations.

The oscillator delay time for samples with different air-gap sizes is shown in Figure 5(a). The results show that the oscillator delay time of YY0, YY1, and YY2 with a large air-gap is 43.4, 133.6, and 174.8ps, respectively. While the oscillator delay time of YY0, YY1, and YY2 with a small air-gap is 44.8, 158.9, and 211.9ps, respectively. By similar manipulation as in Equations 4 to 6, we could obtain the following equations:

$D = - 0.0006X + 0.4156$ (7) (large air-gap sample)
$D = - 0.0008X + 0.523$ (8) (small air-gap sample)

It is obvious that the oscillator delay time increases at a faster rate as metal wire load increases for samples with small air-gap, compared to samples with large air-gap. In fact, for the conventional HDP samples, the size of the air-gap can be regarded as the smallest (i.e., zero in effect), and from Equation (5), i.e., $D = -0.001X + 0.6165$, the delay time for HDP samples indeed shows the largest increase as metal wire load increases. By plotting Equations (5), (7), and (8), as shown in Figure. 5(b), it can be seen that the oscillator delay time per unit wire load gradually decreases when the length of the metal wire load increases.

The effects of the air-gap positions on the delay time were also studied. Figure 6(a) displays the air-gap structures with three different positions with respect to the metal lines, namely, middle (left figure), shift-up (center figure), and extended (right figure), respectively. Figure 6(b) shows that the delay time of YY2 (i.e., metal wire load = 570 μm) oscillators for air-gap samples with three different positions. The delay time is 175.9, 185.1, and 170.8ps, for air-gaps with middle, shift-up, and extended positions, respectively. Figure 6(c) displays SEM photographs of air-gap samples with three different positions. The delay time increases by almost 10ps by shifting the air-gap up by 10 nm (i.e., shift-up position), compared to the air-gap with middle position. On the other hand, the delay time decreases by almost 5ps by extending the air-gap by 10nm (i.e., extended position). This is because the fringing fields between metal lines can be effectively reduced by the air-gap in the extended structure. Obviously, fringing field effect is more pronounced for the shift-up structures.

CONCLUSION

We have successfully developed an air-gap-filling method to effectively reduce the parasitic interconnect capacitance for deep sub-half-micron technologies. Ring oscillators with the air-gap structure indeed display the smallest delay time, compared with that using either conventional

HDP oxide or low-k HSQ. In addition, we have also derived fitting equations to calculate the oscillator delay time for different gap-fill materials and various air-gap sizes, thus provide a handy method for RC delay time estimation. We also found that the oscillator delay time is affected not only by the size, but also the position of air-gap. Optimum results in delay time reduction can be obtained when the air-gap is extended both above and below the metal lines, thus effectively cut down the fringing capacitance.

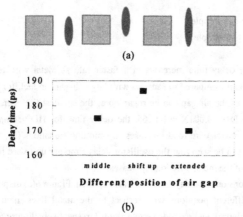

(a)

(b)

Figure. 6 (a) Schematics of air-gap structures with middle, shift up, and extended position. (b) Oscillator delay time of YY2 for air-gap samples with different positions.

REFERENCEES

1. C.Y. Chang, and S. M. Sze, "ULSI TECHNOLOGY", **pp. 371-376**, Mcgraw-Hill, 1996.

2. C. H. Hsieh, and C. C. Hsu, "Implant Treatment of Hydrogen Silsesquioxane Spin-on polymer as an Interlayer Dielectric Material for New SOG and W-Via Process", *VMIC Conference*, **pp. 583-587**, 1997.

3. B. Shieh, K. C. Saraswat, J. P. Mcvittie, S. List, S. Nag, M. Islamraja, and R. H. Havemann, "Air-Gap Formation During IMD Deposition to Lower Interconnect Capacitance", *IEEE Electron Device Lett.*, **pp. 16-18**, 1998.

4. M. T. Bohr, "Interconnect Scaling – The Real Limiter to High Performance ULSI", *Tech. Dig. IEDM*, **pp. 241-244**, 1995.

Mat. Res. Soc. Symp. Proc. Vol. 612 © 2000 Materials Research Society

Fabrication of Air-Gaps Between Cu Interconnects for Low Intralevel k.

Dhananjay M. Bhusari, Michael D. Wedlake, Paul A. Kohl, Carlye Case[1], Fred P. Klemens[1], John Miner[1], Byung-Chan Lee[2], Ronald J. Gutmann[2], J.J.Lee[3], Robert Shick[4] and L. Rhodes[4]

School of Chemical Engineering, Georgia Institute of Technology, Atlanta, GA.
[1]Lucent Technologies, Murray Hill, NJ.
[2]Rensselaer Polytechnic Institute, Troy, NY.
[3]Motorola Inc., Austin, TX.
[4]BF Goodrich Company, Brecksville, OH.

ABSTRACT

We present here a method for fabrication of air-gaps between Cu-interconnects to achieve low intralevel dielectric constant, using a sacrificial polymer as a 'place holder'. IC compatible metallization and CMP processes were used in a single damascene process. The air-gap occupies the entire intralevel volume between the copper lines with fully densified SiO_2 as the planer interlevel dielectric. The width of the air-gaps was 286 nm and the width of the copper lines was 650 nm. The effective intralevel dielectric constant was calculated to be 2.19. The thickness of the interlevel SiO_2 and copper lines were 1100 nm and 700 nm, respectively. Further reduction in the value of intralevel dielectric constant is possible by optimization of the geometry of the metal/air-gap structure, and by use of a low k interlevel dielectric material.
In this method of forming air-gaps, the layer of sacrificial polymer was spin-coated onto the substrate and formed into the desired pattern using an oxide or metal mask and reactive-ion-etching. The intralevel Cu trench is then inlaid using a damascene process. After the CMP of copper, interlevel SiO_2 is deposited by plasma-CVD. Finally, the polymer place-holder is thermally decomposed with the decomposition products permeating through the interlevel dielectric material. The major advantages of this method over other reported methods of formation of air-gaps are excellent control over the geometry of the air-gaps; no protrusion of air-gaps into the interlevel dielectric; no deposition of SiO_2 over the side-walls, and no degradation of the interlevel dielectric during the formation of air-gap.

INTRODUCTION

The need for developing new low dielectric constant (low-k) materials as intralevel dielectrics in the integrated circuits (ICs) arises from the continuing shrinkage in the size of the features in the ICs. The increased proximity of the metallic interconnects introduces propagation delays, crosstalk noise, and increases power dissipation as a result of higher resistance-capacitance coupling[1]. The National Technology Roadmap for Semiconductor, therefore, calls for qualification/preproduction of materials with k = 2.5 - 2.0 in 2000; and k = 1.5 – 2.0 in 2003[2]. As a consequence, several inorganic as well as polymeric materials are being developed to achieve low k [3-5]. However, since these materials are required to satisfy several stringent requirements such as low moisture absorption, excellent adhesion to a variety of materials, high Tg, thermal stability, good mechanical properties and also be able to be integrated in the IC manufacturing process, many materials have shortfalls. In addition, there are in-use requirements such as reliability and manufacturability. For low k applications, there are very few fully-densified materials which have a dielectric constant of < 2.0. None of these materials are acceptable for use. One alternate approach to achieving low-k is to use air or vacuum as the dielectric material. The intra-level (between Cu lines) is the most important location for the low k material. Preliminary work [6-8] has demonstrated the feasibility of the creation of such 'air-gaps' between the metallic conductor lines and indeed a lowering of the intra-level k.

Two distinct approaches have been reported in the literature for creation of air-gaps. Previous studies [7,9] use PECVD growth of SiO_2 over closely spaced metal lines (typically 0.3 μm) to create air-gaps between the lines. The gaps are created due to the anisotropy of the SiO_2 growth at the bottom of the trenches and on top of the metal lines, with the air-gaps extending above the copper lines. This approach, although straight forward to implement, the shapes and sizes of the air-gaps are limited and there is SiO_2 in the intralevel region. Another approach reported by us previously [6] involves use of a sacrificial polymer, which can be patterned to any desired geometry by the conventional photolithography and etching techniques, and encapsulated by a suitable material. This polymer, upon decomposition at elevated temperatures, produces gaseous products that can diffuse through the encapsulating material, thus creating air-gap in the entire volume previously occupied by the polymer. Thus, the exact size and shape of the air-gap can be controlled. However, this approach also suffers from a few drawbacks such as possible limitations on the choice of encapsulation material (so as to allow diffusion of the gaseous decomposition products at moderately high temperatures), and a small amount of residue left after the polymer decomposition. Nevertheless, if these disadvantages can be overcome or mitigated, this method yields high quality air-gaps with excellent geometry control. A similar approach of using sacrificial material for making air-gaps has also been used by Anand et. al [8] who used amorphous carbon as a sacrificial material, which was etched by oxygen at high temperature. However, high temperature oxygen environment may be undesirable.

EXPERIMENTAL

The sacrificial polymer used in this study was a copolymer of butylnorbornene and triethoxy-silyl norbornene[10] (Unity Sacrificial Polymer[TM], BFGoodrich, Cleveland, OH), dissolved as a 12 wt.% solution in mesitylene. The polymer solution was spin coated on Si wafers at 3000 rpm for 30 sec, which yielded a 1.3 μm thick layer. The film was then soft baked at 120 ^{0}C for 3 min. A 500 nm thick layer of SiO_2 was then deposited by RF PECVD at 200 ^{0}C, to serve as a hard mask for subsequent patterning of the polymer film. The gap-fill and planarization test structures were then printed onto this SiO_2 layer. The photoresist pattern was transferred to the SiO_2 layer by reactive ion etching. The exposed polymer was then etched away by reactive ion etching[11], and the SiO_2 mask was removed by dissolving in the buffered oxide etch solution. Blanket layers of tantalum liner and copper were then deposited onto this patterned polymer by DC sputtering. After copper deposition, the structure underwent chemical-mechanical planarization (CMP) using slight modifications to conventional damascene patterning of copper. The CMP was performed using an IPEC-Planar (now Speedfam-IPEC) 372M on Rodel IC1400 pad. The downward pressure of the polish arm was 2.5 psi with 1.5 psi back pressure; the speed of the platen and carrier was 30 rpm; the slurry flow rate was 200 mL/min. A diluted (1:1 by vol.) Rodel metal slurry containing alumina particles and a proprietary liquid, B00045, was used to remove copper; while a diluted (1:1 ratio) Cabot colloidal silica slurry, SS25, was used to remove the Ta liner. A Ta liner is not usually removed with the KOH-based SS25 slurry in conventional damascene structure with oxide, since KOH-based slurry has high removal rate for oxide[12]. However, the removal rate of Unity with this slurry was very low, < 30 nm/min, as measured with blanket Unity films. Care was particularly taken to remove all metal from the top of the Unity polymer with minimal Unity erosion. After CMP, the wafers were cleaned by OnTrak DSS-200 double side brush scrubber with PVA brushes. After CMP of copper, a 1.0 µm thick layer of SiO_2 was deposited by PECVD at 200C to form the overcoat dielectric. The air-gaps were achieved by heat treatment at 425 ^{0}C for 2 hrs in N_2 purged furnace.

The effective dielectric constant of the composite structure was calculated using Maxwell 2D Field Simulator (Ansoft Corp., Pittsburgh, PA). The dielectric constant could not be directly measured because the structures were built on silicon.

RESULTS AND DISCUSSION

The thermal decomposition characteristics of the sacrificial polymer are of paramount importance to successful fabrication of air-gaps, and were hence studied by thermogravimetric analysis. A TGA spectrum of Unity^{TM} with 10% TES groups, recorded at the ramp rate of 3 C/min, is presented in Fig.1. It can be noted from this plot that this polymer is stable and exhibits no significant weight loss up to about 370 ^{0}C, while a sharp decomposition occurs at 425 ^{0}C. Both these properties are desirable, since the stability at temperatures below 370 C allows several process steps to be performed after the polymer coating, such as PECVD coating of the masking and encapsulant materials etc., without causing any degradation of the polymer. The sharp transition to gaseous products at higher temperatures minimizes the temperature window required for decomposition.

A typical copper/air-gap structure produced is shown in Fig. 2. The copper lines in this figure are 0.65 µm wide and 0.73 µm high. The width of the air-gap between copper lines is 0.28 µm. The copper/SiO_2 overcoat layer can be seen to possess very good planarity. The interlevel SiO_2 layer was permeable to the gaseous products formed at elevated temperatures and also withstood the pressure developed inside the cavity. The shape of the sidewalls of the copper lines was verticle and reflect the high anisotropy of reactive ion etching of Unity, which is advantageous for creation of high aspect ratio structures. Fig. 3 shows a cross section of another test structure, wherein 0.8 µm wide copper lines are separated by a 0.7 µm wide air-gaps and support much wider air-gap (> 12 µm) on both sides. These figures clearly show the control that

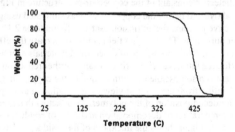

Figure 1: *Thermogravimetric spectrum of Unity^{TM} recorded at the ramp rate of 3 ^{0}C/min.*

the present method offers over the geometry of air-gap structures. In fact, air-gaps of as large as 10 μm in height, 50 μm in width and several millimeters in length have been found to be possible without any structural distortion of the overcoat material.

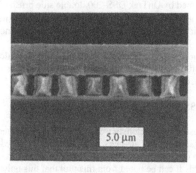

Figure 2: *Scanning electron micrograph of the cross section of copper/air-gap structure. The width of the Cu lines is 0.65 μm and that of air-gaps is 0.28μm. The overcoat is 1.1 μm thick SiO₂.*

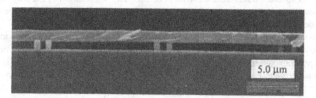

Figure 3: *0.8 μm wide Cu lines separated by 0.7 μm air-gap supporting much wider (12 μm) air-gaps on their sides. The overcoat is 1.1 μm thick SiO₂.*

The effective intralevel dielectric constant of the copper/air-gap structure in Fig. 2, considering 1.1 μm of SiO₂ ($k = 4.1$) above and below the copper layer, was simulated using Maxwell 2D Field Simulator. The effective in-plane dielectric constant was found to be 2.19, which is a significant reduction over pure SiO₂. The fringing fields in SiO₂ increase the effective dielectric constant above the value of that of air. Thus, use of other low k materials as interlevel dielectric will further reduce the effective intralevel dielectric constant. Further reduction in the value of intralevel dielectric constant is also possible by optimization of the copper/air-gap geometry.

An important consideration from the point of view of application of air-gaps in ICs is the amount of residue left after decomposition of the polymer. Two factors that greatly influence the amount of residue are the temperature of decomposition and level of residual oxygen in the furnace during decomposition. Fig. 4 shows the thickness of the residue, as measured by XPS profiling, after decomposition of a 7.5 μm thick film of PNB as a function of decomposition temperature as well as residual oxygen. The rate of Ar-ion sputtering of the residue was calibrated by measuring the thickness of the residue of a patterned PNB film by atomic force

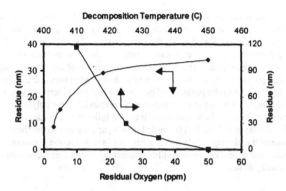

Figure 4: *Thickness of the residue after decomposition of 7.5 μm thick Unity^{TM} film as a function of residual oxygen in the furnace and decomposition temperature.*

microscopy. The PNB films were coated on gold for the XPS studies in order to measure the residue thickness. It can be seen that the amount of residue is less at lower residual oxygen content in the furnace and also when the decomposition temperature is increased. It may be noted that the values plotted in Fig. 4 are for a 7.5 μm thick polymer film, while those in actual practice for typically < 1.0 μm thick polymer will be still lower.

Another important concern is the possible contamination of the interlevel dielectric due entrapment of the decomposition products during diffusion. Such contamination might lower the electronic quality of the interlevel dielectric. In order to check this possibility, carbon content of the glass overcoat layer was studied by secondary ion mass spectroscopy. The measured carbon content before and after decomposition of the polymer are compared in Fig. 5. The conts/sec

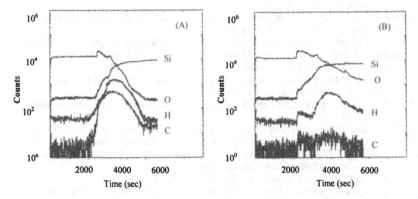

Figure 5: *Secondary ion mass spectra recorded on (A) encapsulated polymer and (B) air-gap formed under the encapsulation.*

for O, Si, H and C are shown as a function of sputtering time. Fig. 5(a) presents the data for SiO_2 without an air-gap (no polymer was decomposed), and Fig. 5(b) shows the mass spectrum for SiO_2 with an air-gap. The initial flat portion of the curves in both Fig. 5(a) and 5(b) (up to 2000 sec) correspond to the overcoat SiO_2 layer. It can be seen that the concentration of C is quite low in this region in both the spectra, suggesting that the contamination of the overcoat layer with C is quite low even after polymer decomposition. The large and sudden increase in C counts in Fig. 5(a) after the SiO_2 layer is due to the presence of polymer; while such increase is very low in Fig. 5(b) and corresponds presumably to the small amount of residue left after polymer decomposition. The rapid increase in Si and initial increase followed by gradual reduction in O counts after the overcoat layer in Fig. 5(b) (recorded on air-gap) are due to the thermally grown SiO_2 layer at the base of the air-gap, while those in Fig. 5(a) are due to the presence of Si and O in the polymer. Thus there appears to be negligible entrapment of the decomposition products in the overcoat layer during diffusion.

CONCLUSIONS

We have presented here a method of forming air-gaps by using a sacrificial polymer as a 'place holder' and demonstrated the feasibility of fabrication of metal-air-gap structures at very low dimensions, of the order of few hundreds of nanometers, using IC compatible metallization and CMP processes. The major advantages of this method over the all-PECVD-method are excellent control over the geometry of the air-gaps, lack of oxide on the sidewalls of the metal and on the bottom of the air-cavities, and lack of protrusion of air-gaps into the overcoat layer. It has been shown that the intralevel dielectric constant can be reduced quite significantly by incorporation of such air-gaps between the metal interconnect lines in integrated circuits.

REFERENCES

1. J.D.Meindl, *Proc. IEEE*, **83**, 619 (1995).
2. *The National Technology Roadmap for Semiconductors*, Semiconductor Industry Association, San Jose, CA, 1997.
3. N.R.Grove, P.A.Kohl, S.A.Bidstrup-Allen, R.A.Shick, B.L.Goodall and S.Jayaraman, *Mater. Res. Soc. Symp. Proc.* **476**, 3 (1997).
4. P.H.Townsend, S.J.Martin, J.Godschalx, D.R.Romer, D.W.Smith Jr., D.Castillo, R.DeVries, G.Buske, N.Rondan, S.Froelicher, J.Marshall, E.O.Shaffer and J-H.Im, *ibid*, **9** (1997).
5. J.A.Theil, F.Mertz, M.Yairi, K.Seaward, G.Ray and G.Kooi, *Ibid*, **31** (1997).
6. P.A.Kohl, Q.Zhao, K.Patel, D.Schmidt, S.A.Bidstrup-Allen R.Shick and S.Jayaraman, Electrochem. and Solid State Lett. **1**, 49 (1998).
7. B.Shieh, K.C.Saraswat, J.McVittie, S.List, S.Nag and M.Islamraja, IEEE Electron Device Letters **19**, 16 (1998).
8. M.B.Anand, M.Yamada and H.Shibata, IEEE Trans. Electron Devices **44**, 1965 (1997).
9. B.Shieh, K.Saraswat, M.Deal and J.McVittie, Solid State Technol.(February), 51 (1997).
10. R.A.Shick, B.L.Goodall, L.H.McIntosh, S.Jayaraman, P.A.Kohl, S.A.Bidstrup-Allen and N.R.Grove, *Proc. IEEE Multichip Module Conf.*, 182 (1996).
11. Q.Zhao and P.A.Kohl, J. Electrochem Soc. **145**, 1257 (1998).
12. J.M.Steigerwald, S.P.Murarka and R.J.Gutmann, *Chemical Mechanical Planarization of Microelectronic Materials* (John Wiley & Sons Inc.), p.150 (1997).

Poster Session
Low-k Dielectrics

Mat. Res. Soc.Symp. Proc. Vol. 612 © 2000 Materials Research Society

PROCESSING, PROPERTIES, AND CMP CHARACTERISTICS OF A SPIN-ON POLYMER: HSQ

Wei-Jung Lin, Chang-Jong Yang, Wen-Chang Chen*
Department of Chemical Engineering, National Taiwan University, Taipei, Taiwan

ABSTRACT

The structures and properties of the HSQ film during the cage/network transformation were studied by curing at 300^0C for 1 hour. The experimental results show that the ratio of the network/cage structure of the cured HSQ film increases from 0.21 to 0.39 by curing. The porosity of the cured film increases from 10.0% to 12.3 %, while the refractive index decreases from 1.413 to 1.376 during curing. These results suggest that the dependence of the structure and properties of the HSQ film by curing. The CMP characteristics of HSQ were studied by using different kinds of slurries and surfactant. The CMP results of polishing HSQ suggest that the hardness and charge status of the abrasive, the interaction of the surfactant with the abrasive and film surface significantly affect the polishing results.

INTRODUCTION

Research interest in low dielectric constant materials for deep sub-micro IC devices remains very high. Poly(silsesquioxane) such as hydrogen silsesquioxane (HSQ) has been recognized early as a potential candidate as a low dielectric constant material. It has a cage structure before curing and part of the cage structure is transformed to a network structure. Curing studies of HSQ have suggested that there are basically four possible stages during curing: (1) room temperature up to 200^0C: solvent loss; (2) 250^0C-350^0C: cage-network redistribution; (3) 350^0C-435^0C (or 450^0C): Si-H thermal dissociation and network redistribution; (4) > 435^0C(or 450^0C): collapse of porous network[1-4]. However, most of current studies focus on the region of 350^0C-435^0C. Quantitative characterization on the transformation of the cage structure to the network structure has not been fully explored. In this study, the structures and properties of the HSQ film cured at 300 ^{0}C were investigated with different curing time, including network/cage ratio, refractive index, and porosity. Chemical-mechanical polishing (CMP) is one of the major technologies for developing multi-level IC structures consisting of metal/low k material. Here, a summary on the CMP characteristics of the HSQ film polished with different slurry formulations is reported.

EXPERIMENTAL

HSQ (Fox-15) was obtained from Dow Corning, Inc.   The obtained liquid was spun coated on a four inch wafer at 3000 rpm for 20 seconds.   The coated film was then cut into 13 pieces and baked at 300^0C with a different baking time. The refractive index and the film thickness of the cured films were then measured using a Nanospec 210XP thickness measurement system. The chemical structure of the cured films was characterized by the

Fourier Transform Infrared (FTIR) spectrophotometer using a Bio-Rad QS300 FTIR. The ratio of the network structure to the cage structure was determined by the ratio of the peak area of the 1130 cm^{-1} peak to the 1070 cm^{-1} peak. Assuming the IR peak shape is a Gaussian-type and then the Bio-Rad built-in program de-convolutes the overlapping peaks based on the algorithm of nonlinear least squares fitting known as the Levenberg-Marquardt method. After the calculation, the information about peak area, cen ter height, width and position of each IR peak was obtained and thus the ratio of the network/cage structure was determined.   The porosity of the cured HSQ film was determined by the ellipsometric method. The CMP experiment was carried out on a Westech Model 372M CMP Processor by polishing the HSQ film on a six inch wafer. The details on the film preparation and the CMP experiments have been described in recent publications [5,6].Three different kinds of slurries based on various abrasives were used for the experiment, including Nissan Chemicals:A-1 slurry(ZrO_2, PH=4.6), SS-25(SiO_2, PH10.9-11.20), and WA 400(Al_2O_3, PH=4.0-4.5). The surfactant, Triton X-100 was purchased from Aldrich Co.

RESULTS AND DISCUSSION

Figure 1 shows the FTIR absorption peaks at 1070 and 1130 cm^{-1} of the HSQ films cured at 300^{0}C at different curing times. The peak intensity at 1070 cm^{-1} increases with increasing curing time while the peak at 1130 cm^{-1} shows a reverse trend. Note that the peaks at 1070 and 1130 cm^{-1} represent the network and cage structures of the HSQ film, respectively[1-4]. Figure 2 exhibits the variation of the ratio of network/cage with curing time. It increases from 0.21 to 0.39 for one hour curing at 300^{0}C. Hence, these results suggest that the cage structure of the HSQ film gradually transforms to the network structure during curing. Such kind of structural transformation results in the variation of film properties. Figure 3 shows that variation of film porosity with curing time. The porosity increases from 10% to 12.3% by curing and this causes the volume increment of the HSQ film. Figure 4 shows the variation of the film refractive index with curing time. It decreases from 1.413 to 1.376 during curing. Since there is no significant composition change for curin g at 300^{0}C, the volume increment results in the decrease of the refractive index after curing. From the above results, it is clear that the cage-network transformation results in the volume and refractive index change. Thus, it is possible to control the film properties by curing temperature and curing time.

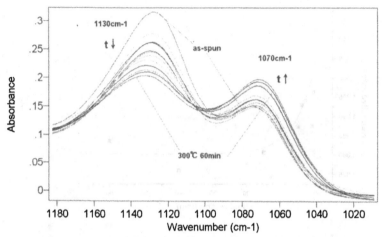

Figure 1. *FTIR spectrum of the HSQ thin film baked at 300 • with different curing time from as-spun to 60 min*

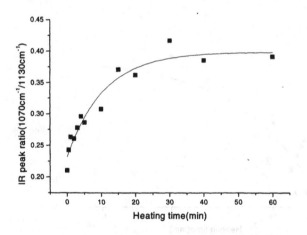

Figure 2. *FTIR peak area ratio(1070 cm^{-1}/ 1130cm^{-1}) of the HSQ thin film baked at 300• as a function of curing time*

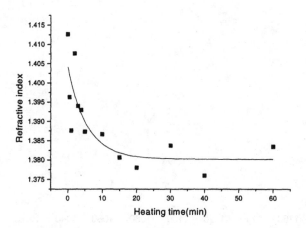

Figure 3. *The refractive index of the HSQ thin film baked at 300 • with different heating time from as-spun to 60 min.*

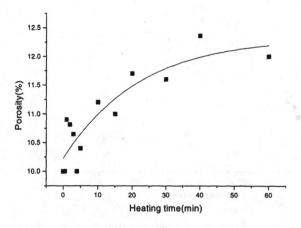

Figure 4. *The Porosity of the HSQ thin film baked at 300• with different heating time from as-spun to 60 min.*

The CMP characteristics of the HSQ film was performed on the different abrasive based slurries and a surfactant by our laboratories[5,6].   Here, summary on the CMP characteristics of the polished HSQ film is described as below. First, the order of the removal rate for different abrasives was A-1 (ZrO_2) > SS-25 (SiO_2) > WA 400 (Al_2O_3). It can be explained from the abrasive hardness and the electrostatic interaction between the abrasive and film surface. The comparison between the slurry PH value and the abrasive isoelectric point (IEP) makes the electrostatic attraction force existed between the ZrO_2 abrasive and the HSQ film. However, the repulsive force is existed for the SiO_2 abrasive. Hence, the removal rate by the A-1 slurry is larger than the SS-25 slurry. The low hardness of the Al_2O_3 abrasives in the WA 400 slurry results in the lowest polishing rate among the studied slurries. Secondly, the removal rate of the polished HSQ film increases as increasing the solid content of the three abrasives. This suggests that the mechanical force plays an important role for polishing the HSQ film Thirdly, the removal rate of the polished HSQ film is significantly reduced by adding the Triton X-100 into the slurries. The OH group of Triton X-100 probably interacts with the silanol group of the Si-O-Si backbone of the HSQ surface by forming hydrogen bonding, inhibiting the abrasive interaction with the silanol group of the film surface. Hence, the removal rate of polishing HSQ film is reduced by adding the surfactant Triton X-100.   Although the previous study showed that the removal rate of the HSQ reached 2000-4000 Å/min by varying different slurry chemistry, the scratches on the polished HSQ are still the major drawbacks. The addition of the lubricant, glycerol, might be a solution for avoiding scratching [7].

CONCLUSIONS

The structural transformation and properties of the cured HSQ film at 300^0C for a different curing time are analyzed.   The ratio of the network /cage structure and the porosity of the cured HSQ film increase with increasing curing time.   However, the variation of the refractive index with curing time shows a reverse trend. These results suggest that the structures and properties of the HSQ film are significantly modified by curing.   The CMP results of polishing HSQ film suggest that the hardness and charge status of the abrasive, the interaction of the surfactant with the abrasive and film surface significantly affect the polishing results.

ACKNOWLEDGENTS

The authors thank the National Science Council of Taiwan for financial support of this work and the Dow Corning Co. for providing the Fox-15 samples.

REFERENCES
1. J. N. Bremmer, Y. Liu, K. G. Gruszynski, and F. C. Dall, *MRS. Res. Soc. Symp. Proc.*, **476**, 37 (1997).
2. M. J. LoLoboda, C. M. Grove, and R. F. Schneider, *J. Electrochem. Soc.*, **145**, 2861(1998).
3. M. G. Albrecht and C. Blanchette, *J. Electrochem. Soc.*, **145**, 4019(1998).
4. Y. K. Siew, G. Sarkar, X. Hu, A. See, and C. T. Chua, *J. Electrochem. Soc.*, **147**, 335(2000).
5. W. C. Chen, S. C. Lin, B. T. Dai, and M. S. Tsai, *J. Electrochem. Soc.*, **146**, 3004(1999).
6. W. C. Chen and C. T. Yen, *J. Polym. Res.*, **6**, 197(1999).
7. S. P. Murarka, *MRS. Res. Soc. Symp. Proc.*, **511**, 277(1998).

Mat. Res. Soc. Symp. Proc. Vol. 612 © 2000 Materials Research Society

MATERIAL PROPERTIES OF A SiOC
LOW DIELECTRIC CONSTANT FILM
WITH EXTENDIBILITY TO k < 2.7

Eugene S. Lopata[1], Lydia Young[2], and John T. Felts[3]

[1]Silicon Valley Group/Thermal Systems, Scotts Valley, CA 95066
[2] now at KLA-Tencor/Viper Division, San Jose, CA 95134
[3]Nano Scale Surface Systems, Inc., Alameda, CA 94501

ABSTRACT

A plasma deposited SiOC very low k (VLK) interlayer dielectric (ILD) film has been developed which can be tuned to $2.5 \leq k \leq 3.0$, demonstrates very good thermal stability, excellent adhesion properties, acceptable hardness, and an indication that it may be extendible to k < 2.5. This paper will disclose properties of this SiOC film which are important to a VLK ILD application.

INTRODUCTION

One of the great material challenges for the semiconductor industry in recent years has been the search for very low dielectric-constant (VLK) materials to replace silicon dioxide as an intrametal and intermetal dielectric for back-end-of-the-line applications.[1,2] While many candidates have been considered as low-k materials, each presents significant trade-offs and none have emerged as a universal solution.

In addition to having the lowest dielectric constant possible, VLK candidate materials must provide adequate performance for basic properties associated with thermal stability, chemical stability and adhesion. An attractive film would ideally have material and processing properties (with the exception of dielectric constant) so similar to silicon dioxide, that the candidate can be dropped into current processes and integration schemes with only minor modifications. Short of this ideal, the trade-offs to performance, material properties, ease or cost of integration, reliability, and the maturity of the technology needs to be considered.

The above requirements for VLK thin films for future generations of semiconductor devices provides an excellent platform to use some of the plasma polymerization process technology used since the early 70's and 80's in other industries to make silicon based thin-films that have organic polymer-like content- applications including optics, magnetic disks, plumbing fixtures and plastic food packaging. Such an approach offers the promise of *tailoring* hybrid materials to provide the best properties from silicon dioxide (thermal stability, hardness, processability, etc...) with the benefits of organic functionality (low dielectric properties)[3].

Experimentation reveals that simply revisiting the prior work does not produce thin-film materials that meet the goals of today's semiconductor environment. Instead, approaches incorporating the basic philosophy of building highly crosslinked hybrid materials through plasma deposition had to be redesigned for VLK applications. By selecting the proper precursor(s), plasma conditions and equipment configurations, thin films that incorporated silicon, oxygen, carbon and hydrogen were deposited. The SiOC(H) thin film materials exhibited some of the properties of silicon dioxide while providing the dielectric property of a polymeric structure. The material properties from the SiOC films deposited from our proprietary approach will be presented herein.

RESULTS

Composition of Film

The composition of the as-deposited SiOC film was determined by FTIR spectroscopy and RBS analysis. As can be seen in Figure 1, the FTIR spectra show absorption characteristic of Si-O bonds ($\sim$1030 cm^{-1}), Si-CH$_3$ bonding (2969 cm^{-1} stretching, 1264 cm^{-1} deformation, and 800 cm^{-1} rocking mode), as well as Si-H groups (2200 cm^{-1}).[4] Note the absence of significant -OH polar groups in the as-deposited film($\sim$3600 cm^{-1}).

Figure 1   FTIR Spectra of SiOC(H) Films

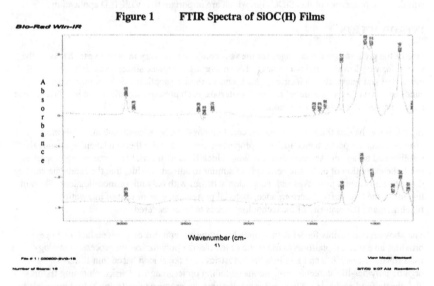

Table 1 gives the results of RBS/HFS analysis of an as-deposited film. The elemental analysis is consistent with the composition of the film deduced from FTIR spectroscopy and demonstrates that this truly is an SiOC(H) film. **A range of film compositions is achievable, even for a given organosilicone precursor, dependent upon the specifics of the plasma used for the deposition. As expected, the range in composition is reflected in a range of film properties, including the dielectric constant.** Table 2 reflects the range of compositions as revealed by ESCA analysis observed during this investigation.

TABLE 1

RBS/HFS ANALYSIS
SVG/THERMAL SYSTEMS VLK FILM

H	C	O	Si	C/Si	O/Si
42.0 %	18.0 %	24.0 %	16.0 %	1.12	1.50

TABLE 2

ESCA RESULTS
SVG/THERMAL SYSTEMS SiOC FILMS
COMPOSITIONAL RANGE

C	O	Si	C/Si	O/Si
40 – 50 %	27 – 33 %	20 – 27 %	1.1 – 2.4	1.0 – 1.5

Electrical Measurements

The dielectric constant of the as-deposited film was determined in-house by evaporating Al dots, then performing C-V measurements @ 1 Mhz. Thermal oxide wafers were used as controls, and comparable results were obtained for samples submitted to Sematech. 4 dots/wafer were measured, and the average dielectric constant for a cassette of 25 wafers deposited in a demonstration tool is given in Table 3.

Most of the characterization of the SVG/Thermal Systems SiOC VLK film was done for a k = 3.0 POR. Except for the electrical properties, it is expected that the other attributes of the films with 2.5 ≤ k ≤ 3.0 will be comparable to those reported below.

Dissipation was measured @ 1 Mhz for 12 dots, leakage current was measured @ 1×10^5 V/cm, and breakdown strength exceeded 1.6 MV/cm (constrained by test system).

TABLE 3

ELECTRICAL MEASUREMENTS
SVG/THERMAL SYSTEMS VLK FILM
(3.0 POR)

k (@ 1 Mhz)	Dissipation (@ 1 MHz)	Leakage Current (@ 1×10^5 V/cm)	Breakdown Strength
3.0	0.0186	$< 1 \times 10^{-9}$ A/cm^2	> 1.6 MV/cm

D5.3.3

Film Stability

Stability under thermal cycling and moisture gain after exposure to ambient conditions were investigated. Thermal testing consisted of TGA analysis under N_2 with a ramp rate of 10 °C/minute from ambient to 400 °C, isothermal weight loss at that temperature for 4 hours, followed by cool down to ambient temperature. *Total* weight loss was reported for the entire thermal cycle. Results are given in Table 4. Long term thermal stability was evaluated by isothermal weight loss measurements by TGA for 8 hours. As can be seen in Table 4, the weight loss asymptotically declines over time, so that the average rate of weight loss over 8 hours is less than over 4 hours.

During development of the k = 3.0 process, film stability was gauged also by measuring k before/after annealing @ 400 °C for 1 hour. As indicated in Table 4, the dielectric constant of this film decreased by ~9% after annealing – the decrease was independent of whether vacuum or N_2 anneals were performed.

In addition, SiOC film stability was gauged by measuring film thickness on a Rudolph spectroscopic ellipsometer before/after annealing @ 400 °C for 1 hour. A 2.1% *increase* in thickness was observed after anneal averaged over 6 wafers (4 locations/wafer).

Moisture stability was evaluated by exposing film scrapped off a wafer to 85% R.H. @ 85 ° C for 24 hours. Gas chromatography/mass spectrometry revealed 0.3% weight gain due to moisture. **No change in the FTIR spectrum was observed after 14 days ambient exposure, indicating low affinity for moisture.**

TABLE 4

STABILITY OF
SVG/THERMAL SYSTEMS VLK FILM
(3.0 POR)

Avg. Wt. Loss (4hrs @ 400 °C)	Avg. Wt. Loss (8 hrs @ 400 °C)	Thickness Change (1 hr @ 400 °C)	k @ 1 Mhz Before/After Anneal (400 ° C, 1 hr)	Wt. gain after 24 hrs @ 85 % R.H. & 85 °C
0.7%/hr	0.5%/hr	+2.1%	3.2/2.9	0.3%

Integration Indicators

Although a formal integration effort has not been made for these SiOC films, a number of performance indicators have been investigated. Adhesion of the film was explored using a simple tape test before/after annealing @ 425 °C for 1 hour. Results are summarized in Table 5. In addition to the adhesion data presented therein, a 5 layer stack of SiOC film (1 μ thick) on SiN (0.1 μ PVD SiN) was found to pass the tape test as deposited.

TABLE 5

ADHESION OF
SVG/THERMAL SYSTEMS VLK FILM
(3.0 POR)

Substrate	Si	SiO$_2$ (th. ox.)	SiN (PVD)	SiN over SiOC	TiN	Cu	Al
As Deposited	pass	pass	pass	pass	pass	pass	pass
Annealed 425 °C, 1 hr	pass	pass	pass	pass	*	*	*

*not tested for adhesion

Other integration parameters investigated include Young's modulus, hardness, stress before/after anneal, and thickness crack limit. Data are presented in Table 6.

TABLE 6

INTEGRATION INDICATORS
OF SVG/THERMAL SYSTEMS VLK FILM
(3.0 POR)

Young's Modulus	Hardness	Stress, As Deposited	Stress After 400 °C, 4 hrs	Thickness Crack Limit
8.7 +/- 0.1 Gpa	1.7 +/- 0.01 Gpa	-5.5 Mpa	+148.5 Mpa	> 1.5 μ on SiO$_2$

Extendibility to Lower Dielectric Constant

As indicated above, careful selection of the proper plasma configuration and process window permits deposition of films with different SiO$_x$C$_y$ compositions, which results in improved electrical performance for VLK applications. To date, films with **k = 2.5 (as deposited)** have been achieved. Some of the relevant performance characteristics of this SiOC(H) film are given in Table 7. While we have not had time to characterize these k = 2.5 films for film stability and integration factors (as we have reported for the k = 3.0 POR) we do expect those properties to be comparable to the latter.

TABLE 7

PERFORMANCE OF
SVG/THERMAL SYSTEMS VLK SiOC FILM
(k = 2.5 process)

k (@ 1 Mhz, as deposited)	k @ 1 MHz Before/After Anneal (400 °C, 1 hr)	Thickness Change (4 hr @ 400 °C)	Stress Before/After Anneal (400 °C, 4 hr)
2.5	2.5/2.5	+2.4%	-9 Mpa/+7 MPa

SUMMARY

By tuning the process with a given precursor, SiOC(H) films can be deposited with a range of compositions and with $2.5 \leq k \leq 3.0^+$. The material properties of these SiOC films make them strong candidates for VLK IMD applications having $k \geq 2.5$, with possible extendibility to $k \leq 2.5$. To date, one U.S. patent has been issued, a second patent has had claims allowed, and a third patent is pending on the technology involved in depositing these films.

ACKNOWLEDGMENTS

The authors acknowledge valuable technical discussions and assistance from Peter Rose and Dr. Josh Golden. Dr. Sanjeev Jain, Eiji Hayashi, and Sydney Fernandez assisted with the k = 3.0 POR development. Rick Matthiesen is acknowledged for vacuum chamber design.

REFERENCES

1. J. A. Thiel, J.Vac. Sci. Technol. B 17 (6), 2397 (1999).

2. W. Chen and C. Yen, J. Vac. Sci. Technol. B 18 (1), 201 (2000).

3. A. Grill, L. Perraud, V. Patel, C. Jahnes, and S. Cohen in Low-Dielectric Constant Materials V, edited by J. Hummel, K. Endo, W. Lee, M. Mills and S. Wang (Mater. Res. Soc. Symp. Proc. 565,

Warrendale, PA, 1999) pp. 107-116.

4. P. G. Pai, S. S. Chao, Y. Takagi and G. Lucovsky, J. Vac. Sci. Technol. A **4** (3), 689 (1986).

Mat. Res. Soc. Symp. Proc. Vol. 612 © 2000 Materials Research Society

EFFECTS OF CURING TEMPERATURE ON THE MECHANICAL
RELIABILITY OF LOW DIELECTRIC-CONSTANT SPIN-ON-GLASSES

Yvete Toivola, Robert F. Cook and Chandan Saha[1]
Department of Chemical Engineering and Materials Science
University of Minnesota
421 Washington Avenue SE, Minneapolis, MN 55455
[1]Dow Corning Corporation
2200 W. Salzburg Road, Midland, MI 48686

ABSTRACT

The variations in the mechanical properties of a commercial low-k silsesquioxane material with curing temperature are examined, focusing on the transition from the low modulus, high stress, under-cured state to the high modulus, low stress, over-cured state. Film modulus and hardness are determined by instrumented nanoindentation and film dielectric constant is determined by ac capacitance measurements of metal dot structures. The mechanical behavior is correlated with changes in molecular structure via infrared spectroscopy. An implication of the results is that there is an intermediate curing temperature for optimum silsesquioxane interconnection performance.

INTRODUCTION

Spin-on glass films, formed by the polymerization of silsesquioxane (SSQ) oligomers [1], have great potential as semiconductor interconnection materials due to their low dielectric-constants $k = (2.5 - 3.3)$, tunable properties, compatibility with silica chemistry and extendibility to even lower dielectric constant [2] via increased porosity. The mechanical properties of SSQ materials, however, are inferior to those of silica, particularly the resistance to moisture-assisted, residual-stress driven stress-corrosion cracking, leading to interconnection yield and reliability concerns [3, 4]. In addition, the underlying mechanical properties controlling cracking – modulus, hardness, toughness and film stress – are extremely sensitive to the time, temperature and environment used during the polymerizing curing process [3-5]. Here we extend our previous work [3-6] characterizing the *processing*-properties linkage in SSQ materials to the *structure*-properties linkage. Infrared spectroscopy will be used to characterize the development of the polymerized three-dimensional -Si-O-Si- bridged network from the initial oligomeric non-bridging -Si-H terminated structure.

EXPERIMENTAL

Commercial low-k Hydrogen Silsesquioxane (HSSQ) solutions (FOx, Dow Corning Corporation, Midland, MI) were spin coated onto 150 mm diameter Si wafers to form films 0.5 and 1 μm thick. The films were cured for 1 hour in a N_2 environment at temperatures of 375, 400, 425 and 450 °C to yield bulk polymerization of the precursor oligomers and hard "spin-on-glass" films. Dielectric constants of the films increased with curing temperature over the range 2.5-3.1, similar to previous observations [3, 4].

RESULTS

Infrared Spectroscopy

Infrared spectroscopy was used to track chemical changes as a function of curing temperature. On curing, HSSQ films polymerize via a chemical reaction in which networking occurs via Si-O bridging. Quantitative information about changes in chemistry as a function of curing temperature was obtained via infrared absorbance peaks associated with Si-O and Si-H.

Figure 1a shows infrared spectra of the Si-H stretch at 2250 cm^{-1} and the Si-H$_2$ stretch at 2200 cm^{-1} of HSSQ films cured at four temperatures, with the highest curing temperature spectrum in bold. Figure 1b shows infrared spectra of the high energy Si-O stretch associated with the cage structure at 1140 cm^{-1} and the low energy Si-O stretch of the network at 1070 cm^{-1}.

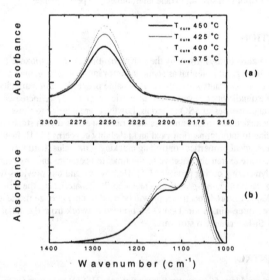

Figure 1. (a) A decrease in peak intensity at 2250 cm^{-1} corresponding to the Si-H stretch and an increase in peak intensity at 2200 cm^{-1} corresponding to the Si-H$_2$ stretch as curing temperature increases. (b) A decrease in peak intensity at 1140 cm^{-1} corresponding to a high energy Si-O stretch and an increase in peak intensity for the lower energy Si-O stretch at 1070 cm^{-1} as curing temperature increases.

A ratio of the baseline-corrected integrated peak area of the Si-H stretch at 2250 cm^{-1} and Si-O stretch at 1070 cm^{-1} was determined to quantify the amount of network formation on curing. A plot of this ratio as a function of curing temperature is shown in figure 2.

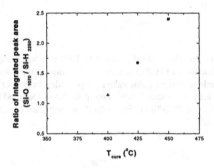

Figure 2. *A plot of the of integrated absorbance area ratio Si-O/Si-H as a function of curing temperature for HSSQ. This ratio is used to quantify the amount of network formation on curing.*

Figure 2 shows explicitly the relationship between curing temperature and the ratio Si-O/Si-H as a quantification of network formation. As curing temperature increases, the amount of networking increases and the HSSQ film tends toward that of SiO$_2$. This is further shown in figure 3 with a comparison of infrared spectra of the lowest and highest temperature cured HSSQ films relative to a plasma enhanced chemical vapor deposited (PECVD) oxide film.

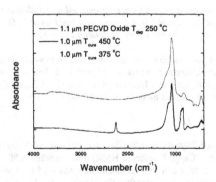

Figure 3. *A comparison of HSSQ to PECVD oxide shows the tendency of HSSQ toward fused silica chemistry as curing temperature increases.*

Curing temperature variation is simply the means for altering the microstructure of HSSQ films; it is thus more instructive to separate out curing temperature as a variable when presenting HSSQ properties information. Therefore, the mechanical properties of HSSQ films will be discussed as a function of the Si-O/Si-H ratio where Si-O represents the Si-O stretch of a network-type bond and Si-H is representative of an HSSQ cage structure.

Nanoindentation

Nanoindentation over the load range of 0.1 to 3000 mN was performed to measure hardness and modulus of HSSQ films cured at various temperatures. Substrate effects were seen in hardness and modulus values for peak depths greater than 20% of the film thickness (peak loads greater than 1 mN). Figure 4a shows the trend in load-displacement as a function of Si-O/Si-H. Figure 4b shows substrate dominated behavior at a 3000 mN peak load.

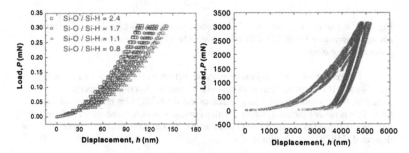

Figure 4. *Load-displacement traces from instrumented nanoindentation of 1.0 μm thick HSSQ films at low loads, 0.3 mN, show a trend as a function of Si-O/Si-H until very large loads, such as 3000 mN, where the substrate dominates the behavior.*

Film hardness and modulus were determined from low-load indents where the calculated values were invariant as a function of peak load. The average hardness and modulus are plotted in figure 5 as a function of Si-O/Si-H. (A Poisson's ratio of 0.25 was assumed to determine the biaxial modulus of the HSSQ films.)

Crack Velocity

Crack extension from a controlled flaw was tracked in 0.5 μm thick capped and uncapped HSSQ films. "Capped" refers to HSSQ films on which a 50 nm SiN_x layer was deposited. In each sample, a 2 N Vickers indentation was used to create a controlled flaw with initial crack lengths much greater than the film thickness. Figure 5 is a plot of the crack velocity of capped and uncapped HSSQ films as a function of the ratio Si-O/Si-H.

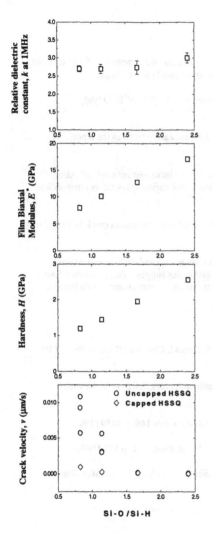

Figure 5. Combination plot showing the relative changes in dielectric constant, modulus, hardness and crack velocity as a function of the ratio Si-O/Si-H for HSSQ.

There is a slight increase in dielectric constant as a function of networking. Both film biaxial modulus and film hardness exhibit a linear increase as a function of Si-O/Si-H.

Crack velocity decreases non-linearly as a function of the ratio Si-O/Si-H. At a single curing temperature, the 50 nm SiN$_x$ cap significantly decreased crack velocity at all four curing temperatures.

CONCLUSIONS

1. On curing, HSSQ $[(HSiO_{1.5})_n]$ films tend toward fused silica chemistry (SiO_2) that can be quantified by the Si-O/Si-H ratio determined by infrared spectroscopy.

2. The dielectric constant of HSSQ films increases as a function of Si-O bridged – network formation.

3. Modulus and hardness of HSSQ films exhibit an increasing trend as the amount of networking increases.

4. There is a decrease in crack velocity as a function of increased networking. Also, scatter in crack velocities for low Si-O/Si-H ratios (low curing temperature) may indicate a more inhomogeneous microstructure.

5. Capping HSSQ films with 50 nm SiN_x layers significantly decreases crack velocity for all amounts of networking.

6. The small change in dielectric constant from changes in (molecular-level) structure compared with the large change in mechanical properties suggests that an intermediate structure optimizes the performance of HSSQ materials as interconnection dielectrics.

REFERENCES

[1] R. H. Baney, M. Itoh, A. Sakakibara, and T. Suzuki, *Chemical Reviews* **95**, p. 1409 (1995).

[2] The National Technology Roadmap for Semiconductors, *Semiconductor Industry Association*, p. 99 (1997).

[3] R.F. Cook and E.G. Liniger, *J. Electrochemical Society* **146**, p.4439 (1999).

[4] R.F. Cook and E.G. Liniger, *Mat. Res. Soc. Symp. Proc.* **511**, p.171 (1998).

[5] R.F. Cook, E.G. Liniger, D.P. Klaus, E.E. Simonyi and S.A. Cohen, *Mat. Res. Soc. Symp. Proc.* **511**, p.33 (1998).

[6] R.F. Cook, *Mat. Res. Soc. Symp. Proc.* **576**, p.301 (1999).

Mat. Res. Soc. Symp. Proc. Vol. 612 © 2000 Materials Research Society

Deposition of Fluorinated Amorphous Carbon Thin Films with Low Dielectric Constant and Thermal Stability

Sang-Soo Han and Byeong-Soo Bae
Laboratory of Optical Materials and Coating(LOMC), Department of Materials Science and Engineering, Korea Advanced Institute of Science and Technology(KAIST), 373-1, Kusongdong, Yusonggu, Taejon, 305-701, Korea

ABSTRACT

Fluorinated amorphous carbon (a-C:F) thin films were deposited by inductively coupled plasma enhanced chemical vapor deposition (ICP-CVD) with increasing $CF_4:CH_4$ gas flow rate ratio, and then annealed with increasing annealing temperature (100, 200, 300, and 400°C). We have found the reduction mechanism of the dielectric constant and the thermally stable condition for the a-C:F films. On the basis of the results, the optimal condition to satisfy both the low dielectric constant and the thermal stability is followed as ; the a-C:F films have to have the compatible F content to make a compromise between the two properties ; the $C-F_x$ bonding configuration has to exist as a form of $C-F_2$ & $C-F_3$ instead of $C-F$; The films should be somewhat cross-linked structure.

INTRODUCTION

Decrease of the feature size in integrated circuits makes interconnect lengths long and contributes to propagation delays due to the increase in RC time constant. Because highly conductive metals and low dielectric materials can reduce propagation delays, the low dielectric interlayer materials are required to improve the circuit performance. Promising candidate materials are fluoropolymers such as polytetrafluoroethylene (PTFE) since their dielectric constants are low (k~1.9) and mechanical strength is reasonably high [1-3]. However, polymer thin films have a difficulty to satisfy desired characteristics such as thermal stability, adhesion, and gap-filling property during integration processing in microelectronics. Thus, the fluorinated amorphous carbon thin films deposited using HDP-CVD (High Density Plasma Chemical Vapor Deposition) have been investigated as new interlayer dielectric materials since they may have similar composition to PTFE with cross-linked structures.[4-8] However, it has been reported that the low dielectric constant and the thermal stability are incompatible properties. Thus, the compromise between the two properties is required and the optimal condition to satisfy the two properties is more favorable to use the a-C:F films as an interlevel dielectrics. In this study, we investigated the reduction mechanism of dielectric constant in the a-C:F thin films and the thermally stable condition. From the results, we will present the optimal conditions to satisfy both the low dielectric constant and the thermal stability.

EXPERIMENTAL

a-C:F thin films were deposited on (100) p-type Si substrate by inductively coupled plasma enhanced chemical vapor deposition (ICP-CVD) with increasing CF_4 gas flow rate (4, 8, 20, and 40 sccm) at constant flow rate of CH_4 gas (4sccm), deposition temperature (25°C) and RF power

(400W). More detailed experimental was described in the previous report.[8] The sample with 10:1 of $CF_4:CH_4$ flow rate ratio was chosen for annealing process since it is the lowest dielectric constant at as-deposited. The thickness and refractive index of the films were measured using an ellipsometer (Rudolph, auto-EL2). From the refractive index(n), electronic polarization was obtained, since $k_e = n^2$. Dielectric constant of the film was calculated from the C-V plot measured at 1MHz in MIS structure (Al/a-C:F/Si). $C-F_x$ bonding configurations of the films were analyzed using hot-stage X-ray photoelectron spectroscopy (XPS, VG ESCALAB 200R, Al K_α radiation). Quantitative analysis of all elements including hydrogen in the films was performed by elastic recoil detection-time of flight (ERD-TOF) measurement [9]. By the inverse reaction scheme, the recoil cross sections induced by ^{35}Cl ion below 10 MeV have been measured. Therefore, we could obtain the energy-counts spectra by plotting with different times of flight.

RESULTS AND DISCUSSION

Figure 1 shows that the dielectric constants of the films vary from 3.2 to 2.4 with increasing $CF_4:CH_4$ flow rate ratio(R). It is observed that the reduction mechanism of the dielectric constant of the film due to the polarization consists of two regimes as represented in Figure 1. In regime (1), the reduction of both electronic polarization and ionic & bond dipole polarization contribute to that of the total dielectric constant. However, in regime (2), electronic polarization remains constant; thus, the reduction of ionic & dipole polarization contributes to the reduction of total dielectric constant.

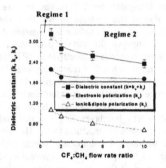

Figure 1. *Variation of dielectric constants in a-C:F thin films with increasing $CF_4 : CH_4$ flow rate ratio.*

The reduction of electronic polarization in the regime (1) can be explained by the variation of atomic ratio obtained by ERD-TOF analysis (Table I). In the regime (1), atomic percent of F content increases from 12 to 23.3 ; in the regime (2), F content remains almost constant. This variation is consistent with k_e variation as expected. It was already observed that the incorporation of F with high electronegativity reduces the refractive index and dielectric constant of the films [4-8]. To find out the origin of the reduction of the dielectric constant in the regime (2), we studied the variation of bonding configuration with increasing R using XPS analysis.

Table I. *Atomic ratio, dielectric constant and area density of the a-C:F samples, where, R is $CF_4:CH_4$ flow rate ratio, A-D is as-deposited, A.T. is annealing temperature, k is dielectric constant and ρ is area density(10^{17} atoms/cm^2)*

	R=1, A-D	R=2, A-D	R=5, A-D	R=10, A-D	R=10, A.T.=100°C	R=10, A.T.=200°C	R=10, A.T.=300°C	R=10, A.T.=400°C
C (%)	78	71.7	70.1	70	74	80	81	82
F (%)	12	23.3	24	25	22	17	15	14
H (%)	10	5	5.9	5	4	3	4	4
k	3.24	2.80	2.58	2.39	2.5	2.75	2.82	2.85
ρ	5.0	3.9	2.8	2.2	2.5	3.3	4.0	5.4

In Figure 2, (a) shows the C 1s binding energy spectra of a-C:F thin films; each peak is deconvoluted and then integrated; (b) represents variation of the integrated area of the peaks depending on the R. In the regime (2), F content remains constant at 25 %, and the C-F$_2$ & C-F$_3$ bondings increase with R while the C-C (or C-H) & C-F bondings decrease. Thus, the dielectric constant reduction in the regime (2) is owing to two polarization terms. One is the ionic polarization and the other is the bond dipole polarization. For polyvalent atom such as carbon, the partial charge builds up when highly electronegative fluorine is added. Since this partial charge induces the extra bonding energy, the ionic polarization decreases as the C-F$_x$ bonding configuration in the a-C:F thin film changes from the C-C & C-F bonds to the C-F$_2$ & C-F$_3$ bonds. The bond dipole (or orientation) polarization decreases with increasing C-F$_2$ & C-F$_3$ bonds since the bond dipole moment of the molecule decreases. Also, the orientation of bond dipole is disturbed by the influence of neighboring F atom. The calculated dipole moments in various molecule structures, CH$_3$F, CH$_2$F$_2$ and CHF$_3$ are 1.81, 1.45, and 1.22, respectively [10].

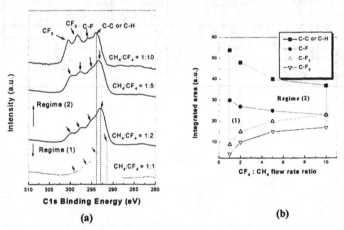

(a)   (b)

Figure 2. *(a) C1s binding energy XPS spectra and (b) multiple Gaussian integration of bonding configuration of a-C:F thin films with increasing CF$_4$:CH$_4$ flow rate ratio. Only the relative position of the peaks is relevant since the spectra are shifted due to the charging effect.*

In addition to polarization, density is another effective factor on dielectric constant. From Clausius-Mossotti eq. [11],

$$\alpha = \frac{M}{\rho}\left(\frac{k-1}{k+2}\right)\frac{3\varepsilon_0}{N_0}$$

where α is an polarizability, M is the molecular weight, ρ is the density, k is the relative dielectric constant, ε_0 is the dielectric constant of vacuum, and N_0 is Avogadro's number, the dielectric constant is proportional to the density of the film. Thus, the additional reduction of the dielectric constant of the a-C:F thin film is expected by the decrease of the density of the film through both regime(1) and regime(2) since the the area density of the film decreases with increasing R as shown in Table I.

Figure 3 shows the variation of the bonding configuration with increasing annealing temperature. As annealing temperature increases, in C-F$_x$ bonding configuration, C-F bondings decrease and C-F$_2$ & C-F$_3$ bondings are almost constant. It means that the C-F$_2$ & C-F$_3$ bonding configurations are thermally stabler than the C-F bonding configuration. As discussed earlier, C-F$_2$ & C-F$_3$ bondings reduce the dielectric constant due to their low ionic polarization. Thus, the C-F$_2$&C-F$_3$ bonding configurations satisfy the condition for low dielectric constant and thermal stability, simultaneously.

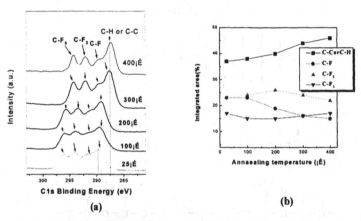

Figure 3. *(a) C1s binding energy XPS spectra and (b) multiple Gaussian integration of bonding configuration in a-C:F thin films with increasing annealing temperature. Only the relative position of the peaks is relevant since the spectra are shifted due to the charging effect.*

Figure 4 (a) is simple structure evolution model of a-C:F thin films with increasing R and (b) is that with increasing annealing temperature. In Table I, the area density decreases with increasing R and increases with increasing annealing temperature. From the results of the atomic ratio in Table I, it is found that the substitution of F for H occurs dominantly in the regime (1). In the regime (2), C-F$_x$ bonding configurations vary without the change of F content. In this regime, excess F ions for high CF$_4$ flow rate break the C-C bonds, and then C-C molecule structure is shortened by the creation of C-F$_3$ bond. In Figure 3, it is observed that the C-F bonding decreases with increasing annealing temperature. It implies that the C-F bonding supplies the site required for cross-linking the structure. Thus, the structure is cross-liked with the sacrificing the C-F bonding with increasing annealing temperature.

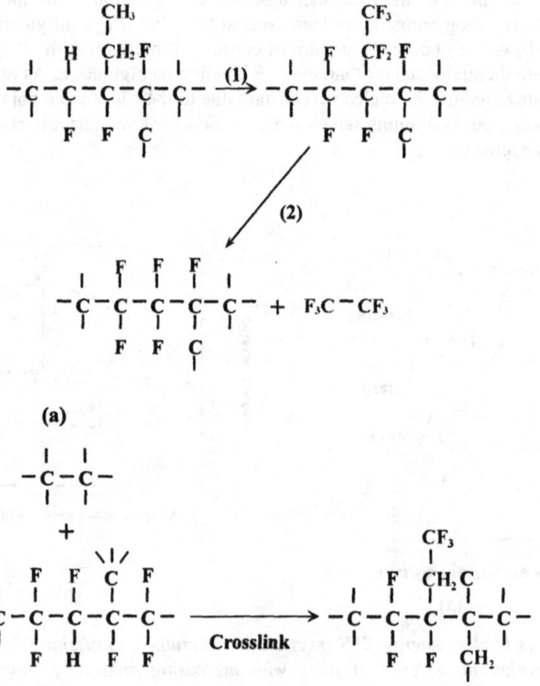

Figure 4. *Structure evolution models of a-C:F thin films (a) with increasing $CF_4 : CH_4$ flow rate ratio and (b) with increasing annealing temperature.*

CONCLUSIONS

Optimal conditions to satisfy both low dielectric constant and thermal stability of a-C:F thin film is followed ; (1) The a-C:F film should have compatible F content (~20%) to make a compromise between the low dielectric constant and thermal stability. (2) The $C-F_x$ bonding configurations must exist in the film as a form of $C-F_2$ & $C-F_3$ instead of C-F. It is the desirable condition for obtaining both the low dielectric constant and thermal stability. (3) The film should have somewhat cross-linked structure. The cross-liked structure is evoluted by sacrificing the C-F bonding as annealing the film

ACKNOWLEDGEMENTS

This work has been financially supported by International Collaborative Research Program in Ministry of Science and Technology in Korea.

REFERENCES

1. T-M. Lu and J. A. Moore, MRS Bulletin **22(10)**, 28 (1997).
2. T. Nason, J. A. Moore and T. M. Lu, Appl. Phys. Lett. **60**, 1866 (1992).
3. G. B. Blanchet, Appl. Phys. Lett. **62**, 479 (1993).
4. K. Endo and T. Tatsumi, Appl. Phys. Lett. **68**, 2864 (1996).
5. K. Endo and T. Tatsumi, J. Appl. Phys. **78**, 1370 (1995).
6. Y. Ma, H. Yang, J. Guo, C. Sathe, A. Agui, and J. Nordgren, Appl. Phys. Lett. **72**, 3353 (1998).
7. H. Yang, D. J. Tweet, Y. Ma and T. Nguyen, Appl. Phys. Lett. **73**, 1514 (1998).
8. S-S. Han, H. R. Kim and B-S. Bae, J. Electrochem. Soc. **146**, 3383 (1999).
9. S. S. Klein, Nucl. Instrum. Methods, **B15**, 464 (1986).
10. C. C. Ku and R. Leipins, *Electrical Properties of Polymers* (Hanser Publishers, New York, 1987), p. 41.
11. W. D. Kingery, H. K. Bowen and D. R. Uhlmann, *Introduction to Ceramics* (John Wiley & Sons, New York, 1991), p. 921.

Mat. Res. Soc. Symp. Proc. Vol. 612 © 2000 Materials Research Society

HIGH-TEMPERATURE MECHANICAL BEHAVIOR AND PHASE MORPHOLOGY OF POLY(TETRAFLUOROETHYLENE)/SILOXANE NANOCOMPOSITES USED AS ULTRA LOW-*k* DIELECTRICS

PING XU [1], SHICHUN QU [2], TOM ROSENMAYER [2] AND MIN Y. LIN [3]
[1] W. L. Gore & Associates, Inc., 2401 Singerly Road, Elkton, MD 21921
[2] W. L. Gore & Associates, Inc., 1414 West Hamilton Avenue, Eau Claire, WI 54703
[3] National Institute of Standards and Technology, React E 151, Gaithersburg, MD 20899

ABSTRACT

Poly(tetrafluoroethylene) (PTFE)/siloxane nanocomposites have been prepared as ultra low-*k* dielectrics. These new nanocomposites show excellent high-temperature mechanical properties compared to unfilled PTFE while their dielectric constant almost remains unchanged. Specifically, the data from the dynamic mechanical study indicates that these nanocomposites have the mechanical behavior similar to that of crosslinked polymers. Small-angle neutron scattering (SANS) has been carried out to characterize the phase morphology of the PTFE/siloxane nanocomposites and the size of the inorganic networks. It has been shown that no phase separations or orientations appear in these nanocomposites in the range of 12 to 469 nm. These SANS results suggest that these materials are single-phase nanocomposites that are very homogeneous and isotropic. They are basically PTFE-based molecular composites.

INTRODUCTION

It is well known that PTFE is an excellent electrical insulator. Its physical and chemical properties have been well documented.[1] Because of its excellent electrical property, this material has been widely used as a dielectric material. To further improve its high-temperature mechanical strength, PTFE/siloxane nanocomposites have been developed. Speedfilm™ BX known as an ultra low-*k* dielectric developed by Gore is a new material.[2,3] This material is a hybrid nanocomposite of PTFE and siloxane, and has a dielectric constant of 2.1. Since this material is an organic-inorganic nanocomposite, several issues need to be studied, for example, inorganic domain sizes, phase separations and orientations, etc. Any submicron-scale inhomogeneity will influence material performance in the microchip application.

This study was carried out to investigate the two issues as to whether PTFE/siloxane nanocomposites have improved high-temperature mechanical strength, and contain any submicron-scale phase separations and orientations.

EXPERIMENTAL

PTFE/siloxane nanocomposites were prepared by mixing a PTFE nanoemulsion and a siloxane as described previously.[2] Mixtures were first dried and then heated at 400 °C for 2 hours in air. The same PTFE nanoemulsion sample without a siloxane was treated under the same condition and used as reference.

Dynamic mechanical analysis (DMA) measurements were made on an ARES rheometer (Rheometric Scientific) in a parallel-plate geometry. Samples were first pressed into 25-mm diameter disks of about 0.8-mm thickness and then loaded under air. Dynamic rheology measurements were carried out in a temperature sweep mode at a fixed frequency of 10 rad/second at temperatures from 250 to 400 °C at a heating rate of 5 °C/minute.

SANS measurements were carried out at room temperature on a 30-m SANS facility at the National Institute of Standards and Technology. Neutron wavelengths of 5, 10 and 12 Å were used, respectively, and the Q range was covered from 0.00134 to 0.5 Å^{-1}. All measurements were calibrated by transmission and with known standards.

RESULTS AND DISCUSSION

POLYMER CHAIN STRUCTURE AND DYNAMIC RHEOLOGY

Polymers consist of long molecular chains. Unlike small molecules, long chain molecules take time to relax once a force is applied. Therefore, polymers give a unique property, i.e., viscoelasticity. The principle of dynamic rheology measurements involves the study of various polymer chain motions with temperature and frequency. Interpretations of viscoelasticity of polymers on a molecular scale are extremely important, which can lead one to correctly understand polymeric materials used and properly utilize them.[4]

Figures 1 illustrates the types of viscoelastic responses most often observed as a function of temperature at a fixed frequency. Specific viscoelastic behavioral patterns for linear or crosslinked polymeric materials are interpreted in terms of the accepted molecular mechanisms.[4] Figure 1-A shows four regions of the viscoelastic behavior of the linear amorphous polymer. At low temperatures below T_g, the polymer is hard and brittle. This is a glassy region (A-B) and T_g is a glass transition temperature of the polymer. In this glassy region, thermal energy is insufficient to surmount the potential barriers for rotational and translational motions of segments of polymer chain molecules. The molecular chain segments are essentially frozen in fixed positions on the sites of a disordered quasi-lattice with their segments vibrating around these fixed positions much like low molecular weights in a molecular crystal. With increasing temperature, the storage modulus slowly decreases and the amplitude of vibrational motion becomes bigger. In this temperature region (B-C), the polymer undergoes a glass transition where short-range diffusional motions begin. Chain segments become free to jump from one lattice site to another, so the brittle polymer becomes a resilient leather.

As the temperature is further increased, the storage modulus reaches a rubbery plateau region (C-D). In this region, the short-range diffusional motions of molecular chain segments that just gave the glass transition occur much faster than the dynamic measurement. On the other hand, the long-range motion of chain molecules that would result in translational motions or the disentanglement of chain molecules is still largely restricted by the presence of strong local physical interactions between neighboring chains. In the case of a crosslinked polymer, these interactions consist of primary chemical bonds as well as physical interactions. However, the viscoelastic responses of linear and crosslinked polymers through the rubbery plateau region are essentially identical. As the temperature is further increased (beyond D), differences between linear and crosslinked polymers become clear as shown in Figures 1-A and 1-B. Chemical crosslinking can prevent molecular chains from sliding by one another. The storage modulus can maintain for a crosslinked polymer up to temperatures where molecular chain scissions start. For a linear polymer, increasing temperature makes chain molecular motions become more and more significant, so that large-scale chain motions take place and eventually entire molecular chains begin to translate. When chain molecules have enough thermal energy, local physical interactions no longer hold chain molecules together. As a result, a melt transition occurs at the melt flow temperature, T_f. In this region, molecular chains disentangle and the storage modulus quickly decreases.

Longer linear molecular chains create more chain entanglements and have stronger physical interactions. These effects extend the rubbery plateau region as shown in Figure 1-C. It is clear that the higher the linear chain molecules, the higher the melt flow temperature. For a linear crystalline polymer, there is one more transition, i.e., a melting transition, as illustrated in Figure 1-D. Crystallites can hold chain molecules together when a polymer is in the crystalline state. Thus, long-range chain molecular motions are prohibited. However, as the temperature is increased beyond the melting temperature, T_m, crystallites melt and thereby chain molecules become mobile. In the E to F region, chain molecules become viscoelastic resulting from chain entanglements and at temperatures above T_f, they become viscous and start flow. For a crosslinked crystalline polymer, except a melting transition, the storage modulus will still maintain until it thermally degrades as described in the situation for a crosslinked amorphous polymer. It should pointed out that for a highly crosslinked polymer that is initially crystalline before crosslinking, the melting transition could disappear.

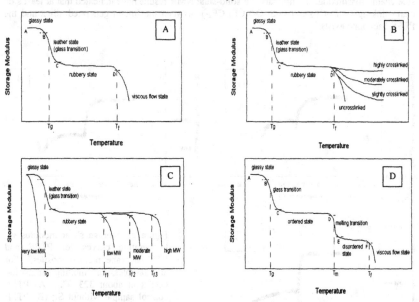

Figure 1. Schematic storage modulus-temperature curves showing various regions of viscoelastic behavior of polymers. (A) linear amorphous polymers; (B) crosslinked amorphous polymers; (C) linear amorphous polymers having different molecular weights; (D) linear crystalline polymers.

Figures 2 shows the viscoelastic responses of the PTFE/siloxane nanocomposites and a PTFE control sample. It is interesting to see these materials have very different viscoelastic behaviors. The control PTFE sample goes through a melting transition at about 323 °C and then the storage shear modulus significantly decreases. And finally, it flows at about 360 °C. This indicates that this control material cannot be used at temperatures higher than 360 °C above which it does not have any mechanical strength. However, this is not a case for the PTFE/siloxane nanocomposites. These composites also go through a melting transition, but the storage shear modulus can well maintain up to the temperature limit, 400 °C. Particularly, the nanocomposite having 1.62% Si shows excellent high-temperature mechanical strength, which indicates the molecular chain mobility at high temperatures have been greatly reduced. Since the nanocomposites and the control sample have the same molecular weight, ca. $M_n = 4 \times 10^5$ g/mol, the effect of molecular weight on viscoelasticity as described earlier can be excluded. Clearly, high thermal mechanical strength results from crosslinking. Based on the sol-gel process,[5] it was speculated that both primary chemical crosslinking and physical crosslinking could exist. In the case of physical crosslinking, PTFE molecular chains could be very tightly embedded in the inorganic silica phase so that molecular chains cannot move on a long-range scale. In the case of chemical crosslinking, primary chemical crosslinks could be created by chain end linking and side chain crosslinking. Preliminary ^{19}F solid-state NMR results have indicated that at least side chain linking ($-CF_2-CF_2-CF(-O-Si-O-)-CF_2-CF_2-$) between the in-situ generated silica and the PTFE backbone exists.[6]

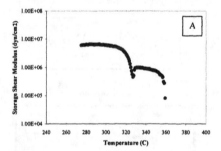

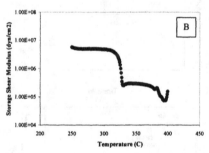

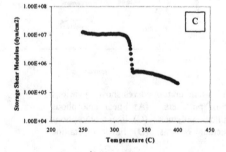

Figure 2. Storage shear modulus vs. temperature curves for PTFE and PTFE/siloxane nanocomposites. For all samples, a melting transition occurs at about 323 °C. (A) PTFE control sample without Si; (B) PTFE nanocomposite containing 0.35% Si; (C) PTFE nanocomposite containing 1.62% Si.

PHASE MORPHOLOGY AND SMALL-ANGLE NEUTRON SCATTERING

In a scattering experiment, the scattered intensity as a function of the length scale is measured. The relation between the scattering angle and the length scale can be described by the Bragg law

$$\lambda = 2D\sin\theta \tag{1}$$

where D is the distance between crystallographic planes, λ is the wavelength of the radiation, and θ is the scattering angle. The intensity is usually measured as a function of the momentum transfer, Q, which is related to θ via

$$Q = (4\pi/\lambda)\sin\theta \tag{2}$$

Combining Equation (1) and (2) gives

$$D = 2\pi/Q \tag{3}$$

Equation (3) indicates the distance scale probed by a measurement at a given value of Q. Although Bragg's law does not apply to amorphous materials, the Fourier relationship between the structure in real-space and the scattering in Q space means that Equation (3) may be applied to first order for all types of scattering. Thus, data at lower Q probes longer length scales. Small-angle X-ray or neutron scattering has been widely used to determine polymer chain dimensions and phase morphologies of materials.[7]

SANS generates information on structure over dimensions of approximately 10 – 1,000 Å. It is clear that the PTFE control sample and the PTFE/siloxane nanocomposite sample have different scattering profiles as shown in Figure 3. For the former, there is a broad peak in the range of 83 – 161 nm, while for the latter, there is no scattering at all related to silica particles and there is no phase separation on the length scale determined by SANS. It is believed that the phase separation in the control sample results from the crystalline and amorphous domain separation whereas the size of the crystalline domain has been significantly reduced in the nanocomposite. This interesting observation shows the PTFE/siloxane nanocomposite is basically a molecular-based composite that is very homogenous across the entire bulk material.

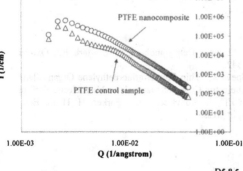

Figure 3. SANS profiles of PTFE control sample and PTFF nanocomposite.

Figure 4 shows that both the control sample and the nanocomposite do not have any orientations on the length scale studied. This result clearly gives evidence that the nanocomposite is isotropic on the submicron scale.

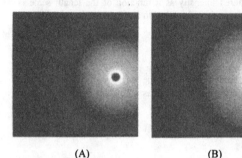

Figure 4. SANS orientation profiles. (A) PTFE control sample; (B) PTFE/siloxane nanocomposite.

(A)   (B)

In fact, in this dimensional regime, many materials display some degree of randomness. The concept of fractal geometry is central to understand small-angle scattering profiles from materials. Fractal geometry provides not only a quantitative measure of disorder, but also insight into the origin of that disorder in terms of growth models.[8] Results regarding this issue will be published elsewhere.

CONCLUSIONS

This study has shown that the PTFE/siloxane nanocomposites have excellent high-temperature mechanical strength relative to the pure PTFE sample. Strong interactions between the PTFE chains and the in-situ generated silica greatly reduce the PTFE chain mobility. The SANS data indicates that such nanocomposites do not have any phase separations and orientations on the length scale of 12 – 469 nm and basically are molecular-based composites, while the pure PTFE sample has a phase segregation in the range of 83 - 161 nm.

ACKNOWLEDGEMENTS

It is a great pleasure to acknowledge that the NIST provides the SANS facility for this study.

REFERENCES

1. D. Kerbow, "Polytetrafluoroethylene" in "Polymer Data Handbook", J. E. Mark, Ed., Oxford University Press, New York, NY, 1999. pp. 842.
2. S. Qu, T. Rosenmayer, P. Xu and P. Spevack, "Thin Polytetrafluoroethylene Organosilane Nanocomposite Films Used as Ultra Low Dielectric Constant Materials in Microelectronics" in "Nanophase and Nanocomposite Materials III", S. Komarneni, J.C. Parker, H. Hahn, Eds., Materials Research Society, Warrendale, PA, in press.

3. T. Rosenmayer, J. Bartz, S. Qu and P. Xu, "A Novel Dual-Damascene ETCH Process Utilising a High-Selectivity, Ultralow-k Dielectric" in "Semiconductor Fabtech", 11th ed., ICG Publishing, London, UK, 2000. pp. 251.

4. J. Ferry, Viscoelastic Properties of Polymers, 3rd ed., Wiley, New York, NY, 1980.

5. P. Xu, "Polymer-Ceramic Nanocomposites: Ceramic Phases" in "Encyclopedia of Materials: Science and Technology", K. H. J. Buschow, R. W. Cahn, M. C. Flemmings, B. Ilschner, E. J. Kramer, S. Mahajian, Eds., Elservier Science, Oxford, UK. in press.

6. P. Xu, unpublished results.

7. R.-J. Roe, Methods of X-Ray and Neutron Scattering in Polymer Science, Oxford University Press, New York, NY, 2000.

8. D. Schaeffer, Science, **243**, 1023 (1989).

Mat. Res. Soc. Symp. Proc. Vol. 612 © 2000 Materials Research Society

STRUCTURAL ANALYSES OF FLUORINE-DOPED SILICON DIOXIDE DIELECTRIC THIN FILMS BY MICRO-RAMAN SPECTROSCOPY

Jeffery L. Coffer*
Department of Chemistry
Texas Christian University, Ft. Worth, Texas 76129

T. Waldek Zerda
Department of Physics
Texas Christian University, Ft. Worth, Texas 76129

Kelly J. Taylor and Scott Martin
Texas Instruments, Kilby Center, Dallas, Texas 75243.

Abstract

Fluorine-doped silicon dioxide, a dielectric material compatible with copper integration, has received considerable attention for applications requiring a k value in the 3.5 to 4.0 range. Given the influence of structure on desired properties, convenient experimental structural probes of this type of material are of widespread interest. This work focuses on Raman spectroscopic analyses of ring defects in fluorine-doped silicon dioxide films prepared by plasma enhanced chemical vapor deposition (PECVD) as well as high density plasma methods (HDP). These measurements are complemented by *ab initio* computational simulations of the ring defects in these films and the impact of nearby fluorine on their stability. The impact of aging on these structures and correlations of observed trends with experimental techniques such as X-ray fluorescence are also described.

Introduction

There is an ever-pressing need for rapid, non-destructive methods in the structural characterization of silicon dioxide-based low k dielectric films, including fluorine-doped variants. The vibrational spectroscopic technique of Raman scattering provides unique information with regard to the presence of strained silicon-oxygen ring defects in these materials.[1,2] Specifically, these defects are noted by characteristic bands attributed to so-called 3-member ring defects (Si_3O_3) at 608 cm[-1] as well as a 4-member ring mode (Si_4O_4) near 503 cm[-1].[1,3,4]

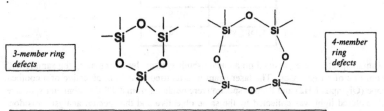

Figure 1. Structures of ring defects in plasma-grown CVD films.

This study focuses on an evaluation of the effect of fluorine incorporation on the relative number of such defects in these films, the type of plasma deposition technique employed, as well as the impact of aging on the evolution of these defects. These observations are complemented by a series of *ab initio* simulations evaluating the effect of fluorine incorporation on the ground state energy of three and four member silica ring structures. Given the sensitivity of these defects to adsorbed water, these studies have important implications for possible interfacial reactions and subsequent device breakdown.

II. Experimental

Four classes of oxide-based films were prepared on 200 mm Si (100) wafers for Raman analyses by standard plasma enhanced CVD (PECVD) and high density plasma CVD (HDPCVD). All PECVD films were deposited in a Novellus Systems Concept 2 Sequel reactor of standard configuration. All HDPCVD films were deposited in a Novellus Systems Speed reactor. Table 1 shows a summary of the different films studied. Undoped SiO_2 films were prepared by PECVD from both $SiH_4/N_2O/N_2$ (PESILOX) and $TEOS/O_2$ (PETEOS) gas mixtures. Three fluorine-doped PECVD SiO_2 films (PEFSG) of concentration 3.8, 4.5, and 4.9 at%,F were made from $SiF_4/SiH_4/N_2O/N_2$ gas mixtures. Fluorine-doped HDPCVD SiO_2 films (F-HDPCVD) were made from $SiF_4/SiH_4/O_2/Ar$ mixtures and had a concentration of 5.4 at%,F.

Prior to the oxide depositions, a 4000 Å thick Al film was deposited directly on the silicon wafer to improved the intensity of the backscattered Raman signal. All oxide films were deposited at about 1 μm and their thickness verified by spectroscopic ellipsometry. Fluorine composition was determined by calibrated x-ray fluorescence measurements with a measurement system error of less than 0.05 at%,F. There may exist a systematic bias of up to 1.0 at%,F applied to all fluorine measurements, but their relative differences will be accurate to the stated measurement system error.

Table 1. Composition of selected dielectric films analyzed by Raman Measurements.

Film	at%, F
PESILOX	-
PETEOS	-
PEFSG	3.8
PEFSG	4.5
PEFSG	4.9
F-HDPCVD	5.4

Raman spectra were obtained on a custom-built system[5] by illuminating the sample with the 514 nm line of an Ar^+ laser. The laser beam was focused by a 100x objective of a confocal microscope (Olympus BH2) and Raman spectra were obtained from a 0.5 ☐m diameter spot size. The backscattered light was collected by the same objective and the spectral analysis was done using an axial transmissive spectrograph (Kaiser Optical Systems, HoloSpec) equipped with a Princeton Instruments CCD camera. Spectra of the films in the frequency range between 200 cm^{-1} and 1700 cm^{-1} were obtained using a 30 second integration time. Raman measurements were conducted at room temperature.

Equilibrium geometries of the silica ring defects were calculated utilizing *ab initio* methods of molecular structure calculation (within the Hartree-Fock approximation) with a commercially-available software package (*Spartan*, from Wavefunction, Inc.) operating on a desktop Wintel platform. In terms of basis sets, a 'split valence' 3-21*G was selected, since it incorporates extra polarization functions (i.e. d orbitals) on the Si atoms. The starting geometry for the three dimensional structure of each ring was first refined by a quick semi-empirical calculation (employing an AM1 Hamiltonian) prior to initiation of the *ab initio* calculation.

III. Results and Discussion

Raman Measurements. Figure 2 illustrates several Raman spectra for a series of PESILOX, PEFSG, and F-HDPCVD films prepared according to the above methods. Two general trends can be discerned qualitatively upon inspection. The first involves a clear diminution of the intensity of the 3-member defect ring mode at 608 cm^{-1} as the %F content increases in a given film. It is also evident from these spectra that films grown under high density plasma conditions do not yield a measurable number of such defects (as reflected by an absence of a detectable Raman band at this frequency).

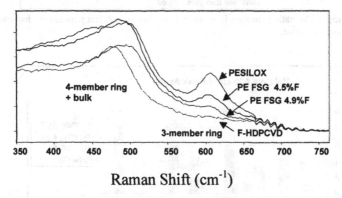

$$\text{Raman Shift (cm}^{-1}\text{)}$$

Figure 2. Raman spectra of selected plasma-grown SiO$_2$ and fluorine-doped SiO$_2$ dielectric films.

To quantify the relative concentration of such defects between film types, we calculate the integrated area under the peak at 608 cm^{-1} and normalize this value by dividing it by the integrated area of the Raman band centered at 490 cm^{-1} (a convolution of 4-member ring structures and bulk Si-O modes). Such an evaluation provides a quantitative verification of the above trends and also permits an assessment of how such defects evolve as a function of film aging in an ambient atmosphere.

In general, PETEOS or PESILOX films possess the largest number of 3-member ring defects, followed by the F-containing SiO_2 films (PEFSG). The relative intensity of this 608 cm^{-1} mode is observed to scale with the relative fluorine percentage; it is also found that the use of a high density plasma process (F-HDPCVD) yields a negligible number of such defects. This observation is also consistent with previous comparisons of strained ring defects in SiO_2 films deposited by PECVD with those prepared in a high density plasma reactor.[1] For undoped SiO_2, high density plasma methods also apparently produce films with relatively fewer three member ring defects, as gauged by these diagnostic Raman modes.

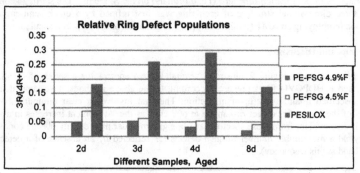

Figure 3. Graphs illustrating changes in relative 3-member ring defect populations for selected SiO_2/FSG films over time.

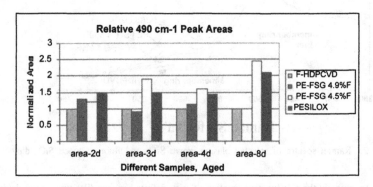

Figure 4. Changes in relative peak areas in the 490 cm^{-1} Raman band for selected dielectric films as a function of time. Values are normalized to F-HDPCVD.

For those films with appreciable intensity at 608 cm^{-1}, it is evident that the relative intensity of this mode diminishes slowly over time, presumably as a consequence of the reactions of these strained silica rings with water to produce open chain species. These 3- member rings are considered to be more strained than the four member ring analogs, and hence more reactive to adsorbed H_2O.[6]

With regard to assessing the relative concentration of 4-member ring defects in this series of fabricated films, the measured linewidth of the broad Raman mode centered at 490 cm^{-1} was used. In essence, the contribution of the lower energy bulk silica modes remain invariant in each case, and thus the narrower the linewidth of this feature the fewer number of the Si_4 ring moieties in a given film.

Computational Studies. From first principles, it is important to understand why the presence of fluorine should strongly affect the stability (and hence presence) of the ring structures in the film. The goal is to use *ab initio* quantum mechanical methods to calculate the changes in ring structure which occur when a nearby Si-F or Si-OH group is brought close to a 3-member or 4-member Si ring defect.

For modeling the 3-member ring defects in FSG, structures of composition R = H, $Si(OH)_3$, or $Si(OH)_2F$ were selected. Since in a real film the stability and structure of a given ring will be strongly affected by its nearest chemical neighbors, the idea here is to assess the role of a nearby Si- or Si-F species on the parent ring structure by replacing a hydrogen of one of the terminal silanol groups with $Si(OH)_3$ or $Si(OH)_2F$.

After initial calculation of the parent Si_3O_3 ring system with terminal silanol groups, the next calculated structure entailed substitution of a terminal silanol hydrogen (emanating from Si(6)) with the more complex $Si(OH)_3$ species, a structure with the empirical formula $[Si_3O_3(OH)_5(OSi(OH)_3]$. The geometry-optimized structure of this derivatized ring system reveals that one of the hydrogens attached to the pendant $Si(OH)_3$ species (Si(17)) positions itself in such a way that it is engaged with a stable intra-molecular hydrogen bond with an adjacent silanol oxygen (O(12), attached to ring Si(7), with a calculated bond length of 1.7960 Å). There are a number of very slight asymmetric distortions of the Si-O bond lengths present in the ring (as compared to the parent $[Si_3O_3(OH)_6]$ species) arising as a consequence of this hydrogen bond.

Replacing one of the OH moieties in the terminal -OSi(OH)$_3$ group lying above the ring with a fluorine atom [i.e. $[Si_3O_3(OH)_5(OSi(OH)_2F]$ (compound c) produces a geometry optimized structure with several interesting points. First, it should be noted that the starting geometry of this structure at the outset of the calculation involved placing the Si-F species close to the nearby silanol (O (11)), where it would have an opportunity for stabilization by hydrogen bonding. Somewhat surprisingly, the lowest energy conformation obtained for this molecule upon geometry optimization resulted in rotation of the Si-F moiety out of the way, and instead it is found that a Si –OH...OSi linkage is preferred (similar to that observed in $[Si_3O_3(OH)_5(OSi(OH)_3]$). However, for $[Si_3O_3(OH)_5(OSi(OH)_2F]$, the calculated intra-molecular hydrogen bond has a measurably shorter distance of 1.753 Å. This, in turn, has marked consequences for the relevant distortions of the Si-O bonds in the ring.

The most striking difference induced by the presence of the fluorine in this specific derivative appears in the bond between Si(7) and O(2). Rather than a slight shrinking (-0.006 angstroms) of this bond when the $Si(OH)_3$ is introduced (compound b), there is a more substantial elongation (+0.015 Å) which is manifested when the Si(OH)$_2$F species is present (compound c). This infers a weakening of that specific bond, thereby enhancing its tendency to react with an adsorbed impurity or alternatively, making it prone to thermal dissociation during the film growth process.

From the above calculations, it is also important to recognize that in a real film it is unlikely that a Si-F species will possess the conformational flexibility to permit its rotation within a rigid silica framework. Thus an additional calculation of the structure of [Si$_3$O$_3$(OH)$_5$(OSi(OH)$_2$F)] was carried out, this time "freezing" the position of the (OSi(OH)$_2$F) species while allowing the rest of the ring to alter its geometry to reach a potential energy minimum. In this calculation, the Si-F moiety is locked into a hydrogen bonding environment with the adjacent ring silanol. The resulting geometry-optimized structure is interesting, as the marked distortions in this particular structure lie not in terms of Si-O bond lengths but rather in terms of bond angles. Specifically, the O(3)–Si(7)-O(12) bond angle increases to 113.4 degrees, while the O(2)-Si(7)-O(12) value distorts to 110.1 degrees. In essence, the O(12)-Si(7)-O(9) moiety is distorting itself as far from the fluorine atom as possible. This distortion is also reflected in the substantial elongation of the intra-molecular distance between the F and the H of the adjacent silanol (from an initial value of less than 2 Å to a value of 5.099 Å for the optimized structure).

IV. Conclusions

Spectroscopic analysis of the structure of several different SiO$_2$ and F-doped SiO$_2$ dielectric thin films by micro-Raman spectroscopy have been completed. The results are consistent with the fact that the 3-member silica ring population mirrors the relative F concentrations according to the trend : PESILOX ~ PETEOS > PEFSG (3.8 %F) > PEFSG (4.5%F) >> PEFSG (4.9%F) > F-HDPCVD (5.4%F). Some diminution of the 3-membered ring population is also observed as a given sample ages. The broader linewidth of the 490 cm^{-1} mode for PEFSG samples and from PESILOX/PETEOS infers a larger 4-member ring population (relative to F-doped samples grown in a high density plasma reactor). Computational studies of the 3-member ring structures suggest that one possible driving force for the relatively fewer 3-member ring defects in F-doped SiO$_2$ films lies in the tendency of these rings to undergo structural distortion(s) to distance itself from the fluorine moieties.

V. Acknowledgements

Financial support by the Texas Advanced Technology Program and the Robert A. Welch Foundation to JLC are gratefully acknowledged.

VI. References

1. S. Okuda, Y. Shioya, and H. Kashimada, J. Electrochem. Soc., 145, 1338 (1998).
2. M. Yoshimaru, S. Koizumi, and S. Shimokawa, J. Vac. Sci. Tech. A, 15, 2908 (1997).
3. A.M . Mulder, J. Non-Cryst. Solids, 95/96, 303 (1987).
4. C.A.M . Mulder and A.A.J.M. Damen, J. Non-Cryst. Solids, 93, 387 (1987).
5. J. Coffer, R. Appel, T.W. Zerda, J. Janik, and R. Wells, Chem. Mater., 11, 20 (1999).
6. B.J. Hopkins and T.W. Zerda, J. Non-Cryst. Solids, 149, 269 (1992).

Mat. Res. Soc. Symp. Proc. Vol. 612. © 2000 Materials Research Society

STUDY OF DRY PHOTORESIST STRIPPING PROCESSES FOR HYDROGEN SILSESQUIOXANE THIN FILMS

HUEY-CHIANG LIOU[1], JERRY DUEL[1], VICTOR FINCH[1], QINGYUAN HAN[2], PALANI SAKTHIVEL[2], AND RICKY RUFFIN[2]
[1]Semiconductor Fabrication Materials, Dow Corning Corporation, Midland, Michigan 48686-0994; [2]Fusion System Division, Semiconductor Equipment Operation, Eaton Corporation, Rockville, MD 20855-2798

ABSTRACT

The impact of dry stripping process chemistries on the selective removal DUV photoresist (PR) in the presence of hydrogen silsesquioxane (HSQ) have been studied along with HSQ film properties in order to develop a new, effective process to minimize changes in HSQ during the PR stripping processes. The results show that oxygen-free gas mixtures, specifically H_2/N_2 gas mixtures, have the best combination of PR:HSQ ash selectivity and minimized changes in HSQ films. However, gas mixtures containing CF_4 or O_2 greatly reduce PR/HSQ ash selectivity. The process temperature is another parameter that strongly influences ash selectivity. While the higher ash temperature greatly enhances selectivity in oxygen-free gas mixtures, the ash selectivity is only marginally enhanced with increasing ash temperature in the presence of O_2. Furthermore, the k-value of HSQ suffers in the presence of O_2 due to the oxidization of HSQ films. The data also shows that lower pressure will help to increase ash selectivity. In this study, processes have been demonstrated, which yield a PR:HSQ selectivity greater than 150, while maintaining the dielectric constant of HSQ at 2.8.

INTRODUCTION

As the minimum geometry of ultra large scale integrated (ULSI) devices moves toward below 0.2 μm, implementation of low k dielectric materials in the fabrication of these devices is needed to reduce the intraline capacitance between metal lines and to increase the signal propagation speed [1]. Among all available low k materials, hydrogen silsesquioxane (HSQ) has attracted attention in the semiconductor industry due to its lower dielectric constant (k ≤ 3.0, after cure), ease of processing, excellent gap fill and excellent planarization capabilities. In addition, HSQ has been used in production for 0.35-0.5 μm devices [2]. However, one of the challenges in applying HSQ in < 0.20 μm devices is the photoresist (PR) stripping step due to possible interactions between stripping chemistries and HSQ films and, therefore, change its dielectric properties.

The general chemical composition of HSQ before curing is $(HSiO_{3/2})_n$ and a representative molecular structure is shown in Figure 1. From the infrared spectrum, it has been shown that HSQ consists of random network linkages based on the eight-corner $H_8Si_8O_{12}$ structure [3]. When thermally processed, the bonds in HSQ break and its molecular structure is rearranged/redistributed into an amorphous film with more network bonding.

It has been known that HSQ films are sensitive to the amine based PR stripper used in wet PR stripping processes and to O_2 used in the traditional O_2 downstream plasma stripping processes [4-5]. The HSQ film shrinkage after the PR stripping step can lead to via bowing and higher via resistance. Therefore, for applications in deep submicron devices, there is a need to develop a better dry PR stripping process for HSQ without changing the HSQ film properties. In this study, the impact of gas mixture, process pressure, and process temperature for photo resist strip have been investigated.

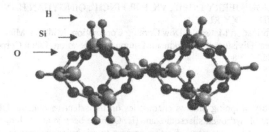

Figure 1. Simulation of HSQ molecule.

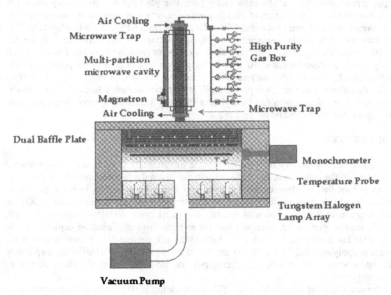

Figure 2. Fusion Gemini ES plasma Asher.

EXPERIMENTAL

HSQ films with an approximate thickness of 4000 Å were prepared by spin-coating Dow Corning FOx®-15 solution onto Si wafers. This was followed by heating on three hot plates at 150°C, 200°C, and 350°C for one min. each. The films were then cured in a quartz furnace at different temperatures and different gas ambients as listed in Table 1. 8500Å of DUV photoresist (PR) (Shipely, UV-6.08) were processed on a Flexfab spin-coater and baked at 120°C for one min. These PR wafers and HSQ wafers were exposed to the same plasma conditions to calculate the PR:HSQ selectivity. The ashing process was performed in a downstream microwave plasma asher (Eaton Corp, Fusion System Division, Model #Gemini ES) as shown in Figure 2. The ash conditions are listed in Table 1. Wafer #12 is the control wafer, which was not treated by ashing. The thickness and the refractive index (R.I.) of the HSQ films before and after ashing were measured by a Tyger Thin Film Analyzer. The Si-H bond density remaining in the HSQ films was measured by Transmission Fourier Transform Infrared (FTIR) spectroscopy. The dielectric constant of the HSQ films was evaluated at 1 MHz using metal-insulator-semiconductor capacitors with Al gate electrodes

Table 1. The ash conditions for HSQ and photoresist.

Run #	Time (seconds)	Pressure (torr)	Temperature (°C)	Process Gas
1	60	Low	270	GasMix 1
2	60	High	270	GasMix 1
3	60	Low	270	GasMix 2
4	60	Low	200	GasMix 2
5	90	Low	80	GasMix 3
6	30	Low	140	GasMix 3
7	30	Low	120	GasMix 3
8	60	Low	100	GasMix 3
9	40	Low	140	GasMix 4
10	20	Low	200	GasMix 4
11	Same as run # 10, with 60-second pre-treatment*			
12	Control Wafer			

GasMix 1: H_2/N_2
GasMix 2: $H_2/N_2/CF_4$
GasMix 3: H_2 rich, $H_2/N_2/CF_4/O_2$
GasMix 4: O_2 rich, $H_2/N_2/CF_4/O_2$
*Pretreatment under H_2/N_2 plasma

Results and Discussions

The thickness shrinkage rate and the refractive index (R.I.) of HSQ films after ashing are shown in Figures 3 and 4, respectively. In general, HSQ films ashed in GasMix 1 and GasMix 2 had relatively smaller thickness changes compared to those ashed in GasMix 3 and GasMix 4. The data shows that HSQ films ashed in H_2/N_2 GasMix 1 have negligible changes in thickness and R.I. compared to control wafer. The data also shows that HSQ films ashed in GasMix 3 (runs # 5-8) have larger thickness loss compared to the control. However, the R.I. of the HSQ films is lower compared to the control, which may be due to these HSQ being exposed to the CF_4 plasma and a reduction of the dipolar polarization through fluorine incorporation. Under the GasMix 3 condition, HSQ films were etched away by the plasma. HSQ films ashed in GasMix 4 (runs # 9-11) show slightly smaller thickness loss compare to those ashed in the GasMix 3 but show a dramatic increase in R. I. to higher than 1.4. The results suggest that ashing in O_2 rich GasMix 4 may oxidize the HSQ film. Apparently, O_2 is detrimental to the HSQ films ashed in both GasMix 3 and GasMix 4 ambients. However, it has been reported that the change of the

HSQ films in the via sidewall is expected to be much less than the blank films due to less plasma exposure and the vertical directionality of plasma components [6].

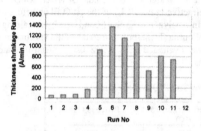

Figure 3. Thickness shrinkage rate of HSQ films under different conditions.

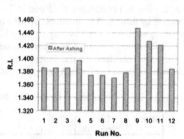

Figure 4. R.I. of HSQ films after ashing process.

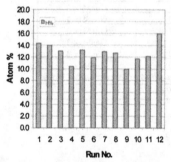

Figure 5. Atomic percent of H in HSQ films after ashing.

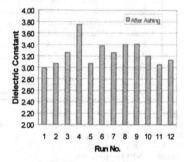

Figure 6. Dielectric constant of HSQ films after ashing.

HSQ film cured in 400°C N_2 ambient contains approximately 16 % H, 31% Si and 63% O atoms [3,7]. The H percentage in HSQ films after ashing is shown in Figure 5. HSQ films can lose a certain percentage of H atoms if the Si-H bonds are broken during the PR stripping process. The HSQ film processed in GasMix 1 has the lowest H loss. In general, there is less H loss if HSQ is exposed to GasMix 3 rather than in GasMix 4. This is speculated to be due to the O_2 rich ambient in GaxMix 4 which expedites the oxidation of HSQ films and leads to larger amount of H loss. However, there are large variations in H loss in GasMix 2, 3, and 4 depending on other process parameters.

The dielectric constant of HSQ films after ashing is shown in Figure 6. It shows that the HSQ film exposed to GasMix 1 has a lower k value than other films. It has been reported that the k value of HSQ film can be maintained when treated by H_2 plasma [8]. Therefore, H_2/N_2 GasMix will help to maintain the k value for HSQ films during ashing processes. However, the k value of HSQ films will increase due to the generation of –OH group when HSQ is treated by

O_2 plasma [9]. For HSQ exposed to GasMix 3, run # 5 has lower k value than other films. It indicates that a lower ash temperature should be used to maintain the k value of HSQ film if O_2 is present in the gas mixtures, such as GasMix 3 and 4. The k value of run # 11 also showed little change compared to the control. Since the k of HSQ can be stabilized by H_2 plasma treatment, the k value of HSQ films with H_2/N_2 plasma pre-treatment remains low compared to films run without pretreatment. Run #4 has the highest k value due to large percentage of H loss in the HSQ film after ashing.

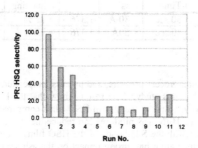

Figure 7. The PR:HSQ ash selectivity.

The PR: HSQ ash selectivity is shown in Figure 7. Among all the GasMix ambients used in the ashing processes, H_2/N_2 GasMix 1 shows ash selectivity close to 100:1 with minimal impact on the film thickness, R.I., and k value. Comparing the PR:HSQ ash selectivity between run # 1 and run #3, shows that adding CF_4 into the H_2/N_2 mixtures will reduce the ash selectivity. All HSQ film treated in GasMix 3 and GasMix 4 have ash selectivity less than 30, which can not be used in production. HSQ films treated in GasMix 3 have more film etched than in GasMix 4. However, there was more oxidization of HSQ films treated in GasMix 4.

The impact of pressure during ash on HSQ can be observed in runs #1 and #2. Lower ash pressure leads to slightly lower H loss and K for HSQ films but much larger PR/HSQ selectivity. This is mostly due to the enhanced PR strip rate at lower pressures.

The impact of ash temperature for GasMix 2 can be observed from runs #3 and #4. There is more HSQ thickness loss when using a lower ash temperature. It has been reported that the etch rate of PR increases with temperature [8]. With higher loss of H, the R.I. of run # 4 is larger than run #3. In addition, the dielectric constant of run # 4 is 0.5 higher than run # 3. The PR:HSQ ash selectivity is higher for run #3, which has higher ash temperature.

However, the HSQ thickness shrinkage rate increases with increasing the ash temperature in an O_2 environment. The H% loss also increases with increasing the ash temperature and the dielectric constant of HSQ films follows the same trend. In this case, the PR:HSQ selectivity is very low and it shows that higher temperature has slightly higher selectivity. The same phenomena were observed on tests # 9 and #10. Higher ash temperature will lead to larger loss of HSQ films and H %, which results in a higher dielectric constant.

One interesting finding is that a 60-second H_2/N_2 pretreatment will slightly improve the resistance of HSQ in an O_2 plasma environment and the PR/HSQ selectivity also increases. It is consistent with Liu's results on HSQ that H_2/N_2 plasma treatment can enhance the resistance of HSQ to O_2 plasma processes [7].

Finally, optimization of the recipes using GasMix 1 and GasMix 2 for HSQ has been explored. The results shows that the PR:HSQ selectivity is greater than 150 while the k value of HSQ remained less than 2.9 as listed in Table 2.

Table 2 Optimized ash processes for HSQ films

GasMix	Pressure	Ash Time	HSQ loss	R.I.	SiH remaining	K	PR/HSQ selectivity
1	Low	30 sec.	< 30 Å	1.376	> 72 %	2.8	> 150
2	Low	30 sec.	< 50Å	1.376	> 72 %	2.8	> 150

SUMMARY

The impact of gas mixture, process pressure, and process temperature on PR strip over HSQ films has been investigated in this study. The results show that the H_2/N_2 gas mixtures have better PR:HSQ ash selectivity with less changes in HSQ films compared to other gas mixtures. In addition, adding CF_4 to the H_2/N_2 gas mixtures can reduce the PR:HSQ ash selectivity. Gas mixtures containing O_2 react with the SiH bonds of the HSQ films and greatly reduce PR:HSQ ash selectivity. The ash temperature has strong impact on the ash selectivity. It is found that higher ashing temperature enhances the ashing selectivity in gas mixtures without the presence of O_2, while maintaining the k value. On the other hand, a slight enhancement in selectivity was observed by increasing the ashing temperature in O_2 gas mixtures but it compromises the k value, due to the oxidization of HSQ films. The data also shows that lower ash pressure will help to increase the ash selectivity. Finally, optimized processes have been demonstrated to have a PR:HSQ selectivity greater than 150 while the k of HSQ is less than 2.9.

REFERENCES

1. The National Technology Roadmap for Semiconductors, Semiconductor Industry Association, San Jose, CA, 1997.
2. S. P. Jeng, K. Taylor, T, Seha, M. Chang, J. Fattaruso, and R. Havemann, Proceedings of VLSI symposium, Kyoto, Japan, 1995, pp. 15-21.
3. M. Loboda, C. M. Grove, R. F. Schneider, J. Electrochem. Soc., **145**, 1998, p. 2861.
4. H. Meynen, R. Uttecht, T. Gao, M. V. Hove, S. Vanhaelemeersch, and K Maex, Proceeding of 3rd Int. Symp. on low and high dielectric constant materials, San Diego, CA, May 5, 1998.
5. D. Louis, E. Lajoinie, F. Pires, W. M. Lee, and D. Holmes, Microelectronic Engineering, 41/42, 1998, p. 415-418.
6. T. Gao, B. Coenegrachts, J .Waeterloos, G. Beyer, M. Meynen, M. Van Hove, and K. Maex, Proceedings of 2nd IITC, San Francisco, CA, 1998, P. 53.
7. H. C. Liou, E. Dehate, J. Duel, and F. Dall, submitted to The Journal of Electrochemical Society.
8. P. T. Liu, T. C. Chang, S. M. Sze, F. M. Pan, Y. J. Mei, W. F. Wu, M. S. Tsai, B. T. Dai, C. Y. Chang, F. Y. Shih, and D. H. Huang, Thin Solid Films, 332, 1998, pp. 345-350.
9. P. Gillespie, R. Mohondro, M. A. Jones, W. L. Krisa, and T. Romig, Semiconductor International, July 1997, p. 161.

Mat. Res. Soc. Symp. Proc. Vol. 612. © 2000 Materials Research Society

CURING STUDY OF HYDROGEN SILSESQUIOXANE UNDER H_2/N_2 AMBIENT

HUEY-CHIANG LIOU, EVAN DEHATE, JERRY DUEL, AND FRED DALL
Dow Corning Corporation, Semiconductor Fabrication Materials, Midland, MI 48686-0994

ABSTRACT

Thin film properties of hydrogen silsesquioxane (HSQ) cured at different temperature under N_2 and H_2/N_2 ambients have been studied. In this study, it was found that compared to an N_2 ambient, film curing in an H_2/N_2 ambient will impact HSQ properties when the temperature is 400°C – 500°C. H_2/N_2 ambient can be used to minimize the dielectric constant while increasing modulus of the films. The data indicates that H_2 can minimize the oxidation of the HSQ films and maintain the dielectric properties.

INTRODUCTION

As the geometry of microelectronic devices continues to shrink, new materials with low dielectric constant (k) and gap fill capability are being sought after by the semiconductor industry [1]. These new materials are used as an interlayer dielectric to enhance the performance of integrated circuits (ICs) by reducing the resistance-capacitance (RC) delay and which, in turn, increases the signal propagation speed.

Recently, hydrogen silsesquioxane (HSQ) films have attracted attention in the semiconductor industry due to low dielectric constant (k < 3.0), ease of processing, good planarization properties and excellent gap fill capability in deep submicron features [2]. The general chemical composition of HSQ before curing is $(HSiO_{3/2})_n$ and a representative molecular structure is shown in Figure 1. From the infrared spectrum, it has been shown that HSQ consists random network linkage based on the eight-corner $H_8Si_8O_{12}$ structure [3]. When thermally processed, the bonds in HSQ break and its molecular structure is rearranged/redistributed into an amorphous film with more network bonding.

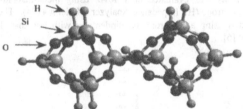

Figure 1. The simulation of HSQ molecule.

A large reduction of SiH bonds occurrs when the cure temperature is greater than 350°C in an N_2 ambient or lower than 350°C in an O_2 ambient [3-4]. The reaction is faster at higher cure temperatures. It was also found that when the loss of hydrogen exceeds a certain percentage of the initial value, polar -OH groups are incorporated into the film and increase the dielectric constant [5]. In addition, increasing the concentration of O_2 in the curing ambient will magnify this effect. Therefore, HSQ might achieve more desirable film properties, e.g. k < 3, if the

Table 1. The curing conditions of HSQ in a quartz furnace

Gas Ambient	Start Temp.(°C)	Cure Temp. (°C)	Cure Time (hr.)
A1, A2, A3	40	350	1.0
A1, A2, A3	40	400	1.0
A1, A2, A3	40	450	1.0
A1, A2, A3	40	500	1.0

Ambient: A1-N_2, A2-5%H_2/95%N_2, A3-10%H_2/90%N_2

presence of H_2 can react with O_2 to reduce the oxidization of HSQ. It might also be possible that the presence of H_2 can participate in the reactions whose net effect is to minimize the loss of Si-H bonds while still allowing film densification, thus improving the mechanical strength. Therefore, in an attempt to produce low k films with higher density and better mechanical properties, H_2/N_2 gas was used as the ambient gas during curing/annealing to verify this hypothesis.

EXPERIMENTAL

Approx. 6000 Å thick HSQ films were prepared by spin-coating Dow Corning FOx®-16 solution onto Si wafers. This was followed by heating on three hot plates at 150°C, 200°C, and 350°C for one min. each. The films were then cured in a quartz furnace at different temperatures and different gas ambients as listed in Table 1. The N_2 is supplied from liquid N_2 boiloff lines which may contain O_2 as an impurity. The H_2/N_2 gas was supplied by gas cylinders containing less than 5 ppm of O_2. The thermal stability tests were performed by annealing HSQ films under 5%H_2/95%N_2 gas ambient in the quartz furnace as listed in Table 2. The thickness and the refractive index (R.I.) of the HSQ films before and after cure were measured by a Tyger Thin Film Analyzer. The Si-H bond density remaining in the HSQ films was measured by Transmission Fourier Transform Infrared (FTIR) (Nicolet, Model# 5SXB). The dielectric constant of the HSQ films was evaluated at 1 MHz using metal-insulator-semiconductor capacitors with Al gate electrode (HP Impedance Analyzer, Model # 4194). The modulus of the HSQ films was evaluated using a Hysitron® nanoindentator with less than 10% of the film thickness being indented [6].

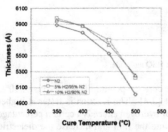

Figure 2. The film thickness of HSQ films at
different cure temperatures and gas ambient.

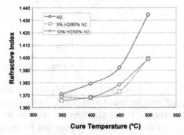

Figure 3. The refractive index of HSQ films at
different cure temperatures and gas ambient

RESULTS AND DISCUSSIONS

The film thickness of HSQ cured at different temperatures and gas ambients is shown in Figure 2. The data shows that the final film thickness decreases with increasing cure temperature in N_2 and H_2/N_2 ambients. However, thicker films and less film shrinkage were observed for HSQ films cured in an H_2/N_2 ambient when compared with those cured in an N_2 ambient. There is quite a difference in the thickness changes between N_2 and H_2/N_2 cured films but there is not much difference in thickness between films cured in 5% H_2 and 10% H_2 ambients.

The R.I. of the HSQ films cured in different ambients are shown in Figure 3. The data shows that the HSQ films cured in the H_2/N_2 gas ambient have a smaller R.I. than those cured in the N_2 ambient. Within measurement error, the difference in R.I. becomes significant when the cure temperature is greater than 400°C. Combining the trends in thickness and R.I. vs. temperature, the data suggests that the structure of the HSQ films might be different when cured in N_2 and H_2/N_2 ambients at different temperatures.

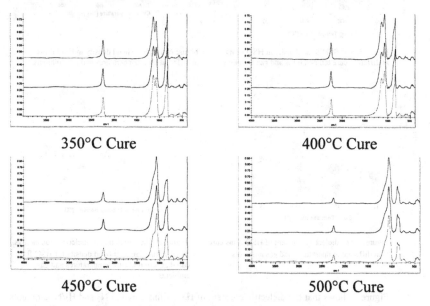

350°C Cure   400°C Cure

450°C Cure   500°C Cure

Figure 4. FTIR spectra of HSQ films cured in N_2 and H_2/N_2 ambient

The FTIR spectra of HSQ films cured at different temperatures are shown in Figure 4. There is little change in the Si-H bond density at 2250 cm^{-1} and in the Si-O bond density at 1070 cm^{-1} in the FTIR spectra at 350°C and 400°C in all gas ambients. However, the different gas ambients result in larger variations in bond density of Si-H bonds and Si-O bonds at 450°C and 500°C. Comparing with the Si-H bond density of as-spun HSQ films, the remaining Si-H bond density in the film decreases with increasing cure temperature both in N_2 and H_2/N_2 ambients as shown in Figure 5. Figure 6 shows that the Si-O bond density increases with increasing the curing temperature. It indicates that the HSQ films were densified at higher curing temperatures. Comparing films cured in the N_2 ambient, there is a higher percentage of hydrogen for HSQ films cured above 450°C in the H_2/N_2 ambient. It is proposed that the H_2/N_2 gas can minimize the percentage of hydrogen dissociated from HSQ at higher curing temperatures, e.g. $\geq$ 450°C.

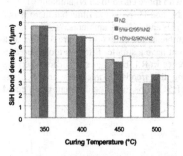

Figure 5. The Si-H bond density in HSQ films at different cure temperatures and gas ambients.

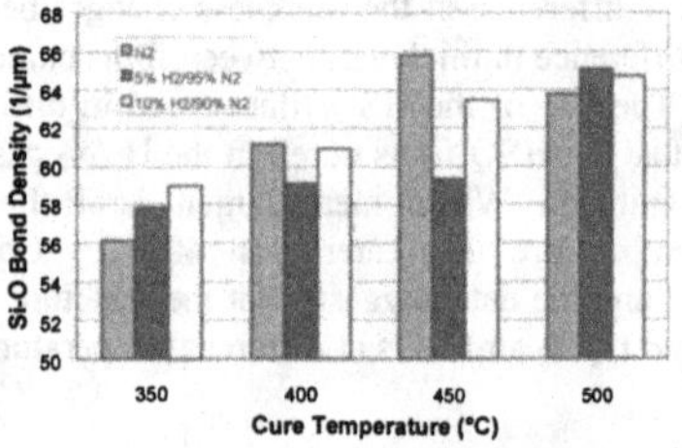

Figure 6. The Si-O bond density in HSQ films at different cure temperatures and gas ambients.

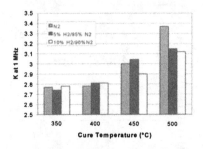

Figure 7. The dielectric constant of HSQ films cured at different temperatures and gas ambients.

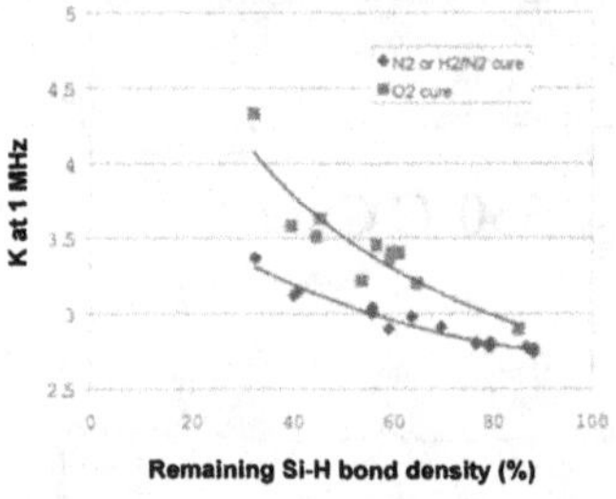

Figure 8. The correlation of dielectric constant vs. % of remaining Si-H bond density for HSQ films cured at different cure temperatures and gas ambients.

Figure 7 shows that the dielectric constant of HSQ films cured in N_2 and H_2/N_2 ambients increases with increasing curing temperature. The dielectric constant of the HSQ film is less than 2.8 when cured at temperature between 350°C and 400°C. There is not much difference in the dielectric constant between N_2 cured and H_2/N_2 cured films. However, the dielectric constant of HSQ films cured in an H_2/N_2 ambient is significantly lower than that cured in an N_2 ambient when the cure temperature is above 450°C. As shown in Figure 5, there is a higher percentage of Si-H bond density in HSQ films cured in the H_2/N_2 ambient at 450°C and 500°C. It is suspected that the H_2 plays an important role in the reaction kinetics at these high temperatures since less Si-H bond dissociated from HSQ. Combining the data of HSQ films cured in O_2 [5], the correlation between dielectric constant and the percentage of hydrogen remaining compared to the as-spun HSQ film is shown in Figure 8. The data shows that the correlation of dielectric constant vs. percentage of hydrogen remaining for HSQ cured in N_2 and H_2/N_2 ambients is quite different from that those cured in an O_2 ambient. It has been reported that there is an oxidation reaction when there is more than 50 ppm of O_2 present during the curing process [7]. The oxidation of HSQ will break the Si-H bonds at a lower temperature and lead to higher dielectric constant due to the formation of Si-OH and H_2O in the films. The data shows that curing HSQ in the N_2 and H_2/N_2 ambients can reduce the occurrence of oxidation, which leads to the decrease in

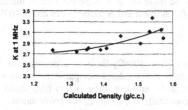

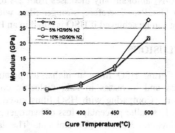

Figure 9. The correlation between dielectric constant and calculated density for HSQ films cured at different cure temperatures and different gas ambients.

Figure 10. The modulus of HSQ films cured at different cure temperatures and gas ambients.

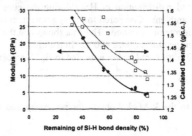

Figure 11. The correlation of modulus and density vs. % of remaining Si-H bond density for HSQ films cured at different cure temperatures and gas ambients.

dielectric constant. H_2 also inhibits the loss of hydrogen in the HSQ films with less network formation. The density of HSQ films can be calculated using the elemental composition from bond density data [3]. The correlation between density and the dielectric constant of HSQ films is shown in Figure 9. It indicates that the dielectric constant of HSQ increases with increasing film density.

The modulus of HSQ films cured in the N_2 and H_2/N_2 ambients measured by the indentation test is shown in Figure 10. The data shows that there is little modulus change between 350°C and 400°C cured HSQ films in all gas ambients, and between N_2 and H_2/N_2 cured HSQ films at these temperatures. However, significant changes in the modulus were observed for HSQ films cured at 450°C and 500°C. The data shows that HSQ films cured in an H_2/N_2 ambient have a lower modulus than those cured in an N_2 ambient at the same cure temperature. Possibly, the increase in modulus is due to more oxidation and film densification as the cure temperature increases. The data in Figure 11 shows that both the modulus and density of HSQ films increase with decreasing percentage of hydrogen. These results are similar to the results reported by Liou et al. [4].

In comparison, HSQ films with a dielectric constant of 2.9 and modulus of 11 GPa were observed when cured at 450°C in an 10% H_2/ 90% N_2 ambient while HSQ films with dielectric constant of 3.0 and modulus of 12 GPa were observed when cured in an N_2 ambient. Compared with HSQ cured at 400°C in an N_2 ambient, the dielectric constant of HSQ films cured at 450°C in an H_2/N_2 ambient only increases from 2.8 to 2.9 but the modulus greatly increases from 6.5 GPa to 11 GPa. Therefore, stronger films with the trade off of a slight increase in dielectric constant can be achieved if HSQ is cured in an H_2/N_2 ambient and at temperatures above 400°C.

CONCLUSIONS

Curing HSQ films at higher temperatures (T > 400°C) leads to a decrease in film thickness and increases in R.I., dielectric constant, and modulus compared to its as spun condition. Comparing with HSQ films cured in an N_2 ambient, there is less film shrinkage, lower R.I., smaller percentage loss of hydrogen, lower dielectric constant, and lower modulus for HSQ films cured in an H_2/N_2 ambient. The data shows that the dielectric constant and elastic modulus increase with decreasing the percentage of hydrogen in the HSQ films. However, the correlation between dielectric constant and hydrogen loss for HSQ cured in N_2 and H_2/N_2 ambients is quite different from those cured in an O_2 ambient. This is because the mechanisms for the oxidation and bond redistribution of HSQ during the cure process depend on the O_2 concentration of the cure ambient. Compared to an N_2 ambient, curing HSQ with H_2 present in the ambient reduces the degree of oxidation. The data shows that the impact of H_2 on the cure of HSQ films increases at temperatures above 450°C. In addition, the thermal stability of dielectric constant is better for HSQ films cured and/or annealed in H_2/N_2 ambient than N2 ambient alone.

ACKNOWLEDGEMENT

The authors would like to thank Mark Loboda for reviewing and making suggestions to this manuscript.

REFERENCES

1. The National Technology Roadmap for Semiconductors, Semiconductor Industry Association, San Jose, CA, 1997.
2. S. P. Jeng, K. Taylor, T, Seha, M. Chang, J. Fattaruso, and R. Havemann, Proceedings of VLSI symposium, Kyoto, Japan, 1995, pp. 15-21.
3. M. Loboda, C. M.Grove, R. F. Schneider, J. Electrochem. Soc., 145, 1998, p. 2861.
4. H.-C. Liou and J. Pretzer, Thin Solid Films, 335, 1998, pp. 186-191.
5. J. Bremmer, Y. Liu, K. Gruszynski, and F. Dall, Mater. Res. Symp. Proc. 476, 1997, p. 37.
6. G. M. Pharr and W. C. Oliver, MRS Bulletin, July, 1992, p. 28-33.
7. J. Bremmer, K. Gruazynski, C. Saha, and K. Chung, VMIC 1998, P. 631.

Mat. Res. Soc. Symp. Proc. Vol. 612 © 2000 Materials Research Society

Characterization and integration in Cu damascene structures of AURORA, an inorganic low-k dielectric

R. A. Donaton, B. Coenegrachts, E. Sleeckx, M. Schaekers, G. Sophie[1], N. Matsuki[1], M. R. Baklanov, H. Struyf, M. Lepage, S. Vanhaelemeersch, G. Beyer, M. Stucchi, D. De Roest, K. Maex[2]
IMEC, Kapeldreef 75, B-3001 Leuven, Belgium
[1] ASM Japan, 6-23-1, Nagayama, Tama, Tokyo, Japan
[2] IMEC and E.E. Department, K. U. Leuven, Belgium

ABSTRACT

AURORA films, which have a Si-O-Si network with $-CH_3$ terminations, were characterized and integrated into Cu single damascene structures. The relatively low carbon concentration ($\sim 20\%$) and the very small pore size (~ 0.6 nm) found could be advantageous during integration of AURORA. Integration of AURORA into Cu single damascene structures was successfully achieved. Suitable resist strip processes, which are critical for Si-O-C type materials, were developed, resulting in trenches with satisfactory profiles. After a complete single damascene process, a interline dielectric constant value of 2.7 was found for line spacing down to 0.25 μm.

INTRODUCTION

The replacement of silicon dioxide by materials with lower dielectric constant in metallization schemes becomes mandatory as the technologies migrate towards the 130 nm (and below) technology nodes. As the dimensions decrease, the line to line capacitance dominates the parasitic capacitance of the interconnects and, in long parallel lines, cross talk becomes an issue due to coupling capacitance [1]. Even though a long list of candidates, deposited either by spin-on or vapor deposition techniques, has been investigated over the last years [2,3], it is not clear at this moment which material will be the selected one. Among them, there is a myriad of materials with dielectric constant between 2.6 and 3.0, including diamond-like carbon films, organic polymers, hybrid organic siloxane polymers and organosilicate glasses (OSG), the later also known as carbon-doped SiO_2 or simply Si-O-C films.

A long residence plasma CVD (LR-CVD) method to deposit inorganic low-k materials has been shown recently by Matsuki et al. [4]. The key feature of this process is the enhanced dissociation and polymerization of source gas molecules in the vapor phase, followed by a completion of the reaction on the substrate after deposition. As the residence time increases, the dielectric constant is reduced from 3.6 to 2.6 and the deposition rate is increased. AURORA, a material with Si-O-Si network with $-CH_3$ terminations, is therefore formed by optimizing the LR-CVD technique and using dimethyldimethoxysilane $(CH_3)_2Si(OCH_3)_2$ as precursor.

In this paper we discuss the basic characteristics of AURORA films deposited at IMEC and the integration of this material in Cu single damascene structures.

RESULTS AND DISCUSSION

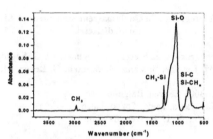

Fig. 1 – FTIR spectrum measured from a 500nm thick AURORA layer.

Film deposition

Deposition of AURORA films at IMEC was done in a single-wafer, parallel plate plasma CVD system based on an ASM Eagle-10 system. Dimethyldimethoxysilane is used as precursor and He as additive gas. Temperature during deposition was 400 °C.

The deposited films have a refractive index of 1.41, as measured by spectroscopic ellipsometry.

Film structure

Fourier Transform Infrared Spectroscopy (FTIR) is used to characterize the AURORA film structure. Fig. 1 shows an infra-red spectrum obtained from a 500nm thick AURORA film. The strong absorption peak observed at 1040 cm^{-1} results from the presence of Si-O bonds (stretch) in the film. Also observed in the spectrum are peaks corresponding to CH$_3$ (2970 cm^{-1}), CH and CH$_2$ asymmetric bonds at 2910 cm^{-1}, a small peak at 2170 cm^{-1} that corresponds to Si-H bonds, the antisymmetric and symmetric deformation of C-H in a Si-(CH$_2$)$_n$-Si at 1412 and 1360 cm^{-1} respectively [5], and a strong peak at 1280 cm^{-1} corresponding to CH$_3$-Si bonds. A peak around 800 cm^{-1} is also observed, and it can be assigned to Si-C, Si-CH$_2$ and Si-CH$_3$ bonds. FTIR spectra measured at center and edge of the wafer, for both 250 and 500nm thick AURORA films showed the exact same features, indicating a good uniformity in the film structure. No indication of moisture uptake is observed from the FTIR spectrum, even after a month of storage in clean room environment.

Film composition and surface roughness

Rutherford Backscattering Spectrometry (RBS) and Elastic Recoil Detection (ERD) are used to study the composition of AURORA films. ERD is used for hydrogen detection. Energies of 3.045 MeV and 4.26 MeV are used for resonance measurements of oxygen and carbon, respectively. A composition of 25% Si, 50% O, 20% C and 5% H is found.

The surface roughness of a 500 nm thick AURORA film was analyzed by atomic force microscopy (AFM). Roughness was measured over different spots on the sample and the scans were done over different areas (1 x 1, 2 x 2 and 5 x 5 μm^2). Fig. 2 shows a 3D image of the film surface. An average surface roughness below 0.5 nm is found, and this value is independent of the area measured.

Porosity

Ellipsometric porosimetry (EP), which is a new version of adsorption porosimetry, is a powerful technique to characterize the porosity of microporous materials such as AURORA. In this

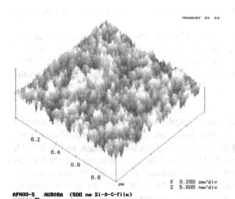

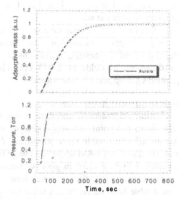

Fig. 2 – 3D AFM image showing the surface roughness over an area of 1 x 1 μm² of a 500 nm thick AURORA film.

Fig. 3 - Kinetics of toluene adsorption in AURORA

method, the mass of an adsorptive condensed / adsorbed in the pores is determined by changes in the optical characteristics of the porous film during vapor adsoption and desorption. A more detailed explanation of the technique and the models used for calculation of porosity and pore size distribution can be found in references [6,7]. Fig. 3 shows the kinetic evolution of the adsorptive mass into AURORA. Toluene, which has a molecule kinetic diameter around 0.6 nm, is used as adsorbate vapor. In the same figure the increase of the pressure in the chamber, due to the introduction of toluene vapor, is shown as a function of time. If the film had a pore diameter larger than the kinetic diameter of the toluene molecule, the adsorbate amount in the film would increase in the same rate as the vapor pressure in the chamber [7]. It can be seen that adsorption of toluene into the AURORA film occurred in a much slower way than the change of the vapor pressure. This suggests that the pore diameter of AURORA is comparable to the kinetic diameter of the toluene molecule, i.e. $d_{AURORA} \approx 0.6$ nm. This reduced pore size of AURORA can be advantageous during integration of this material in damascene structures.

Electrical properties – blanket films

Dielectric constant, leakage current and breakdown field were measured using a metal / insulator / metal (MIM) structure, with the top electrode patterned by lithography and dry etch for accurate area determination. Capacitance vs. frequency (C-F) measurements were performed. The values reported here were taken at a frequency of 100 kHz. Capacitors of different sizes were measured. At least 5 different sites on the wafers were probed. The thickness of the AURORA layers used in the calculation of the dielectric constant was measured by spectroscopic ellipsometry. The capacitors measured were located as close as possible to the thickness measurement sites.

A dielectric constant of 2.9 was found for the AURORA films deposited in IMEC. The dielectric constant was the same for all the various film thicknesses, and did not change during a period of a month (maximum time interval measured). Annealing the samples at 400°C for 20 minutes did not result in any change of the dielectric constant. A standard deviation of the measurements of around 1% is found, indicating that the dielectric constant does not vary significantly over the wafer.

Leakage current was measured at a field of 1 MV/cm. The measurements were carried out in a nitrogen environment. The leakage current values are below 10^{-11} A/cm^2 and are independent of processing conditions and thickness. The AURORA films did not show breakdown up to fields of 2 MV/cm. The measurements were limited to this field by the equipment set-up and film thickness.

<u>Single damascene integration</u>

Single damascene structures using 500 nm AURORA with a 100 nm α-SiC:H (referred as SiC from this point on) etch stop layer were fabricated. Two different hard masks were used: 100 nm SiC or a double layer 50 nm SiC / 150 nm SiO$_2$. Etching of hard mask and AURORA was carried out in a LAM 4520XL. The 100nm SiC hard mask was etched using a Ar/N$_2$/CF$_4$/O$_2$ chemistry, which resulted in good CD-control and etch uniformity. Due to differences in etch rates of oxide and SiC, the same etch time used for etching 100nm SiC could be used for opening of the double hard mask, with similar characteristics for CD-control and uniformity. Etching of the trenches was done using a Ar/CF$_4$/CHF$_3$/O$_2$ chemistry, resulting in trenches with straight profiles and no faceting of the hard mask was observed. Various stripping processes were evaluated in order to remove the remaining photoresist and polymers formed during the trench etch. A downstream O$_2$ plasma (high temperature, high pressure) and a O$_2$/CF$_4$ plasma (lower pressure and temperature) resulted in undercut and bowing, as can be seen in fig. 4a. The severe conditions during the strip caused a modification of the film leading to shrinkage of the layer, as can be observed at the side walls of the trenches. On the other hand, a LAM 4520XL in-situ O$_2$ strip performed at low pressure and low temperature resulted in reasonable profiles, with practically no undercut or bowing of the trenches. However, from patterning point of view, the best strip process was the one carried out in a LAM TCP9100 chamber at low temperature and low pressure using N$_2$/O$_2$ chemistry. Fig. 4b shows a cross-section SEM image of AURORA trenches after stripping in N$_2$/O$_2$. No bowing or undercut is present, and some rounding of the hard mask is observed due to this RIE-type strip.

Integration of AURORA in single damascene structures was successfully achieved. Fig. 5 shows a cross-section image of 0.25 μm Cu trenches of a meander/fork structure made in AURORA with the 50nm SiC / 150nm SiO$_2$ double hard mask. The resistance of 1.8 cm long meanders is also shown in the figure. The metallization scheme consisted of PVD TaN followed by PVD Cu seed layer and subsequent Cu electroplating fill in an ECD chamber from Semitool.

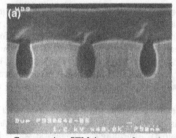 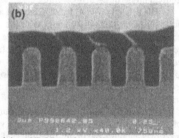

Fig. 4 – Cross-section SEM images of trenches etched in AURORA after different strips: (a) high pressure and high temperature O$_2$ plasma and (b) N$_2$/O$_2$ plasma (low temperature and low pressure).

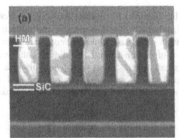
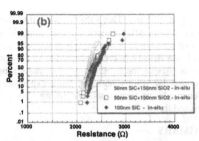

Fig. 5 – FIB image of 0.25 μm Cu trenches in AURORA made with a double hard mask (50nm SiC / 150nm SiO₂) (a), and resistance of 1.8 cm long meanders (0.25 μm wide) for different strips and hard masks (b). 56 chips per wafer were measured.

The wafer with 100nm SiC hard mask stripped with the O_2 in-situ process shows slightly higher values than the ones with the double hard mask and same strip. The small difference comes from the fact that the total trench depth is smaller (due to the thinner hard mask) and less hard mask is removed during the CMP process when 100 nm SiC is used.

Interline capacitance measurements are very useful to investigate the effect of processing steps on the dielectric constant of the low-k material being integrated. Fig. 6 shows interline capacitance as a function of various line spacings for wafers with different hard masks and different strip processes. Interline capacitance of single damascene oxide structures is plotted for comparison. The capacitance of the structures prepared with the double hard mask (SiC/SiO₂) is higher than the ones prepared with a single SiC hard mask due to the higher thickness of the trenches (~ 60 nm thicker). No significant difference is observed between the two different strip splits: O_2 in-situ and N_2/O_2. In order to evaluate the effect of these two strip processes on the dielectric constant of AURORA, simulations using TMA Raphael software were done. The actual profile of the measured structures, determined from FIB cross-section images (Fig. 7), was used in the simulations. The line spacing was not dependent on the strip used, according to these images, which is in agreement with the capacitance measurements. From the simulations and electrical measurements, a dielectric constant of 2.7 was found for AURORA, down to line

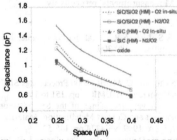
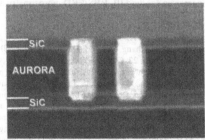

Fig. 6 – Interline capacitance of AURORA trenches as a function of the line spacing for different hard masks and strips.

Fig. 7 – Cross-section FIB image of a 0.25 μm interline capacitance structure (100nm SiC hard mask and O_2 in-situ strip)

spacing of 0.25 μm. This value is lower than the value obtained for blanket AURORA wafers, extracted from MIM capacitors. One possible explanation for the lowering of the dielectric constant between lines upon single damascene integration is the formation of a modified material with lower density (more porous), due to the removal of carbon from the film during the strip process. Similar mechanism was observed during oxidation of a-SiC:H films [6]. This idea is supported by results obtained from experiments performed on blanket wafers exposed to different plasma treatments, and taking into account the differences between etching blanket films and trenches.

SUMMARY

AURORA films deposited at IMEC were characterized and successfully integrated into Cu single damascene structures. AURORA, which has a Si-O-Si network with $-CH_3$ terminations, showed to be a very stable film. No moisture uptake and no change in the dielectric constant were observed. A composition of 25% Si, 50% O, 20% C and 5% H was found. A pore size of 0.6 nm was determined from ellipsometric porosimetry. The relatively low carbon concentration ($\sim 20\%$) and the very small pore size could be advantageous during integration of AURORA. A dielectric constant of 2.9 and leakage current values below 10^{-11} A/cm^2 were measured from MIM capacitors.

As any Si-O-C type material, AURORA showed to be sensitive to resist strip processes during patterning of damascene structures, since it involves the use of plasma with oxygen chemistries. However, low temperature and low pressure processes were found to result in acceptable trench profiles. AURORA was integrated into Cu single damascene structures having different hard masks. Good continuity in meander / fork structures was achieved. The interline dielectric constant of AURORA after single damascene integration was found to be 2.7, down to 0.25 μm spacing. We believe that the lower k-value found is due to the modification of the material at the trench side wall, resulting in a less dense film with lower dielectric constant.

ACKNOWLEDGEMENTS

The authors would like to acknowledge C. Alaerts, T. Dupont, G. Mannaert and IMEC's pilot line for the technical support, H. Bender and C. Drijbooms, B. Brijs, D. Vanhaeren and D. Shamiryan for the FIB, RBS/ERD, porosimetry and AFM measurements, respectively.

REFERENCES

[1] S.W. Russel, A.J. McKerrow, W.-Y. Shih, A. Singh, R.S. List, A.R.K. Ralston, W.W. Lee, K.J. Newton, M.-C. Chang, and R.H. Havemann, Proc. ULSI XIII, 1998, pp.289-300.
[2] MRS Bulletin 22, Oct 1997.
[3] L. Peters, Semiconductor International 22, 1999, pp.56-64.
[4] N. Matsuki, A. Matsunoshita, J.S. Lee, Y. Morisada, Y. Naito, and C. Merritt, 2000 Proceedings Sixth International Dielectrics for ULSI Multilevel Interconnection Conference (DUMIC), pp. 151 to 160.
[5] M. J. Loboda, J.A. Seifferly, R.F. Schneider, and C.M. Grove, Proceedings of the Symposia on Electrochemical Processing in ULSI Fabrication I and Interconnect and Contact Metallization: Materials, Processes, and Reliability. Electrochem. Soc, Pennington, NJ, USA; 1999; viii+274 pp. 145-52.
[6] M.R. Baklanov, K.P. Mogilnikov, V.G. Polovinkin, and F.N. Dultsev, J. Vac. Sci. Technol. B 18(3), May/June 2000.
[7] M.R. Baklanov and K.P Mogilnikov, to be published in Mat. Res. Soc. Symp. Proc., Symposium D: "Materials, Technology, and Reliability for Advanced Interconnects and Low-k Dielectrics", 2000 MRS Spring Meeting.

Mat. Res. Soc. Symp. Proc. Vol. 612 © 2000 Materials Research Society

Low Dielectric Constant Porous Silsesquioxane Films: Effect of Thermal Treatment

Y. K. Siew[*],G. Sarkar[*],X. Hu[*],Y. Xu[**] and A. See[**]
[*]Division of Materials Engineering, School of Applied Science, Nanyang Technological
University Nanyang Avenue Singapore 639798 (E-mail: siewyk@charteredsemi.com)
[**]R&D Dept. Chartered Semiconductor Manufacturing Ltd., 60 Woodlands Industrial Park D,
Street 2, Singapore 738406

ABSTRACT

Low dielectric constant, nanoporous hydrogen silsesquioxane (HSQ) films were prepared
using hybrid templating method by addition of a sacrificial labile polymer, polybutadiene (PB).
Thermal gravimetric analyzer (TGA) was employed to monitor the decomposition behavior of
HSQ-PB films. Curing of HSQ and decomposition of PB were performed by a two-stage thermal
treatment to obtain a defect-free film. The porosity level and pore morphology of the resultant
porous films were found to be dependent upon thermal treatment conditions applied.

INTRODUCTION

As metal pitch of multilevel interconnect continues to shrink, the interconnect RC delay
becomes larger than the intrinsic gate delay. To maintain high signal propagation speed in IC
devices, it will be mandatory to reduce dielectric constants (k) of intermetal dielectric. When
choosing low k materials with k below 3.0, not only are the material properties of great concern,
their extendibility to future generations also affects material selection. It is well agreed that only
porous materials can deliver the lowest required dielectric constants (k<2.0) [1]. In other words,
there is no true dielectric generational extendibility, i.e. similar materials for multiple device
generations differing primarily in k, without embracing the concept of porosity.

A desired porous dielectric should give much lower k while having thermal and mechanical
properties required for subsequent processing. The advantage of a porous dielectric approach can
be explained by figure 1a, which shows a Maxwell-Garnett modeling of porous structures based
on a matrix material with an initial k of 2.8 [2]. Logarithmic mixture rule was also shown [3].
Incorporation of air-filled pores with a k of 1.0, causes a dramatic reduction in the k. However,
porous materials pose some problems for dielectric applications. Pore size must be much smaller
than both the film thickness and minimum device feature size. Moreover, it is desired that the
porosity be in the form of closed cell in order to prevent absorption and diffusion of
contaminants. Although porosity lowers k, it normally has negative impacts on the other material
properties. Therefore porosity should be no higher than needed to achieve the dielectric goals.

Conventionally, effort to develop porous dielectrics has focused on sol-gel derived porous
silica [4, 5]. However, many alternative approaches have been explored due to process
complexity, open-cell morphology and poor mechanical properties of porous silica. An
increasingly viable alternative is the hybrid templating technique. Besides process simplicity,
templated porous materials posses more regular structures, leading to mechanical stability. In
addition, closed-cell pores and intrinsically hydrophobic matrix minimize the films' absorption
of water, a characteristic that distinguishes these dielectrics from porous silica [1, 6]. Materials
under research include inorganic-organic hybrids [7], porous polymers [8-10] and porous
silsesquioxane films [1, 10-11]. This approach involves use of polymeric organic-inorganic

hybrids or block copolymer with controlled morphologies, where the function of the thermally labile macromolecule or pore generator (porogen) is to template the vitrification of the matrix material. The porogen is then removed by thermolysis to leave behind a nanoporous structure.

Hydrogen Silsesquioxane (HSQ), commercially available in solution form as FOx®, offers many properties of silica such as high thermal stability, high hardness and low outgassing. Without Si-C or C-C bonds, HSQ does not cause via poisoning. Since k value of HSQ (2.8-3.0) is smaller than that of silica, smaller porosity is needed for porous HSQ to achieve low k level. Moreover, pore surface of porous HSQ film is hydrophobic due to the hydrophobic nature of Si-H group in HSQ [12]. For these reasons, HSQ is a good candidate for preparation of nanoporous dielectric films using templating technique. This work focuses on the extendibility of HSQ by preparing nanoporous HSQ films via organic/inorganic hybrid templating process using thermally labile macromolecule as porogen. HSQ-polybutadiene matrix-porogen system was explored and the effect of porogen loading and thermal treatment on the pore size and morphology were investigated.

EXPERIMENTAL

Commercially available HSQ was the base materials in our study. FOx® solution (15wt% HSQ in methyl isobutyl ketone) was mixed with polybutadiene (PB), the porogen in this study. A clear solution was obtained following complete mixing of HSQ and PB. The weight percentage of porogen was varied to control the volume fraction of porosity in the resulting films. Spin coating was carried out on bare silicon wafer before baking at 100°C on a hotplate for 3 minutes to evaporate the organic solvent. Spin speed of about 2700rpm was used to obtain film thickness of about 800nm. Subsequently, the films were thermally treated in N_2 ambient to induce curing of HSQ and to remove the porogen by decomposition. A major goal in preparation of porous films is to achieve controllable pore size. In order to obtain evenly distributed pores, it is important to ensure that morphology of the porogen molecules is locked in before their decomposition. This involves two competing processes from vitrification of the matrix and diffusion of the porogen molecules which leads to porogen agglomeration. Therefore, different thermal treatment conditions were explored to evaluate their effect on the resulting films.

The refractive indices (RI) of the films before and after thermal treatment were measured using ultra-violet (UV) interferometry. Porosity of the film (p) can be calculated using the Lorentz-Lorenz equation [13],

$$p=3(n_s^2-n^2)/(n_s^2-1)(n^2+2) \tag{1}$$

where, n_s and n are the RI of dense and porous HSQ films, respectively. This method was confirmed to be effective by Hong and co-workers using both ellipsometry and Rutherford backscattering spectrometry [14]. The thickness of the film was measured by alpha-step techniques and confirmed by XSEM. The film morphology was also assessed by XSEM. In addition, reflection mode Fourier Transform Infrared (FTIR) spectroscopy was employed to verify loss of PB after thermal treatment.

The residue of the mixture solution was prepared by drying the mixture solution and then characterized using thermogravimetric analyzer (TGA) to understand the interaction between HSQ and PB. Different cure conditions can be simulated using TGA. The thermogravimetric analysis allowed us to follow the decomposition of PB under varying heating conditions. TGA experiments were performed in N_2 ambient.

RESULTS AND DISCUSSIONS

Figure 1b shows the decomposition behavior of Polybutadiene (PB). The decomposition of PB starts at around 300°C and ends before 500°C with substantial weight loss occurring after 400°C. Previous studies on thermal curing of HSQ suggested that vitrification of HSQ network took place between 250°C and 435°C [15, 16]. Thus the thermal treatments of HSQ-PB were designed at temperatures between 350 to 450°C.

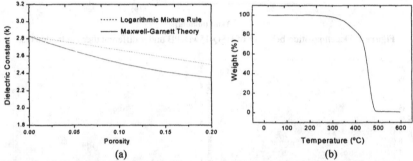

(a)   (b)

Figure 1 (a) Plot of dielectric constant (k) versus porosity as modeled by the Logarithmic mixture rule and Maxwell-Garnett thoery. (b) Decomposition behavior of polybutadiene.

The decomposition behavior of HSQ-15wt%PB under different thermal treatment conditions is given in figure 2. HSQ-15wt%PB was heated at 10°C/min to 350, 400, 425 and 450°C and held isothermally for 60 minutes to simulate the decomposition of PB. Besides initial solvent loss, HSQ experienced less than 3% weight loss during curing [18]. Hence any weight loss above 300°C was considered as a result of PB decomposition. It is clearly shown in figure 2 that only 1, 10, 10.7 and 14.3wt% of PB were driven out of HSQ-15wt%PB under the four thermal treatments; in other words, about 6.67, 66.7, 71.3 and 95.3% of the 15wt% PB were decomposed. Meanwhile, instantaneous weight loss can be observed at about 410°C for thermal treatments at 425 and 450°C. This observation suggested a time independent weight loss in HSQ-15wt%PB after it was heated above 400°C. High ramp rate (10°C/min) employed might have promoted instantaneous weight loss. There is no significant further weight loss 30 minutes after holding temperature is reached for all the thermal treatments. Therefore, to reduce thermal budget and avoid overcuring of HSQ, HSQ-PB films were treated for only 30 minutes.

Figure 3a and 3b show the porosity levels and thickness reduction of porous HSQ films as a function of treatment condition for HSQ-PB of different porogen loading respectively. Similar trend as observed in TGA simulation can be found. Film porosity was not optimized owing to incomplete decomposition of porogen. Recorded FTIR spectra of the treated films supported this finding as well. This demonstrates that TGA can be employed to monitor the decomposition behavior of HSQ-PB films. Treated HSQ films did not suffered significant film shrinkage with less than 5% thickness reduction.

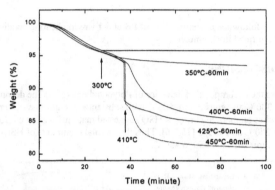

Figure 2 Decomposition behavior of HSQ-15wt%PB under different thermal treatments.

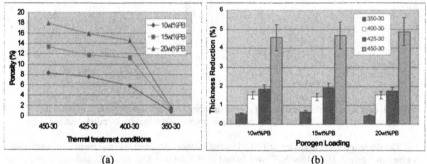

(a)   (b)

Figure 3 (a) Porosity levels (b) Thickness reduction of porous HSQ films.

The morphology of the porous films was assessed by XSEM. There was no evidence of pores for films treated at 350°C. For films treated at 400°C, pores were observed in the form of micro-domain in which pores of about 50nm in size were densely distributed. The uniformly distributed micro-domains are elliptical in shape and have a lateral dimension of about 400nm. These micro-domains tend to scatter near the bottom of the porous film. These micro-domains could be a result of PB agglomeration. Despite similar pore size and morphology, percolation pathways, which rendered the films defective, were observed in films treated at 425 and 450°C. Figure 4a shows the XSEM micrograph of HSQ-15wt%PB film treated at 425°C. The morphology in these porous films seems to be unique to this matrix-porogen system and process.

Formation of percolation pathways in films treated above 425°C can be correlated to the instantaneous weight loss observed in TGA simulation (figure 2). Formation of pores involves generation of low molecular weight decomposition products which diffuse out of the film. There are basically three processes taking place during thermal treatment, namely, decomposition of porogen, diffusion of the decomposition by-products through the matrix and thermal curing of the matrix material. When decomposition rate of the porogen is greater than the diffusion rate of the decomposition products, the gas pressure in the pores builds up. If gas pressure exceeds the strength of the matrix, percolation pathways will form to facilitate escape of the decomposition by-products. When HSQ-PB films were heated to above 400°C, tremendous gas pressure would build up in the pores due to excessive decomposition of PB. On the other hand, HSQ was still under-cured and did not possess sufficient mechanical strength as curing was a time dependent diffusion-controlled process. Hence the gaseous by-products were forced to find a path to escape. This leads to the formation of percolation pathways resulting in instantaneous weight loss.

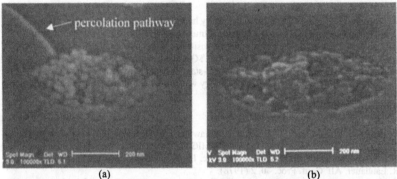

(a)   (b)

Figure 4 Cross-sectional SEM micrograph of HSQ-15wt%PB film (a) heated to 425°C and held for 60 minutes, (b) after a 2-stage 400-450°C thermal treatment.

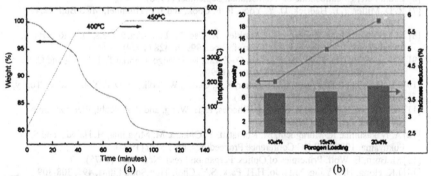

(a)   (b)

Figure 5 (a)Decomposition behavior of HSQ-15wt%PB during a two-stage thermal treatment (b) Porosity and thickness reduction of resulting films.

In order to prevent formation of percolation pathway, a two-stage thermal treatment is proposed. Figure 5a shows the decomposition behavior of HSQ-15wt%PB monitored by TGA. The temperature was increased at 10°C/min to 400°C and held for 30 minutes before being increased again to 450°C at 5°C/min and held for 60 minutes. During the 30-minute isothermal holding at 400°C, moderate decomposition rate allows the decomposition products to diffuse out slowly without pressure build-up while HSQ continues to cure leading to greater mechanical strength. This eliminates formation of percolation pathway and hence no evidence of instantaneous weight loss was found when the film was heated to 450°C and held for 60 minutes. HSQ-PB films treated by this two-stage thermal treatment exhibited the targeted porosity level (figure 5b) and similar pore size and morphology without evidence of percolation pathways (figure 4b).

CONCLUSIONS

Nanoporous HSQ films were demonstrated by hybrid templating method using polybutadiene as porogen. One stage thermal treatments were found to lead to unoptimized porosity level due to incomplete porogen decomposition or defective films due to percolation pathway formation. It was also demonstrated that TGA can be employed to monitor the decomposition behavior of HSQ-PB films. A two-stage thermal treatment was proposed to resolve the issues. Micro-domain pore morphology was observed with pore size of about 50nm.

REFERENCES

[1] W. Volksen, R.D. Miller, J.L. Hedrick, C.J. Hawker, J. F. Remenar, P. Furuta, C. Nguyen, D.Y. Yoon and M. Toney, Sept. 7-9, 1999 VMIC Conference. P.407 (1999).
[2] J.C. Maxwell-Garnett, Philos. Trans. Roy. Soc. 15, 2033 (1982).
[3] R. Landauer, AIP Conf. Proc. 40, 2 (1978).
[4] C. M. Jin, J. D. Luttmer, D. M. Smith, and T. A. Ramos, MRS Bull. 22, 39 (1997).
[5] C. H. Ting and T. E. Seidel, Mater. Res. Soc. Symp. Proc. 381, 3 (1995).
[6] Laura Peters, Semiconductor international, September (1998).
[7] R. D. Miller, J. L. Hedrick, D. Y. Yoon, R. F. Cook, and J. P. Hummel, MRS Bull. 22,44 (1997).
[8] M.L. O'Neill, L.M. Robeson, T.J. Markley, X. Gao, M. Langsam, J. Stets, D.A. Roberts, S. Motakef, and P.R. Sierocki, VMIC Conf. Proc.1999, p. 428 (1999).
[9] Y. Xu, Y.-P. Tsai, K. N. Tu, B. Zhao, Q.-Z. Liu, M. Brongo, George T. T. Sheng and C. H. Tung, Applied Physics Letters 75, 6, 853 (1999).
[10] J.L. Hedrick, R.D. Miller, C.J. Hawker, K.R. Carter, W. Volksen, D.Y. Yoon and M. Toney, Advanced Materials 10, 13, 1049 (1998).
[11] A. T. Kohl, R. Mimna, R. Shick, L. Rhodes, Z. L. Wang, and P. A. Kohl, Electrochem. Solid-State Lett. 2,77 (1999).
[12] A. Nakashima, R. Muraguchi, M. Komatsu, Y. Ohkura, M. Miyajima, H. Harada, and S. Fukuyama, 1998 DUMIC Conference Proceedings, pg. 25-30 (1998).
[13] M. Born, E. Wolf, Principles of Optics, Pergamon Press, New York (1975).
[14] J.K. Hong, H.S. Yang, M.H. Jo, H.H. Park, S.Y. Choi, Thin Solid Films, 495, 308-309 (1997).
[15] D. Tobben, P. Weigand, M.J. Shapiro, and S.A. Cohen, Mat. Res. Soc. Symp. Proc. 443, p 195 (1997)
[16] Y.K. Siew, G. Sarkar, X. Hu, J. Hui, A. See and C. T. Chua, Journal of Electrochemical Society, 147 (1) 335-339 (2000).

Mat. Res. Soc. Symp. Proc. Vol. 609 © 2000 Materials Research Society

Microstructure and Electronic Properties of Thin Film Nanoporous Silica as a Function of Processing and Annealing Methods

Christine Caragianis-Broadbridge[1*], John R. Miecznikowski[1], Wenjuan Zhu[2], Zhijiong Luo[2], Jin-ping Han[2] and Ann Hein Lehman[1]
[1]Department of Engineering; Trinity College, Hartford, CT, 06106; [2]Department of Electrical Engineering; Yale University, New Haven, CT, 06520; *Visiting Fellow, Yale University.

ABSTRACT

Alcogels, aerogel precursors, were prepared by hydrolysis and condensation of the metal alkoxide tetraethylorthosilicate and were catalyzed by both acids and bases, according to a standard reaction. Alcogel solution was spin coated onto p-type silicon wafers and fluid extraction was achieved in an uncontrolled (room temperature, atmospheric pressure) environment. Film porosity was retained through surface modification and/or low vapor pressure solvent techniques. The microstructure and electronic properties of the resulting films were evaluated using non-contact atomic force microscopy (nc-AFM), cross sectional scanning electron microscopy (SEM), and transmission electron microscopy (TEM). Metal insulator semiconductor (MIS) devices were prepared and current-voltage and capacitance-voltage measurements were obtained from these devices. Annealing studies reveal a dramatic temperature dependent effect on both the microstructure and electronic properties of the porous silica films.

INTRODUCTION

Aerogels are highly porous materials with unique optical, thermal, mechanical and electrical properties. Bulk nanoporous silica has been studied and has been shown to exhibit the lowest refractive index (n), thermal conductivity, sound velocity, and dielectric constant (k) of any known material. The potential applications for these materials include sensors, photonic, and electronic devices. Specifically, the low-k properties of thin film silica aerogels (k=1-3) make them highly attractive as a replacement for the conventional insulator (silicon dioxide, k=3.9) which is now utilized for interlevel dielectrics (ILDs) in integrated circuits [1]. The introduction of low-k ILDs into ICs will potentially result in decreased power dissipation, reduced cross-talk and faster achievable switching speeds [2]. The k of silica aerogel is, however, process dependent, while the reliability is similarly determined by such properties as porosity, film uniformity and adhesion. The long-term stability of these films is also an issue, with such environment parameters as annealing response critical to subsequent process integration. Before semiconductor manufacturers can implement thin film nanoporous silica, these issues must be examined in detail.

The focus of this research was the fabrication of thin film nanoporous silica while utilizing methods that do not require pore-fluid extraction under supercritical conditions. Standard aerogel preparation techniques require supercritical extraction of pore fluid to prevent pore collapse. Supercritical extraction retains porosity by eliminating the capillary stress exerted by the solvent upon removal from the microscopic pores of the gel [3]. Several groups have recently demonstrated alternative methods of fluid extraction that require less stringent environmental control [1-6]. According to the process patented by Smith *et al.*, a second lower vapor pressure solvent is introduced [1]. Surface modification (silylation, before drying) has also been utilized both for control of film porosity and for improvement of dielectric properties. In this study, we have applied Fourier Transform Infrared Spectroscopy (FTIR), non-contact atomic force microscopy (nc-AFM), cross-sectional scanning electron microscopy (SEM), and transmission electron microscopy (TEM), coupled with capacitance-voltage (C-V) and current-voltage (I-V) measurements to characterize thin films prepared by techniques requiring limited process control. We have also examined the impact of subsequent annealing conditions (temperature and environment) on microstructure and electronic properties.

EXPERIMENTAL METHOD

For the preparation of nanoporous thin silica films a sol composed of tetraethylorthosilicate (TEOS), ethanol, ethylene glycol, and water was converted to a gel composed of a solid skeleton of Si-O-Si bonds, ethylene glycol, and ethanol according to the process patented by Smith *et al* [1]. Fabrication of the gel was accomplished by combining TEOS, $Si(OC_2H_5)_4$, ethylene glycol, $C_2H_6O_2$, ethanol, C_2H_5OH, water, H_2O, and 1 M nitric acid, HNO_3, in the following molar ratios: 1.00 (TEOS): 2.43 (ethylene glycol): 1.48 (ethanol): 0.946 (water): 0.0420 (nitric acid). These molar ratios were refluxed for 1.5 hours at about 60-65°C. During this reflux, reaction (1), a hydrolysis reaction, occurred. Following the initiation of hydrolysis and condensation reactions a gel was formed. The hydrolysis reaction (1) is shown below:

$$Si(OC_2H_5)_4 + H_2O \Leftrightarrow Si(OC_2H_5)_3 \text{-OH} + C_2H_5OH \qquad (1).$$

After the addition of a base catalyst (0.25 M NH_4OH) the following condensation reactions (2 and 3) occurred:

$$Si(OC_2H_5)_3 \text{-OH} + HO\text{-}Si(OC_2H_5)_3 \Leftrightarrow (OC_2H_5)_3Si\text{-}O\text{-}Si(OC_2H_5)_3 + H_2O \qquad (2).$$

$$Si(OC_2H_5)_3 \text{-OH} + (OC_2H_5)\text{ - }Si(OC_2H_5)_3 \Leftrightarrow (OC_2H_5)_3 Si\text{-}O\text{-}Si(OC_2H_5)_3 + C_2H_5OH \qquad (3).$$

The gel solution formed by reactions 1, 2 and 3 was subsequently spin coated [0.8 mL at 5000 rpm] onto RCA cleaned p-type silicon (Si) wafers that were pretreated with an adhesion promoter solution [1]. The adhesion promoter solution consisted of 3-aminopropyl triethoxysilane and solvent (60 % ethylene glycol and 40 % ethanol) in a 1:25 volume ratio. Immediately following deposition, the wafers were aged in an ethylene glycol environment at room temperature for 24 hours. Selected wafers were surface modified before drying utilizing a solution of 90 % n-hexane and 10 % trimethylchlorosilane (TMCS).

MIS capacitors were prepared by evaporation of ~300 nm Al for both the front and back contacts. A shadow mask with a variety of gate sizes was utilized. A post-deposition (pre-metal) annealing/post-metal annealing study was performed on selected wafers. The annealing conditions utilized were 400°C for 30 minutes in forming gas (FG - 95/5%: N_2/H_2) for the post-deposition annealing (PDA) and 400°C, 30 minutes in N_2 for the post metal annealing (PMA). To examine microstructure, nanoporous silica coated silicon wafers were characterized using nc-AFM and compositional information was obtained with FTIR Spectroscopy. SEM samples were prepared by the focused ion beam technique [7]. TEM studies were performed with nanoporous silica samples that were deposited onto poly-vinyl formal coated copper grids.

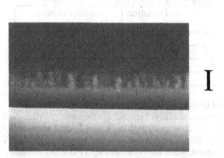

Figure 1. *Cross-sectional SEM image of nanoporous silica on Si with scale marker (**I**) indicating ~600 nm film thickness.*

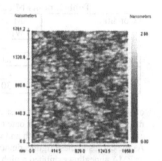

Figure 2. *nc-AFM image of nanoporous silica with mean pore width = 39nm.*

RESULTS AND CONCLUSIONS

A representative SEM image of nanoporous silica on Si is presented in Figure 1 revealing a nanoporous film of approximately 600 nm thickness. Close examination of SEM data revealed pores distributed throughout the film thickness with pore sizes ranging from 7.5 to 37.5 nm and a mean pore width of 15.1 nm. With TEM, for samples that were deposited onto polymer coated copper grids, pores of smaller width were observed. Specifically TEM data revealed a pore width range of 1 to 10 nm with a mean value of 4 nm. To further examine *relative* pore size and size distribution at the surface of the films, nc-AFM was employed. Throughout this study AFM data was collected from 3-5 samples of each type with at least 10 different areas tested per sample. A representative nc-AFM image of a porous silica film is presented in Figure 2 revealing a mean pore size of 39 nm. When interpreting AFM data, tip effects must be considered [8]. Specifically the finite radius of the tip (~15 nm) determines the minimum pore size resolvable, thus relative pore size comparisons are of greatest value and were utilized for this study.

The impact of annealing temperature and environment was examined next. MIS devices were prepared and selected samples were exposed to post-deposition annealing (PDA, in forming gas) and/or post-metal annealing (PMA, in N_2) treatments. The AFM data in Table I was obtained from regions not exposed to metal (between the metal dots). These results indicate that PDA and PMA have a significant impact on the mean pore width. PMA alone (without PDA) yielded the lowest mean pore widths for the nanoporous silica films studied. The impact of the PDA alone (without PMA) also resulted in a lower mean pore width when compared to wafers that were not subjected to either type of annealing. However, the combination of both PDA and PMA yielded a higher mean pore width than that recorded for wafers that were subjected to either a PDA or PMA alone. This result is important because the relative pore size and size distribution are parameters that must be optimized for electronic applications. Specifically, pore sizes well below IC feature size (<50 nm) must be achieved consistently -- independent of post deposition processing conditions employed [1].

The electronic properties of the MIS devices were subsequently examined with I-V and C-V data revealing relative accumulation capacitance and leakage current values as a function

Table I. nc-AFM data: Mean pore width from PDA and PMA study

Condition	Mean Pore Width (nm)	Standard Deviation	Number of pores measured
no PDA, no PMA	32.33	11.64	113
no PDA, PMA	21.99	8.72	121
PDA, no PMA	22.39	8.55	107
PDA and PMA	30.25	15.56	138

of annealing temperature. At least 10 different devices were tested per wafer and devices were mapped and probed at a later date to confirm reproducibility. Typical high frequency (1MHz) C-V and I-V plots for devices with and without annealing are presented in Figures 3 and 4 respectively. As the curves in these figures demonstrate, the MIS capacitors that were not subjected to either PDA or PMA annealing exhibited a high leakage current, while the capacitors with PDA had several orders of magnitude lower leakage current values and reduced accumulation capacitance. The leakage current was also reduced when wafers were subjected to PMA alone (without PDA). However, fabricated capacitors exposed to both PDA and PMA exhibited decreased capacitance but a slightly *higher* leakage current when compared to those receiving one annealing treatment alone. These results again suggest that an optimal annealing sequence, that is highly dependent on film properties, may exist for these films. The issue of film stability must also be considered, charging effects in MIS devices are strongly dependent on film composition both initially and after probing.

To further examine the role of annealing conditions on final film microstructure and electronic properties the impact of PDA temperature (300 vs. 400°C; 5/95%: H_2/N_2) was studied. Pore width data and electronic measurements were acquired for samples processed with several annealing temperatures. In addition, to study the impact of surface modification on the annealing response, selected wafers underwent surface modification. The results of the AFM study are presented in Table II while representative I-V curves and a nc-AFM image are presented in Figures 5 and 6 respectively. The results in Table II indicate that surface modification and annealing temperature do play a role in determining the final pore size for the nanoporous silica films studied. The wafers prepared without surface modification exhibited the lowest mean pore widths when annealed at 400°C. The mean pore width for the non-surface modified wafers was highest when annealed at 300°C. Surface modified wafers exhibited the smallest mean pore width when annealed at 300°C.

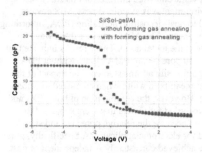

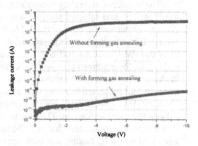

Figure 3. *1 MHz C-V characteristics for an Al/Sol-gel/Si capacitor with and without PDA.*

Figure 4. *I-V characteristics for an Al/Sol-gel/Si capacitor with and without PDA.*

Table II. Mean pore width data for silica aerogel as a function of PDA temperature

Condition	Mean Pore Width (nm)	Standard Deviation	Number of Pore Widths Measured
Without surf. mod., unannealed	22.77	9.15	139
Without surf. mod., 300°C	28.43	8.67	129
Without surf. mod., 400 C	20.07	7.65	110
With surf. mod., unannealed	29.27	16.09	122
With surf. mod., 300°C	25.88	9.14	142
With surf. mod., 400°C	31.02	14.40	104

Electronic measurements obtained from these films revealed a similar trend in terms of reduced leakage current and accumulation capacitance. Specifically, as demonstrated in Figure 5, 300°C was the optimal annealing temperature for surface modified films while a higher 400°C anneal resulted in improved properties for films that were not surface modified. Several groups [e.g., 2] have hypothesized that the presence of surface organic groups plays an important role in determining the electronic properties of surface modified films with higher annealing temperatures resulting in the removal of surface terminating organic groups (e.g., $-CH_3$, $-OC_2H_5$). For the current study FTIR data obtained from films representative of all types (with and without surface modification) revealed similar spectra - all indicating the primary presence of Si-O-Si bonding (~1080 cm^{-1}) as well as a minor contribution from OH absorption in the 3400 – 3600 cm^{-1} range. More detailed FTIR studies coupled with other surface sensitive techniques are currently underway to study the role of starting film composition and chemical stability in the determination of electronic behavior.

To further examine the impact of aerogel preparation method on annealing response, data was obtained from samples that were prepared using the method described by Nitta et al [3]. This method does not include ethylene glycol but instead uses surface modification to control film densification. All of the wafers in this part of the study underwent surface modification, with the extent of surface modification varied to produce films of varying degrees of porosity [3]. Pore width data was obtained on unannealed samples and samples that were annealed at two different temperatures (300 and 400°C, PDA, in forming gas). The index of refraction was also obtained for these wafers with ellipsometry. This data is shown in Table III. The results in Table III indicate that the smallest mean pore width for each degree of porosity occurred when the wafers were annealed at 300°C. Furthermore, the smallest mean pore width for the unannealed samples occurred for the samples that exhibited a low relative degree of porosity.

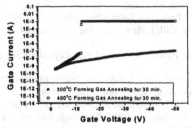

Figure 5. *I-V of Al/nanoporous silica/Si capacitor after PDA at 300°and 400° With surface modification.*

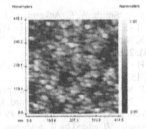

Figure 6. *nc-AFM image of nanoporous silica film after 400°C PDA; mean pore width = 19nm. Without surface modification.*

Table III: Pore width data and refractive index measurements as a function of annealing temperature for surface modified films prepared by method described by Nitta *et al.*

Condition	Mean Pore Width (nm)	Std. Dev.	Number	Mean Ref. Index
Low Porosity, no anneal	22.75	6.59	131	1.332
Low Porosity; 300°C	18.61	7.86	151	1.296
Low Porosity; 400°C	26.51	12.05	132	1.262
High Porosity, no anneal	28.34	9.28	130	1.068
High Porosity, 300°C	21.46	7.81	132	1.057
High Porosity, 400°C	28.35	15.86	117	1.068

SUMMARY

The results of this study reveal that the annealing response of thin film nanoporous silica is a function of starting film microstructure, chemistry and the annealing conditions (i.e., temperature, environment and annealing sequence). Specifically, for films subjected to surface modification, the annealing temperature utilized (300 vs. 400°C) plays an important role in determining the final film microstructure and electronic properties. Future studies will be directed toward determining the optimal annealing conditions necessary given starting material microstructure and processing history. Specifically, the impact of such factors as initial surface chemistry on annealing response will be examined in greater detail.

ACKNOWLEDGEMENTS

We acknowledge Prof. T. P. Ma, Yale University, Department of Electrical Engineering for many helpful discussions. We acknowledge Mr. Feng Wang and Dr. William Gill, Department of Chemical Engineering, Rensselaer Polytechnic Institute, Troy, N.Y, for the contribution of high porosity films obtained for comparative purposes. This work was partially funded by the National Science Foundation under grant numbers ECS-9753096, BIR-9512508 and DBI-9512508.

REFERENCES

[1]. D.M. Smith, W. C. Ackerman, U. S. Patent No. 5 736 425 (1998).
[2]. H –S. Yang, S –Y. Choi, S –H. Hyun, C –G.Park, Thin Solid Films, **348**, 69-73 (1999).
[3]. S. Nitta, V. Pisupatti, A. Jain, P. Wayner, W. Gill, J. Plawsky, J. Vac. Sci. Technol. B., **17(1)**, 205-212 (1999).
[4]. H –S. Yang, S –Y. Choi, S –H. Hyun, H –H. Park, J –K. Hong, J. Non-Crystalline Solids, **221**, 151-156 (1997).
[5]. P. Mezza, J. Phalippou, R. Sempere, J. Non-Crystalline Solids, **243**, 75-79 (1999).
[6]. S. Prakash, C. Brinker, A. Hurd, J. Non-Crystalline Solids, **190**, 264-275 (1995).
[7] Young, R. J. et al. *in Specimen Preparation for Transmission Electron Microscopy of Materials – II*, edited by R. Anderson. (Mater. Res. Soc. Proc. 199, Pittsburgh, PA), pp.205-216.
[8] D. Keller, Surface Science, **253**, 353-364 (1991).

Mat. Res. Soc. Symp. Proc. Vol. 612 © 2000 Materials Research Society

a) E-mail j.zhang@ee.ucl.ac.uk, Tel: +44 (0) 20 7419 3196, Fax: +44 (0) 20 7388 9325

Photo-induced growth of low dielectric constant porous silica film at room temperature

Jun-Ying Zhang[a] and Ian W. Boyd
Electronic and Electrical Engineering, University College London, Torrington Place, London WC1E 7JE, UK

ABSTRACT

We report low temperature (25-200°C) photo-assisted sol-gel processing for the formation of porous silicon dioxide films on Si (100) substrates using 172 nm radiation from an excimer lamp. The effects of substrate temperature and irriadation time on the properties of the films formed have been studied using ellipsometry, Fourier transform infrared spectroscopy (FTIR), and electrical measurements. The FTIR spectra revealed the presence of a Si-O-Si stretching vibration peak at 1070 cm^{-1} after UV irradiation at 200°C. This is similar to that recorded for oxides grown by thermally oxidation of silicon at temperatures between 600-1000°C. Capacitance measurements indicated that the dielectric constant values of the films, found to be between 1.7-3.3, strongly depended on the substrate temperature during irradiation. Dielectric constant values as low as 1.7 were readily achievable at room temperature. These results show that the photochemical induced effects initiated by the UV radiation enable both reduced processing times and reduced processing temperatures to be used.

PACS: 42·78·Hk, 61.80.Ba, 81.60.Cp, 85.50.Na
Keywords: Photo-assisted sol-gel processing, excimer lamp, porous SiO$_2$ film, low dielectric constant

1. Introduction

As microelectronic device densities increase and chip dimensions shrink, propagation delay, crosstalk noise, and power dissipation become significant due to resistance-capacitance (RC) coupling. Integration of low-dielectric-constant (k) materials, as means of reducing these RC time delays, has been identified for 0.1 μm technology and beyond. Current low-k commercialization emphasizes spin-on glasses (SOGs) and fluorinated SiO$_2$ with k>3, and a number of polymers are under development with k values in the range of 2-3 [1-3]. These suffer from potential problems including thermal stability, mechanical and electrical properties, low thermal conductivity, and reliability. However, low-k dielectric nanoporous silica with tuned k values from 1-4 have the advantage of facilitating manufacture of higher performance integrated-circuit (IC) devices because of compatibility with standard microelectronic processing and the ability to tune k over a wide range [4-6].

Low temperature deposition has been required for nanoporous SiO$_2$, because thermal stress degrades device characteristics and reliability of interconnection. Low temperature growth techniques, such as plasma-enhanced chemical vapor deposition (PECVD) [7], atmospheric pressure CVD (APCVD) [8], and electron cyclotron resonance (ECR) CVD [9], have been investigated. However, these cannot reduce the deposition temperature sufficiently, because substrate temperatures above 300°C are required to initiate dehydration and dissociation reactions. In recent work, we have reported the use excimer lamps to form high quality tantalum pentoxide films at low temperatures by a photo-induced sol-gel process which avoids dopant diffusion and defect generation caused by high temperature processing [10]. In this paper, we report the formation of low dielectric porous silica films using photo-induced sol-gel processing with a 172 nm excimer lamp. With this method, high energy photons are utilized to decompose the source materials. The effects of substrate temperature and lamp exposure time on the properties of the films formed have been studied using ellipsometry, Fourier transform infrared spectroscopy (FTIR), and electrical measurements. The results show that the photo-induced sol-gel process is a very promising technique since the pores and nanoscaled particles of porous silica can be tailored at the chemical solution stage.

2. Experimental details

A solution of tetraethoxysilane, $Si(OC_2H_5)_4$ (TEOS), C_2H_5OH, H_2O and HCl was stirred at room temperature and the progress of the gelation (to form a sol-gel solution) process in the solution was monitored by FTIR. The silica sol-gel solutions were formed by the hydrolysis-condensation reaction of mixing the solution. Films of 10-200 nm thickness were then spin-coated on p-type (100) Si substrates cleaned by ultrasonic washing in a propanol solution. These layers were subsequently mounted into a vacuum chamber and exposed to the 172 nm UV (ultraviolet) radiation from a dielectric barrier discharge operating in pure xenon. Fig. 1 shows the flow chart for the growth of porous SiO_2 films by this photo-induced processing route. More details of the excimer UV source and experimental apparatus used have been described in a previous paper [11-13]. The substrate temperatures between 25 and 200°C were measured with a thermocouple attached to a heater stage in the chamber while total gas pressure fixed at 1 mbar.

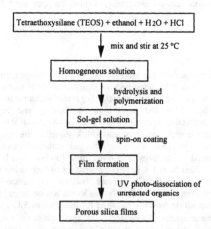

Fig. 1 Flow chart for growth of porous SiO_2 films by photo-induced processing.

The film thickness and refractive index were determined by ellipsometry (Rudolph AutoEL II) whilst an FTIR spectrometer (Perkin Elmer, Paragon 1000) was used to examine the dependency of the film composition and structure on processing parameters. The capacitance and dielectric constant of the films were determined by using metal-insulator-semiconductor (MIS) (Al/SiO$_2$/p-Si) test structures with aluminum electrodes of area 7.2 x 10^{-4} cm^2 evaporated through a metal contact mask. The capacitance-voltage (C-V) measurements were carried out at a frequency of 1 MHz.

3. Results and discussion

Sol-gel processing generally involves three reactions from the synthesis of an inorganic network by a chemical reaction in solution at low temperatures. The reactions leading to gelation are hydrolysis and condensation (water and alcohol condensation) of metalorganic compounds in a solution. For example, to form silica sol-gel from tetraethoxysilane (TEOS) in ethanol and water employing an acid as a catalyst the following reactions take place:

1) Hydrolysis

$$Si(OC_2H_5)_4 + H_2O \rightarrow Si(OC_2H_5)_xOH_y + yC_2H_5OH \qquad (1)$$

2) Alcohol condensation

$$Si(OC_2H_5)_xOH_y + Si(OC_2H_5)_4 \rightarrow (HO)_{y-1}(OC_2H_5)_xSi-O-Si(OC_2H_5)_3 + C_2H_5OH \qquad (2)$$

3) Water condensation

$$Si(OC_2H_5)_xOH_y + Si(OC_2H_5)_xOH_y \rightarrow$$
$$HO)_{y-1}(OC_2H_5)_xSi-O-Si(OC_2H_5)_x(HO)_{y-1} + H_2O \qquad (3)$$

The hydrolysis reaction (Eq. (1)) replaces alkoxide groups (OC$_2$H$_5$) in the TEOS with a hydroxyl group (OH) to form OH-substituted polymers. Subsequent two condensation reactions involving the silanol groups in OH-substituted polymers react with each other or with TEOS to produce siloxane (Si-O-Si) bridging bond contained polymers plus the by-products alcohol (C$_2$H$_5$OH) (Eq. 2) or water (Eq. 3). By the hydrolysis and condensation of TEOS a polymeric product is formed, leading to increased viscosity of the solution. Fig. 2 shows the FTIR spectra of sol-gel films spin-coated on Si wafers before and after an irradiation time of 5 min for wavenumbers between 400 and 4000 cm^{-1}. As can be seen, the O-Si-O stretching mode (1075 cm^{-1}), Si-O rocking vibration (470 cm^{-1}), silicon-hydroxyl Si-OH (950 cm^{-1}) and broad OH-group (around 3400 cm^{-1}) absorption peaks in the non-irradiated film can be identified. It was found that the intensity of the Si-O, Si-OH and OH-group bands significantly increased with gelation time due to the hydrolysis and condensation of the TEOS [14]. After an irradiation time of 5 min, the silicon-hydroxyl Si-OH (950 cm^{-1}) (see the inset of Fig. 2) and broad OH-group (around 3400 cm^{-1}) absorption peaks diminished, compared with those for the non-irradiated film. Fig. 3 shows the thickness and refractive index of the sol-gel films as a function of exposure time at spin-on speeds of 2000rpm. Clearly the thickness of the layers decreased with irradiation time while the refractive index slightly decreased due to decomposition. The refractive index values of the films formed at 200°C were found to be around 1.44, which is very close to those expected for stoichiometric SiO$_2$ layers (n=1.46).

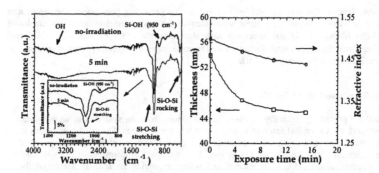

Fig. 2 FTIR spectra of sol-gel films on Si wafer before and after UV irradiation. speeds The inset shows 950 and 1070 cm⁻¹ bands

Fig. 2 FTIR spectra of sol-gel films on Si wafer before and after UV irradiation.
The inset shows 950 and 1070 cm^{-1} bands

Fig. 3 Thickness and refractive index of films as a function of irradiation time at spin-on speeds of 2000rpm.

The evolution of the FTIR spectra of the sol-gel films irradiated at different temperatures and at a fixed irradiation time of 10 min is presented in Fig. 4. As can be seen, after UV irradiation the intensity of the Si-OH and OH-group absorption bands decreased considerably, indicating that many of the hydroxyl groups had decomposed and been removed. Complete removal can be achieved either by increasing the UV exposure times or the substrate temperature. In fact removal in films of other thicknesses can be also achieved by optimizing exposure time and temperature. After 10 min exposure to UV light at a temperature of 200°C the Si-O-Si stretching vibration peak appeared at around 1070 cm^{-1}, a feature often associated with silicon dioxide, especially for oxides thermally grown at temperatures between 600-1000°C [15].

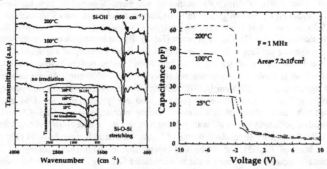

Fig. 4 FTIR spectra of layers irradiated at different temperatures for a fixed irradiation time of 10 min.

Fig. 5 High frequency C-V traces of Al/SiO₂/Si MIS test structure fabricated using films formed at various temperatures with irradiation time of 10 min.

It was found that the decomposition rate of the sol-gel films was dependent upon total gas pressure and increased as pressure decreased [14], which is similar to the situation with photo-induced decomposition of palladium acetate and tantalum oxide films [16-17].
Figure 5 shows high frequency C-V traces of Al/SiO$_2$/Si MIS capacitors fabricated using silica films formed at temperatures between 25-200°C with an irradiation time of 10 min. A typical C-V hysteresis effect, i.e. flat-band shift, was observed in the films formed at different temperatures. The dielectric constant was calculated from the C-V measurements using the well known expression k=C·d/ε_0A, where k, C and d are the dielectric constant, capacitance and thickness of the layer, ε_0 is the permittivity in vacuum, and A is the capacitor area. Table 1 summarizes the capacitance, dielectric constant and flat-band shift values of the SiO$_2$ films formed at substrate temperatures between 25°C and 200°C.

Table I Capacitance, dielectric constant, and flat-band shift of SiO$_2$ films grown at different substrate temperatures.

Temperature (°C)	Capacitance (pF)	Dielectric constant	Flat-band shift (V)
200	61.0	3.30	-1.0
100	48.4	2.73	-2.2
25	22.5	1.70	-1.3

It is clearly seen that the dielectric properties strongly depend on the substrate temperature during irradiation. As the organic compounds are decomposed and the water content removed from the sol-gel films during UV irradiation, the film structure becomes more porous unless a densification process occurs. After irradiation at temperatures between 25 and 200°C, the layers exhibit dielectric constants ranging from 1.7 - 3.3. These values are considerable lower than for silicon dioxide (k=3.9) indicating the lower density (higher porosity) of the films. A dielectric constant of 1.7 is ready obtained at room temperature which can be associated with a porosity of approximately 75% [4,6,18]. Thus a tuneable dielectric constant can be achievable in these porous silica films by a simple change of processing temperature.

4. Conclusions

A new technique of low temperature photo-induced sol-gel process for the growth of low dielectric constant porous silicon dioxide thin films from TEOS sol-gel solutions with a 172 nm excimer lamp has been successfully demonstrated. The FTIR spectral analyses revealed that the silicon-hydroxyl Si-OH (950 cm^{-1}) and OH-group (around 3400 cm^{-1}) absorption peaks decreased after irradiation and indicated that the UV light decomposes the organic compounds in the sol-gel to convert it into a silicon dioxide film at low temperatures. The properties of the processed layer strongly depend on the operating parameters. Refractive index values were found to be around 1.44 for the films formed at 200°C, while dielectric constant values as low as 1.7 could be achieved at room temperature. These results indicate that this low temperature photo-induced sol-gel technique is very promising for the preparation of tuneable low-k dielectric porous silica films and possibly other interlayer dielectrics in future ULSI multilevel interconnections.

Acknowledgement:
This work is partly supported by the Engineering and Physical Science Research Council of UK (grant No. GR/190909).

References
1. P. Pai, A. Chetty, R. Roat, N. Cox, and C. Ting, J. Electrochem. Soc. 134 (1987) 2829.
2. T. Homma, R. Yamaguchi, Y. Murao, J. Electrochem. Soc. 140 (1993) 687.
3. J.-Y. Zhang and I.W. Boyd, Optical Materials, 9 (1998) 251.
4. J. Hong, H. Yang, M. Jo, H. Park, and S. Choi, Thin Solid Films 308 (1997) 495.
5. L.W. Hrubesh and J.F. Poco, J. Non-Cryst. Solids 188 (1995) 46.
6. M. Jo, H. Park, D. Kim, S. Hyun, S. Choi, and J. Paik, J. Appl. Phys. 82 (1997) 1299.
7. S. Nguyen, D. Dobuzinsky, D. Harmon, R. Gleason and S. Fridmann, J. Electrochem. Soc. 137 (1990) 2209.
8. M. Matsuura, Y. Hayashide, H. Kotani and H. Abe, Jpn. J. Appl. Phys. 30 (1991) 1530.
9. C.S. Pai, J.F. Miner, P. Foo, Int. VLSI Multilevel Interconnection Conf. Proc., Santa Clara, CA, 11-12 June 1991, p442.
10. J.-Y. Zhang, L. Bie and I.W. Boyd, Jpn. J. Appl. Phys. 37 (1998) L27.
11. J.-Y. Zhang and I.W. Boyd, J. Appl. Phys. 80 (1996) 633.
12. I.W. Boyd and J.-Y. Zhang, Nucl. Instr. Methods in Phys. Res., B121 (1997) 49.
13. J.-Y. Zhang, L. Bie, V. Dusastre and I.W. Boyd, Thin Solid Films 318 (1998) 252.
14. J.-Y. Zhang and I.W. Boyd, E-MRS99 Spring Meeting, June 1-4, 1999, Strasbourg, France.
15. M.P. Woo, J.L. Cain, and C. Lee, J. Electrochem. Soc. 137 (1990) 196.
16. J.-Y. Zhang and I.W. Boyd, Appl. Phys. A, 65 (1993) 379.
17. J.-Y. Zhang and I.W. Boyd, Ultrathin SiO_2 and high-k materials for ULSI gate dielectrics. ed. by M. Green, H. Huff, MRS, Warrendale, Pennsylvania, 567 (1999) 495.
18. T. Homma, Material Science and Engineering, R23 (1998) 243.

Mat. Res. Soc. Symp. Proc. Vol. 612 © 2000 Materials Research Society

ULTRA LOW-K INORGANIC SILSESQUIOXANE FILMS
WITH TUNABLE ELECTRICAL AND MECHANICAL PROPERTIES

Thomas A. Deis, Chandan Saha, Eric Moyer, Kyuha Chung, Youfan Liu, Mike Spaulding, John Albaugh, Wei Chen, and Jeff Bremmer
Dow Corning Corporation, Midland, MI 48611 U.S.A.

ABSTRACT

Low-k dielectric films have been developed using a new silsesquioxane based chemistry that allows both the electrical and mechanical properties to be tuned to specific values. By controlling the composition and film processing conditions of spin-on formulations, dielectric constants in the range 1.5 to 3.0 are obtained with modulus values that range from 1 to 30 GPa. The modulus and dielectric constant are tuned by controlling porosity, which varies from 0 to >60%, and final film composition which varies from $HSiO_{3/2}$ to $SiO_{4/2}$. The spin-on formulation includes hydrogen silsesquioxane resin and solvents. Adjusting the ratio of solvents to resin in the spin-on formulation controls porosity. As-spun films are treated with ammonia and moisture to oxidize the resin and form a mechanically self-supporting gel. Solvent removal and further conversion to a more "silica-like" composition occur during thermal curing at temperatures of 400 to 450°C. The final film composition was controlled through both room temperature oxidation and thermal processing. Final film properties are optimized for a balance of electrical, mechanical and thermal properties to meet the specific requirements of a wide range of applications. Processed films exhibit no stress corrosion cracking or delamination upon indentation, with indenter penetration exceeding the film thickness, and followed by exposure to water at room temperature. Films also exhibit high adhesive strength (> 60MPa) and low moisture absorption. Processing conditions, composition and properties of thin are discussed.

INTRODUCTION

Multilevel interconnect technology beyond the 0.18 micron technology node requires interlayer dielectric (ILD) materials with low permitivity values. An inorganic material and a unique process method have been developed to meet this need. A commercially viable process has been developed that produces controlled void volume in hydrogen silsesquioxane (HSQ) based films and results in a more "silica" like film. New advances in the formulation and processing allow for fast, simple processing that significantly improved both the mechanical and electrical properties of hydrogen silsesquioxane resin. In addition the dielectric constant value and mechanical properties of films processed using this method can be tuned over a wide range to meet specific needs.

EXPERIMENTAL

Thin films are processed by spin coating a solution onto a silicon wafer, treating the as-spun film with moist ammonia, and then thermally curing the film. The initial solution was comprised of hydrogen silsesquioxane (HSQ) resin and two solvents. One solvent, a low boiling solvent, was used to control the thickness of the as-spun film. The second solvent was a high boiling solvent, which remained with the resin in the as-spun film. The ratio of the resin to high boiling solvent was used to control the density of the as-spun film. After spin coating, the as-spun film was exposed to moist ammonia to cause the film to gel in the presence of the high boiling solvent. The remaining high boiling solvent was then removed by subsequent heating steps, which included a final cure in the range of 400 to 450°C. Non-porous HSQ films were prepared from solutions with only a low boiling solvent.

Fourier transform infrared spectroscopy (FTIR) was used to characterize the structure and composition of thin films. Figure 1 shows the FTIR structure of dense HSQ and also a porous, more silica like, film. The relevant absorptions are the Si-H stretching region near 2250 cm^{-1}, and the Si-O stretching region near 1000 cm^{-1}. Electrical properties were measured by evaporating Al electrodes onto processed films to form MIS structures. X-ray reflectivity measurements were used to determine thickness and density. The overall film density was calculated from the critical edge of the x-ray reflectivity curve. A combination of x-ray reflectivity and small angle neutron scattering (SANS) was used to obtain nano-porosity information. Debye's equation was used to fit neutron scattering profile. From the data an average pore chord length and film density are calculated. Modulus values were determined by nano-indentation and film toughness was measured using the modified edge liftoff technique.

RESULTS

Figure 1 shows FTIR spectra from a standard processed HSQ film (k=3) and a low-k processed film. The standard process, used to form non-porous HSQ films, involves a thermal cure at 400°C in N_2, resulting in a film composition that had ~80% of the original hydrogen content in the as-spun film. In contrast, the low-k process resulted in a final composition that had 30-40% of the original hydrogen content. The lower hydrogen content is a result of the hydrolysis and condensation that occurred during the moist ammonia treatment in addition to hydrogen loss, which occurred during a higher 450°C thermal treatment. The final composition of low-k films has a significant impact on both electrical and mechanical properties.

Figures 2 shows the change in modulus, for a low-k film with a k value of 2.0, as a function of composition. The composition was altering by varying the time that the as-spun film was exposed to moist ammonia before curing at 450°C for 1 hr. As shown in Figure 2, the film modulus increased as the composition of the film became more "silica-like." Figure 3 shows the effect of cure temperature on observed modulus values of a film with a k value of 2.0. Both composition and cure temperature significantly influence the mechanical properties of HSQ based films.

Table 1 lists the properties of HSQ films, both porous and nonporous, that have been processed to achieve a wide range of properties. The most porous films were observed to have a k value of 1.6 and a modulus value of 1 GPa. By controlling the amount of porosity, dielectric constants in the range of 1.6 to 3.0 were achieved. Pore dimensions, obtained from SANS measurements, varied in the range of 17 to 24 Å for the k=2.0 and 2.2 films, and a single data point for k=1.6 film was 39 Å. While sufficient data is lacking to be conclusive, pore sizes observed decrease as the amount of porosity was reduced. Films with k values of 2.5 have been analyzed by SANS, but scattering was insufficient. A possible reason for this could be that the pore size was too small (< 10 Å) to be adequately detected by SANS. Figure 4 shows that the hydrophobic nature of HSQ films was maintained even after introducing >50% porosity.

As porosity was reduced in a controlled manner, the modulus values increased from a low of 1 GPa, for a k=1.6 film, to 30 GPa for a supposedly nonporous HSQ film treated with ammonia. In contrast, HSQ films processed in the standard manner (i.e. no ammonia treatment, 400°C 1hr cure in N_2) had a modulus value of 6 GPa.

In all cases, the HSQ based films on bare Si were observed to have adhesive strength values in excess of 60 MPa. This conclusion was based on stud pull failure occuring at the epoxy film interface. However, ammonia treated porous films were found to have higher K_{lc} values when compared to standard processed HSQ films,. K_{lc} values were observed to increase as the level of porosity decreased. The difference in toughness values, with respect to porosity, could also have been influenced by pore size differences.

CONCLUSIONS

Hydrogen silsesquioxane films have been processed in a manner that allowed both the electrical and mechanical properties to be tuned to specific values. Gelling as-spun films in the presence of a solvent with moist ammonia created porosity, which could be controlled by the ratio of resin to solvent. The size of pores introduced has been observed to be in the range of 20 to 40 Å. Composition was controlled by the duration of exposure to moist ammonia and by the duration and temperature of the thermal curing. Films processed with moist ammonia were more silica-like in composition compared with typical HSQ films. Compared to standard processed HSQ films, which resulted in a k value of ~3 and a modulus of ~6 GPa, ammonia treated samples had significantly higher modulus values with a modest increase in k value, 30 GPa and k=3.5. By introducing porosity, into the ammonia treated films, the dielectric constant of films was reduced to as low as 1.6, while the film modulus value was 1.5 GPa. Excellent adhesion and mechanical toughness were also observed with the ammonia treated HSQ based films.

Table 1. Properties of HSQ based films

Property	Material					
	Porous HSQ (NH₃ process)	Porous HSQ (NH₃ process)	Porous HSQ (NH₃ process)	Porous HSQ (NH₃ process)	HSQ (std process)[†]	HSQ (NH₃ process)
Target k	1.5	2	2.2	2.5	3	—
Measured k	1.6 - 1.8	1.9 - 2.1	2.0 - 2.3	2.5 - 2.6	2.6 - 3.0	3.4 - 3.5
Density (g/cc)	0.62	0.9 - 1.0	1.1	1.2	1.4 - 1.5	1.6 - 1.7
Pore Size (Å)	39	20-24	17-20	poor scattering	—	—
Porosity	>60%	56 - 60%	45 - 55%	25%	—	—
Modulus (GPa)	0.5 - 1.7	2.0 - 3.0	3.0 - 5.0	6.0 - 10.0	6	27 - 30
K_{Ic} (MPa·m$^{1/2}$)	0.23	0.28 - 0.38	0.34 - 0.38	0.37 - 0.46	0.11 - 0.17	0.21
Adhesive Strength (MPa)	> 60	> 60	> 60	> 60	50 - 60	> 60

[†]standard process: 150/200/350°C hotplates 1min each, 400°C 1hr cure in N_2 atm

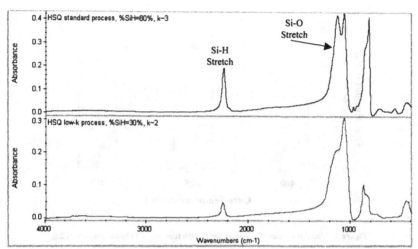

Figure 1. FTIR spectra of HSQ films, top: nonporous standard process, bottom: porous film from low-k process.

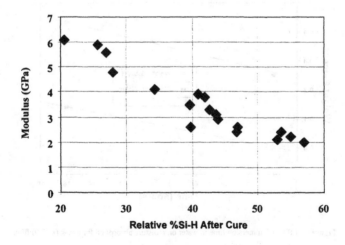

Figure 2. Relationship observed between modulus and %Si-H for low-k processed films (k=2.0). Range of %SiH obtained by varying aging time of low-k films.

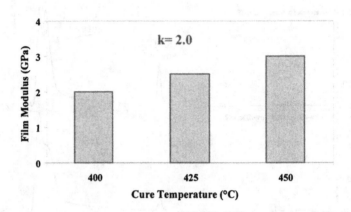

Figure 3. Modulus vs. cure temperature for 5,000Å film with a dielectric constant of 2.0.

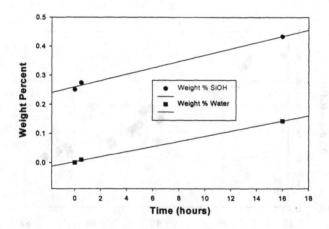

Figure 4 Effect of immersion time in water on moisture absorption for low-k (k=2.0) film.

Mat. Res. Soc. Symp. Proc. Vol. 612 © 2000 Materials Research Society

A Three-phase Model for the Structure of Porous Thin Films Determined by X-ray Reflectivity and Small-Angle Neutron Scattering

Wen-li Wu, Eric K. Lin, Changming Jin[1], and Jeffrey T. Wetzel[1]
Polymers Division, Materials Science and Engineering Laboratory
National Institute of Standards and Technology
100 Bureau Drive, Stop 8541, Gaithersburg, MD 20899-8541 USA
[1]SEMATECH
2706 Montopolis Drive, Austin, TX 78741-6499 USA

ABSTRACT

A methodology to characterize nanoporous thin films based on a novel combination of high-resolution specular x-ray reflectivity and small-angle neutron scattering has been advanced to accommodate heterogeneities within the material surrounding nanoscale voids. More specifically, the average pore size, pore connectivity, film thickness, wall or matrix density, coefficient of thermal expansion, and moisture uptake of nanoporous thin films with non-homogeneous solid matrices can be measured. The measurements can be performed directly on films up to 1.5 μm thick while supported on silicon substrates. This method has been successfully applied to a wide range of industrially developed materials for use as low-k interlayer dielectrics.

INTRODUCTION

Porous materials with nanoscale voids are leading candidates for next-generation low dielectric constant or low-k thin films to be used as an interlayer dielectric (ILD) materials. The introduction of voids in the material can effectively lower the dielectric constant, k, of the base material. For example, when the film porosity approaches 75% by volume, the dielectric constant of silica, nominally about 4, can be reduced to about 2.0 and below [1]. A low dielectric constant for ILD is needed to decrease the energy needed to propagate a signal and to decrease crosstalk between adjacent conductors. Many techniques have been established to characterize the performance properties for interlevel dielectric thin films [2]. However, few techniques are able to measure the structural properties of porous films ~1 μm thick supported on silicon substrates. Recently, Gidley et al. used positronium annihilation lifetime spectroscopy to measure void sizes and distributions [3]. Baklanov et al. use ellipsometric porosimetry to also measure void sizes and distributions [4]. Wu et al. [5] have demonstrated that the average void size, porosity, film density, coefficient of thermal expansion, connectivity among voids and moisture uptake can be measured using a combination of high resolution specular x-ray reflectivity [6-8] (SXR) and small angle neutron scattering [9] (SANS) techniques.

In the use of SXR and SANS to characterize nanoporous thin films, an underlying hypothesis of this methodology is the homogeneity of the material surrounding the voids. It invokes a simple two-phase model in which the material around the voids is assumed to be homogeneous. The only heterogeneity in the film comes from the voids. For many materials from different sources, the two-phase methodology has been successfully applied. However, there exist samples where the SANS intensities are too high for any two-phase material given its measured elemental composition. As a result, the two-phase model results in unrealistically high values for the density of the material surrounding the voids. To ensure more reasonable values, a new three-phase model methodology is developed here that is capable of dealing with deficiencies in the two-phase methodology. The underlying principles for both the three phase and the two-phase methodology [5] are identical; they both rely on the complementary nature of the results from SANS and SXR. The difference lies in the model used to interpret the SANS results. Consequently, the results on CTE, film thickness, and the electron density are unchanged with regard to the specific methodology used because these quantities are determined solely from SXR data.

The rest of this manuscript will be arranged as follows. The fundamentals of SANS, SXR and the two-phase methodology will be briefly reviewed, the shortcoming of the two-phase model will then be discussed and followed by outlining the rationale for introducing three-phase model. An example of the application of the three-phase model will be provided at the end of the manuscript.

TWO-PHASE METHODOLOGY

The novelty of our approach to characterizing porous thin films is two fold: the use of a new high-resolution x-ray reflectometer to accurately characterize films up to 1.5 μm thick and the use of complementary data obtained from both SXR and SANS as a set of simultaneous equations to quantitatively determine structural parameters of porous thin films. In the two-phase method [5], we used the simplest description of a porous material; a two-phase model where one phase is comprised of the voids and the other is comprised of the connecting material. The connecting material (the pore wall material) is assumed to be uniform in composition and in density. With this assumption, the average density of the film can be parameterized with two unknowns, the porosity, P, and the wall density, ρ_w. These two variables cannot be independently determined from either SXR or SANS data alone. By using both techniques and solving simultaneous equations, specific to each technique and involving these two variables, the values of the unknowns can be determined. In order to perform this analysis, we must also know the chemical composition of the film. The chemical compositions were determined using a combination of Rutherford back scattering (RBS) (for silicon, oxygen, and carbon) and forward recoil elastic scattering (FRES) (for hydrogen). The film composition is used to convert electron density to mass density in the SXR data analysis and to determine the scattering contrast between the connecting material and pores in the SANS analysis. In addition to P and ρ_w, another parameter, the correlation length, ξ, is determined from the SANS data. These three parameters are widely used to characterize two-phase materials. Most of the low k ILD films analyzed by us are well characterized using the two-phase model.

Many of the ILD samples, however, are found to contain a significant amount of hydrogen atoms. Hydrogen represents a special case in neutron scattering because the neutron scattering length, b_H, of hydrogen is negative. As a result, the spatial distribution of hydrogen has a major impact on the observed SANS intensity. The three-phase model is developed to help interpret the inhomogeneous distribution of hydrogen with the thin film.

THREE-PHASE METHODOLOGY

For a two-phase system, the neutron scattering contrast, η^2, can be expressed as

$$\eta^2 \propto P(1-P)(\rho_w \sum (n_i b_i / m_i))^2 \tag{1}.$$

P again is the porosity or the volume fraction of the voids, and b_i, n_i and m_i denote the neutron scattering length, the number fraction and the atomic weight of element i, respectively. The summation in equation (1) is over all the elements present in samples. The observed SANS intensity dictates the magnitude of η^2 and the SXR data determine the value of the product $\rho_w (1-P)$. In samples with significant amount of hydrogen, the observed SANS intensity (and hence η^2) can only be accounted for with an unrealistic value of ρ_w. For example, a sample with the following composition by number, Si (16%), O (26%), C (19%) and H (39%) was measured and SXR results indicated that the film density was (0.72 ± 0.01) g/cm^3. Using a combination of SANS and SXR data, the calculated value of ρ_w was (3.30 ± 0.2) g/cm^3, a value higher than that of quartz and thus an unrealistic one. The corresponding porosity was (78 ± 1.5) %. This high wall density value can be regarded as a definitive sign that the hydrogen atoms are not uniformly distributed within the wall or matrix material. In the two-phase methodology, the overall density of the matrix material was forced to adopt a high value in order to provide sufficient scattering contrast between the matrix material and the wall and compensate for the inhomogeneous distribution of hydrogen.

The three-phase model is developed as a reasonable model to represent the heterogeneous distribution of hydrogen in the porous thin film. Since it is unphysical to have all the hydrogen atoms to form a phase or clusters by themselves, we assume that all the hydrogen and carbon atoms exist as hydrocarbons and are segregated from the silicon and oxygen atoms. The nanoporous thin film mentioned above is then assumed to be comprised of three phases, the hydrocarbon phase (phase 1), the silicon and oxygen or silica phase (phase 2) and the voids (phase 3). The corresponding contrast factor of the three phase material can be expressed as

$$\eta^2 \propto \Phi_1(1-\Phi_1)(B_2-B_3)^2 + \Phi_2(1-\Phi_2)(B_3-B_1)^2 + \Phi_3(1-\Phi_3)(B_1-B_2)^2 \tag{2}$$

where Φ_j stands for the volume fraction occupied by phase j [10]. By definition, Φ_3 is equal to P, the porosity of the film. B_j is the neuron scattering length of phase i and is defined as $\rho_j \sum (n_i b_i / w_i)$ where ρ_j is the mass density of phase j and the summation is over all the elements existed in phase j. Also by definition, B_3, the neutron scattering length of voids, is zero. Within the above equation there are a total of five unknowns and they are Φ_1, Φ_2, Φ_3, ρ_1, and ρ_2. There are two obvious constraints or relations for these unknowns and they are;

$$\Sigma \Phi_i = 1 \tag{3}$$

for the sum of all three volume fractions to be unity, and

$$\Phi_1\rho_1 / \Phi_2 \rho_2 = (n_c m_c + n_H m_H) / (n_{Si} m_{Si} + n_o m_o) \tag{4}$$

where the mass ratio between phases 1 and 2 is equal to the ratio of the total mass of the constituents. At this point, there are still three unknowns and the SXR data provide an additional relation or constraint for these unknowns. More explicitly,

$$Q_c^2 \propto (\Phi_1 \rho_1 + \Phi_2 \rho_2)/(\Phi_1 + \Phi_2) \qquad (5).$$

Where Q_c^2 stands for the critical angle expressed in the Fourier space measured by SXR, and the right hand side of the above equation is, by definition, the matrix material or the wall density. The SANS data provide a measure of η^2 of the equation (2), but there is still a need for one more experimental measurement of some of the five variables. In this work, we instead assume that the density of hydrocarbon phase, ρ_1, to be unity. This is believed to be a reasonable assumption because the bulk density of many hydrocarbons is close to one.

The sample mentioned above was reanalyzed using the three-phase model and the matrix density was found to be (1.71 ± 0.05) g/cm^3, a value close to that of a thermally grown silicon oxide and its porosity was (58 ± 1.5) %. This example does not necessarily prove that the matrix material surrounding the voids is indeed made of two phases, a hydrocarbon phase and a silica phase. This result does provide strong evidence that the matrix is not a homogeneous one-phase material. The three-phase model discussed here is the simplest extension of the two phase model and its application shall be limited to cases where the two-phase methodology fails to provide physically meaningful results.

In addition to all six Φ_i and ρ_i parameters, three correlation lengths, ξ_i, one for each phase i, are needed to fully characterize a three-phase system. It is noteworthy that there is only one correlation length for a two-phase system and its value can be deduced directly from the SANS data using Debye, Porod, or other analysis schemes [11]. For a three-phase system, all three correlation lengths manifest themselves in the SANS results via a relation similar to that of equation (2). Each correlation length is weighted by the neutron scattering contrast factor of that particular phase. Conveniently, the neutron contrast factor of hydrocarbons, especially for those with a 1:2 carbon to hydrogen ratio, is nearly zero because the scattering length of hydrogen is -3.74×10^{-13} cm and 6.65×10^{-13} cm for carbon. Accordingly, the measured correlation length from SANS is dominated by that of the phase composed of silicon and oxygen. The correlation length in our example can be treated as if the system is a two-phase system with a silicon-oxygen phase and a voids-hydrocarbon phase. The chord length of each phase can be deduced with the two-phase scheme [11].

After their structural parameters being determined with the three-phase scheme, the pore connectivity and moisture uptake can be determined using the methodology developed for two-phase system. The pore connectivity and moisture uptake were measured by conducting SANS measurement on samples immersed respectively in either a deuterated organic solvent or in deuterated water. Organic solvents with low interfacial tension can readily fill interconnected pores having a passage to the exterior surface to cause a scattering contrast change. Deuterated toluene has been used for all of the samples tested because it spreads readily on surfaces of those samples. Once the pores are filled, the scattering contrast changes dramatically depending on the neutron scattering length of the solvent used. The percentage of the pores filled by solvent or water can be determined from the difference in SANS intensities between thin films before and after immersion.

CONCLUSION

A combination of SXR and SANS data from some nanoporous thin films indicate that the matrix material surrounding the voids is not homogeneous and that hydrogen within the film must be segregated in some way. By assuming that of all the hydrogen and carbon are segregated into a phase with a mass density of unity, a three-phase methodology has been developed and applied successfully to samples that previously could not be described with a simple two-phase model.

ACKNOWLEDGMENT

The authors would like to thank Professor Russ Composto and Howard Wang of the University of Pennsylvania for performing the high-energy ion scattering experiments.

REFERENCES

1. L. W. Hrubesh, L. E. Keene, and V. R. Latorre, J. Mater. Res. **8**, 1736 (1993).
2. M. Grasserbauer and H. W. Werner, *Analysis of microelectronic materials and devices* (Wiley, New York, 1991).
3. D. W. Gidley, W. E. Frieze, T. L. Dull, A. F. Yee, C. V. Nguyen and D. Y. Yoon, Appl. Phys. Lett., **76 (10)**, 1282 (2000).
4. F. N. Dultsev and M. H. Baklanov, Elec. Solid State Lett., **2**, 192 (1999).
5. W. L. Wu, W. E. Wallace, E. K. Lin, G. W. Lynn, C. J. Glinka, E. T. Ryan and H. M. Ho, J. Appl. Phys., **87**, 1193 (2000).
6. J. Lekner, *Theory of Reflection* (Nijhoff, Dordrecht, 1987).
7. S. Dietrich and A. Haase, Phys. Rep. **260**, 1 (1995).
8. E. Chason and T. M. Mayer, Crit. Rev. Solid State Mat. Sci. **22**, 1 (1997).
9. J. S. Higgins and H. C. Benoit, *Polymers and Neutron Scattering* (Oxford University Press, Oxford, 1994).
10. W. L. Wu, Polymer, **23**, 1907 (1982).
11. P. Debye, H. R. Anderson, and H. Brumberger, J. Appl. Phys. **28**, 679 (1957).

Mat. Res. Soc. Symp. Proc. Vol 612 © 2000 Materials Research Society

Grazing Incidence Small Angle X-Ray Scattering Study on Low Dielectric Thin Films

C.-H. Hsu [1], Hsin-Yi Lee [1], K.S. Liang [1], U-Ser Jeng [2], D. Windover [3], T.-M. Lu [3], and C. Jin [4]
[1.] Synchrotron Radiation Research Center, Hsinchu, Taiwan.
[2.] Department of Engineering and System Science, National Tsing-Hua University, Hsinchu, Taiwan.
[3.] Center for Integrated Electronics, Electronics Manufacturing and Electronic Media, Rensselaer Polytechnic Institute, Troy, NY, U.S.A.
[4.] Texas Instruments, Dallas, TX, U.S.A.

ABSTRACT

Highly porous silica films with pore size in the nanometer scale are being extensively studied as potential candidates for interlevel dielectrics. Because these dielectric materials appear in the form of thin films with a thickness of only several thousand Angstroms, conventional techniques are difficult to be readily applied to study their structure and porosity. We employed small angle scattering in the grazing incidence geometry in this study. Using high resolution x-ray beamline with synchrotron radiation source, we demonstrate that the small angle x-ray scatteirng (SAXS) data of the porous films can be obtained. The structure of sol-gel derived silica - xerogel films on silicon substrate studied by specular reflectivity and grazing incidence small angle x-ray scattering (GISAXS) will be presented.

INTRODUCTION

One of the principal considerations in the selection of an interlevel dielectric material for integrated circuit applications is the low dielectric constant. However, the relatively high dielectric constant of SiO_2 will eventually preclude its use as an interlevel dielectric due to an increased RC delay, as well as increased cross talk, noise and power dissipation. Hence materials with a lower intrinsic dielectric constant need to be developed and evaluated as potential replacements for SiO_2. Porous materials have been demonstrated to be a potential candidate of the interlevel dielectric materials [1]. It is well known that the microscopic structure of the porous materials is strongly related to its mechanical strength and electric properties. Moreover, as an interlevel dielectrics, the pore size must be uniformly distributed and small compared to the feature size of the devices. Small angle x-ray scattering (SAXS) has already been demonstrated very useful for studies of the micro-structure of complex disordered, porous materials. Therefore, in this study we employ x-ray reflectivity [2] and SAXS to investigate the microscopic structure of a sol-gel derived silica - xerogel thin film.

EXPERIMENT

The xerogel sample is grown on a Si(001) wafer, which has a SiO_2 layer about 10,000 Å thick on top of it. A 5,000 Å film is made via spin-coating using a sol-gel process and dried under atmospheric pressure. Xerogel composition was determined using Rutherford backscattering spectrometry to be approximate $SiO_{2.2}C_{0.6}$. A sample of size 92 x 64 mm^2 is used in this work.

X-ray scattering measurements were conducted at a wiggler beamline BL17B of Synchrotron Radiation Research Center (SRRC). The beamline is equipped with a double crystal monochromator (DCM) with sagittal focusing to provide monochomatic beam. Photons with an energy of 8 keV was used in this study. In addition to beamline slits, two pairs of slits, 1 m apart, are used inside the experimental hutch as collimator. X-ray reflectivity is performed to find the density and surface/interface roughness of the sample. SAXS is conducted to investigate the structure of the porous xerogel. Because xerogel exists in the form of thin film in microelectronic applications, conventional SAXS measurements performed in transmission geometry is not applicable. We adopt the grazing incident geometry to significantly increase x-ray travel distance in xergoel and thus enhance its scattering intensity. X-ray incidence angle is fixed at 0.2°, which is larger than the critical angle θ_c of xerogel but is smaller than that of silicon oxide. An 8-circle diffractometer, which provides a detector with both horizontal and vertical degree of freedom, was used. During x-ray measurements, incident and reflected x-rays were in a horizontal plane, and the SAXS signals were collected via a vertical scan away from the reflected beam. A scintillation counter was used in all the measurements.

RESULTS

A scanning electron microscopy (SEM) image of xerogel sample with a magnification of 50,000 is displayed in figure 1. It clearly shows the topology of the xerogel surface. Figure 2 illustrates the x-ray reflectivity curve of the xerogel sample. The open circles denote the experimental data. The two maxima at $q_z = 4\pi sin(\theta)/\lambda$, where 2θ is the scattering angle and λ is the x-ray wavelength, equal to 0.014Å^{-1} and 0.03Å^{-1} are associated with the total external reflection occurred at air/xerogel and xerogel/SiO$_2$ interface, respectively. Due to the finite size of sample, the footprint effect results in the initial intensity rising before the critical angle of xerogel. Two distinct period of intensity oscillations- small period at low q_z's, as shown in the inset, and large period at high q_z region - are observed in the reflectivity curve, implying the presence of at least two layers of different density on top of the substrate.

Figure 1. SEM image of surface morphology of xerogel film.

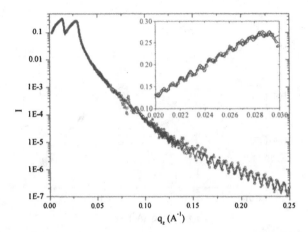

Figure 2. Experimental data and fitted results of the specular reflectivity of xerogel/Si sample: (o) the experimental data; (—) the fitted results using parameters of the three-layer best fit listed in table I.

We adopt a three-layer model to describe the sample. Layer 1 with an infinite thickness represents the substrate. Because of the very thick SiO_2 and the small difference in density between SiO_2 and Si, we do not distinguish SiO_2 from the Si substrate in the model. We also assume the composition of $SiO_{2.2}C_{0.6}$ for the top two layers. The interface roughness is simulated by a density variation of an error function profile [3], i.e.,

$$\rho(z) = \frac{(\rho_{n+1} - \rho_n)}{2}\left[1 + erf(\frac{z - z_n}{\sqrt{2}\sigma_n})\right] \qquad (1)$$

where ρ_n denotes the density of the nth layer, z_n is the location of the top interface of the nth layer, and σ_n represents the width of interface between the nth and the $(n+1)^{th}$ layers. The Parratt recursion formalism [4] modified to incorporate the interface roughness is applied to calculate reflectivity curve. The calculated intensity is convoluted with a Gaussian function to simulate the instrument resolution. There are totally nine varying parameters, i.e., the density and thickness of each layer as well as the interface roughness and a scaling constant. The best-fit result of

Table I. Parameters from the best fit to the reflectivity data using a three-layer model.

Layer	Density (g/cm³)	Thickness (Å)	Interface roughness (Å)
1 (substrate)	2.19	∞	16.0 ($\sigma_{sub-xerogel}$)
2	.47	5525	3.0 ($\sigma_{xerogel-top layer}$)
3	.53	915	5.5 ($\sigma_{top layer-air}$)

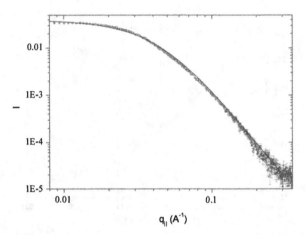

Figure 3. GISAXS data of xerogel/Si sample: (o) the experimental data; (——) the best-fitted results using a fractal model [5].

experimental data is displayed by a solid line in figure 1 and the best-fit value of parameters are listed in table I. Though the best-fit result matches both the high-q and low-q regions as well as the overall profile of the reflectivity curve very well, the intensity modulations of 0.075 $Å^{-1} < q_z$ < 0.15 $Å^{-1}$ cannot be fitted satisfactorily. There are some minor features still missing in the model. The 16 Å roughness of the xerogel/SiO$_2$ interface is unreasonably large. We speculate that it is associated with a density gradient between the SiO$_2$ and the porous xerogel material. The second layer has a thickness of 5,525 Å, which is close to the expected value of xerogel film thickness. Therefore, we assign the 2nd layer as the xerogel film. Assuming the skeletal density of xerogel is close to that of conventional SiO$_2$ film, approximate 2.27g/cm^3, the porosity of the film is
estimated to be about 79%. The topmost layer, which has a density slightly larger than that of the 2nd layer, has a thickness of 915 Å. The origin of the top layer structure may be related to the surface treatment to the xerogel film to cut down on moisture uptake but is not clear at this point.

Figure 3 is the log-log plot of the SAXS signals of xerogel thin film, where $q_{||}$ denotes the momentum transfer parallel to surface. The open circles depict the scattering intensities measured with incidence angle fixed at 0.2°. The SAXS intensities of a substrate prepared by same way as that used in xerogel sample are recorded as background and have been scaled and subtracted from the raw data. To obtain a detailed information about the microscopic structure of the xerogel, we fit scattering intensity $I(q)$ to a fractal model, which is commonly adapted to describe porous materials made of aggregates. The characteristic function $\gamma(r)$ of a fractal object in 3-dimension is of the form of $r^{-(3-D)} e^{-r/\xi}$ [5], where D is the fractal dimension and ξ is the cut-off length which corresponds to the finite length scale of fractal objects. Fourier transform of the correlation function yields the structure factor $S(q_{||})$

$$S(q_\parallel) = 1 + \frac{C\ \Gamma(D-1)\xi^D}{q_\parallel\xi\ (1+q_\parallel^2\xi^2)^{(D-1)/2}}\sin[(D-1)\arctan(q_\parallel\xi)] \quad (2)$$

where C is a constant and Γ is the gamma function. The SAXS intensity is fitted by $S(q_\parallel)$ without introducing any additional form factor. There are four parameters : D, ξ, C and a scaling constant I_0, in the model. The calculation results using the best-fitted parameters is displayed by the solid line in figure 3. The least-square fitting of the experimental data yields $\xi = 22.4\text{Å}$. It is particularly noteworthy that the best-fit value of D is 3, which is the limit of a mass fractal dimension and the fractal model reduces to the Debye-Bueche (D-B) model [6, 7], i.e., $\gamma(r) = e^{-r/\xi}$ and $I(q) \propto (1+q^2\xi^2)^{-2}$. The D-B model was developed for materials consisting of two strongly segregated, interpenetrating phases and it assumes a random distribution of pores both in size and shape at any scale. It is expected for an aggregated system that fractal dimension D increases toward the dimensionality of space, d = 3, while the aggregates get more and more entangled. It is speculated that the aggregates in xerogel are tightly entangled that a complete random distribution can well describe it.

Based on the D-B model several characteristics of the xerogel thin film can be readily obtained. The specific surface S of xerogel can be calculated from the porosity ϕ and correlation length ξ using $S = 4\phi(1-\phi)/\xi$. With the estimated porosity 0.8 determined from the reflectivity data, the specific surface is approximate 0.029 Å^{-1}. Moreover, the average size of the aggregates is obtained from the cut-off length, $l_c = 2\int_0^\infty \gamma(r)dr = 2\xi$, is about 45Å.

CONCLUSIONS

X-ray reflectivity and grazing incidence SAXS are measured on a xerogel thin film grown on a Si wafer. Specular reflectivity reveals that the xerogel film has a two-layer structure. The microscopic structure of xerogel can be described by a D-B model - a random distribution of of pores both in size and shape.

ACKNOWLEDGEMENTS

The authors would like to acknowledge the technical support of the SRRC staffs, in particular, Mr. Tang-Eh Dann. We also like to thank Prof. T.E. Hsieh of National Chiao-Tung University for taking SEM micrographs and Prof. T.L. Lin of National Tsing-Hua University for very useful discussion.

REFERENCES

1. Changming Jin, Scott List, and Eden Zielinski, Mat. Res. Soc. Proc. **511**, 213, (1998).
2. D. Windover, T.-M. Lu, S.L. Lee, A. Kumar, H. Bakhru, C. Jin, W. Lee, Appl. Phys. Lett. **76**, 158 (2000).
3. S.D. Kowowsky, C.-H. Hsu, P.S. Pershan, J. Bevk, and B.S. Freer, Appl. Surf. Sci. **84**, 179 (1995).
4. L.G. Parratt, Phys. Rev. **95**, 359 (1954).
5. T. Freltoft, J.K. Kjems, and S.K. Sinha, Phys. Rev. B. **33**, 269 (1986).
6. P. Debye, H.R. Anderson, Jr., and H. Brumberger, J. Appl. Phys. **28**, 679 (1957).

7. E.Z. Valiev, S.G. Bogdanov, A.N. Pirogov, L.M. Sharygin, and V.I. Barybin, JETP **76**, 111 (1993).

Mat. Res. Soc. Symp. Proc. Vol. 612 © 2000 Materials Research Society

PROCESSING, CHARACTERIZATION AND RELIABILITY OF SILICA XEROGEL FILMS FOR INTERLAYER DIELECTRIC APPLICATIONS

Anurag Jain, Svetlana Rogojevic, Feng Wang, William N. Gill, Peter C. Wayner Jr., Joel L. Plawsky*
Department of Chemical Engineering, Rensselaer Polytechnic Institute, Troy NY- 12180.
Arthur Haberl, William Lanford
Department of Physics, University at Albany, Albany, NY-12222.

Abstract

Surface modified silica xerogel films of high porosity (25-90 %) and uniform thickness (0.4-2 μm) were fabricated at ambient pressure on silicon and other substrates. Mechanical reliability of the films was determined by measuring fracture toughness (adhesive) as a function of aging time and temperature using the modified edge-lift-off technique. There is an optimum aging time at 60 °C aging to obtain maximum fracture toughness for the procedure used here.

Cu/xerogel/Si and Ta/xerogel/Si structures were annealed at different temperatures and in different ambient environments were analyzed using RBS and optical microscopy to assess the extent of interaction with the xerogel film. When annealed in N_2 with trace amounts of O_2 (equivalent to 10^{-7}-10^{-6} Torr vacuum), RBS analysis does not show diffusion of Cu or Ta through the xerogel up to 450 °C. At higher temperatures, or in the presence of larger concentrations of O_2, Cu and Ta oxidize. Cu oxidation leads to significant diffusion through the xerogel. Ta oxidation also results in diffusion-like RBS spectra. Using the micron-size ion beam to probe the Ta surface, this was found to be solely due to buckling of Ta films on xerogel. A thin SiN_x layer on top of Cu and Ta prevents metal oxidation up to 640 °C, Cu diffusion, and Ta buckling.

Introduction

Silica xerogels offer promise as ultra-low K dielectric materials. Fabrication methodologies under ambient conditions have recently been developed[1, 2, 3,] to replace impractical supercritical drying methods[4, 5]. Integration studies of these films using Al and Cu as the metallization layers have also been reported[6]. The porosity of these materials could cause reliability problems if they are used in the future interconnect schemes. Mechanical stability of the films and adverse interactions with other materials (metals, barriers, hard masks) are the main concerns. In this work, we study the effect of processing conditions on the fracture toughness, which is measure of the crack resistance of the films. To study the interaction with other materials, multilayer structures were fabricated and annealed at various temperatures and in different ambient environment and the extent of interaction was analyzed by Rutherford Backscattering Spectroscopy (RBS).

Experimental

Xerogel films were fabricated on silicon wafers with a native oxide layer, using the procedure described in previous publications [7, 8]. By controlling the amount of solvent evaporation from the xerogel film during spin coating, films of wide range of porosity were produced. The final films were produced by aging, surface modification, and ambient drying steps [7, 8]. The thickness and refractive index (R. I.) of the films were measured by ellipsometry at λ = 632.8 nm, θ = 70°. Interferometry and profilometry were used to obtain an estimate of the film thickness to determine the correct ellipsometric period. The porosity of the films was obtained from the R. I. using effective medium theory[7].

Fracture Toughness Measurements: Maintaining the mechanical integrity of multilayer coatings is an important requirement for successful integration in back end of the line processing. The materials must possess both cohesive and adhesive fracture toughness. Aging processes affect the microstructure (and hence the mechanical strength) of the film as dissolution and reprecipitation reactions cause neck formation between the particles. The time and temperature of the aging, the pH of the medium, and the solvent influence these reactions[9].

The modified-edge lift off test (m-ELT)[10, 11] was used to quantify the toughness of the xerogel coatings. It consists of applying a thick backing layer (e.g. epoxy) to the test film. The test film is supported on a rigid substrate (Si). The backing layer materials must have a known stress temperature profile, higher fracture toughness than the test material and excellent adhesion to the test material. The wafer is diced so that 90° edges to the substrate are formed. The sample is then cooled until debonding is observed. If the backing layer material is much thicker than the test material, the stored energy is approximately that in the backing layer, so, the applied fracture intensity, K_I, is given by the stress in the backing layer times the square root of half the backing

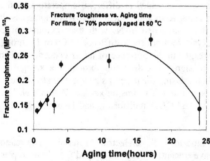

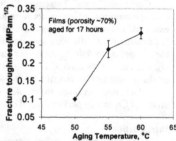

Figure 1: Fracture toughness of films aged at 60 °C. Each data point represents a separate film. The whole wafer was broken into small pieces. The values represent the average toughness and the error bars are the standard deviation

Figure 2: Fracture toughness of different xerogel films aged at different temperatures for aging time of 17 hours. The aging should be done at highest possible temperature to achieve the required strength

layer thickness. To perform the m-ELT tests, several films aged at different conditions were fabricated. The aging temperature was varied from 50 to 60 °C for a fixed aging time of 17 hours and the aging time was varied from 0.5-24 hours for a fixed aging temperature of 60 °C. Thus the effect of both aging time and temperature could be determined separately. They were then coated with 0.1 μm of PECVD oxide. This ensured that resist used in the test would not wick through the pores of the xerogel during curing. The multilayer structure was again annealed at 400 °C for 1 hour in vacuum. Before doing the m-ELT test, a scotch tape test was performed and all samples survived. An epoxy was applied to the xerogel and cured for one hour at 177 °C. In a separate experiment, the residual stress versus temperature was measured for the epoxy on a bare silicon substrate. The samples were then cleaved and observed for debonding as described above.

Figure 1 shows the fracture toughness of films aged for different times at 60 °C. The toughness first increases with increased aging time reaches a maximum and then decrease for long aging times (24 hours). The film aged for 17 hours shows the maximum fracture toughness. Figure 2 shows the fracture toughness versus aging temperature of the films. The results show that the toughness increases almost three times with an increase in the aging temperature of just 10 °C. Higher aging temperature lead to higher strength. This is in agreement with Smith et al.[12] that at temperatures in the range 100-150 °C the aging time required to achieve significant mechanical strength is few minutes. However, these types of conditions can only be achieved by super-saturation in pressurized vessels.

Thus results from figure 1 and 2 suggest that the optimum aging conditions for the experiments done so far a temperature of 60 °C and an aging time of 15 hours. The locus of failure for the samples from the visual observation was the silicon-xerogel interface, which means the measured fracture toughness may be adhesive, rather than cohesive. The exact locus of failure has to be identified after doing careful surface analysis of failed samples. Efforts to improve the adhesion are underway so that we can measure the cohesive fracture toughness. The mechanical properties and adhesion also vary with porosity of the xerogels and the hence toughness of films of different porosities aged at optimum conditions will be measured and compared in future. We have developed a method to fabricate films of a wide range of porosity (25-90 %) and thereby improve the thermal and mechanical properties including strength.

Xerogel Interaction with Cu and Ta: In an interconnect scheme, the dielectric material is in contact with other materials such as metals (Copper, Cu), barriers (e.g. Ta, or TaN), or hard masks (SiN_x). It is important that any adverse interactions between these different materials be minimized, although some interaction is necessary to provide adhesion. Cu, the material of choice for IC metallization, is known to be a fast diffuser through both SiO_2 and Si, and requires a diffusion barrier. Tantalum is one of the materials often investigated as an adhesion layer/diffusion barrier between copper and SiO_2 or Si substrates[13, 14].

Thermal oxide (SiO_2) films grown using wet oxidation at 1000 °C were used for comparison with the silica xerogel. Cu was deposited on all films in a magnetron sputterer. Deposited Cu films were 500-950Å thick. Ta films were deposited at conditions similar to Cu, with an Ar pressure in the range of 2.8 to 17mTorr. A series of annealing experiments was done in a vacuum furnace with a lowest achievable base pressure of 8-10 mTorr. The samples were loaded at 100°C, furnace was then evacuated to ~ 20 mTorr. Following evacuation, we maintained a pressure of 100 Torr using a N2 purge. We estimate that the O_2 concentration during annealing was less than $5X10^{-8}$ mol/m^3. ("trace of O_2" in further text). During several of the experiments the O_2 concentration increased due to pump failure. These conditions are marked with an "air leak" label. Using RBS [15], the atomic composition of the major constituents can be determined from the heights of the peaks in the RBS spectra. RBS spectra were obtained using the dynamitron ion accelerator at U. Albany's Accelerator Lab., using 2 MeV $4He^+$ ions. The detection sensitivity for Cu was estimated to be $1x10^{20}$ atoms/cm^3 of SiO_2. This is equivalent to $0.5 X 10^{20}$ atoms/cm^3 of 50% porous xerogel. Similarly, the detection sensitivity for Ta is estimated at $2.5X10^{19}$ atoms/cm^3 of SiO_2. The detection of Cu or Ta diffusion within the first 0.5μm of the xerogel layer is difficult, due to the presence of the thick layer of dense metal film immediately above this layer. Therefore, xerogel films around 1μm thick were used.

Copper: Cu/xerogel samples annealed with the N_2 purge up to 450°C for 11 hours show no oxidation or diffusion of Cu through xerogel (no change in the RBS spectrum upon annealing). A Cu/xerogel sample annealed at 550 °C for 2 hours at the same conditions (Figure 3) starts showing Cu oxidation (decreased peak height) and diffusion though the xerogel (Cu peak tail on the left edge) due to trace amounts of air (O_2) that enter the furnace under normal operating conditions. One Cu/xerogel sample was annealed at 500°C for 11 hours with a substantial

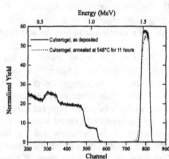

Figure 3: Cu starts oxidizing and diffusing into xerogel at 550°C when annealed in an ambient containing trace O_2 (< 5×10^{-8} mol/m³)

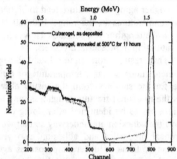

Figure 4: Cu annealed at 500 °C in ambient containing substantial O_2 shows oxidation and enhanced diffusion through xerogel.

amount of O_2 in the furnace. The RBS spectrum, Figure 4, shows that Cu diffused through the xerogel film and into the Si substrate substantially. The Cu layer is also oxidized. In contrast, Cu/thermal SiO_2 samples show oxidation, but no Cu diffusion through SiO_2. Thus we conclude that Cu diffusion is faster through xerogel than through thermal SiO_2, probably due to the prevalence of the surface diffusion mechanism. A piece of Cu/xerogel sample was coated with 500Å of plasma deposited SiN_x. The samples were annealed up to 650 °C. Neither Cu oxidation nor diffusion into the xerogel was observed. SiN_x is thus a good barrier to oxidation of Cu in ambient containing trace amounts of O_2. Thus Cu in the elemental state does not diffuse into xerogel at temperatures up to 650 °C.

Tantalum: Ta could be deposited on xerogels as a continuous film at Ar pressures in excess of 10 mTorr. Below 10 mTorr or after annealing, the Ta film buckled when deposited on xerogel. An optical micrograph of the buckled Ta surface is shown in Figure 5. This is consistent with the

Focusing micron beam at the intersection of buckled plates showed diffusion like tails (Figure 7)

Focusing micron beam in this region showed no diffusion like tails

Figure 5: Optical micrograph of the surface of Ta films deposited on 70% porous xerogel shows Ta buckling

reports [16, 17, 18] stating that Ta has a very high intrinsic compressive stress (in GPa) when deposited at conditions similar to ours. The film buckles to relieve the stress. Buckling increases as the annealing time is increased and is amplified if the sample is annealed in an O_2 containing environment. The intrinsic stress in the Ta film increases when the film is partially oxidized, due to the volume increase caused by oxidation. Similar to the case of Cu, Ta showed no diffusion through the xerogel or thermal oxide films when annealed at 450°C for up to 11 hours in N_2. To

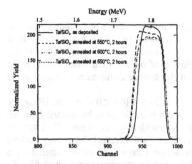

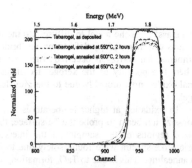

Figure 6: Ta deposited on SiO₂ is progressively more oxidized as the annealing temperature is increased, but no diffusion through SiO₂ is observed.

Figure 7: Ta is progressively more oxidized and diffusion-like tails through xerogel are more pronounced as the annealing temperature is increased. These tails were found solely due to the irregular Ta topography.

determine whether Ta diffusion occurred at higher temperatures, the samples were annealed at temperatures ranging from 450 to 650°C.

Figure 6 shows that the Ta/SiO₂ samples annealed up to 650°C (Ta does not buckle on SiO₂) show progressively increased Ta oxidation, but no noticeable Ta diffusion. In contrast (Figure 7), Ta/ xerogel samples annealed at 550, 600 and 650°C show progressively more Ta oxidation and diffusion-like tails as the annealing temperature is increased. Using the micron-size ion beam to probe the Ta/xerogel sample surface, the diffusion- like RBS spectra were found to be due solely to the buckling of Ta films on xerogel. The RBS signals corresponding to the energies of the Ta tails are emitted only from the localized regions of the surface, resembling the shape of the buckles (figure 5). When capped by SiNₓ, and annealed at temperatures up to 650 °C, all Ta/xerogel RBS spectra are indistinguishable - there is no Ta oxidation or diffusion. It is important to note that the SiNx capped samples showed no Ta buckling. This confirms that Ta buckling is related to Ta oxidation.

To show that Ta buckling and diffusion-like spectra are caused by the oxidation of Ta, Ta/xerogel samples were annealed in air at 304 °C and at 450 °C. Ta on xerogel annealed at 304 °C in air and the as-deposited Ta shows no difference. A temperature of 304 °C is not sufficient to cause significant oxidation of Ta in air. Ta annealed at 450 °C in air for 2 hours shows a significant amount of Ta oxidation (to Ta₂O₅) as well as Ta diffusion-like tails. This confirms our previous observation that whenever a small extent of Ta oxidation occurs, Ta buckling causes Ta tails in the RBS spectra. Even in its oxidized state, the Ta^{+5} charge prevents thermal diffusion into oxide.

Conclusions

The fracture toughness of xerogel films is affected by aging conditions. To obtain high toughness films should be aged at the highest possible temperature. There may be an optimal aging time and temperature for maximum fracture toughness.

In a N₂ environment with O₂ < 5X10^{-8} mol/m³ neither Cu nor Ta show diffusion through the xerogel (beyond the detection limit of RBS), when annealed at temperatures up to 450 °C. At this concentration of O₂ (~ 5X10^{-8} mol/m³) oxidation of Cu and Ta thin films occurs at temperatures higher than 450 °C. Cu oxidation causes significant Cu diffusion into the xerogel. The Cu diffusion results agree with a model we had previously developed for diffusion of Silver ions in glass[19]. For the metal to diffuse into the SiO_2, it must be oxidized or corroded first. The metal ions can then occupy defects, or non-bridging oxygen sites in the silica matrix. In these experiments, Cu corrosion occurred whenever air was introduced or leaked into the furnace chamber. The porosity of the xerogel facilitates diffusion of oxidized Cu when compared to thermal oxide, most probably due to the prevalence of the surface diffusion mechanism in the xerogel.

Ta oxidation at higher temperatures also causes diffusion-like RBS spectra. Using the micron-size ion beam to probe the Ta surface, these signals were found to be emitted from the localized regions of the sample (at the intersection of buckled plates, figure 5). Thus the diffusion-like spectra were found solely due to buckling of Ta on xerogel. Ta always buckles upon annealing. This is due to TaO_x formation, which increases the volume of the material up to 150%. A 500Å SiN_x layer prevents the oxidation of Cu and Ta at temperatures up to 650 °C. Under these conditions, Cu does not diffuse into the xerogel (beyond the RBS threshold). SiN_x cap layer prevented oxidation, buckling and diffusion of Ta up to 650 °C. Thus high quality dielectric xerogel films are compatible with the adjacent layers of Cu or Ta in an inert environment.

Acknowledgements

The authors wish to thank Semiconductor Research Corporation for providing the financial support for the project. The authors thank Dr. E.O. Shaffer II of Dow Chemical Company for the edge-lift off measurements and S. Ponoth (RPI) for CVD oxide and nitride deposition.

References

1. S. S. Prakash, S. S., Brinker, C. J., Hurd, A. J., J. Non-Cryst. Solids, 190, 264, (1995).
2. C. Jin, J. D. Luttmer, D. M. Smith, T. A. Ramos, MRS Bulletin, pp 39-42, Oct. (1997).
3. H-S.Yang, S-Y. Choi, S-H. Hyun, H-H. Park, J-K. Hong, J. Non-Cryst. Solids, 221, 151, (1997).
4. L. H. Hrubesh, and J. F. Poco, J. Non-Cryst. Solids, 188, 46, (1995).
5 M. - H Jo, H.-H. Park, D-J. Kim, S-H. Hyun, S-Y. Choi, J-T. Paik, J. Appl. Phys., 82, 1299, (1997).
6 Jin, C., List, S., Zielinski, E., Mat. Res. Soc. Symp. Ser., 511, 213, (1998).
7. S. V. Nitta, V. Pisupatti, A. Jain, P. C. Wayner, Jr., W. N. Gill, and J. L. Plawsky, J. Vac. Sci. Tech. B, 17(1), 205, (1999).
8. S. Nitta, A. Jain, V. Pisupatti, W. N. Gill, P. C. Wayner, Jr., and J. L. Plawsky in Low Dielectric Constant Materials-IV (Mat. Res. Soc. Symp. Proc. 511, Pittsburgh, PA, 1998) pp 99-103.
9. C. J. Brinker and G. W. Scherer, Sol Gel Science (Academic Press, 1990).
10. E. O. Shaffer, F. J. McGarry, and L. Hoang, Polym. Sci. & Eng., 36(18), 2375, (1996).
11. E. O. Shaffer, P. H. Townsend, and J-H. Im, Conf. Proc. ULSI XII, 429, (1997).
12. D. M. Smith, G. P Johnston, W. C. Ackerman, and S. - P. Jeng, U. S. Patent # 5753305, May (1998).
13. P. Gallais, J. J. Hantzpergue, and J. C. Remy, Thin Solid Films 165, 227-236 (1988).
14. L. P. Buchwalter, J. Adhes. Sci. Tech., 9(1), 97-116 (1995).
15. W.-K. Chu, J. W. Mayer, and M. -A. Nicolet, Backscattering Spectrometry, (Academic Press, New York, 1978).
16. H. Windischmann, Crit. Rev. Solid. State Mat. Sci., 17(6), 547-596 (1992.)
17. D. W. Hoffman, J. Vac. Sci. Tech. A, 20(3), 355, (1982).
18. D. T. Price, Ph.D. Thesis, Rensselaer Polytechnic Institute, (1999).
19. D. Kapila and J. L. Plawsky, AIChE Journal, 39(7), 1186, (1993).

Mat. Res. Soc.Symp. Proc. Vol. 612 © 2000 Materials Research Society

Sol-Gel Derived Silica Layers for Low-k Dielectrics Applications

Sylvie Acosta[1], André Ayral[1], Christian Guizard[1], Charles Lecornec[2], Gérard Passemard[3] and Mehdi Moussavi[2]

[1]LMPM, UMR CNRS 5635, ENSCM, 8, rue de l'Ecole Normale, F34296 Montpellier cedex 5, France
[2]LETI/DMEL/TCI, CEA-Grenoble, 17, rue des Martyrs, F38054 Grenoble cedex 9, France
[3]ST Microelectronics, Central R&D, CEA-Grenoble, 17, rue des Martyrs, F38054 Grenoble cedex 9, France

ABSTRACT

Porous silica exhibits attractive dielectric properties, which make it a potential candidate for use as insulator into interconnect structures. A new way of preparation of highly porous silica layers by the sol-gel route was investigated and is presented. The synthesis strategy was based on the use of common and low toxicity reagents and on the development of a simple process without gaseous ammonia post-treatment or supercritical drying step. Defect free layers were deposited by spin coating on 200 mm silicon wafers and characterized. Thin layers with a total porosity larger than 70% and an average pore size of 5 nm were produced. The dielectric constant measured under nitrogen flow on these highly porous layers is equal to ~ 2.5, which can be compared to the value calculated from the measured porosity, ~ 1.9. This difference is explained by the presence of water adsorbed on the hydrophilic surface of the unmodified silica.

INTRODUCTION

The continuous decrease of the integrated circuit dimensions has led to prospect for new low dielectric-constant materials for interlayer dielectric application [1]. Among the various dielectrics, porous silica is particularly attractive for low -permittivity electronics applications [2]. Various studies were carried out in order to prepare such dielectric layers by the sol-gel route [2-6]. The larger the porosity, the lower is the dielectric constant. Aimed K values lower than 2 require a silica porosity larger than 70% [7].The porosity of the sol-gel derived silica thin layers depends strongly on the synthesis, deposition and drying conditions [8,9]. Various strategies were applied in order to promote the preparation of highly porous silica layers : sol deposition near the gelation point [2,3], supercritical drying of the layers [4], use of low volatility solvents [5] or of the templating effect by liquid crystal mesophases [6,10], additional ammonia post-deposition treatment [5].

The synthesis route investigated for this study is based on the use of common and low toxicity reagents and on the development of a simple process without gaseous ammonia post-treatment or supercritical drying step. The optimization of the synthesis parameters in order to obtain highly porous silica layers is first described. The layer characteristics are then described and discussed.

EXPERIMENTAL

The chemical products used to prepare the sols were common and low toxicity compounds: Tetraethoxysilane ($Si(OC_2H_5)_4$) as silica precursor ; acidic (10^{-2} M HCl) and basic (NH3) aqueous solutions to hydrolyze the silicon alkoxide. The molar ratio water/alkoxide in the final sol, h_t, was ranging from 12 to 28 ; Polyethylene glycol 300 ($HO(CH_2CH_2O)_{5.4}H$), organic additive was introduced to control the sol viscosity and to limit the water evaporation during deposition and drying, by hydrogen bonding of water molecules with the oxyethylenic units. The added amount of polyethylene glycol was defined by the weight percentage in the final sol, w%PEG. Octylphenyl polyether alcohol (C_8H_{17}-C_6H_4-$(OCH_2CH_2)_{9-10}OH$; Triton X100™), a non-ionic surfactant was added to decrease the solid-liquid and solid-vapor surface tensions and by this way to act as wetting agent on the silicon wafers and to decrease the capillary stresses and related shrinkage during drying. The weight percentage of surfactant in the final sol was kept equal to 1w%.

The silicon alkoxide was first hydrolyzed by addition of a 10^{-2} M HCl aqueous solution ($2h_t/3$) under vigorous stirring and maintained under reflux at 78°C for 1 hour. After cooling the organic additives (PEG and surfactant) were added. The resulting sol was stable and could be stored during several days. Before deposition, the pH of the solution was increased by addition of a NH_3 aqueous solution ($h_t/3$). The pH of the final sol was equal to 5 (excepted later specified case), leading to gelation times, t_G, ranging from 1 to several hours as a function of the h_t values. The deposition of the layers was always carried out at t<0.5 t_G, corresponding to a period of time during which the viscosity does not vary. For the first step of optimization of the synthesis parameters and in order to obtain large amounts of sample, sols were poured in wide beakers resulting in cracked thick layers. The sols were then deposited as thin layers on 200 mm silicon wafers by spin coating (4000rpm). The aging conditions were defined by the aging time, t_a, the aging temperature, T_a, and the relative humidity, HR, of the oven. After aging the layers were thermally treated up to 450°C for 2 hours.

The thermal removal of residual water and organic additives was studied from thermogravimetric measurements. The porosity P_{N2} of the calcined samples was analyzed using nitrogen adsorption-desorption isotherms at 77 K [11]. For the thin layers deposited on silicon wafers, specific conditions were applied which were detailed in [12]. The thickness and the refractive index, n, of the thin layers were measured by ellipsometry. From n, a second value of the porosity, P_e, was calculated using the Lorentz-Lorenz relation [13]. The thin layers were observed by scanning electron microscopy (SEM) and analyzed by Fourier Transform IR spectroscopy (FTIR). The stress was measured using a Flexus stress system. The dielectric constant was determined, before and after N2 purge, by C-V measurement using a SSM Mercury Probe system.

RESULTS AND DISCUSSION

A preliminary study was carried out on unaged thick layers. In Figure 1 is reported the thermogravimetric curve obtained for a sample dried for several hours at room temperature. The first weight loss between 60 and 200°C is associated to the water departure and the second weight loss up to 450°C corresponds to the elimination of the organic additives. From these data it can be calculated that 60% of the water initially introduced in the sol is kept in the material after drying. This result evidences the positive effect of the polyethylene glycol addition for the water retention. The evolution of the porosity as a function of the weight percentage of PEG 300 is shown in Figure 2. It appears that a porosity larger than 60% requires an important weight percentage of PEG 300 (> 60%). From this first set of data, we decided to fix the weight percentage at a low value : 20% and to add a supplemental stage in the process : a aging treatment at controlled temperature and relative humidity in order to favor the silica polymerization and the related strengthening of the inorganic network before drying and calcination. The results reported on Table I evidence that the increase of the temperature and of the relative humidity during aging induces a strong increase of the porosity.

In a second stage, we focused our investigations on thin layers. The PEG 300 weight percentage and the hydrolysis ratio h_t were kept equal to 20% and 28 respectively. For these conditions, the viscosity of the sol was equal to 8 MPa.s and crack free thin layers were deposited on 200 mm silicon wafers. We increased the aging temperature up to 75°C and we tried to adjust the other aging parameters : HR and t_a.

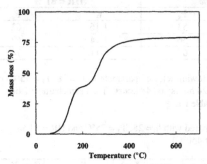

Figure 1. *Thermogravimetric analysis of a dried thick layer ($h_t = 12$ and w%PEG = 20).*

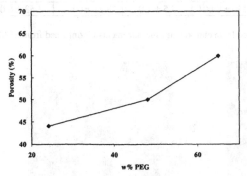

Figure 2. *Variation of the porosity for calcined thick layers vs. w% PEG ($h_t = 16$; final pH = 3.5).*

Table I. Porosity P_{N2} of calcined thick layers ($h_t = 28$ and $t_a = 64$ hours).

HR (%)	$T_a = 50°C$	$T_a = 60°C$
no humidity regulation	44	48
78	53	75

The results reported on Table II show that an increase of HR from 75% to 81% enables to obtain an equivalent final porosity for a shorter aging time. For instance, the aging time required to obtain a porosity of 67% is divided by a factor 2.5. Attempts of aging at higher relative humidity gave rise to less porous layers than for HR = 81% and also to defects at the surface of the layers probably related to water condensation phenomena.

Table II. Refractive index n and porosity P_e of calcined thin layers ($h_t = 28$ and $T_a = 75°C$).

t_a (hour)	HR = 75 %		HR = 81 %	
	n	P_e	n	P_e
24	1.20	53	1.16	62
48	1.19	56	1.14	67
120	1.14	67	1.11	74

Thin layers were prepared using the following selected parameters : $h_t = 28$, $T_a = 75°C$ and HR = 81 % and two different aging times : 24 hours and 48 hours. The main characteristics measured on these layers are reported on Table III.

Table III. Characteristics of thin layers prepared with $h_t = 28$, $T_a = 75°C$ and HR = 81 % (measurements performed after annealing at 450°C)

t_a (hour)	thickness (nm)	n	stress (Mpa)	K_{air}	K_{N2}
24	630 ± 25	1.155 ± 0.005	12.5 ± 4.0	4.22 ± 0.20	-
48	640 ± 15	1.140 ± 0.001	9.5 ± 1.0	3.21 ± 0.05	2.53 ± 0.05

The thickness of the layers determined from ellipsometry measurements is confirmed from SEM observations (Figure 3).

(a)   (b)

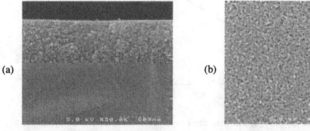

Figure 3. Scanning electron microscope observations of a calcined thin layer ($h_t = 28$, $T_a = 75°C$, HR = 81 % and $t_a = 48$ hours). (a) cross section image; (b) surface image.

The porosity P_e calculated from the refractive index is equal to 66.8 ± 0.3 % for $t_a = 48$ hours. The nitrogen adsorption-desorption isotherm obtained for layers prepared using the same conditions is presented on Figure 4. The hysteresis loop is characteristic of a mesoporous material [10] and the calculated mean pore size is equal to 5 nm. The porosity measured from the N_2 isotherm is equal to 72 ± 1 %. The lower value derived from the ellipsometric measurements carried out at ambient conditions can be explained by the presence of adsorbed water in the layer This adsorbed water is not taken into account in the calculation of P_e and the refractive index of water (1.33) is higher than that of air (1.00). The presence of adsorbed water is confirmed by the FTIR analyses. In Figure 5, are reported the IR spectra of the layer aged 48 hours, recorded before and after the last step of calcination. On the spectrum of the calcined layer, the adsorbed water is evidenced by the presence of large absorption bands centered around 3500 cm^{-1} and 1600 cm^{-1} and assigned to the stretching vibration $\nu(OH)$ and to the deformation vibration $\delta(H_2O)$ respectively. Moreover, for the calcined layer, it can be noted the disappearance of the absorption band located at 2900 cm^{-1} and associated to the stretching vibration $\nu(CH)$ of the organic compound. In the low wave number range (<1200 cm^{-1}) absorption bands assigned to amorphous silica are observed.

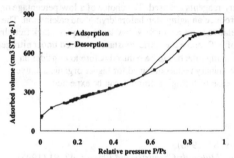

Figure 4. Nitrogen adsorption-desorption isotherm for a calcined thin layer ($h_1 = 28$, $T_a = 75°C$, $HR = 81$ % and $t_a = 48$ hours).

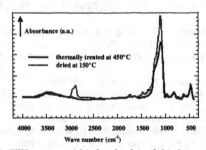

Figure 5. FTIR spectra of dried and calcined thin layers ($h_1 = 28$, $T_a = 75°C$, $HR = 81$ % and $t_a = 48$ hours).

Low values of residual tensile stress were measured on the calcined layers (Table III). The first measurements of the dielectric constant carried out in air gave K values larger than 3 (Table III). A new measurement under dry nitrogen flux was done on the calcined layer for $t_a = 48$ hours. The corresponding value is equal to 2.53 ± 0.05. This value can be compared to the value estimated from the layer porosity assuming ideal pure silica [7] : 1.9. This difference and also the higher values measured in air can be explained by the contribution of the adsorbed water to the dielectric properties of the layers. The dry nitrogen flux reduced the amount of adsorbed water in the layer but it was not sufficient to remove all the adsorbed water. The produced silica layers are highly hydrophilic and a hydrophobic post-treatment is so required to prevent water adsorption. This kind of treatment, usually silylation, has been applied in the processes previously described in the literature [2, 3,5] .

CONCLUSION

A new route of preparation of highly porous silica layers by the sol-gel process was investigated. The effect of different synthesis parameters was studied. The developed process is simple and only common and low toxicity reagents are used. The choice of a low percentage of organic additive (%PEG = 20) led to introduce an aging step before drying and calcination. The aging time of 48 hours at 75°C with a relative humidity of 81% enables to prepare crack free thin layers exhibiting a total porosity larger of 72% and a dielectric constant measured under nitrogen flow equal to ~ 2.5. The duration of the aging time has to be reduced before to consider an industrial application of this process. Preliminary results obtained for higher organic additive percentages show that an important decrease of the aging duration can be expected.

REFERENCES

1. S.P. Murarka, *Solid State Technology*, 83 (1996).
2. C. Cho, D. M. Smith and J. Anderson, *Materials Chemistry and Physics*, **42**, 91 (1995).
3. H. S. Yang, S. Y. Choi, S. H. Hyun, H. H. Park and J. K. Hong, *J. of Non-Cryst. Solids*, **221**, 151 (1997).
4. M. H. Jo, H. H. Park, J.J. Kim, S. H. Hyun, S. Y. Choi and J.T Paik, *J. Appl. Phys.*, **82**, 1299, 1997.
5. D. M. Smith, G. P. Johnston, W. C. Ackerman, S. P; Jeng, B. E. Gnade, R. A. Stoltz, A. Maskara and T. Ramos, European Patent 0 775 669 A2, (1997).
6. P.J. Bruinsma, N.J. Hess, J.R. Bontha, J. Liu and S. Baskaran, *Mat. Res. Soc. Symp. Proc.* **443**, 105 (1997).
7. L.W. Hrubesch, L.E. Keene and V.R. Latorre, *J. Mat. Res.*, **8**, 1736 (1993).
8. C. J. Brinker, G. C. Frye, A. J. Hurd and C. S. Ashley, *Thin Solids Films*, **201**, 97 (1991).
9. M. Klotz, A. Ayral, C. Guizard and L. Cot, *Bul. Korean Chem. Soc.*, **20**, 879 (1999).
10. M. Klotz, A. Ayral, C. Guizard and L. Cot, *J. Mat. Chem.*, **10**, 663 (2000).
11. S. Lowell and J.E. Shields, *Introduction to Powder Surface Area* (J. Wiley & Sons,1984).
12. A. Ayral, A. El Mansouri, M.P. Vieira, C. Pilon, *J. Mat. Sci. Lett.*, **17**, 883 (1998).
13. M. A. Fardad, E. M. Yeatman, E. J. C. Dawnay, M. Green and F. Horowitz, *J. Non Cryst. Solids*, **183**, 260 (1995).

Barrier and Seed
Layer-Deposition Techniques

Mat. Res. Soc. Symp. Proc. Vol. 612 © 2000 Materials Research Society

Seed-layer Deposition for Sub 0.25 μm Cu Metallization Using a Line Cusp Magnetron Plasma Source

Sunil Wickramanayaka, Hanako Nagahama, Eisaku Watanabe, Toshihiko Hayashi, Makoto Sato, Yukito Nakagawa, Shinya Hasegawa, Shigeru Mizuno and Yoichiro Numasawa
Anelva Corporation, Yotsuya 5-8-1, Fuchu, Tokyo 183-8508, Japan

Abstract

A magnetically enhanced capacitively coupled plasma source was developed for sputter deposition of Cu seed layers on sub 0.25μm via or contact holes. The plasma source is of planer parallel plate configuration where the Cu target plate is one of the electrodes. For the generation of plasma, 60 MHz rf power is selected in order to increase the plasma density. Additionally, a line cusp magnetic field is used to further increase the plasma density. The film deposition rate and uniformity obtained with this plasma source is ~200 nm/min and ~±5%, respectively. The Cu film resistivity lies around 2 μΩcm. This sputtering system yields good film coverage on bottom and sidewalls of via holes with an aspect ratio > 5; therefore, a perfect Cu filling could be realized by electroplating process.

Introduction

Recently Cu interconnects technology came into application as an alternative to Al metallization in the production of integrated circuits in semiconductor devices.[1,2] To date, there are several methods investigated for Cu fillings in via or contact holes, such as CVD, ECD (electro-chemical deposition) and PVD processes.[1-5] In CVD, via holes are filled by chemical vapor deposition (CVD) processes where a plasma may or may not be used. In ECD processes, electrodeless or electroplating may be employed in gap filling. In PVD, Cu seed layer is first deposited on the sidewalls of via holes and then filled by electroplating process.[1,2,5] Here, the Cu seed layer is deposited by a plasma assisted sputtering process. The PVD process is seen to be more attractive compared to other alternatives because this process is a dry process and carried out at low temperatures. In this paper the development of a plasma assisted sputtering system for Cu seed layer deposition on 200 mm diameter wafers is reported.

There are several facts to be considered when developing a sputter deposition system for Cu seed layer deposition. 1. The seed layer must deposit on the side and bottom walls of via hoes of sub-0.25 μm with an aspect ratio >5. 2. The seed layer must be thick enough for the electroplating process. (Usually about 20 nm thick sidewall coverage is needed for electroplating process.) 3. The over-hang of film at the mouth of via-holes must be minimized in order to eliminate formation of voids during the electroplating process. 4. The film morphology must be good enough for better electrical conductance, which is also of importance for electroplating. 5. The nonuniformity of the Cu film deposited on the wafer surface must be < ±5%. 6. The plasma should not cause damages on the integrated circuits on the wafer surface.

Considering all these facts a new type of magnetically enhanced capacitively coupled plasma source is developed for Cu seed layer deposition on 200 mm diameter wafers. Even though conventional plasma sources are operated with 13.56 MHz rf power, a 60 MHz rf power is selected for this sputtering system in order to increase the plasma density. In addition, the plasma density is further improved by applying a line cusp magnetic field.

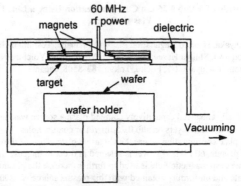

Figure 1. *A schematic diagram of sputter deposition plasma source.*

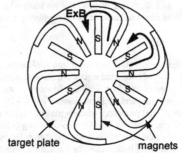

Figure 2. *A schematic diagram of magnet arrangement on the upper surface of Cu target plate.*

Experimental details

A cross sectional diagram of the developed plasma source is shown in Fig. 1. This sputtering system is based on parallel plate capacitively coupled plasma configuration. The upper electrode is the target plate and is made of Cu. This Cu target plate is simply a planar Cu plate with a diameter of 280 mm. The target plate was given a 60 MHz rf power through a matching circuit. Except for this 60 MHz rf power, no other rf or DC power was employed in this sputtering system to control the process parameters.

On the upper surface of the target plate a magnet arrangement was placed in order to enhance the plasma density. A schematic diagram of the magnet arrangement is shown in Fig. 2. Magnets were arranged in radial lines with alternate polarity. Those magnets, of which N-polarity faces towards plasma, were slightly bent along the perimeter of the target plate. In this configuration, magnets were not placed in the central region covering a 70 mm diameter. These magnets were arranged on a separate metal plate (which is called the magnet-plate hereafter) and placed at a close proximity to the upper surface of the Cu target plate so that the magnet plate can be rotated with the use of an electrical motor.

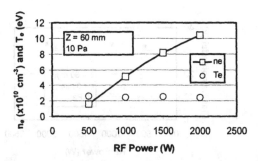

Figure 3. Plasma density (n_e) and electron temperature (T_e) as a function of applied rf power.

With the magnet arrangement explained above, several line-cusp magnetic fields are generated below the target plate. That is, the magnetic field generated from a line of magnet passes through the target plate and bends towards the nearest opposite polarized magnets. Therefore, a strong magnetic field can be observed at the surface of Cu target plate. The strength of the magnetic field decays rapidly with the increase of distance from the target plate, thus at the wafer surface a magnetic field-free environment is obtained.

The sputtering process was carried out at the pressure region of 10 Pa to 14 Pa with the use of Ar as the process gas. Once the plasma is ignited, the electrons in the plasma are trapped and undergo cyclotron rotation due to the presence of magnetic field. This results in an increase of electron path length and thereby a higher degree of ionization. Accordingly, this magnetic field causes a confinement of plasma and enhancement of plasma density. Further, when the plasma is generated, a negative self-bias voltage is generated on the target plate. Owing to this negative bias, Ar ions in the plasma gain energy by accelerating towards the target plate and bombard on the target which result in the sputtering of Cu atoms. In addition to this process, the self-bias voltage causes a drift of plasma due to the ExB force, where **E** and **B** are the dc electric field and magnetic field at the surface of target plate, respectively. The direction of plasma drift between each two magnetic lines is also shown in Fig. 2. With this magnet arrangement plasma drifts inward and outward alternately. The plasma drifted towards the outside of the target plate bends and flows towards the center of the target plate as shown in Fig. 2. This causes an increase of plasma density in the central region.

The plasma parameters were measured by using a single probe (Langmuir probe). These probe measurements were performed at 60 mm below the target plate as a function of applied rf power.

Films were deposited on bare or patterned Si wafers at 70 mm below the target plate. A 200 nm thickness Cu films were deposited on bare Si wafers containing a 2 μm thick SiO_2 layer for the electrical resistivity measurements. When patterned wafers were used, a TaN film of 35 nm thickness was first deposited on the wafer as a barrier layer. Thereafter, the wafer was cooled and transferred to the sputtering module. Film depositions were carried out at –20 °C though out this study. During the sputter deposition process, wafer was electrostatically chucked to the wafer holder and cooled by maintaining a higher Ar gas pressure between the wafer and the holder. The electroplating of Cu seed layer deposited wafers were carried by a different institute.

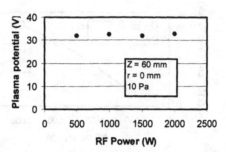

Figure 4. Variation of plasma potential as a function of applied rf power.

Results and discussion

The variation of plasma density (n_e) and electron temperature (T_e) as a function of applied rf power are shown in Fig. 3. The plasma density is seen to increase almost linearly with rf power. The electron temperature of plasma is estimated as around 2.5 eV. This lower electron temperature compared to those of conventional plasma sources is attributable to the increase of excitation frequency. The variation of plasma potential as a function of rf power is shown in Fig. 4. The plasma potential lies around 32 V and shows no variation with rf power. The plasma generated below the target plate is nonuniform since the plasma density between two magnetic lines gets increased due to the magnetic field effect. However, this nonuniform plasma diffuses and makes a uniform plasma about few centimeters below the target plate. Therefore, even though the above magnet arrangement is not rotated, a uniform plasma can be obtained at the wafer surface. The above argument can extend in explaining the uniform film deposition as well. At the places where the plasma density is higher, the sputtering rate is higher. These higher density plasmas lie close to each other, that is between every two magnetic lines, thus sputtered atoms diffuse and make a uniform film deposition on the wafer surface. This is confirmed by calculating the uniformity of electrical resistance of the film on the wafer surface. Figure 5 shows the radial profile of film resistance. The nonuniformity of the film resistance is estimated as ±5.1%. Since the electrical resistance is a function of film thickness, the nonuniformity of the film thickness is also considered to be equal to ±5.1%. The electrical resistivity of deposited film is around 2 µΩcm.

If the magnet-plate is not rotated the etch rate profile on the target plate becomes non-uniform. In order to make a uniform etch rate on the target plate the magnet arrangement is rotated with a rate of 0.1Hz. When the magnets are rotated, an etch rate profile that is almost uniform could be observed on the target plate.

The deposition characteristics of Cu seed layers in 0.25 µm via holes are shown in Figure 6. Here, SEM photographs of via holes after the film deposition at 10 Pa, 12 Pa and 14 Pa are shown. For the clarity of figures only the lower parts of via holes are shown in Fig. 6. It is observed that side and bottom coverage increase with an increase of pressure. The calculated bottom coverage of seed layers deposited at 10 Pa, 12 Pa and 14 Pa are 35%, 37% and 41%, respectively. Further the deposition rate decreases with an increase of pressure. The variation of deposition rate as a function of applied rf power is shown in Fig. 7. The increase of pressure results in an increase of gas phase collisions and thereby back scattering of Cu atoms. In addition, it seems that increase of gas phase collisions causes an

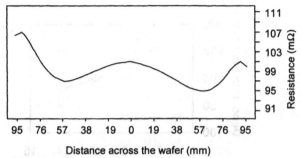

Figure 5. *Resistivity profile of Cu seed layer across the wafer.*

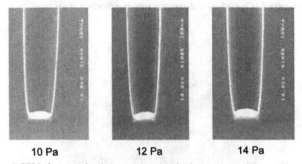

| 10 Pa | 12 Pa | 14 Pa |

Figure 6. *SEM photographs showing bottom and side coverage of Cu seed layer in 0.18 μm via holes.*

increase of ionization of sputtered atoms. These ionized Cu atoms accelerate towards the wafer due to the potential difference between the plasma and the wafer. This improves the bottom coverage.[6,7] Figure 8 shows a SEM photograph of via holes after the electroplating process. The diameter and aspect ratio of these via holes are 0.25 μm and 5, respectively. This photograph shows good Cu filling characteristics. The same filling characteristics are observed throughout the wafer. These results point out that the sputtering source developed in this project can be used for Cu seed layer deposition on sub-micron via holes for the purpose of electroplating and thereby for obtaining good filling characteristics.

Conclusion.

A plasma assisted sputtering system with a very simple configuration is developed for the purpose of Cu seed layer deposition on sub 0.25 μm via holes on 200 mm diameter Si wafers. The plasma is generated by employing a 60 MHz rf power applied to the Cu target plate. Except for this rf power, no other rf or DC power source is applied to the Cu target. This configuration of sputtering system yields a higher density plasma and thereby a higher sputtering rate. It is assumed that a fraction of film deposition is occurred by ionized Cu atoms. The deposition rate and uniformity obtained are ~ 200 nm/min and

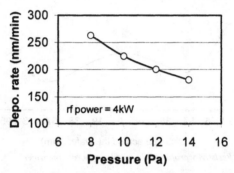

Figure 7. Cu seed layer deposition rate as a function of pressure.

Figure 8. A SEM photograph of 0.25 μm via holes after electroplating process.

~±5%, respectively. Electroplating results show that via holes of 0.25 μm with an aspect ratio of 5 can be filled perfectly.

References

1. M. Biberger, K. Ashtiani, M. Hamed, L. Hartsough, S. Jackson, E. Klawuhn and L. Tam, Proc. of the 5th International Symposium on sputtering and plasma process 1999 (ISSP-99), p. 51
2. J. Jorne, in extended abstract – Electrochemical Society Meeting , Hawaii (1999)
3. N. Awaya and Y. Arita, *Jpn. J. Appl. Phys.* 32, 3915 (1993)
4. H. J. jin, M. Shiratani, T. Fukuzawa, and Y. Watanebe, Proc. of the 4th International Conference on Reactive Plasmas - p. 95 (1998)
5. P. Gopalraja, J. Forster, A. Chan, J. van Gogh, Z. Xu, F. Chen, Proc. The 5th International Symposium on Sputtering and Plasma processes 1999, p 45
6. S. M. Rossnagel and J. Hopwood, *J. Appl. Phys. Lett.*, 63, 3285 (1993)
7. S. M. Rossnagel and J. Hopwood, *J. Vac. Sci. Technol.*, B12, 449 (1997)

Mat. Res. Soc. Symp. Proc. Vol. 612 © 2000 Materials Research Society

Atomic Layer CVD for Continuously Shrinking Devices

Suvi Haukka, Kai-Erik Elers and Marko Tuominen
ASM Microchemistry Ltd., Kutojantie 2B, P.O. Box 132, FIN-02631 Espoo, Finland

ABSTRACT

This paper will review the basics of the atomic layer chemical vapor deposition (ALCVD) thin film growth technique. The emphasis is on the ALCVD metal nitride growth and dual damascene barrier requirements.

INTRODUCTION

The shrinkage of IC device dimensions makes great demands on the thin film processing techniques. The processing technique should enable the growth of uniform ultra-thin films with atomic layer accuracy on large surface areas. Furthermore, the atomic layer accuracy should be realized in high aspect ratio vias and on irregular shaped surfaces to attain the best device performance. One possible technique to meet this challenge is atomic layer chemical vapor deposition (ALCVD).[1,2,3]

The ALCVD technique was developed in Finland in the early 1970's by Dr. Tuomo Suntola. His very simple and revolutionary idea was to introduce the metal compound and non-metal compound precursors sequentially to the surface and allow the reactive sites at the surface to control the film growth. This was contrary to conventional CVD where the precursors are introduced at the same time to the surface and the growth is controlled by the precursor flux intensity or the time of the growth.

The basics of the ALCVD technique as well as the factors affecting the growth of ALCVD high-k gate dielectrics on differently treated silicon surfaces have previously been covered.[4] Here the ALCVD growth basics are reviewed, emphasizing the metal nitride growth and dual damascene barrier requirements.

DUAL DAMASCENE BARRIER REQUIREMENTS

The barrier film should meet a great number of requirements (see Fig. 1). The film should have a low resistivity and almost no precursor residues should be incorporated into the film. Conformality is also crucial and the film should cover not only the bottom surface but also the side walls and bottom corners as well. Furthermore, in the dual damascene structure various surfaces are simultaneously present and the growth of the barrier film on these surfaces must equally be realized. The thickness of the barrier film will decrease with the decreasing feature size, which means that highly conformal ultra thin films (< 20 Å) will be needed in the future.

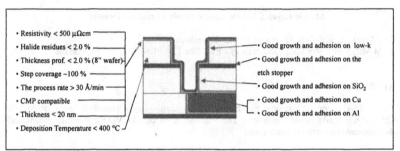

Figure 1. Dual damascene barrier film requirements.

ALCVD TECHNIQUE

The basic principle of the ALCVD technique is simple: metal compounds and non-metal compounds are reacted alternatively with the growing surface with an inert gas purging step in between the reactions. However, to achieve the excellent ALCVD features such as good step coverage and uniformity etc. the simultaneous fulfilment of the following growth conditions must take place:[4]

- availability of reactive volatile metal and non-metal compounds used as precursors,
- a substrate surface with proper reactive (adsorption) sites,
- a reaction temperature, at which a chemical reaction between the precursor and the reactive sites take place (chemisorption) without any decomposition or condensation of the precursors,
- a precursor dose high enough to saturate the surface,
- a sufficient inert gas purge after each reaction step to remove the surplus precursor molecules as well as the reaction by-products.

ALCVD PRECURSORS IN METAL NITRIDE GROWTH

Metal halides, alkyls and alkyl amides can be used as metal compound precursors and ammonia (NH_3) as a ligand removing or ligand exchange agent in the ALCVD growth of nitrides.[5-11] The growth mechanism of AlN from $Al(CH_3)_3$ and NH_3 has been studied in detail[5] and therefore it is used in Fig. 2 to describe the ALCVD growth basics, although AlN is not a suitable barrier material. Fig. 2 also clearly illustrates how the NH_x and OH groups as well oxygen bridges control the growth.

The starting surface for the AlN growth in Fig. 2 is a silicon oxynitride with both OH and NH_2 groups and oxygen bridges as reactive sites (Fig. 2A). In the first step $Al(CH_3)_3$ is reacted with the starting surface until surface saturation is achieved (Fig 2B). The reaction temperature is selected so that no decomposition or condensation takes place. Besides the reaction with the OH and NH_x groups also the reaction with the oxygen bridges occurs on the starting surface as illustrated by the formation of Si-CH₃ surface species. After saturation the reaction space is purged with inert gas to remove the surplus $Al(CH_3)_3$ molecules and methane that is formed as a by-product in the reaction between the CH₃ ligands and the hydrogen atoms in the OH and NH_x groups. To form AlN the methyl ligand terminated surface is reacted with ammonia (NH_3) until surface saturation is achieved (Fig. 2C). In the NH_3 reaction the methyl groups are replaced with NH_x groups which serve as reactive sites for the next $Al(CH_3)_3$ reaction (Fig. 2D). After NH_3 reaction the reaction space is again carefully purged with inert gas.

In ALCVD the four steps – reaction of the metal compound precursor, purging, reaction of the non-metal compound precursor and purging – forms one reaction cycle. In the thin film growth the cycle is repeated as many times as needed to grow a film with the desired thickness. This stepwise nature of the process differentiates ALCVD from conventional CVD, where continuous film growth takes place due to the simultaneous introduction of the metal and non-metal compounds to the growing surface.

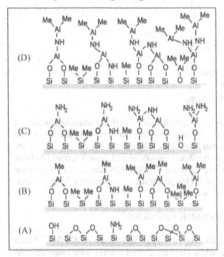

Figure 2. Growth mechanism of AlN from Al(CH₃)₃ and NH₃ (CH₃=Me); (A) starting surface, (B) first Al(CH₃)₃ reaction, (C) NH₃ reaction and (D) second Al(CH₃)₃ reaction.[5]

REACTIVE SITES

In ALCVD the nature of the reactive sites on the starting surface should be well characterized. This is because the growth base plays an important role by largely determining the quality of the interface and the film itself. A substrate surface with reactive sites uniformly distributed results in uniform film and good adhesion of the film to the surface. An example of a substrate with various starting surfaces is presented in Fig. 3. The oxide, nitride and Cu surfaces can be simultaneously present in dual damascene structure.

The best growth bases in ALCVD are the oxide surfaces terminated with OH groups and oxygen bridges and the nitride surfaces terminated with NHₓ groups and nitrogen bridges. As explained below and shown in Fig. 2 the presence of proper reactive sites leads to the formation of strong bonding of the precursor to the surface i.e. to good adhesion. Most low-k and etch stop materials contain silicon oxide and silicon nitride, which ensures a good adhesion of the ALCVD barrier. SiC and organic based low-k materials do not contain silicon oxide or nitride, but for instance the SiC surface might be somewhat oxidized when exposed to air making the growth possible. In addition, the ammonia used in the growth can also modify the surface suitable for the growth. However, to obtain uniform films either OH or NHₓ group termination should if possible be formed in a controlled way prior to the growth.

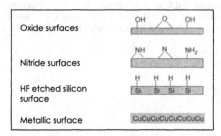

Figure 3. Starting surfaces with different surface sites.

Besides low-k materials and etch stop layers also the metal surfaces must be considered when ALCVD is used in the processing of Cu barriers. Pure copper surface is present on the bottom of the via and this copper surface should stay intact during the Cu barrier deposition. Copper pitting has, however, been observed in the direct growth of TiN from TiCl$_4$ and NH$_3$ on the Cu surface.[11] The pitting most probably occurs because HCl formed in the reaction etches the Cu surface through the formation of volatile CuCl. This problem can be overcome by processing for instance a protective layer of W$_x$N$_y$ on the Cu surface.[11]

GROWTH TEMPERATURES AND PRECURSOR RESIDUES

Metal nitrides can be grown at temperatures of 180 to 500 °C by ALCVD.[5-12] Metal halides can withstand elevated temperatures, while alkyl amide compounds with lower thermal stability start decomposing above 230 °C.[7] The low temperature requirement in case of alkyl amide compounds also leads to high resistivity of the as-deposited films.

In the ALCVD metal nitride growth from metal halides the amount of halide residues is dependent on the temperature, length of ammonia pulse and purging. At the growth temperatures required in Cu metallization (< 400 °C) lower chloride residues have been detected (< 1.5 at.%)[12] in the TiN growth from TiCl$_4$ and NH$_3$ than in the conventional CVD. This is because the ligand removal and purging steps can quite efficiently remove the HCl and NH$_4$Cl that are formed as by-products. The formation of NH$_4$Cl from NH$_3$ and HCl has been verified by FTIR measurements.[13] In conventional CVD where the metal halide and ammonia are introduced at the same time to the surface and no purging is possible, typically quite high temperatures (>600 °C) are required to get the halide residuals to acceptable level.

SURFACE SATURATION AND STEP COVERAGE

In Cu metallization 100% step coverage of the Cu barriers in the dual damascene structures is a requirement in the near future. In ALCVD the 100% step coverage has repeatedly been shown,[14] the growth on the porous high surface materials exhibiting the most extreme case.[3] In ALCVD the possibility of depositing highly conformal, ultra thin coatings on irregular surfaces is a result of the systematic utilization of the surface saturation through chemisorption in the growth. From a process standpoint this requires proper removal of the surplus precursor molecules with inert gas purge after each reaction step. Improper purging, as well as decomposition or condensation of metal precursors, would lead to uncontrolled CVD growth and eventually to closing up of the extremely narrow vias.

ALCVD GROWTH PARAMETERS

In ALCVD the growth rate can be expressed as growth rate per reaction cycle and thereby the thickness of the films can simply be determined by counting the number of the cycles. Typically the ALCVD growth rate varies from 0.2 to 1.5 Å per reaction cycle. The growth rate i.e. the surface coverage attained per cycle in the saturating reaction conditions is to a large extent dependent on the following:

1) reactivity of the precursors,
2) number of reactive sites available
3) size of the precursor molecule.

In the ALCVD metal nitride growth the amount of Al atoms deposited on the silica surface from $Al(CH_3)_3$ is high, which is due to the high reactivity of $Al(CH_3)_3$ and its small size.[5] On the contrary, the ALD growth rate of TiN thin film is relatively low (0.2 Å/cycle). This is most probably caused by the incapability of $TiCl_4$ to utilize other reactive sites than NH_x groups. Furthermore, the ALCVD TiN has almost a bulk density,[12] which suggests that the number of the NH_x groups on the TiN surface cannot be very high.

In the growth of metal nitrides from alkyl amide compounds that have quite large ligands $(-N(C_xH_y)_2)$ the effect of the size of the precursor on the growth rate can be expected. J.-S. Min has reported an ALCVD growth rate of 0.5 Å/cycle for tetrakis(dimethylamido) titanium precursor. Large ligands can cover part of the reactive sites very efficiently, thus preventing them from participating in the growth. In addition, the quite a low growth temperature (< 230 °C) that must be used for alkyl amides to prevent the thermal decomposition of the precursor can also partly explain the growth rate achieved. It must be emphasized here that because of the three factors mentioned above a growth rate corresponding to a full monolayer is generally not achieved.

The ALCVD cycle time in general varies from 0.5 s to 5 s and the growth rate per minute from 10 Å to 200 Å. These two parameters are dependent on the process and also on the ALCVD reactor design.

ALCVD REACTION MECHANISMS

In ALCVD the following reaction mechanisms have experimentally been identified:[1,3,5]
1) Ligand exchange reaction with the OH and NH_x groups.
2) Dissociative adsorption on the oxygen and nitrogen bridges.
3) Agglomeration through the formation of highly reactive volatile intermediate species.

A true atomic layer-by-layer growth can only be achieved when the bonding of the metal compound to the surface takes place directly through the OH or NH_x groups or as a result of a dissociation on the oxygen or nitrogen bridges. When agglomeration takes place the film growth cannot obviously be considered an atomic layer-by-layer growth. Despite agglomeration the growth is highly reproducible and uniform films are obtained.[4]

Examples of the ligand exchange and dissociation reactions can be found in Fig 2. Agglomeration reaction that is typical for metal chlorides has been thoroughly studied in the case of the TiO_2 growth from $TiCl_4$ and H_2O and ZrO_2 growth from $ZrCl_4$ and H_2O.[3] In the agglomeration reaction highly reactive volatile intermediate $M(OH)_xCl_y$ species form. According to the TEM results similar agglomeration seems to take place when TiN is grown from $TiCl_4$ and NH_3.[12] Whether corresponding volatile intermediate species $(M(NH_x)_yCl_z)$ form in the interaction of metal halides with NH_x groups has not yet been experimentally verified. However, the structure and the high density of the films suggest that the reaction mechanism is similar to that observed in the oxide growth from metal chlorides.

CONCLUSIONS

Atomic layer CVD is a highly potential technique for future Cu barrier deposition. With good knowledge of the surface chemistry involved in ALCVD, 100% step coverage along with good adhesion to different surfaces present in dual damascene structure are possible to achieve.

REFERENCES

1. T. Suntola, Atomic Layer Epitaxy, in: *Handbook of Crystal Growth 3, Thin Films and Epitaxy, Part B: Growth Mechanisms and Dynamics*, Chapter 14, Elsevier, (1994).
2. S. Haukka and T. Suntola, *Interface Sci.*, **5**, 119 (1997).
3. S. Haukka, E.-L. Lakomaa and T. Suntola, *Stud. Surf. Sci. Catal.* **120**, 715 (1998).
4. S. Haukka, M. Tuominen and Ernst Granneman, paper presented in High-k Dielectrics Session organized by Semieducation on the 5th of April, 2000 in Munchen.
5. R. L. Puurunen, A. Root, S. Haukka, E. I. Iiskola, Marina Lindblad and A. O. I. Krause, *J. Phys. Chem*, in press.
6. M. Ritala and M. Leskelä *J. Phys. IV*, **5**, C5-937 (1995).
7. J. S. Min, Y.-W. Son, W.-G. Kang, S.-S. Chun and S.-W. Kang, *Jpn. J. Appl. Phys.* **37**, 4999 (1998).
8. M. Ritala, P. Kalsi, D. Riihelä, K. Kukli, M Leskelä and J. Jokinen, *Chem. Mater.* **11**, 1712 (1999).
9. M. Ritala, M. Leskelä, E. Rauhala and J. Jokinen, *J. Electrochem. Soc.*, **145**, 2914 (1998).
10. M. Ritala, M. Leskelä, E. Rauhala and P. Haussalo, *J. Electrochem. Soc.* **142** 2731 (1995).
11. K.-E. Elers, V. Saanila, P. Soininen and S. Haukka, unpublished results.
12. A. Satta, G. Beyer, K. Maex, K.-E. Elers, S. Haukka and A. Vantomme, submitted to MRS Symp. Vol. 612.
13. E. Iiskola, personal communication.
14. M. Ritala, M. Leskelä, J.-P. Dekker, C. Mutsaers, P. J. Soininen and J. Skarp, *Chem. Vap. Deposition*, **5**, 7 (1999).

Mat. Res. Soc. Symp. Proc. Vol. 612 © 2000 Materials Research Society

PROPERTIES OF TiN THIN FILMS DEPOSITED BY ALCVD AS BARRIER FOR CU METALLIZATION

Alessandra Satta, Gerald Beyer, Karen Maex[1]
IMEC, Kapeldreef 75, B-3001 Leuven, Belgium

Kai Elers, Suvi Haukka
ASM Microchemistry, Kutojantie 2B, P.O. Box 132, FIN-02631 Espoo, Finland

A. Vantomme
IKS, K. U. Leuven, Celestijnenlaan 200 D, B-3001 Leuven, Belgium

[1]Also at E. E. Dept., K. U. Leuven, Belgium

ABSTRACT

In advanced multi-level metallization schemes, the application of copper as interconnect metal requires the prevention of Cu diffusion into the active area and into interlevel dielectrics by total encapsulation of Cu with barrier films. Critical requirements for diffusion barriers are very small thicknesses, low resistivity, low deposition temperature and conformality on high aspect ratio trenches and vias. For this application, we have studied TiN films deposited by atomic layer chemical vapour deposition (ALCVD) at 400°C and 350°C. This paper discusses the ALCVD TiN films properties and compares them to the properties of TiN deposited by ionized physical vapour deposition (I-PVD).
The ALCVD TiN deposited at 400°C exhibits a resistivity comparable to I-PVD TiN resistivity. However, the ALCVD films deposited at 350°C show higher resistivity. The Cl residue in ALCVD films is 1.5% at 400°C and 3% at 350°C. The microstructure is fine-grained. A very high level of conformality on trenches characterizes the ALCVD TiN films. We believe this property gives a clear advantage over the sputtered I-PVD TiN since its coverage in high aspect ratio trenches and vias is expected to be limited for the future devices interconnection scheme.

INTRODUCTION

Nowadays copper is accepted as interconnect metal in advanced metallization schemes, because it offers a lower resistivity and a superior electromigration resistance compared to Al and its alloys. The major drawbacks of the Cu metallization are the fast Cu diffusion and drift into interlevel dielectrics. Consequently, in order to benefit from the advantages offered by Cu as an interconnect metal, a high quality and high performance diffusion barrier is necessary. Therefore, the choice of the diffusion material as well as the choice of the deposition technique are important key Cu interconnect issues.[1] Hence, the need for an alternative deposition technique for metal barriers is progressively increasing. The new deposition technique should ensure several critical properties of the deposited barrier film, like ultra thin and uniform thickness, low resistivity, low deposition temperature, as well as conformality on high aspect ratio trenches and vias, and barrier effectiveness.

We propose the atomic layer chemical vapour deposition (ALCVD)[2] as a technique to deposit TiN as diffusion barrier for Cu interconnect. Though ALCVD may be considered a special mode of a conventional CVD process, the characteristics of the two techniques are substantially different. ALCVD is based on the alternate supply of the precursors and the saturation of each individual surface reaction between the growing surface and the reactant.

This paper discusses the structural, chemical and electrical properties of TiN thin films deposited by ALCVD and compares them to the properties of layers deposited by I-PVD technique, which is currently used for diffusion barrier deposition.

EXPERIMENTAL DETAILS

The ALCVD TiN films have been deposited in the Pulsar 2000 reactor on 100 nm of PECVD SiO_2 on top of <100> Si substrates. $TiCl_4$ and NH_3 have been used as precursors to grow the TiN films and N_2 as a carrier and purging gas. An ALCVD full cycle consisted of a $TiCl_4$ pulse and a purge pulse, followed by a NH_3 pulse and a purge pulse. The deposition temperatures of the ALCVD TiN films are 400°C and 350°C. For comparison, TiN films have also been deposited on 100 nm PECVD SiO_2 on <100> Si wafers by I-PVD. The pressure during the sputter deposition is 16 mT, the DC power and the RF power are 4 kW and 2.5 kW respectively.

The thickness of the TiN films has been determined by ellipsometry, and the sheet resistance of the TiN layers has been evaluated by four point probe measurements. The film composition and the density have been analyzed by Rutherford backscattering spectroscopy (RBS). In order to determine the surface morphology of the films, analysis by atomic force spectroscopy (AFM) has been done. The microstructure and the conformality of the films have been studied by transmission electron microscopy (TEM). The electrical performance of the layers has been tested by capacitance-voltage measurement in combination with bias-thermal stress.

RESULTS AND DISCUSSION

1. Structural and chemical characterization
In order to calculate the resistivity of the different barrier layers we have used the thickness obtained by ellipsometry and the sheet resistance value measured by the four point probe system. Fig. 1 shows the resistivity obtained for TiN deposited by ALCVD at 400°C and 350°C and for I-PVD TiN. The thickness of the films ranges from 7 nm to 40 nm. Importantly, the process temperature strongly affects the resistivity of ALCVD TiN films: at 400°C, the resistivity of ALCVD TiN layers is in the range of 150-250 μΩcm which is comparable to the resistivity of I-PVD TiN films. The ALCVD films deposited at 350°C show higher resistivity values, in the range of 400-500 μΩcm. The resistivity of the ALCVD layers surprisingly does not increase as the layer thickness gets thinner, in the range of considered thickness. However, the I-PVD films exhibit a clear increase of the resistivity as the layer becomes thinner.

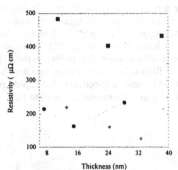

Fig. 1. Resistivity of the TiN films deposited at 400°C and 350°C as a function of the thickness. The resistivity of I-PVD TiN is also reported for comparison.

The deposition temperature also plays a fundamental role in determining the amount of Cl residue. The RBS spectra of ALCVD TiN layers (Fig. 2) show that the Cl content of all TiN films deposited at 400°C is 1.5%, whereas it is 3% for all TiN films deposited at 350°C. The amount of Cl content does not change with the thickness of TiN films.

The dependence of Cl residue on the temperature is related to the different reactivity of the precursors at different temperatures: the TiCl₄/NH₃ reaction is less favourable at lower temperatures. The lower reactivity gives inevitably rise to more precursors residues. In addition, the Cl removal from the surface with NH₃ is kinetically controlled. Therefore, it is slower at lower temperatures than at higher temperatures.

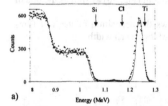 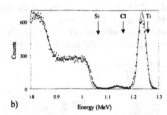

Fig.2. RBS spectra of (a) 7 nm ALCVD TiN deposited at 400°C, containing a Cl content of 1.5%, (b) 11 nm ALCVD TiN deposited at 350°C showing a Cl content of 3%. The arrows indicate the energy of particles back-scattered from the elements at the surface.

The amount of chlorine in the ALCVD films is reasonable: it has been found that the Cl content should not exceed 5% in the case of Al lines.[3] Moreover, attempts to reduce the residual chlorine concentration in TiN films deposited by flow modulation chemical vapour deposition (FMCVD) have resulted in the best case in a residual chlorine percentage equal to 4%.[4]

Using the thicknesses determined by ellipsometry, the density values of ALCVD and I-PVD films can be derived from the RBS measurement. The density of ALCVD TiN processed at 400°C is the same as the bulk density, equal to 10.6×10^{22} at/cm^3. The ALCVD TiN films deposited at 350°C exhibit lower density value (8.6×10^{22} at/cm^3) than the layers deposited at 400°C. The high density values of ALCVD TiN films agree very well with the morphology of the layers. Fig. 3a shows the AFM picture of 7 nm ALCVD TiN deposited at 400°C. The RMS roughness is 0.753 nm. As the AFM image show, the ALCVD TiN consists of a dense packing of grains, with a rather narrow distribution of grain size. In Fig. 3b the AFM picture of 13 nm I-PVD TiN shows that the I-PVD TiN films exhibit a rougher surface than the ALCVD films, with a wider grain size distribution and more distinctively developed grains. These surface characteristics are confirmed by the higher RMS roughness, equal to 1.795 nm.

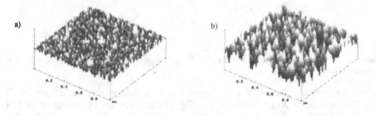

Fig. 3. AFM images of (a) 7 nm ALCVD TiN deposited at 400°C, RMS = 0.753 nm and (b) 13 nm I-PVD TiN, RMS = 1.795 nm.

2. Conformality

Fig. 4a shows cross-sectional TEM images of 26 nm ALCVD TiN deposited at 350°C on a trench of aspect ratio equal to 3.4. The dielectric stack consists of 600 nm of SiO$_2$. The ALCVD layer exhibits a columnar microstructure. Within the accuracy of the images, a conformality of 100% is achieved in the trench. The coverage of the trench bottom (Fig. 4b) as well as the one of the trench top corner (Fig. 4c) is excellent.

The surface-controlled growth mechanism of ALCVD allows the deposition of the TiN films with a very uniform thickness in the trenches, independently of the trench width.

Fig. 4. TEM cross section pictures of a trench of aspect ratio of 3.4 deposited with 26 nm ALCVD TiN at 350°C. a) View of the feature coverage. b) View of the bottom and c) view of the top corner coverage of the same feature.

3. Electrical characterization

The electric performance of ALCVD TiN layers has been tested by high frequency capacitance/voltage (C/V) measurements of Cu gate capacitors. The test structure is a typical MOS capacitor, with 20 nm of thermal SiO_2 as a dielectric. The thickness of the ALCVD TiN barrier layer is 10 and 20 nm, deposited at 400°C and 350°C.

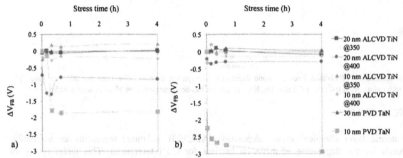

Fig. 5. Flatband voltage shifts as a function of the stressing time for ALCVD TiN diffusion barrier of 10 and 20 nm, deposited at 400°C and 350°C. The behavior of 30 and 10nm of PVD TaN as a barrier is also reported as a reference. a) and b) plots show the behavior of capacitors stressed at 2 MV/cm, 200°C and at 2 MV/cm, 300°C, respectively.

On top of the Cu gate 0.5 μm thick, TaN is deposited as a passivation layer. The structure is stressed at 2 MV/cm for two different temperatures, 200°C and 300°C, up to 4 h. Fig. 5 shows the flatband voltage shifts ΔV_{FB} of the C/V curves after bias temperature stressing (BTS). ΔV_{FB} is defined as the flatband voltage after BTS minus the flatband voltage before stress. As a reference, 30 and 10 nm PVD TaN as a barrier are also reported. For capacitors with ALCVD layers as a barrier, the rapid initial flatband shift is probably related to the annealing out of impurities present at the barrier/oxide interface.[5]

Stressing for longer times does not cause any further shift. This indicates no significant Cu diffusion into the SiO_2. The reference sample with 30 nm TaN as diffusion barrier also does not show any significant change in V_{FB}. However, the capacitor with 10 nm of TaN as a barrier exhibits a continuous decrease of ΔV_{FB}, which can be related to the Cu diffusion into the oxide.

4. ALCVD TiN integration

ALCVD TiN of 10 nm has been integrated in a single damascene interconnection scheme. The dielectric stack has consisted of 700 nm of SiO_2. The resistance of isolated metal lines with linewidth ranging from 0.4 μm to 0.2 μm has been measured. Each line is 300 μm long and it is configured as a Kelvin resistor for four points probe resistance measurement. The resistance of the lines has been measured on 56 different devices over a single wafer. Fig. 6 shows the increase of the resistance with the shrinking linewidth. The distribution of the line resistance is sharp for linewidth down to 0.25 μm. The resistance distribution of 0.2 μm-wide trenches is clearly not as sharp as wider lines, which is attributed to patterning effects.

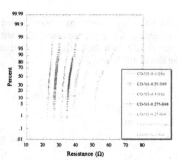

Fig. 6. Resistance of isolated lines in single damascene integration scheme. The lines are 300 μm long. The linewidth ranges from 0.4 μm to 0.2 μm. The lines resistance has been measured on 56 chips over a wafer.

CONCLUSIONS

The atomic layer chemical vapour deposition (ALCVD) technique represents an alternative approach for the deposition of diffusion barriers for Cu interconnect. This technique is an effective way of growing high quality thin films. The ALCVD TiN layers exhibit low resistivity, high density, low surface roughness, good homogeneity and remarkable conformality on trenches. The fraction of chlorine residues of films deposited at 400°C and 350°C is acceptable. The C/V test, performed in combination with BTS, shows a stable behaviour in terms of flatband voltage shift after 4h of stress. The ALCVD process, at the deposition temperature ≤ 400°C, has a large promising potential for the deposition of ultrathin films as diffusion barriers in the future Cu interconnection scheme.

ACKNOWLEDGMENTS

The authors wish to thank D. Vanhaeren for AFM measurements, H. Bender for TEM mesurements, F. Lanckman, IMEC.

REFERENCES

1. P. Murarka, Multilevel Interconnections for ULSI and GSI era, Material Science and Engineering, R19, 87 (1997).
2. T. Suntola, Atomic Layer Epitaxy, Handbook of Crystal Growth, 3, 601, edited by D. T. J. Hurle (1994).
3. P. Felix, Interconnect for ULSI: State of art and future trends, ESSDERC 1995. Proceedings of the 25th European Solid State Device Research Conference. p. 5 (1995).
4. H. Hamamura, R. Yamamoto, K. Takahiro, S. Yamaguchi, H. Komiyama, Y. Shimogaki, Advanced Metallization Conference in 1998, Materials Research Society, 345 (1998).
5. A. L. S. Loke, C. Ryu, C. P. Yue, J. S. H. Cho, S. S. Wong, IEEE Electron device letters, vol.17, No 12, (1996)

Mat. Res. Soc. Symp. Proc. Vol. 612 © 2000 Materials Research Society

A Study on CVD TaN as a Diffusion Barrier for Cu Interconnects

Se-Joon Im*, Soo-Hyun Kim*, Ki-Chul Park**, Sung-Lae Cho*, and Ki-Bum Kim*
*School of Materials Science and Engineering, Seoul National University, San 56-1, Shillim-dong, Kwanak-gu, Seoul, 151-742, Korea
**Samsung Electronics Co. Ltd. Kihung, Korea

ABSTRACT

Tantalum nitride (TaN) films were deposited using pentakis-diethylamido-tantalum [PDEAT, $Ta(N(C_2H_5)_2)_5$] as a precursor. During film growth, N- and Ar-ion beams with an energy of 120 eV were supplied in order to improve the film quality. In case of thermally-decomposed films, the deposition rate is controlled by the surface reaction up to about 350 °C with an activation energy of about 1.07 eV. The activation energy of the surface reaction controlled regime is decreased to 0.26 eV when the Ar-beam is applied. However, in case of N-beam bombarded films, the deposition is controlled by the precursor diffusion in gas phase at the whole temperature range. By using Ar-beam, the resistivity of the film is drastically reduced from approximately 10000 μΩ-cm to 600 μΩ-cm and the density of the film is increased from 5.85 g/cm^3 to 8.26 g/cm^3, as compared with thermally-decomposed film. The use of N-beam also considerably lowers the resistivity of films (~ 800 μΩ-cm) and increases the density of the films (7.5 g/cm^3). Finally, the diffusion barrier properties of 50-nm-thick TaN films for Cu were investigated aftre annealing by X-ray diffraction analysis. The films deposited using N- and Ar-beam showed the Cu_3Si formation after annealing at 650 °C for 1 hour, while thermally-decomposed films showed Cu_3Si peaks firstly after annealing at 600 °C. It is considered that the improvements of the diffusion barrier performance of the films deposited using N- and Ar-ion beam are the consequence of the film densification resulting from the ion bombardment during film growth.

INTRODUCTION

The development of suitable CVD diffusion barrier in manufacturing the Cu-based metallization is critical issue because of increase of aspect ratio and fast migration of Cu into Si or SiO$_2$. From the previous investigation, it was proved that Ta and its nitride such as Ta$_2$N and TaN showed excellent diffusion barrier properties against Cu [1]. Among them, TaN showed the superior barrier properties. For this reason, chemical vapor deposition of TaN films has drawn much attention [2-7]. Two approaches are currently being used for the production of CVD-TaN films. The first approach is based on the reaction of TaCl$_5$ with NH$_3$ or N$_2$ and H$_2$[2]. However, high deposition temperature (>900 °C) is required. And, in case of using NH$_3$ as a reaction gas, Ta$_3$N$_5$, which is dielectric, can be deposited. Other researchers used the TaBr$_5$[3] as a source gas of tantalum in order to lower the deposition temperature. There have been several attempts to develop a CVD-TaN process by using metallorganic source gases, such as (NEt$_2$)$_3$Ta=Nbu^t (tert-butylimido-tris-diethylamido-tantalum, TBTDET) [4], Ta(NEt$_2$)$_5$ (pentakis-diethylamido-tantalum, PDEAT) [5-6], and Ta(NMe$_2$)$_5$ (pentakis-dimethylamido-tantalum, PDMAT) [7]. In this case, the high impurity content and high resistivity can be problem.

In this work, we have studied the new CVD system, which was devised in order to obtain the TaN film with high density. The basic idea is to deposit the TaN film by thermal decomposition and simultaneously bombard the growing film surface using an ion beam source. The plasma was isolated from the deposition chamber in order to prevent the precursor from being exposed to the plasma. Thereby, it was intended that the step coverage of the film was still governed by the thermal growth characteristics. The characterization for the ion beam source was accomplished using a retarding-field ion energy analyzer. PDEAT was used as a precursor and N- and Ar-ion beams were used to enhance the film properties. In addition, the diffusion barrier properties of the films for Cu were also presented.

EXPERIMENTS

Figure 1 shows the ion beam induced chemical vapor deposition (IBICVD) system. The IBICVD system used in this work has been described in detail elsewhere [8]. The upper grid (floating grid) was electrically floated to repel most of the electrons and attract the ions in the plasma. The lower grid (accelerating grid) was negatively biased in order to extract and accelerate the ions. N- and Ar-ion beams were generated by using N_2 and Ar as plasma gases, respectively. Using the retarding-field ion energy analyzer, we can identify that the negative increase in the accelerating grid bias ($V_{acc.}$) of the ion beam source increases the ion beam flux. Also, we can know that the ion energy is little changed around 100 eV with the change of the accelerating grid bias of ion beam source.

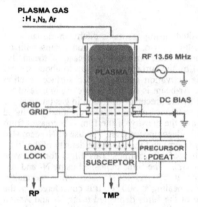

Fig. 1. Schematic diagram of IBICVD system.

Pentakis-diethylamido-tantalum (PDEAT) precursor was carried by 20 sccm of Ar gas through the bubbler maintained at 80 °C. The susceptor was located 7 cm below from the lower grid or ion beam source. TaNx films were deposited using PDEAT as a precursor under the bombardments of N or Ar- beam. For comparison, thermal decomposition of PDEAT was also investigaed. The deposition pressure was fixed to about 10 mTorr and the substrate temperature was varied from 275 °C to 400 °C. In order to obtain maximum ion beam flux, the accelerating grid bias was fixed to – 1kV. All depositions were carried out onto the Si substrates. Before deposition, the Si substrate was cleaned by using the hydrogen ion beams.

The film thickness was measured using a step profilometry. The properties of the deposited films were analyzed by four-point probe for sheet resistance, Auger electron spectroscopy (AES) for composition, Rutherford backscattering spectrometry (RBS) for density, and X-ray diffractometry (XRD) for phase identification. For evaluating the diffusion barrier properties against Cu, 300-nm-thick Cu films are sputter deposited onto 50-nm-thick TaN films. Then these Cu/TaN/Si samples were annealed in vacuum ambient for 1 hour at temperature ranging from 500 °C to 650 °C with a 50 °C interval. The diffusion barrier performance was estimated by XRD analysis.

RESULTS AND DICUSSION

A. Deposition Rate

Figure 2 shows the dependence of the deposition rate on the substrate temperature. As has been reported already [5-6], an Arrhenius plot of the deposition rate of thermally-decomposed films shows a transition temperature from surface reaction controlled regime to diffusion controlled regime at around 350 °C. The activation energy of the surface reaction controlled regime is about 1.07 eV, which is a typical number for the thermal decomposition transition metal dialkylamido metallorganic source gas [5-6, 9]. When we apply Ar-beam, the overall deposition rate is also divided into two regimes. However, the activation energy of the surface-reaction-controlled regime is much smaller (0.26 eV) than that of thermally-decomposed film. Although it is not clear at this stage whether this is due to ions, radicals, or metastables, it shows

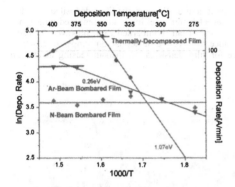
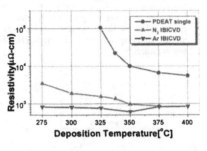

Fig. 2. Arrhenius plot of deposition rates.   **Fig. 3.** As-deposited films resistivity.

that the surface reaction mechanism is significantly changed to have lower activation energy. It is also noted that the deposition rate becomes much faster at the lower temperature region. The lower activation energy means that the surface reaction is enhanced by Ar-beam although it cannot chemically participate in the reaction. These results are similar with our previous investigations that performed IBICVD of TiN using tetrakis-dimethyl-amido titanium (TDMAT) [8]. With the addition of nitrogen ion beam in the source gas, overall deposition rate was significantly decreased and it rarely changes with the substrate temperature. In other words, the deposition rate follows mass-transfer-controlled regime in the temperature ranges of tested. In N_2 plasma similar to our experimental condition, it is known that there are 2-3 % reactive N radicals [10] This is a much higher value compared with the typical ionization efficiency, about 0.1%. These N radicals are neutral and will diffuse from the plasma quartz tube through the grid hole and to the deposition chamber without being influenced by the accelerating grid bias. It is considered that, however, the participation of these N radicals into the deposition reaction also significantly lowers the activation energy for surface reaction.

B. Film Resistivity

The film resistivities are shown in Fig. 3 as a function of deposition temperature. In general, the film resistivity decreases with increasing deposition temperature. The decrease of the film resistivity with the increase of deposition temperature is not a surprising result. It is generally observed that the resistivity of the films deposited by CVD process is decreased as the deposition temperature is increased as far as the films forms a continuous layer and is explained due to the formation of large grain size and the densification of the as-deposited film. However, the effect of ion beam on the film resistivity is more drastic. The effect of ion beam on the resistivity of the film is in the sequence of nitrogen and argon. The strongest effect of Ar-beam demonstrates that the physical bombardment to make the film dense is the most important factor to reduce the film resistivity. The minimum resistivity of TaNx films was about 600 $\mu\Omega$-cm, which was deposited at 350 °C by using Ar-beam.

C. Sheet Resistance Increase after Air-Exposure

Figure 4 shows the sheet resistance of TaNx films before and after air-exposure for 24 hours as a function of temperature. These results support the densification of the film by both increase of deposition temperature and bombardments of ion beam. It is generally believed that the increase of sheet resistance of CVD-TiN film using metallorganic precursor increased by the incorporation of oxygen due to the porous microstructure [9]. Thus, the films deposited at higher temperature show lower resistivity and also reveal better stability in air compared with those deposited at lower temperature. In case of Ar or N-beam bombarded film, however, all films deposited show little increase of sheet resistance irrespective of deposition temperature, which shows applying ion beam is more effective for film densification than increasing deposition temperature.

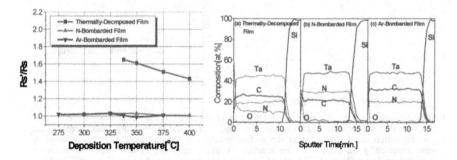

Fig. 4. Sheet resistance changes of as-deposited films after air-exposure of 24 hours.

Fig. 5. AES depth profiles of TaNx deposited at 325 °C (a) thermal CVD, (b) N₂ IBICVD, and (c) Ar IBICVD.

D. Composition

Figure 5 shows AES depth profiles for thermally-decomposed and Ar or N-bombarded films (150 nm) deposited at 325 °C. It is clearly shown that there is about 12 at. % of oxygen in the thermally-decomposed film. On the contrary, the oxygen content of the Ar or N-bombarded films was below detection limit of AES. As explained earlier, this behavior is well consistent with the fact that the thermally-decomposed films show an increase of sheet resistance, while Ar or N-bombarded films do not.(See Fig. 4.) In case of N-bombarded fillms, the lower carbon contents in the film are probably caused by the reaction between these N radicals and PDEAT.

E. Density

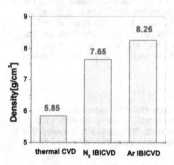

The density of the TaNx films deposited at 325 °C is evaluated by RBS and shown in Fig. 6. As is expected, the density of the TaNx films is significantly increased with the aid of ion beam. However, the overall densities of the films deposited in this experiment (5.85 ~ 8.26 g/cm³) appear to be much lower than that of bulk TaN (16.3 g/cm³) or bulk TaC (13.9 g/cm³). However, it should be pointed out that the stoichiometry of the films is quite different from that of bulk TaN. In other words, the existence of excess light elements results in much lower gram densities relative to that of bulk TaN or TaC.

Fig. 6. Density of TaNx films deposited at 325°C.

F. Diffusion Barrier Property

TaNx films (50 nm) deposited were evaluated as a diffusion barrier between Cu and Si by XRD analysis (not shown). The formation of Cu_3Si is firstly observed at annealing at 600 °C in the thermally-decomposed film. However, in case of Ar or N-bombarded film, the formation temperature of Cu_3Si is increased to 650 °C. Thus, the improvement of diffusion barrier performance is shown when we applied the ion beam during the thermal decomposition, which may be due to the increase of film density.

G. Step Coverage

The step coverages of the films were observed at the 0.5 μm × 1.5 μm contacts and summarized in Fig. 7. As already reported [5], the thermally-decomposed film using PDEAT in surface reaction-controlled regime shows a good step coverage (> 80 % in 0.5 μm × 1.5 μm contact). However, in case of the film deposited using a N-ion beam at 250 °C, step coverage is significantly degraded. This is because deposition is controlled by precursor diffusion from the gas phase as shown by Fig. 1. We speculate that this poor step coverage is caused by a reaction between PDEAT and reactive N radicals. Ar-beam bombarded film shows better conformality compared with N-beam bombarded film at same deposition temperature.. However, the step coverage is poor compared with that of thermally-decomposed film. This is because the Ar-ion beam activates the surface reaction by the transfer of momentum to the adsorbed species on the surface. (see Fig. 1) In our experiments, therefore, it is considered that the Ar-ion beam also activates the surface reaction although it cannot chemically participate in the reaction. In addition, it should be noted that the poor step converage of Ar-beam bombarded film compared with that of thermally-decomposed film is due to non-uniformity of Ar-ion beam flux along the contact geometry.

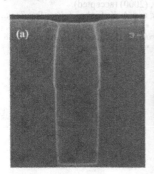 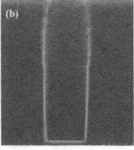

0.5 μm

Fig.7. Step coverage of (a) N-beam bombarded and (b) Ar-beam bombarded film.

CONCLUSION

Tantalum nitride film was deposited by using ion beam induced chemical vapor deposition, which was devised in order to obtain films with different chemistry and with high density. The ions have the energies between 115 eV and 127 eV. Ion current density is about 57 ~ 103 μA/cm^2 as lower grid bias is changed. Pentakis (diethylamido) tantalum was used as a precursor to deposit TaNx film. The use of argon ion beam significantly lowers the resistivity of TaNx film and increases the density of TaNx film (~ 600 μΩ-cm, 8.26 g/cm^3), as compared with the thermally decomposed film (~ 10000 μΩ-cm, 5.85 g/cm^3). The use of nitrogen and hydrogen ion beam also considerably lowers the resistivity of the films ($\sim$ 800 $\mu\Omega$-cm) increases the density of the films (7.37 ~ 7.65 g/cm^3). While the thermally grown TaNx films shows an aging effect, the IBICVD TaNx films do not show aging effect after air-exposure. The XRD analysis shows that the diffusion barrier performance is improved by ion beam bombardment, which may be due to the increase of film density. However, the step coverage was degraded due to the relative enhancement of the reaction rate on the surface by the Ar ion beam bombardment as compared to the reaction rate on the side walls and bottom of the contact.

ACKNOWLEDGMENT

This work has been supported by A Collaborate Project for Excellence in Basic System IC Technology through Consortium of Semiconductor Advanced Research (COSAR) and was partially funded by Applied Materials. Authors are grateful to Brain Korea 21 Scholarship by Ministry of Education, Korea.

REFERENCES

1. K. H. Min, G. C. Jun, and K. B. Kim, *J. Vac. Sci. Tech.*, **B 14**, p. 3263 (1996).
2. K. Hieber, *Thin Solid Films*, **24**, p. 157 (1974).
3. X. Chen, G. Peterson, T. Stark, H. L. Frisch, and A. E. Kaloyeros, *in Proceedings of VLSI Multilevel Interconnection Conference*, p. 434 (1997).
4. M. H. Tsai, S. C. Sun, H. T. Chiu, C. E. Tsai, and S. H. Chuang, *Appl. Phys. Lett.*, **67**, p. 67 (1995).
5. G. C. Jun, S. L. Cho, and K. B. Kim, *Jpn, J. Appl. Phys.*, **37**, L 30 (1998).
6. S. L. Cho, S. H. Min, K. B. Kim, H. K. Shin, and S. D. Kim, *J. Electrochem. Soc.*, **146** 3724 (1999).
7. R. M. Fix, R. G. Gordon, and D. M. Hoffman, *Chem. Mater.*, **5**, p. 614 (1993).
8. K. C. Park, S. H. Kim, and K. B. Kim, *J. Electrochem. Soc.*, (2000) (accepted).
9. M. Eizenberg, K. Littau, S. Ghanayem, M. Liao, R. Mosely, and A. K. Sinha, *J. Vac. Sci. Tech.*, **A 13**, p. 590 (1995).
10. D. L. Smith, *Thin-Film Deposition: Principles and Practice*, Mcgraw-Hill, Inc., p. 458 (1997).

Mat. Res. Soc. Symp. Proc. Vol. 612 © 2000 Materials Research Society

THE 2,2,6,6-TETRAMETHYL-2-SILA-3,5-HEPTANEDIONE ROUTE TO THE CHEMICAL VAPOR DEPOSITION OF COPPER FOR GIGASCALE INTERCONNECT APPLICATIONS.

Rolf U. Claessen, John T. Welch, Paul J. Toscano, Kulbinder K. Banger, Andrei M. Kornilov, Eric T. Eisenbraun and Alain E. Kaloyeros
NYS Center for Advanced Thin Film Technology, Department of Physics and Department of Chemistry, University at Albany, SUNY
Albany, NY 12222, U.S.A.

ABSTRACT

A new class of copper(II) precursors containing silylated β-diketonate ligands has been developed for the chemical vapor deposition (CVD) growth of copper for applications in ultralarge scale integration interconnect schemes, including conformal seed layer for gigascale Cu integration and ultrathin Cu lines with enhanced conductivity characteristics. Cu(tmshd)$_2$ (tmshdH = 2,2,6,6-tetramethyl-2-sila-3,5-heptanedione) has been studied as a representative compound and is appreciably more volatile than nonsilylated compounds such as Cu(tmhd)$_2$ or Cu(tmod)$_2$ (tmhdH = 2,2,6,6-tetramethyl-3,5-heptanedione; tmodH = 2,2,7-trimethyl- 3,5-octanedione). The CVD process employs Cu(tmshd)$_2$ as the metalorganic precursor and hydrogen as the reducing and carrier gas. These films were deposited using a custom made, cold wall, stainless steel CVD. Copper films were produced at a substrate temperature of 250 – 320 °C, hydrogen flow rates of 20 - 100 sccm, deposition pressure of 0.2 - 1 Torr, and a source temperature of 120 – 135 °C. The films were analyzed by X-ray photoelectron spectroscopy, cross section scanning electron microscopy, transmission electron microscopy, four-point resistivity probe, Rutherford backscattering spectrometry and Auger electron spectroscopy.

INTRODUCTION

Low temperature chemical vapor deposition (CVD) is among the methods of choice for the preparation of metallic and metal containing thin films for ultralarge integration (ULSI) of electronic devices.[1-4] The current technology for the deposition of seed layers for electroplated copper employs ionized physical vapor deposition. As continued device scaling demands improved conformality of such layers in high aspect ratio dual damascene structures, the need for alternative techniques to deposit seed layers become increasingly important[5]. In addition to conformality, the primary issues related to seed layer deposition are continuity, resistivity and adhesion. Chemical vapor deposition is one candidate process to achieve these goals.
The structure of ancillary ligands for CVD precursors may profoundly influence the desired physical and chemical properties of the metal complexes, and thus, the performance of these materials in processing. Factors such as the vapor pressure of the precursor, as well as adsorption/desorption behavior, and the purity of the deposited film, are important considerations.[1-4]

Binary and ternary β-diketonate metal complexes possess excellent volatility, decompose in a desirable manner, and are easily prepared and handled.[1-4,6-7] These complexes also offer the possibility to tailor the physical and chemical properties of the precursor via variation of the ligand substituents. For example the volatility of metal β-diketonate complexes can be enhanced by increasing the steric bulk of or by incorporation of fluorine atoms into peripheral β-diketonate substituents.[3,7-9] While fluorine substitution results in significant increases in precursor volatility, fragmentation of the ligand at elevated temperatures may lead to fluorine contamination in the deposited films.[10-11]

Recently, we have developed a method for the preparation of silylated β-diketones and their metal complexes.[12] It is well known that selective silylation of organic compounds can improve volatility. Herein, we report on the suitability of a representative metal complex, Cu(tmshd)$_2$, for the low temperature CVD of copper thin films.

EXPERIMENTAL DETAILS

Cu(tmshd)$_2$ was prepared as reported by the current authors elsewhere.[12] The melting point was determined using a TA Instruments DSC 2920 Differential Scanning Calorimeter to be 163 °C. The suitability of Cu(tmshd)$_2$ for CVD of copper was examined using a custom-made three inch wafer capable, stainless steel CVD reactor. The delivery system included a sublimator with a carrier gas line leading into the reaction chamber. Before a deposition, the delivery system was heated to 150 °C, full vacuum was applied and a flow of 100 sccm hydrogen was used to purge the system. After cooling the system, the sublimator was isolated from the rest of the system and filled with argon. The air stable precursor was placed into the sublimator, and substrates were loaded into the chamber and placed on top of the substrate heater. For this work the substrates included Si/SiO$_2$, Si/SiO$_2$/TaN, Si/SiO$_2$/W$_2$N and patterned Si/SiO$_2$/W$_2$N. After evacuating the sublimator, the delivery system was again purged with hydrogen. The system was then evacuated to a base pressure of about 1 mTorr. Hydrogen flow into the sublimator and the delivery system, regulated by an electronic mass flow controller, was initiated. Actual growth was initiated by opening a valve between chamber and delivery system after heating the precursor to a preset temperature of 120 °C. The temperature of the delivery system was maintained at 130 °C. Typical deposition parameters included a substrate temperature of 300 °C, a chamber pressure of about 600 mTorr and a carrier gas flow of 100 sccm H$_2$.

RESULTS

Characterization of the Precursor

The crystal structure (figure 1) and the packing diagram (figure 2) of Cu(tmshd)$_2$ has been determined by single crystal X-ray diffractometry. Thermogravimetric Analysis has been performed using a TA Instruments TGA 2050 Thermogravimetric Analyzer. The TGA shows a clean weight loss of >98% at a 50% weight loss temperature of 148 °C (figure 3). These thermal properties are indicative of the increased volatility resulting from incorporation of silicon into the molecule. The parent carbon compound, Cu(tmhd)$_2$, has the first 50% weightloss temperature well above 170°C.

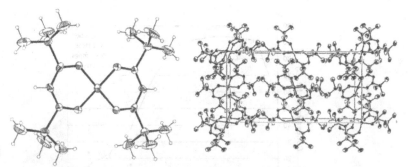

Figure 1. Molecular Structure of Cu(tmshd)₂   Figure 2. Packing Diagram for Cu(tmshd)₂

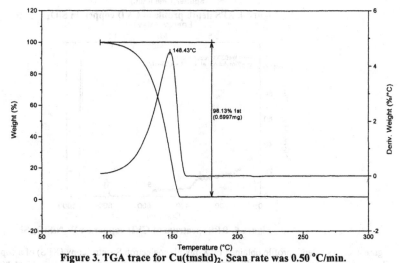

Figure 3. TGA trace for Cu(tmshd)₂. Scan rate was 0.50 °C/min.

Characterization of CVD Grown Copper Films

The composition characteristics of the deposited copper films were determined by X-ray Photoelectron Spectroscopy (XPS), Auger Electron Spectroscopy (AES) and Rutherford Backscattering Spectrometry (RBS).

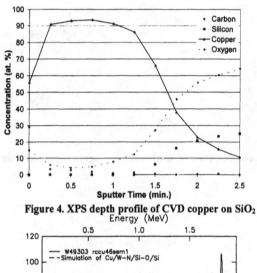

Figure 4. XPS depth profile of CVD copper on SiO$_2$

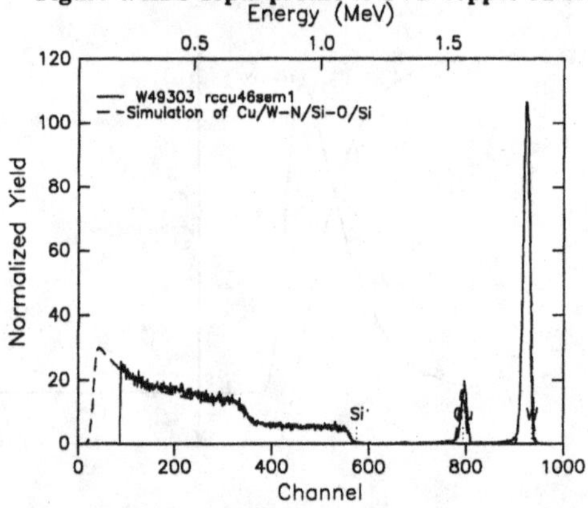

Figure 5. RBS spectrum of CVD copper on W$_2$N/SiO$_2$/Si

Figure 4 shows a depth profile obtained by X-ray Photoelectron Spectroscopy (XPS) of a copper film on a SiO$_2$ substrate. The resulting spectra indicate that the films possess at least 93 at. % copper, with less than 2 at. % carbon contamination. XPS also indicates approximately 5 at. % contamination from oxygen, although the source of the contamination could not be clearly identified. Possible sources include breakup of the precursor during the deposition or oxygen contamination while handling samples *ex situ*.

Rutherford Backscattering Spectrometry (figure 5) obtained from a copper film on a W$_2$N/SiO$_2$/Si stack indicates no heavy element contamination.

The thickness of our films was measured by RBS and Scanning Electron Microscopy (SEM). SEM was also used to measure film microstructure. An SEM image of a 15 nm thick copper film on TiN substrate (figure 6) shows an approximate surface grain size of 20 nm.

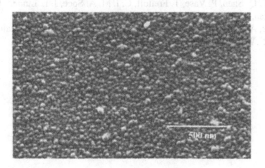

Figure 6 SEM image of CVD copper film from Cu(tmshd)₂ grown on TiN

CONCLUSIONS

A novel precursor for CVD copper has been developed. The synthesis of Cu(tmshd)₂ was demonstrated to be scalable on an industrial scale. The results of theses studies indicate that the films were > 93 at. % pure. Carbon and oxygen contamination were found to be as low as < 2 and < 5 at. % respectively.

ACKNOWLEDGMENT

The authors would like to thank the NYS Center for Advanced Thin Film Technology for generous support of this project.

REFERENCES

1. *The Chemistry of Metal CVD*, ed. T. T. Kodas, M. J. Hampden-Smith (VCH Publishers: New York, 1994).
2. J. T. Spenser, *Prog. Inorg. Chem.*, **41**, 145 (1994).
3. S. P. Murarka, S. W. Hymes, *Crit. Rev. Solid State Mater. Sci.*, **20**, 87 (1995).
4. T. J. Marks, *Pure Appl. Chem.*, **67**, 313 (1995).
5. SIA Roadmap 1999
6. R. C. Mehrotra, R. Bohra, D. P. Gaur, *Metal β-Diketonates and Allied Derivatives* (Academic Press: New York, 1978).
7. R. E. Sievers, J. E. Sadlowski, *Science (Washington, D.C.)*, **201**, 217 (1978).
8. P. J. Toscano, C. Dettelbacher, J. Waechter, N. P. Pavri, D. H. Hunt, E. T. Eisenbraun, B. Zheng, A. E. Kaloyeros, *Coord. Chem.*, **38**, 319 (1996).

9. A. E. Kaloyeros, M. A. Fury, *MRS Bull.*, **18**, 22 (1993).
10. M. L. Hitchman, S. H. Shamlian, D. D. Gilliland, D. J. Cole-Hamilton, J. A. P. Nash, S. C. Thompson, S. L. Cook, *Mater. Chem.*, **5**, 47 (1995).
11. B. C. Richards, S. L. Cook, D. L. Pinch, G. W. Andrews, G. Lengeling, B. Schulte, H. Jürgensen, Y. Q. Shen, P. Vase, T. Freltoft, C. I. M. A. Spee, J. L. Linden, L. Hitchman, S. H. Shamlian, A. Brown, *Physica C (Amsterdam)*, **252**, 229 (1995).
12. K. K. Banger, A. M. Kornilov, R. U. Claessen, E. T. Eisenbraun, A. E. Kaloyeros, P. J. Toscano, J. T. Welch, *Angew. Chem. (submitted)*.

Interconnects

Mat. Res. Soc. Symp. Proc. Vol. 612 © 2000 Materials Research Society

FABRICATION AND PERFORMANCE LIMITS
OF SUB-0.1 μm Cu INTERCONNECTS

T. S. Kuan, C. K. Inoki, and G. S. Oehrlein
Department of Physics, University at Albany, State University of New York, Albany, NY 12222
K. Rose,[a] Y. –P. Zhao,[b] and G. –C. Wang[b]
[a]Department of Electrical, Computer and Systems Engineering, [b]Department of Physics,
Rensselaer Polytechnic Institute, Troy, NY 12180
S. M. Rossnagel and C. Cabral
IBM T. J. Watson Research Center, Yorktown Heights, NY 10598

ABSTRACT

As the on-chip interconnect linewidth and film thickness shrink below 0.1 μm, the size effect on Cu resistivity becomes important, and the electrical performance deliverable by such narrow metal lines needs to be assessed critically. From the fabrication viewpoint, it is also crucial to determine how structural parameters affect resistivity in the sub-0.1 μm feature size regime. To evaluate the scaling of resistivity with thickness, we have fabricated a series of $Ta/Cu/Ta/SiO_2$ thin film structures with Cu thickness ranging from 1 μm to 0.02 μm. These test structures revealed a far larger (~2.3 ×) size effect than that expected from surface scattering. We have also fabricated test structures containing 50-nm-wide Cu lines wrapped in Ta-based liners and embedded in insulating SiO_2 using e-beam lithography, high-density plasma etching, ionized PVD Cu deposition, and chemical-mechanical planarization processes. Direct current (16 nA) resistance measurements from these 50-nm-wide Cu lines have also shown a higher-than-expected distribution of resistivity. Cross-sectional TEM and surface AFM observations suggest that the observed extra resistivity increase can be attributed to small grain sizes in ultra-thin Cu films and to Cu/Ta interface roughness. Monte Carlo simulations are used to quantify the extra resistivity resulting from interface roughness.

INTRODUCTION

To meet the challenges posed by the emerging sub-0.1 μm Si technology [1], a physical model capable of predicting the scaling of resistivity in sub-0.1 μm Cu interconnects is needed. Such a model should be based on experimentally measured parameters and take into account microstructure features imposed by the fabrication process. As the metal film thickness or linewidth approaches the conducting electrons' mean free path l, surface/interface scattering becomes increasingly predominant and the resistivity is expected to rise. This size effect on resistivity has been investigated by analyzing the influence of surface scattering on the rates of change of electron distribution functions in an electric field. The scattering at surfaces or interfaces can be specular ($p = 1$) or diffuse ($p = 0$), and its effect on film or line resistivity is (within a 10% error) inversely proportional to film thickness t or linewidth d [2-5]:

$$\rho_{film}/\rho_{bulk} \approx 1+0.375l(1-p)/t \qquad (1)$$

$$\rho_{line}/\rho_{bulk} \approx 1+0.75l(1-p)/d \qquad (2)$$

The effect of surface scattering on Cu thin films and Cu fine lines, assuming 100% diffuse scattering ($p = 0$) at the surface, is shown in Fig. 1. Experimentally it is much easier to determine the surface effect by measuring a series of thin films of various thicknesses. High-purity Cu whiskers have been reported to exhibit specular scattering at surfaces with $p = 0.6$ [6]. Polycrystalline Cu films, ~0.05 μm in thickness, on the other hand, have shown diffuse scattering ($p = 0$) in an earlier report [7]. In an interconnect structure, the scattering parameter p can be sensitive to the liner materials. In this study we have fabricated a series of Ta/Cu/Ta thin film test structures to measure the size and surface effects. Our experimental results showed, however, a much larger resistivity increase than that predicted in Fig. 1. Transmission electron microscopy (TEM) and atomic force microscopy (AFM) analyses of the test structures suggested that other factors, such as small grain size and interface roughness, are as important in causing the drastic rise in resistivity in ultra-fine structures.

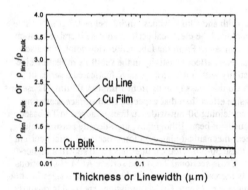

Fig. 1 Resistivity increases in Cu thin films and Cu fine lines due to surface/interface scattering.

EXPERIMENTAL

A. Thin film test structure

Two series of Ta/Cu/Ta test structures, each with Cu thickness ranging from 1 μm to 0.02 μm, have been fabricated to test the size effect on DC electrical resistivity. Ultra-thin Ta liners, ~7 nm in thickness, were included in these test structures for evaluating the carrier scattering at Cu/Ta interfaces. Four Ta and Cu thin film structures were deposited at room temperature by PVD on SiO_2 (5-μm)/Si substrates, followed by a rapid thermal annealing (RTA) in highly purified N_2 from 25° to 400°C at a rate of 3°C/s. Another group of four test structures were deposited at 400°C. Cross-sectional and plan-view TEM micrographs of the test structures showed rougher top Cu/Ta interfaces in films deposited at high temperatures (Fig. 2) and smaller Cu grain sizes in thinner films (Fig. 3). AFM measurements of the room-temperature-deposited films indicated that the rapid thermal annealing process has caused a 10-20% increase in the RMS surface roughness. The grain size distribution and surface roughness as determined by plan-view and cross-sectional TEM are listed in Table 1.

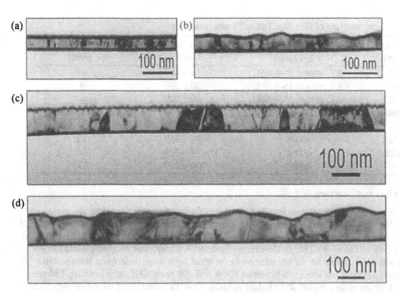

Fig. 2 Cross-sectional TEM images of (a) a Ta(7.6 nm)/Cu(20 nm)/Ta(7.6 nm)/SiO$_2$ test structure deposited at room temperature and annealed at 25-400°C, (b) a Ta(8 nm)/Cu(26 nm)/Ta(8 nm)/SiO$_2$ test structure deposited at 400°C, (c) a Ta(7.6 nm)/Cu(60 nm)/Ta(7.6 nm)/SiO$_2$ test structure deposited at room temperature and annealed at 25-400°C, and (d) a Ta(8 nm)/Cu(78 nm)/Ta(8 nm)/SiO$_2$ test structure deposited at 400°C.

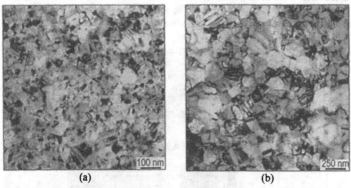

(a)   (b)

Fig. 3 Plan-view TEM micrographs showing grain size distributions in (a) Ta(8 nm)/Cu(26 nm)/Ta(8 nm)/SiO$_2$ test structure and in (b) Ta(8 nm)/Cu(125 nm)/Ta(8 nm)/SiO$_2$ test structure deposited at 400°C.

Table 1. Microstructure of thin film Ta/Cu/Ta test structures

Cu thickness (nm)	Deposition temperature	Grain size distribution (nm)	Surface roughness (nm)
20	RT, anneal 25-400°C	10 – 100	0.7
60	RT, anneal 25-400°C	20 – 200	4
92	RT, anneal 25-400°C	60 – 300	–
980	RT, anneal 25-400°C	200 – 1000	–
26	400°C	25 – 100	6
78	400°C	50 – 350	12
125	400°C	75 – 500	16
1200	400°C	500 – 2000	200

B. Fine line test structure

A van der Pauw test structure [Fig. 4(a)] was designed and fabricated by the damascene method for evaluating the manufacturability and electrical resistivity of fine metal lines [8]. The test structure was written onto a 200-nm-thick 950PMMA resist layer in a JEOL JSM-848 SEM using Nabity's Nanometer Pattern Generation System. A transformer coupled plasma (TCP) reactor was used to transfer the test patterns in the resist layer to an underlying 500-nm-thick SiO_2 layer. Using a low inductive power of 400W and 400 sccm CHF_3 at 5.3 mTorr, 150 nm deep and 50 nm wide trenches were etched in SiO_2. Thin Ta-based liners were then deposited with conventional physical vapor deposition (PVD), followed by Cu trench fill using ionized magnetron sputtering. TEM observations have indicated slightly tapered (~80°) sidewalls, conformal liners with good Cu fill, and an average Cu grain size of ~50 nm in the trenches. The field Cu was removed by a chemical-mechanical polish (CMP) using a 50-nm-sized alumina slurry at ~5 psi pressure. A second short polish using a slurry with 100-nm-sized SiO_2 removed the field Ta-based liner. SEM observations have shown that 50-nm-wide damascene Cu lines were left in the oxide trenches after the polishing [Fig. 4(b)]. However, the patterning, etching and polishing processes were found to produce significant sidewall roughness (~5 nm) in these 50-nm-wide lines.

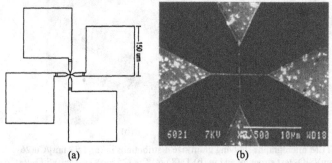

(a)   (b)

Fig. 4   (a) A van der Pauw test structure and (b) a SEM micrograph of the test structure after removal of field Cu by the chemical-mechanical polishing process.

RESULTS AND DISCUSSION

A. Effect of film thickness and grain size on resistivity

The resistivity in the Cu films deposited at room temperature increased from 2.2 $\mu\Omega$-cm to 5.2 $\mu\Omega$-cm as the thickness decreased from 1 μm to 20 nm, which was 2.4 times larger than the expected effect of surface scattering [Fig. 5(a)]. Part of the observed increase can be attributed to small grain sizes in thinner films (Table 1). A combination of larger grain sizes and rougher surfaces in Cu films deposited at 400°C was found to produce an overall ~10% decrease in resistivity [Fig. 5(b)]. Mayadas and Shatzkes had shown that the effect of grain boundary scattering on resistivity is, to a good approximation, inversely proportional to the average grain size g [9]:

$$\rho_{gb}/\rho_{bulk} \approx 1+1.5lR/(1-R)g \approx 1+0.6l/g \qquad (3)$$

where R is the grain boundary reflection coefficient and is estimated to be ~0.3 for Cu. Since the average grain size in Cu films is of the same order as film thickness, the resistivity increase due to grain boundary scattering can be as large or larger than the effect of surface scattering expressed in Eq. (1).

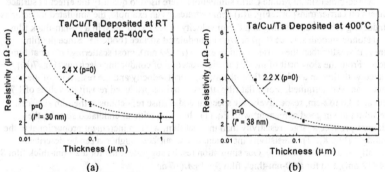

(a)   (b)

Fig. 5 Resistivities of Ta/Cu/Ta test structures measured at room temperature. The solid line depicts the expected effect of Ta/Cu interface scattering, assuming $p = 0$ (100% diffuse scattering).

B. Effects of interface roughness

A resistivity increase larger than in Eq. (1) is expected when the reflecting surface or interface is not smooth. The resistivity increase associated with surface roughness was observed during the rapid thermal annealing (RTA) of the 20-nm-thick and 60-nm-thick Cu films. The RTA of Cu films thicker than 0.1 μm produces a prominent (~30%) resistivity drop at ~150°C through extensive stress relaxation and grain growth [Fig. 6(a)]. In sub-0.1-μm-thick films, the resistivity reduction was much less, and a resistivity rise was detected at higher (~300°C) temperatures, resulting in no improvement in conductivity from annealing [Fig. 6(b)]. This rise in resistivity can be attributed to the anneal-induced 10-20% increase in RMS surface roughness as measured by AFM.

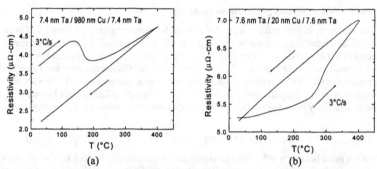

(a)   (b)

Fig. 6 In-situ resistivity measurements during a 25-400°C anneal in purified N_2 for (a) a ~1-μm-
thick Cu film and (b) a 20-nm-thick Cu film.

Three-dimensional Monte Carlo simulations were used to quantify the effect of surface
roughness on carrier transport. Fig. 7(a) demonstrates, as an example, the simulated trajectories
of conducting electrons, traveling at a Fermi velocity of 1.57×10^6 m/s, in a 20-nm-thick film
with an atomic smooth surface (top) or with an undulated surface (bottom). Since the film
thickness is smaller than the carrier mean free path ($l = 39$ nm), most scatterings occur at film
surfaces. From the slow drift of the distribution function of conducting electrons [Fig. 7(b)]
simulated with time in an electric field, the average drift velocity and the resistivity of the
structure can be determined. For a flat film, the simulations produced resistivity values of 1.7
μΩ-cm and 3.0 μΩ-cm, respectively, for specular and diffuse reflections at the film surfaces.
These values are in good agreement with Eq. (1). In a film with an undulated surface, the
simulations produce a larger resistivity than in a flat film, with an increment proportional to the
amplitude A [Fig. 8]. This resistivity increment due to surface roughness can be described
empirically by a surface factor S. Our simulation results suggested that for a 20-nm-thick film S
$\approx 1+A/(12$ nm), and for a 60-nm-thick film $S \approx 1+A/(16$ nm).

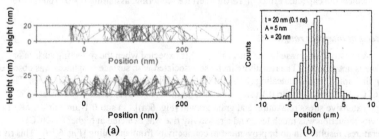

(a)   (b)

Fig. 7 (a) Simulated carrier trajectories in flat and undulated films and (b) the distribution of
10,000 electrons in the undulated film after 0.1 ns.

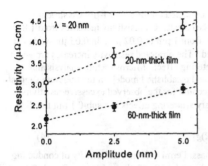

Fig. 8 Simulated resistivities of Cu thin films with an undulated top surface as a function of undulation amplitude. The wavelength of the undulation λ is 20 nm.

C. Resistivity of 50-nm-wide Cu interconnects

The testing of 50-nm-wide structures requires very low current densities, due to the small Cu wire cross section. A 16 nA supply source was used for the resistance measurements. The exact supply current was measured by observing the voltage drop across a 1 MΩ resistor. Cu wire resistance was then determined by the supply current and the voltage across the test structure. Errors in the measurements due to the geometry of the structure's layout in Fig. 4(a) were found to add four squares to an approximately 200 square line (a ~2% negligible error). A large spread in resistivities, ranging from ~7 $\mu\Omega$-cm to more than two orders of magnitude higher, was observed, indicating the effect of process-induced irregularities. Fig. 9 plots the best group of resistivity values we have obtained, indicating that ~7 $\mu\Omega$-cm is achievable in a 0.05-μm-wide Cu interconnect. SEM observations of the test structures suggest that sidewall roughness and surface scratches induced by the patterning and polishing processes may be responsible for the excessive resistivity rise.

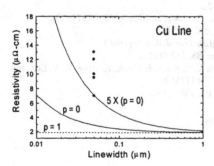

Fig. 9 Resistivity measured from 50-nm-wide Cu interconnects.

CONCLUSION

Room-temperature resistance measurements carried out on sub-0.1-μm Cu structures have indicated larger than expected size and surface effects. Cu resistivity increases from 1.8 μΩ-cm to ~5 μΩ-cm as the film thickness reduces from 1 μm to 0.02 μm. In 0.05-μm-wide Cu interconnects, resistivities ≥7 μΩ-cm are measured. The observed resistivity increases are about 2 to 5 times larger than the size of the surface scattering effect. Microstructure analysis and Monte Carlo simulations carried out in this study and established models in the literature suggest that small grain sizes and surface roughness can account for the observed excessive resistivity increase. The combined surface and grain boundary scattering effects in a sub-0.1 μm fine line can be summarized as:

$$\rho_{line} \approx \rho_{bulk} \, [1+0.75l(1-p)S/d +1.5lR/(1-R)g] \tag{4}$$

where $\rho_{bulk} = mv_F/ne^2l$, and m, v_F, and n are the mass, Fermi velocity, and density of conducting electrons, respectively. By assuming $p = 0$, $R = 0.3$ and $l = 39$ nm, the resistance of a sub-0.1 μm Cu interconnect at 300 K can be expressed as:

$$\rho_{line} \approx (1.8 \ \mu\Omega\text{-cm})[1+(30 \ nm)S/d +(25 \ nm)/g] \tag{5}$$

It is noted that these size effects are difficult to overcome, since the parameters associated with surface roughness (S), linewidth (d), and grain size (g) are fixed with the structure. They cannot be quenched away by cooling either. The high value of ρ_{line} will be a serious concern in the sub-0.1 μm linewidth regime.

ACKNOWLEDGMENT

This work was partially funded by the SRC Center for Advanced Interconnect Science and Technology (Task 448.045).

REFERENCES

1. *The National Technology Roadmap for Semiconductors* (Semiconductor Industry Association, San Jose, CA, 1999).
2. K. Fuchs, *Proc. Cambridge Phil. Soc.* **34**, 100 (1938).
3. M. S. P. Lucas, *J. Appl. Phys.* **36**, 1632 (1965).
4. D. K. C. MacDonald and K. Sarginson, *Proc. Roy. Soc. London* **A203**, 223 (1950).
5. R. B. Dingle, *Proc. Roy. Soc. London* **A201**, 545 (1950).
6. R. V. Isaeva, *Soviet Phys. JETP Letters English Transl.* **4**, 209 (1966).
7. F. W. Reynolds and G. R. Stilwell, *Phys. Rev.* **88**, 418 (1952).
8. Y. Hsu, T. E. F. M. Standaert, G. S. Oehrlein, T. S. Kuan, E. Sayre, K. Rose, K. Y. Lee, and S. M. Rossnagel, *J. Vac. Sci. Technol.* **B16**, 3344 (1998).
9. A. F. Mayadas and M. Shatzkes, *Phys. Rev.* **B1**, 1382 (1970).

Mat. Res. Soc. Symp. Proc. Vol. 612 © 2000 Materials Research Society

Tantalum-Nitride diffusion barrier studies using the transient-ion-drift technique for copper detection

T. Heiser, C. Brochard, M. Swaanen[1]
University Louis Pasteur, Laboratoire de Physique et Applications des Semiconducteurs, CNRS, BP20 F67037 Strasbourg Cedex 2, France
[1]at ST Microelectronics, 850, rue Jean Monnet B.P. 16, F-38921 Crolles cedex, on leave of absence from Philips Electronics, Eindhoven, Netherlands.

ABSTRACT

The permeability of 5 nm thick TaN, Ta and TiN diffusion barriers has been studied by monitoring the bulk copper concentration in silicon after isothermal and isochronal annealing experiments with a transient-ion-drift (TID) technique. The method estimates quantitatively the bulk copper concentration in silicon from capacitance transients of a Schottky barrier which arise when copper ions drift out of the depletion region towards the quasi-neutral region. The correlation between the copper lateral distribution and the position of the copper metal is used to distinguish between background contamination and copper originating from barrier leakage. The TID detection limit is found lower than 10^{12}at/cm^3, which makes this technique particularly well adapted for quantitative diffusion barrier studies. Isothermal and isochronal annealing experiments show that TaN fails at a higher temperature than Ta barriers. The copper concentration does not exceed the solubility limit, indicating that copper precipitates nucleate rapidly at the interface. The opposite is found in TiN covered samples where a large copper supersaturation is obtained even after short annealing times.

INTRODUCTION

Although diffusion barriers are a key part of the copper interconnect technology, a quantitative method to evaluate barrier efficiency against copper diffusion in terms of metal contamination is still missing. Standard analytical tools, such as Secondary Ion Mass Spectroscopy (SIMS) or Auger Electron Spectroscopy (AES) are limited by a detection limit for copper higher than 10^{15}at/cm^3.[1] The presently most sensitive techniques measure electrical parameters of the barrier (i.e. barrier resistivity) or of a test component (oxide breakdown of a MOS capacitor) which do not allow a quantitative estimation of the copper flux through the barrier.[2] The copper concentration needed to affect these electrical parameters is not well established yet. On the other hand, diffusion barriers, less than a few nanometers thick, should allow to keep copper contamination of silicon oxide and bulk silicon below 10^{10}at/cm^2.[3] A quantitative copper detection tool with a low enough unambiguous detection limit would thus be of considerable interest to study diffusion barrier quality versus barrier composition, following mechanical and thermal stress. Validation and calibration of existing techniques, which may be better suited for on-line investigations, could be achieved by such a tool as well. In the present study we show that the recently developed transient-ion-drift (TID) technique,[4,5] which combines both, high sensitivity of electrical measurements and access to quantitative results, is particularly well suited for diffusion barrier investigations. In the present article, the technique is used to investigate time and temperature dependent diffusion barrier failure for TiN, Ta and TaN thin films.

THE TRANSIENT- ION - DRIFT METHOD

Charged defects located in the depletion region of a Schottky barrier tend to drift out of the high electric field region if their mobility is high enough. The resulting change in electric charge distribution induces a transient capacitance signal which carries information about the defect mobility and concentration. The observation and processing of such ion drift induced transient signals are the matter of the so-called Transient Ion Drift technique.[4,5] Recently, diffusion properties and defect reactions of interstitial copper in silicon were investigated by TID and let to a better understanding of the complex behavior of this impurity.[6,7] The possible application of TID for trace analyses of copper in silicon was suggested by Heiser et al. only after they established that most copper impurities in p-type silicon remain interstitially dissolved for a few hours, provided that the thermal treatment, during which copper contamination occurred, is ended by a fast quench.[8,9] As only interstitial copper contributes to the signal, the fraction of interstitially dissolved impurities restrain the TID sensitivity. Further, it was shown that TID transients can be repeated by applying a pulsed voltage to the Schottky barrier, allowing standard signal processing techniques to be used and reach a detection limit close to 10^{11}at/cm^3. The physical origin of the signal, and in particular its differentiation from common hole emission induced transient capacitance signals, can be ascertained by estimating the signal time constant and measuring the signal amplitude dependence on the applied voltage.[5] Drawbacks of the technique are its limitation to p-type silicon, the necessity of a Schottky contact and the thermal treatment required to transfer all copper impurities into the interstitial state.

In the framework of diffusion barrier studies, TID can be easily implemented to monitor the amount of copper impurities that crossed the barrier during a given thermal treatment. Indeed, at temperatures above 600K, it takes less than a few minutes to the copper impurities to distribute uniformly over the silicon substrate thickness.[7] If, following the thermal stress, the sample is quenched to room temperature, the in-diffused copper impurities remain interstitially dissolved. After removing the top layer, including at least the copper metallization, the diffusion barrier and eventually the silicon oxide, the bulk copper concentration can be determined by TID and related to the amount of in-diffused copper.

EXPERIMENTAL DETAILS

The sample structure depicted on figure 1 is used. TID measurements possess a spatial resolution equal to the Schottky barrier diameter (about 1mm). This is low enough to allow the observation of the spatial correlation between bulk copper distribution and the position of the copper metal dots on top of the barrier. This information can be used to distinguish between background contamination and copper impurities which crossed the barrier.

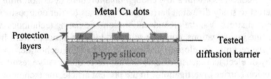

Figure 1. *Test structure used for diffusion barrier investigation.*

For all samples a 25nm TaN protection layer is deposited on the backside of the wafer by ionised metal plasma physical vapour deposition (PVD). After that, 5nm thick barriers are deposited full sheet on the front surface of the wafer. For the Ta and TaN barriers ionised metal plasma PVD has been used as deposition technique and chemical vapour deposition for the TiN barrier layer. The copper dots, which are 6 mm in diameter and spaced 6 mm, have been deposited through a metal hard mask by ionised metal plasma PVD. A second, 25nm thick, TaN layer is deposited by ionised metal plasma PVD on top of the samples to protect them against damage and contamination during transport and to prevent the heating equipment to get contaminated by the samples. The copper backside contamination of the deposition equipment is checked by TXRF on a regular basis.

Samples of 1 x 2 cm^2 are cleaved from the wafers and submitted to isothermal or isochronal heat treatments in a vertical furnace at temperatures between 773K and 873K. The annealing time varies between 5 minutes and 2 hours. The quenching is proceeded by dropping the sample into ethyleneglycole. Next, the front and backside surface layers are lapped off by mechanical polishing and chemical etched in a CP4 solution (mixture of HNO_3/ HF /CH_3COOH). In this way, approximately 50μm are removed from the sample surface. Arrays of Schottky contacts, 1mm in diameter, separated center to center by 1.6mm, are obtained by Al evaporation. The diodes are aligned with the copper metal dots as shown by the "measurement line" in figure 2. The backside ohmic contacts are obtained by covering the sample with an InGa eutectic. The samples are stored in liquid nitrogen in order to avoid interstitial copper room temperature precipitation and outdiffusion.[8,10] The sample preparation time (approximately one and a half hours) is taken into account to estimate the total amount of in-diffused copper (see below).

TID transients are recorded at a reverse voltage of 2V during 900ms following a zero bias pulse of 10 seconds. The TID time constant, defined as in Ref.5, is close to 8ms, which is typical for interstitial copper drift at room temperature in boron-doped material. The ion-drift origin of the signal is ascertained by checking the pulse voltage dependence of the signal amplitude.[5] The TID signal amplitude ΔC is converted into copper concentration using the expression (1) :

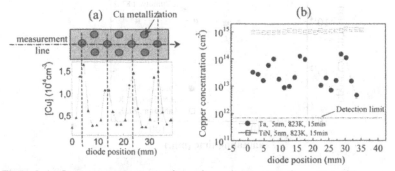

Figure 2. (a) Copper concentration in silicon along a line crossing the position of the copper metal dots. The sample was covered by a 5nm TaN film and was annealed at 873K for 10 minutes. (b) Similar result on 5nm thick Ta and TiN layer after 15 minutes at 823K.

$$[Cu] = 2\frac{\Delta C}{C_o}[Acc] \cdot \exp\left(t_o / \tau_{Cu_i}\right) \qquad (1)$$

where C_o is the quiescent capacitance at the reverse bias, $[Acc]$ is the acceptor concentration, t_o the time that the sample spent at room temperature prior to measurement, and τ_{Cu_i} is the effective lifetime of interstitial copper in silicon estimated to 300 ± 50min.

EXPERIMENTAL RESULTS

Figures 2a and 2b show the Cu concentration, measured along a line crossing the copper metal dots, for various diffusion barriers and annealing conditions. In the case of Ta and TaN barriers, the spatial correlation between the maximum copper amount and the initial position of the metal dots is clearly observed. Note that the signal is far above the detection limit, shown by the dashed line in figure 2b. In the case of a TiN barrier the copper concentration is almost uniform and one order of magnitude higher than the value found in the Ta sample. This surprising result will be discussed below.

Figure 3 summarizes the isothermal annealing experiments done on Ta and TaN diffusion barriers. The copper concentrations shown correspond to the maximum values observed underneath the metal dot position. In all cases, the spatial correlation with the copper dots is observed. The detection limit, defined by a signal to noise ratio of 3, is estimated to $8 \cdot 10^{11}$at/cm^3, which corresponds to a transient capacitance amplitude of about 10fF. For the shorter heat treatments, the copper concentration in between the copper dots is due to a residual copper contamination of the sample surface prior to barrier deposition, and is close to $(5\pm4) \cdot 10^{12}$ at/cm^3,

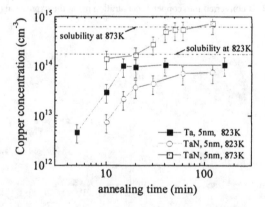

Figure 3. Copper concentration in silicon after isothermal annealing at 82K or 873K for 5 nm Ta or TaN diffusion barriers.

which corresponds to a surface density of $5 \cdot 10^{11} \mathrm{cm}^{-2}$. The error bars are considerably larger than the detection limit of TID measurements and are essentially due to the uncertainty on τ_{Cu_i} and the fluctuations of the copper concentration from one maximum to the other (see figure 2).

Figure 4 shows the results of isochronal annealing experiments for all three barriers. The contamination is below the detection limit only in the TaN sample annealed at 773K.

DISCUSSION

In the case of Ta and TaN barriers, the spatial correlation between the bulk contamination and the position of copper dots (figures 2a and 2b) leaves no doubt that the copper atoms crossed the barriers. Consequently, the annealing experiments can be used to compare both structures and intend the influence of N on the permeability. The isothermal results, shown in figure 3, do not present a significant difference in the characteristic time constant (or slope) of the kinetics, but rather a delayed failure in the case of TaN barrier : the Ta barrier starts to leak earlier than the TaN barrier. Such a behavior is consistent with barrier failure induced by a structural transition (for instance from an amorphous to a polycrystalline state), which would be retarded in the presence of nitrogen.[2] Yet, a full study of the isothermal kinetics over a large temperature range is needed to confirm this assertion. In the opposite case, an activation energy, which varies with the nitrogen concentration, would be expected.

For both barriers, the maximum copper concentration is reached after less than an hour at 823K and is close to the copper solubility in equilibrium with the copper silicide.[11] From these results we may infer that supersaturated copper rapidly forms precipitates in these samples, most probably at the Si/Ta(N) interface. In the TiN sample a supersaturation is observed (see figure 4) indicating that precipitation is slower than for the tantalum samples. This, in turn, may indicate a better quality of the barrier - silicon interface in terms of defects or nucleation centers for copper

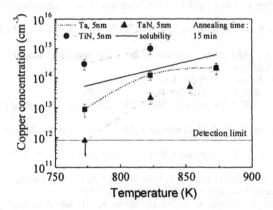

Figure 4. Isochronal anneal experiments on TiN, Ta and TaN diffusion barriers

precipitates. The absence of a spatial correlation in the TiN sample does not allow us yet to conclude about the origin of the copper contamination. Either an uncontrolled contamination, larger than $5\cdot10^{13}$ at/cm^2, happened prior to the TiN deposition or a significant lateral diffusion occurred along the Si/TiN interface during the anneal. The bulk diffusion length of copper in our annealing experiments is indeed less than the distance between the copper plots and cannot, alone, lead to the observed uniform copper concentration.[7]

The isochronal annealing experiment confirms the better resistance against thermal stress of TaN barrier as compared to a Ta barrier. The TaN leakage becomes significant above 823K while, for the Ta sample, a significant copper contamination is observed already at 773K.

CONCLUSION

A comparative study of Ta and Ti based diffusion barriers where performed using the transient-ion-drift technique. The detection limit of the technique is found to be low enough to allow a comparative study of diffusion barriers. In the case of Ta and TaN thin films, the contamination can be related unambiguously to barrier leakage, while for the TiN sample the absence of a spatial correlation between bulk contamination and position of the metal layer, allows no firm conclusion about the origin of the copper. The temperature and time dependent annealing experiments indicate that nitrogen improves the quality of the barrier, by retarding the leakage occurrence. From the present results we may infer that the combination of TID analyses with microscopic analytical characterization techniques will allow us to investigate the relationship between diffusion barrier breakdown kinetics and film microstructure.

REFERENCES

1. M. Stavrev, D. Fischer, C. Wenzel and T. Heiser, Microelectronic Engineering, **37/38**, 245 (1997)
2. Yoon-Jik Lee, Bong-Seok Suh, Sa-Kyun Rha, and Chong-Ook Park, Thin Solid Films **320**, 141 (1998)
3. *The National Technology Roadmap for Semiconductors* (Semiconductor Industry Association - SIA , San José, CA, 1994)
4. T. Heiser and A. Mesli, Applied Physics **A 57**, 325 (1993)
5. T. Heiser and E.R. Weber, Physical Review B, **58**, 3893 (1998)
6. T. Heiser, A.A. Istratov, C. Flink, and E.R. Weber, Material Science Engineering **B58**, 149 (1999)
7. A.A. Istratov, T. Heiser , C. Flink, H. Hieslmair, and E.R. Weber, Physical Review Letters, **81**, 1243 (1998)
8. T. Heiser, S. McHugo, H. Hieslmair, and E.R. Weber, Applied Physics Letters, **70**, 3576, (1997)
9. T. Heiser, S. McHugo, H. Hieslmair, and E.R. Weber, Mat. Res. Soc. symposium *on Defects and Diffusion In Silicon Processing*, San Francisco (USA, 1997)
10. C. Flink, H. Feick, S.McHugo, W. Seifert, H. Hieslmair, T. Heiser, A.A. Istratov, and E.R. Weber, Physica B, **273**, 437 (1999)
11. R.F. Hall and J.H. Racette, J. Appl. Phys. **35**, 379 (1964)

Mat. Res. Soc. Symp. Proc. Vol. 612 © 2000 Materials Research Society

Reliability of Tantalum Based Diffusion Barriers between Cu and Si

Tomi Laurila[1], Kejun Zeng[1], A. Seppälä[2], Jyrki Molarius[3], Ilkka Suni[3] and Jorma K. Kivilahti[1]

[1] Lab. of Electronics Production Technology, P.O. Box 3000, FIN-02015 HUT, Finland.
[2] Accelerator laboratory, P. O. Box 43, FIN-00014 University of Helsinki, Finland
[3] VTT Microelectronics, P.O. Box 1101, FIN-02044 VTT, Finland

ABSTRACT

The reaction mechanisms in the Si|Ta|Cu and Si|TaC|Cu metallization systems are discussed based on the experimental results and the assessed ternary Si-Ta-Cu, Si-Ta-C and Ta-C-Cu phase diagrams. The ternary Si-Ta-N and Ta-N-Cu phase diagrams were also assessed in order to compare the thermodynamic properties of the TaC diffusion barriers to more widely investigated TaN_x diffusion barriers. With the help of the sheet resistance measurements, RBS, XRD, SEM, and TEM the Ta barrier layer was observed to fail above 650 °C due to the formation of $TaSi_2$. This was accompanied by the diffusion of Cu through the silicide layer and the resulting formation of Cu_3Si precipitates. The stability of the TaC layers was better and the failure was observed above 750 °C due to the formation of Cu_3Si and $TaSi_2$. However, interdiffusion of Cu and Si was observed already at lower temperatures due to the presence of pinholes in the TaC layer. This emphasises the importance of the fabrication method and the quality of the TaC layers.

INTRODUCTION

There is a great interest in the use of copper for interconnections in integrated circuits, since it offers many advantages over the currently used Al based materials. Unfortunately, the interaction between Si and Cu is detrimental to the electrical performance of Si even at temperatures below 200 °C.[1-4] Thus, it is necessary to implement a barrier layer between Si and Cu. Tantalum has many good properties from the diffusion barrier point of view and it has been investigated widely.[5-9] However, according to our investigations thin Ta barrier layers may not have sufficient thermal stability, especially as very thin films. Nevertheless, based on the thermodynamic considerations, it is expected that binary tantalum compounds, such as carbides and nitrides, should provide feasible solutions to the diffusion barrier problem. Two stable carbides, Ta_2C and TaC, with melting points of 3330 °C and 3985 °C, respectively, exist in the Ta-C binary system.[10] Despite the many potentially good properties of TaC barriers, only one reported investigation has been published on the Si|TaC|Cu system[11].

MATERIALS AND METHODS

The copper, tantalum, and tantalum carbide films were sputtered onto cleaned and oxide-stripped (100) n-type Si substrates in a dc/rf-magnetron sputtering system. The deposition of TaC was obtained from the TaC-target in argon atmosphere. The thicknesses of the tantalum and tantalum carbide layers were about 10, 50 and 100 nm. The copper films with thicknesses of 100 nm for the thin Ta and TaC films (10-50 nm) and 400 nm for the thick films (100nm) were subsequently sputtered without breaking the vacuum. The samples were annealed in the vacuum of 10^{-4} Pa or better at temperatures ranging from 500 to 800 °C for 30 min.

The sheet resistance measurements at room temperature using a four-point probe were used to detect interfacial reactions after the each annealing step. The reaction products in the Si|Ta|Cu and Si|TaC|Cu metallization schemes were characterized by x-ray diffraction (XRD), Rutherford backscattering spectroscopy (RBS), and transmission electron microscope (TEM). Surfaces of the samples were also examined with an optical microscope and scanning electron microscope (SEM). The results of the investigations were compared with the assessed phase diagrams and to the results obtained from the literature.

Although the complete thermodynamic equilibria are never met in thin film systems - because the materials in the contact regions are under continuous microstructural evolution - the local equilibrium is, however, generally attained at interfaces. Therefore, the phase diagrams provide us an efficient method for designing diffusion barrier layer between various metallizations, especially when they can be combined with kinetic information. Hence, ternary Ta-Si-Cu, Si-Ta-C, Ta-C-Cu, Si-Ta-N and Ta-N-Cu phase diagrams were calculated from the assessed binary thermodynamic data by using the CALPHAD method [12] and compared with the experimental results obtained.

RESULTS AND DISCUSSION

Since the detailed investigation of the Si|Ta|Cu metallization system is described elsewhere[13], only a summary of the results obtained is presented here. Three Ta layer thicknesses of 10, 50 and 100 nm were used to investigate the failure mechanism(s) of the metallization system. The thin (10 and 50 nm) Ta diffusion barriers failed at 550 °C and 600 °C, respectively, due to the diffusion of Cu through the Ta layer and the resulting formation of Cu_3Si. This was followed by the formation of $TaSi_2$ in the 50 nm sample, but in the 10 nm sample Cu_3Si was the only reaction product. The Cu overlayer enhanced the formation of $TaSi_2$, which was not expected to take place before 650 °C[14]. The thicker (100 nm) samples did not fail completely until at 685 °C when the formation of the large Cu_3Si "precipitates" took place (figure1).

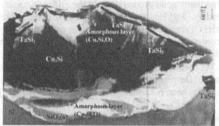

Figure 1. Bright field TEM image from the sample annealed at 685 °C for 30 min.

However, the formation of $TaSi_2$ had already started at 650 °C as revealed by the XTEM investigations (figure 2). It was concluded that the failure mechanisms of thin (10-50 nm) and thick (100 nm) Ta layers were different: as long as Ta layer was thick enough to prevent the diffusion of Cu up to the formation temperature of $TaSi_2$, tantalum silicide was the first phase to form. If the Ta layer was so thin that Cu penetrated through it before the formation temperature of TaSi$_2$ was reached, the Cu$_3$Si was the first phase to form. The resulting reaction structure Si|Cu$_3$Si|TaSi$_2$ was supposed to be the equilibrium structure based on the calculated ternary Si-Ta-Cu phase diagram (figure 3). The role of oxygen was also found to be important and it may have a strong influence to the thermal stability of Ta layers[13].

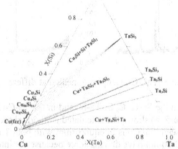

Figure 2. *Bright field TEM image from the sample annealed at 650 °C showing the formation of TaSi$_2$ and its columnar grain structure designated by M.*

Figure 3. *Isothermal section from the calculated ternary Si-Ta-Cu phase diagram at 700 °C.*

The stability of the TaC diffusion barriers was found to be higher in comparison to the Ta layers. The sheet resistance measurements vs. temperature of the Si|TaC(100 nm)|Cu (400 nm) samples showed an abrupt rise in the sheet resistance at 775 °C indicating that reaction had occurred. Above 800 °C the sheet resistance started to decrease, and the structure behaved as an intrinsic semiconductor. The stabilities of the thin 10 nm and 50 nm TaC layers were 625 °C and 675 °C, respectively. RBS analysis revealed that the integrity of the layers was obtained up to 750 °C. However, the analysis also showed that the TaC layers had large amount of pinholes. These pinholes were also detected by optical microscope on as deposited samples. Therefore Cu diffusion towards Si was detected already at lower temperatures. In order to obtain information about the phase formation the XRD analysis were conducted (figure 4). The analysis shoved that up to 725 °C there were only TaC and Cu present in the samples. In 725 °C there were some weak diffractions from the Ta$_2$O$_5$. At 750 °C the formation of Cu$_3$Si took place. At 800 °C, the TaC and Cu peaks had disappeared and only Cu$_3$Si and TaSi$_2$ peaks were present. According to the phase diagram (figure 5(b)) the formation of SiC should have taken place. However, it was not detected in the analysis. The formation of TaSi₂, which took place at 800 °C according to the XRD analysis, could have started at lower temperatures. In the previous case in the Si|Ta|Cu metallization scheme, the formation of TaSi₂ as detected with XTEM, could not be detected with XRD. Hence, a detailed TEM analysis must be conducted in order to reveal the phase formation sequence. Althought the TaC layers seem to be very promising as to be used as diffusion barriers the fabrication of the layers must be conducted with utmost care. The formation of pinholes into the layer must be prevented. We have to investigate the sputter deposition of TaC from TaC-target further to see, if pinholes can be prevented.

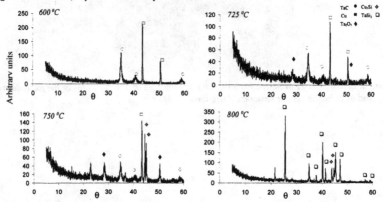

Figure 4. XRD-analysis from the Si|TaC|Cu samples annealed at different temperatures.

The phase relationships in the Si-Ta-C, Si-Ta-N, and in the Ta-C-Cu,Ta-N-Cu phase diagrams are very similar. Figure 5 (a) shows the calculated isothermal section of the ternary Ta-C-Cu phase diagram. It can be seen that Cu is in equilibrium with both TaC and Ta₂C. Thus, the reactions at the interface Cu|TaC do not have driving force because the interface is already in local thermodynamic equilibrium. Nevertheless, the driving force for copper diffusion through barrier towards silicon is expected to be substantial, since copper has a very high affinity towards silicon. Figure 5 (b) shows the isothermal section of the calculated Ta-Si-C ternary phase diagram. There are two important issues which have to be carefully considered in the calculations due to the lack of reliable experimental information. The first one is the existence of ternary phase(s). It is possible, but it has been neglected due to the lack of experimental information. This is also consistent with the general situation in the thin film systems where a ternary phases often have kinetic difficulties to nucleate. Second one is the solubility of carbon into Ta₅Si₃, which is expected to be substantial based on the other Me₅Si₃ silicides (e.g. Ti₅Si₃ ~10 at-%).[15]

Figure 6 (a) shows the isothermal section from the Ta-N-Cu diagram. The same thermodynamic considerations about the barrier|Cu interface, as in the carbide case, are also valid here. Figure 6 (b) shows the isothermal section from the Si-Ta-N phase diagram. Again the similarity is evident and the same considerations about the solubilities and ternary compounds have to be taken into account. Hence, the systems resemble each other very closely. Indirect support to the possible solubility of nitrogen into Ta₅Si₃ can be found in the literature. Holloway *et al.*[6] stated that their Ta$_2$N diffusion barrier failed due to the reaction with the substrate and the resulting formation of Ta$_5$Si$_3$ which was followed by Cu penetration and the formation of Cu$_3$Si. If local equilibrium can be assumed in this system, the formation of Ta$_5$Si$_3$ in the reaction between Si and Ta$_2$N is not thermodynamically possible because there is not Ta$_5$Si$_3$-Ta$_2$N equilibrium in the Ta-Si-N phase diagram(figure 6 (b)), unless some amount of nitrogen could be dissolved into Ta$_5$Si$_3$. However, local equilibrium is not always obtained in thin film systems and therefore additional experiments with bulk materials must be done to find the solubility limits.

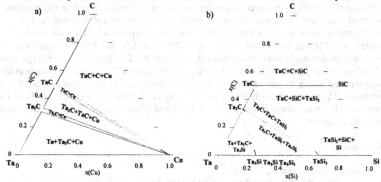

Figure 5. (a) Isothermal section at 700 °C from the ternary Ta-C-Cu phase diagram (b) isothermal section at 700 °C from the ternary Si-Ta-C phase diagram.

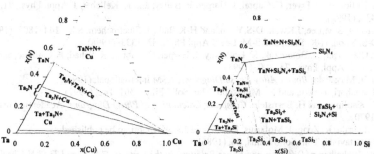

Figure 6. (a) Isothermal section at 700 °C from the ternary Ta-N-Cu phase diagram, (b) isothermal section at 700 °C from the ternary Si-Ta-N phase diagram.

There are similarities in the failure mechanisms of the carbide and nitride barrier layers, as expected. According to Oku *et.al.*[16] the TaN barrier layer failed by Cu grain boundary diffusion through the layer and no reaction between Si and TaN was observed before the failure in the same manner as with our TaC layers. Our results are also in good agreement with Imahori *et al.*[11] whose Ta$_{53}$C$_{47}$ diffusion barrier failed at 800 °C due to formation of Cu$_3$Si and TaSi$_2$

CONCLUSIONS

The interfacial reactions in the Si|Ta|Cu and Si|TaC|Cu systems were investigated. The failure mechanism of the Ta diffusion barrier was found to be dependent on the thickness of the layer. The chemical stabilities of the layered structure with Ta layer thicknesses of 10nm, 50 nm and 100nm, were 550 °C, 600 °C and 685 °C, respectively. The failure mechanism of the TaC layers was found to correspond that of TaN_x layers, since the reaction was induced by the diffusion of Cu through the layer, before the reaction of the barrier layer with the substrate occurred. The reaction sequence is not yet unambiguously solved and therefore further investigations are needed. The chemical stabilities of the layered structure with TaC layer thicknesses of 10 nm, 50 nm and 100 nm, were 625 °C, 675 °C and 750 °C, respectively. Hence, marked improvement in the stability was achieved with the TaC layer. However, large amounts of pinholes were detected in the layer enabling the diffusion of Cu towards Si already at lower temperatures. Hence, the deposition process of TaC layer must be developed to avoid the pinhole formation. The ternary phase diagrams Si-Ta-Cu, Si-Ta-C, Ta-C-Cu, Si-Ta-N, and Ta-N-Cu were assessed. The phase relationships in the Si|TaC|Cu and Si|TaN_x|Cu metallization schemes were found to be similar.

References:

1. E. Weber, Appl. Phys. A, 1, (1983).
2. A. Broniatowski, Phys. Rev. Lett., **62**, 3074,(1989).
3. J. Torres, Appl. Surf. Sci., **91**, (1995).
4. S.P. Murarka, Microelectronic Engineering, **37/38**, (1997).
5. K. Holloway, and P. Fryer, Appl. Phys.Lett., **57**, 1736, (1990).
6. K. Holloway, P. Fryer, C. Cabral, J. Harper, P. Bailey, and K. Kelleher, J. Appl. Phys., **71**, 5433, (1992).
7. B.Kang, S-M.Lee, J.Kwak, D-S.Yoon, and H-K.Baik, J. Electrochem. Soc., **144**, 1807, (1997).
8. D-S. Yoon, H-K. Baik, and S-M. Lee, J.Appl.Phys., **83**, 1333, (1998).
9. L. Clevenger, N. Bojarczuk, K. Holloway, J. Harper, C. Cabral, R. Schad, F. Cardone, and L. Stolt, J. Appl. Phys., **73**, 300, (1993).
10. T.B. Massalski, *Binary Alloy Phase Diagrams*, ASM International, (1996).
11. J. Imahori, T. Oku, and M. Murakami, Thin Solid Films, **301**, 142, (1997).
12. L. Kaufman and H. Bernstein, *Computer Calculation of Phase Diagrams*, Academic Press, (1970).
13. T.Laurila, K. Zeng, J. Molarius, I. Suni, and J.K. Kivilahti, to be published.
14. G. Ottaviani, Thin Solid Films, **86**, (1986).
15. W. Wakelkamp, *Diffusion and Phase Relations in the Systems Ti-Si-C and Ti-Si-N*, Thesis, Tech. University of Eindhoven, Netherlands, (1991).
16. T. Oku, E. Kawakami, M. Uekebo, K. Takahiro, S. Yamaguchi, and M. Murakami, Appl. Surf. Sci., **99**, 265, (1996).

Mat. Res. Soc. Symp. Proc. Vol. 612 © 2000 Materials Research Society

Microstructural Analysis of Copper Interconnections Using Picosecond Ultrasonics

James M.E. Harper, Sandra G. Malhotra, Cyril Cabral Jr.,
Christian Lavoie, Hsin-Yi Hao*, Wadih Homsi*, Humphrey J. Maris*
IBM T.J. Watson Research Center, Yorktown Heights, NY 10598
*Department of Physics, Brown University, Providence, RI 02912

ABSTRACT

We demonstrate that picosecond ultrasonics provides detailed information on the structure and properties of patterned arrays of copper fine lines used in silicon chip interconnections. In this method, the sample surface is momentarily heated several °C using a pump laser beam, and the transient change in the optical reflectivity is measured by a probe laser beam. Measurements of the optical reflectivity are made on time scales ranging from picoseconds to nanoseconds, revealing information on electronic, acoustic and thermal properties. We have applied this method to samples consisting of copper line arrays of 0.4 µm linewidth, 0.65 µm pitch and 0.35 µm depth in SiO_2 on silicon wafers. For comparison, we examined the picosecond ultrasonic response of 200 nm-thick blanket copper thin films. The patterned Cu lines are found to have long-term oscillations at frequencies of 4.39 and 8.29 GHz with lifetimes at least 10 times longer than the oscillations in the blanket Cu film. A two-dimensional mechanical analysis was developed which uses as input parameters the dimensions and sound velocities of the materials in the sample, and finds the normal mode frequencies and displacements. The main vibrational modes are identified and described for the patterned lines, and the simulations confirm that the lowest frequency modes have very small damping coefficients. Also, the time-dependent signal is shown to reveal details of interface layers and integrity of the copper/liner interface.

INTRODUCTION

Picosecond ultrasonics was developed to probe material properties and interfaces at length scales that are inaccessible to lower frequency acoustic techniques [1,2,3]. Since the sound velocity in copper is about 4.5 nm/ps, a 45-nm thick copper film is traversed by an acoustic wave in 10 ps. To examine film thicknesses below the tens of nm range, it is therefore necessary to produce and detect acoustic signals with picosecond time resolution. The development of stable, pulsed, mode-locked lasers has made possible the application of quantitative picosecond ultrasonics as a practical thin film measurement method [4]. In its simplest application, the method provides a film thickness measurement on multilayer samples that need not be transparent. However, as has been previously demonstrated, the sample response covers a wide range of time scales from ps to ns, providing information on electronic [5], mechanical [6] and thermal properties [7,8] of the sample. In this paper, we demonstrate that this method can provide detailed information on the structure and properties of patterned copper fine line arrays in addition to blanket copper thin films. With patterned fine line arrays, we show that the normal modes of oscillation are detected and identified by comparison with a two-dimensional vibrational analysis. To confirm the assignments of vibrational modes, we used some of the many degrees of freedom available in this experiment including varying the angle of incidence of the pump and probe beams, the orientation of the fine line array, the polarizations of pump and probe, and the wavelength of pump and probe. For samples consisting of copper line arrays of 0.4 μm linewidth, 0.65 μm pitch and 0.35 μm depth in SiO_2 on silicon wafers, the most intense vibrational modes occur at 4.39 GHz and 8.29 GHz, and correspond to displacements in which the free surface of the copper lines is moving relative to the surrounding surface of the SiO_2. One interesting observation is that the low-frequency modes of oscillation in the fine line array at 4.39 and 8.29 GHz are found to be very lightly damped, persisting over thousands of picoseconds. These lifetimes are much longer than those observed with the oscillations of a blanket thin film of similar thickness which are damped out within hundreds of picoseconds. We also show that analysis of the time-dependent reflectivity reveals the thickness of the diffusion barrier liner between the copper lines and the surrounding SiO_2 insulator.

EXPERIMENTAL METHOD

The picosecond ultrasonic method has been previously described [1,2,3] and an example is shown schematically in Fig.1.

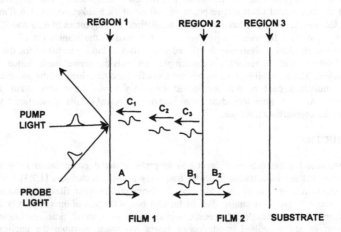

Fig. 1. Schematic diagram of a typical picosecond ultrasonic experiment on planar films. The pump light pulse generates acoustic pulses A, B_1 and B_2. After some time, echo pulses C_1, C_2 and C_3 return to the top surface of the sample.

Here, the pump laser beam is shown striking the surface of a sample which has two thin film layers (film 1 and film 2) on a substrate. The energy of the light pulse is transferred to the conduction electrons in a region near the surface of film 1 (shown as region 1). The hot electrons propagate rapidly throughout the film while losing energy to the thermal phonon system [5,9]. This sets up a thermal stress in the film that is largest near to the upper surface. As a result of this stress, an acoustic pulse A is launched from the near-surface region. If metal films 1 and 2 are not too thick, the heating and thermal stress will be significant throughout the film thickness, and acoustic pulses B_1 and B_2 may be generated from the bottom of the film at the buried interface with film 2 (shown as region 2), or from the interface of film 2 with the substrate (region 3). It is important to note that because the hot electrons propagate with such a high velocity, all of these acoustic pulses are generated at essentially the same time.

Each of the generated acoustic pulses propagates through the thin films at the local sound velocity, and undergoes partial reflections at interfaces between layers and at the bottom interface. As a result, a number of acoustic echoes C_1, C_2, C_3, ..., return to the top surface. These echoes arrive over a time interval that depends on the thickness of the different films, and which may vary from a few picoseconds to several hundreds of picoseconds. Each returning echo results in a change in the optical reflectivity of the sample surface, and this change is detected by a time-delayed probe light pulse. If the thickness of a layer is comparable to the acoustic wavelength, the layer may be set into platelike vibrations, rather than producing distinct echoes clearly separated in time. For patterned thin films, more complex vibrational mode patterns are excited instead of the simple thickness modes observed in blanket films.

For the present studies, a mode-locked Ti:sapphire pump laser was used at 800 nm wavelength (denoted as IR) and also frequency-doubled to 400 nm (denoted as UV). Each light pulse has a duration of 200 fs, and the repetition rate is 76 MHz. The beam size on the sample is approximately 20 μm. The time-delayed probe laser was derived in each case from the pump laser using a beam splitter, and therefore has the same wavelength as the pump beam. The ability to operate at either IR or UV wavelengths is particularly useful in studies of patterned copper samples because different normal modes of vibration tend to dominate the response at the two wavelengths.

A schematic diagram of the experimental configuration is shown in Fig. 2. In the particular configuration shown in Fig. 2, the pump and probe beams are polarized parallel to the plane of the sample, the pump beam direction lies in the plane M that contains the normal to the sample surface and the direction in which the copper lines run, and the probe beam is directed perpendicular to the direction of the lines.

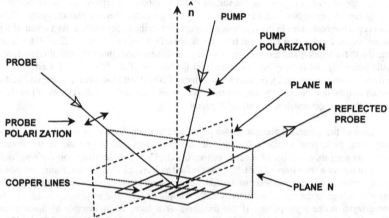

Fig. 2. Diagram of the experimental configuration. Plane M contains the sample normal and the copper line direction. Plane N is perpendicular to plane M and to the lines.

Blanket thin film samples consisted of 200 nm-thick sputter-deposited Cu on 10 nm-thick Ta on SiO₂ on Si. Patterned thin film samples for this study consisted of Cu fine line arrays with 0.4 μm linewidth, 0.65 μm pitch and 0.35 μm thickness in SiO₂ on Si, shown schematically in Fig. 3. These lines were prepared by etching the line patterns into SiO₂. Next, a thin liner and Cu seed were sputter-deposited into the etched line features, followed by electrolytic plating of Cu into the features, and chemical mechanical planarization to remove the excess Cu. Examination of the lines showed the sidewalls to have an angle of a few degrees, i.e. the lines taper slightly from the top to the bottom.

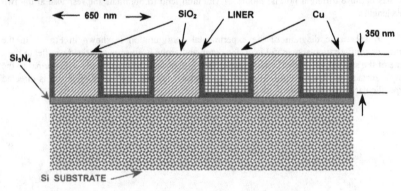

Fig. 3. Cross section of copper line samples.

RESULTS AND DISCUSSION

Blanket copper thin films

Examples of ultrasonic signals from a 200 nm-thick blanket Cu film on 10 nm Ta on SiO_2 on Si are shown in Fig. 4. Only the region extending to 400 ps is shown to illustrate the information contained in the early portion of the signal.

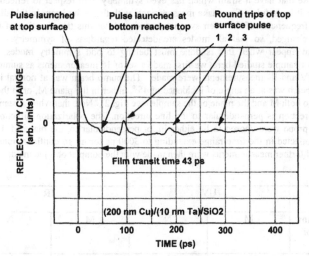

Fig. 4. Ultrasonic signals from a 200 nm-thick blanket Cu film on 10 nm Ta on SiO_2 on Si.

Regions indicated include the initial rapid response in which an acoustic pulse is launched from the top surface, the arrival at the top surface of the acoustic pulse launched at the bottom of the film, and multiple round-trips of the pulse launched at the top surface. The time between successive echoes is used to determine the film thickness, and the ratio of the amplitudes of successive echoes gives information about the extent of the coupling between the film and substrate [10].

Patterned copper thin films

For blanket copper films, it is hard to make measurements using IR light because the piezo-optic coefficient that determines the magnitude of the acoustic response is very small [11]. However, it turns out that acoustic vibrations of patterned copper structures can readily be measured with both UV and IR. The IR response arises because when the copper bars vibrate, the spacing between the bars changes and this gives a change in the optical reflectivity of the sample.

To study patterned films it is advantageous to make measurements for several different combinations of the direction and polarization of the pump and probe beams. In this way it is possible to enhance the response from particular normal modes of vibration so that their frequency can be measured with greater accuracy. Variation of the beam direction also makes it possible to control the symmetry of the normal modes that are excited and measured. When the pump beam direction lies in the plane M of Fig. 2, the vibrational modes of the sample that are excited are those that have a strain which has even symmetry with respect to reflection in the plane M. Thus, it is useful to first make measurements with the pump beam in this direction to determine the frequencies of the even symmetry modes. After this the direction of the pump beam can be changed, so that all modes are detected regardless of symmetry. The new frequencies that appear when this is done must belong to odd symmetry modes. For the patterned copper sample studied here, we first made a set of 16 measurements as summarized in Fig. 5. Both UV and IR measurements were made. The pump beam was at normal incidence, and the probe beam was at an angle of incidence of 45°, either in the plane M, or in the plane N perpendicular to both M and the plane of the sample (see Fig. 2). Note that when a beam is in the plane N, its direction is perpendicular to the line direction. The polarization directions of the pump and the probe were chosen to lie either in, or perpendicular to, the plane M. Ultrasonic signals were collected in the time range extending to 3000 ps for all 16 configurations listed in Fig. 5. Selected further measurements were then made with the pump at oblique incidence.

Light wavelength	UV								IR							
Plane containing the direction of probe beam	M				N				M				N			
Plane containing the pump polarization	M		N		M		N		M		N		M		N	
Plane containing the probe polarization	M	N	M	N	M	N	M	N	M	N	M	N	M	N	**M**	N

Fig. 5. List of 16 configurations, not including variations in the angle of incidence. Planes M and N are shown in Fig. 2. Bold letters show conditions used to obtain data in Fig. 6.

Examples of these ultrasonic signals from the patterned line array are shown in Fig. 6. It can be seen that the signal contains oscillations at well-defined frequencies that decay very slowly. In these examples, the laser wavelength is 800 nm (IR), the polarization of the pump is perpendicular to the lines, and the angle of incidence of the pump is 0° (normal incidence). The probe direction is perpendicular to the lines, and the angle of incidence of the probe is 45°.

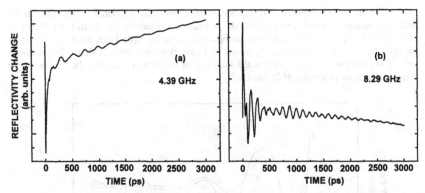

Fig. 6. Examples of ultrasonic signals showing (a) dominant frequency 4.39 GHz and (b) dominant frequency 8.29 GHz. Note the weak damping of the oscillations.

In Fig. 6(a), the probe polarization is perpendicular to the lines, and the acoustic response is dominated by a strong oscillation at 4.39 GHz. In Fig. 6(b), the probe polarization is parallel to the lines. This configuration is preferentially sensitive to a mode of frequency 8.29 GHz. These examples are given to show how different configurations detect different responses in the same sample, providing information which helps to identify the vibrational modes for comparison with the predictions of the mechanical model. Fourier transforms of the data in Fig. 6 are shown in Fig. 7. From these transforms, accurate values for the mode frequencies can be obtained.

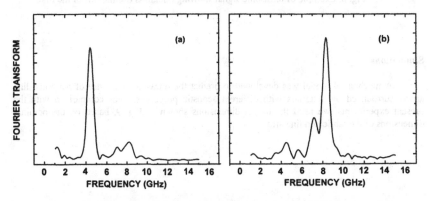

Fig. 7. Fourier transforms of the ultrasonic data shown in Figure 6a and 6b.

The short time region of the ultrasonic signal provides additional information on the individual echoes of sound waves within the structure. For example, the signal extending to 40 ps is shown in Fig. 8 for the same beam configuration as in Fig. 6(b). Here, the data is shown as the solid line, and a fit to the data by means of a numerical simulation is shown by the dotted line. The model was used to estimate the thickness of the liner surrounding the Cu lines to be consistent with the properties of 26 nm-thick Ta.

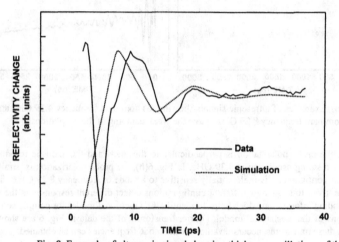

Fig. 8. Example of ultrasonic signal showing thickness oscillations of the liner.

Simulations

A mechanical model was developed to predict the ultrasonic response of patterned line arrays surrounded by materials with different acoustic properties. For comparison with the present experiments, we used the model dimensions shown in Fig. 9, based on the nominal dimensions of the fabricated line array.

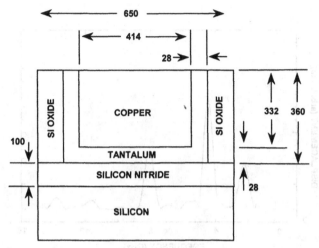

Fig. 9. Diagram showing the sample dimensions used in the simulations. All units are in nm.

Since the length of copper lines that is excited by the pump pulse is large compared to the lateral dimensions of the lines, the motion of the lines occurs entirely in a plane perpendicular to their length. Thus, it is sufficient to perform a simulation of the motion in this plane, i.e. a simulation of the motion in two dimensions. The sample is modeled by dividing the structure into a square array of "atoms" coupled by "springs". The spacing between atoms and the mass of each atom are chosen so as to give the correct local density. The atoms interact by simple springs between nearest and next-nearest neighbors, and also by forces dependent on nearest neighbor bond angles. The strength of the springs is chosen so as to make the material elastically isotropic and to give the correct values for the longitudinal and transverse sound velocity.

The excitation by the pump beam is simulated by expanding the interatomic springs as the starting condition for the model. The volume in which the springs are expanded is selected according to the pump wavelength and the symmetry of excitation (normal or off-normal angle of incidence). The atoms are then released and their motion is calculated over the period of time of the data set (here up to 3000 ps). The simulation program provides the time-dependent displacement of each part of the structure, and the Fourier transforms of the displacements give the frequency spectra of the resulting oscillations. The sound velocity in Cu was varied by 10% above and below its nominal literature value to obtain the best fit to the experimental data such as those shown in Figs. 6-8. An example of a simulated frequency spectrum is given in Fig.10, showing that the model predicts dominant vibrational modes at frequencies of 4.4 and 8.1 GHz, in close agreement with the measured values.

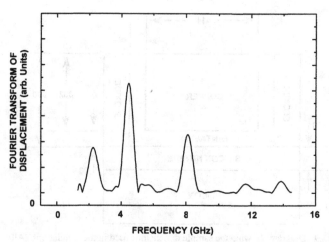

Fig. 10. Fourier transform of the displacement as calculated by numerical simulation.

The simulation does not make allowance for any dissipative process, and so the damping of the vibrations is determined solely by the rate at which acoustic energy is radiated away into the substrate. From the simulation, we find that the rate at which the dominant vibrational modes lose energy by radiation is very small. This is consistent with the experimental results. The computer simulation program was then extended so that the displacement pattern associated with each vibrational normal mode of the structure could also be determined. These displacement patterns corresponding to the 4.4 and 8.1 GHz frequency modes are shown in Fig. 11. The figure shows a sequence of displacements covering one full vibrational cycle.

CONCLUSIONS

We have demonstrated that the picosecond ultrasonic method is capable of resolving structural detail of copper interconnection structures, in addition to the thickness of blanket thin films. In patterned submicrometer copper line arrays, we have identified the main vibrational modes by comparison with a general mechanical model. We found that these vibrations are very lightly damped compared with those of blanket films on similar substrates, and have verified these low damping coefficients with simulations. The variety of physical configurations available with this technique and the frequency-dependent response of the materials were useful in identifying the symmetry of the dominant modes. Finally, the short time resolution of the method was used to reveal the thickness of the metal liner surrounding the copper lines.

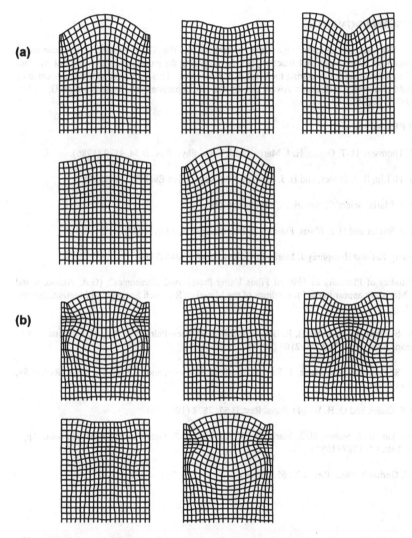

Fig. 11. Schematic diagram of the two low-frequency vibrational modes (a) 4.4 GHz and (b) 8.1 GHz. The displacements have been greatly exaggerated for clarity. The displacement pattern for each mode is shown for phases 0, $\pi/2$, π, $3\pi/2$ and 2π.

ACKNOWLEDGMENTS

The authors acknowledge their colleagues at the IBM Advanced Semiconductor Technology Center, Hopewell Junction, NY for preparing the patterned copper fine line samples and Roy Carruthers for preparing the blanket film samples. This work was partially supported by an IBM University Partnership Award and by the US Department of Energy through Grant No. DE-FG02-86ER45267.

REFERENCES

1. C. Thomsen, H. T. Grahn, H. J. Maris, and J. Tauc, Phys. Rev. B **34**, 4129 (1986).

2. H.-N. Lin, R. J. Stoner, and H. J. Maris, J.Non-Destructive Eval. **9**, 239 (1990).

3. H. J. Maris, Scientific American, **278**, 86 (1998).

4. R. J. Stoner and H. J. Maris, Future Fab International, **1**, 339n (1997).

5. Guray Tas and Humphrey J. Maris, Phys. Rev. B **49**, 15046 (1994).

6. "Studies of Plasticity in Thin Al Films Using Picosecond Ultrasonics", (G.A. Antonelli and H.J. Maris), to appear in the Proceedings of the Materials Research Society Fall Meeting, Boston, 1999.

7. W. S. Capinski, H. J. Maris, E. Bauser, I. Silier, M. Asen-Palmer, T. Ruf, M. Cardona, and E. Gmelin, Appl. Phys. Lett. **71**, 2109 (1997).

8. W. S. Capinski, H. J. Maris, T. Ruf, M. Cardona, K. Ploog, and D. S. Katzer, Phys. Rev. B **59**, 8105 (1999).

9. V.E. Gusev and O. B. Wright, Phys. Rev. B **57**, 2878 (1998).

10. G. Tas, R. J. Stoner, H. J. Maris, G. W. Rubloff, G. S. Oehrlein, and J. M. Halbout, Appl. Phys. Lett. **61**, 1787 (1992).

11. U. Gerhardt, Phys. Rev. **172**, 651 (1969).

Poster Session
Interconnects

Mat. Res. Soc. Symp. Proc. Vol. 612 © 2000 Materials Research Society

Integrated CVD-PVD Al Plug Process for Sub-Quarter Micron Devices: Effects of Underlayer on the Via Filling and the Microstructure of the Al Film

Won-Jun Lee, Jung Joo Kim, Jun Ki Kim, Jin Won Park, Hyug Jin Kwon, Heung Lak Park, and Sa-Kyun Rha[1]
Memory R & D Division, Hyundai Electronics Industries, Ichon, 467-701, KOREA
[1]Department of Materials Engineering, Taejon National University of Technology, Taejon, 300-717, KOREA.

ABSTRACT

The integrated CVD-PVD Al plug process was successfully applied to a sub-quarter micron device for the simultaneous formation of plugs and wires. The effects of the underlayer on the via filling and the microstructure of the CVD-PVD Al films were investigated. Three types of underlayers were examined in this work: the Ti film deposited by the ionized PVD (I-PVD) method, the MOCVD TiN film stacked on the I-PVD Ti film, and the PVD Al film deposited on the I-PVD Ti film. Excellent via filling was achieved by employing the MOCVD TiN/I-PVD Ti or the PVD Al/I-PVD Ti as an underlayer. When only I-PVD Ti film was used as an underlayer, complete via filling was not obtained, because the CVD Al film sealed the top of vias. The CVD-PVD Al film deposited on the PVD Al/I-PVD Ti underlayer also showed excellent crystallographic texture of Al <111> and surface morphology, which is superior to those of the CVD-PVD Al film deposited on the MOCVD TiN/I-PVD Ti underlayer.

INTRODUCTION

As the scaling down of LSI circuits continues, interconnect metallization is playing an important role in determining the performance, reliability and cost of LSI circuits. Currently, W plugging followed by PVD Al wiring is popular, because it shows better results at small contact/via filling compared with Al plugging by PVD Al reflow. However, it is a relatively complex process, and the high cost and low yield of metallization are regarded as disadvantageous for DRAM manufacturing. One of the viable metallization alternatives is a technique which integrates the CVD Al "wetting" layer and the PVD Al(Cu) "reflow" film [1]. With this integrated CVD-PVD Al approach, it was possible to form Al plugs and doped Al interconnects simultaneously for the 0.25-μm technology [2-3]. The CVD-PVD Al process was also adopted in the simultaneous Al filling of vias and trenches with dual damascene structure [4]. In the previous work, we have successfully integrated the CVD-PVD Al plug process into the process flow of a 0.18-μm device [5]. Metalorganic CVD (MOCVD) TiN film stacked on ionized PVD (I-PVD) Ti film was used as the underlayer of CVD Al from dimethylaluminum hydride (DMAH). However, MOCVD TiN is a costly process and it deteriorates the crystallographic texture of the Al film [6].

In this work, three types of barrier metals were examined as the underlayer for CVD-PVD Al plug process: the MOCVD TiN film stacked on the I-PVD Ti film (MOCVD TiN/I-PVD Ti), the I-PVD Ti film, and the PVD Al film deposited on the I-PVD Ti film (PVD Al/I-PVD Ti). The via filling and the microstructure of CVD-PVD Al films were investigated on various underlayers. In order to investigate the effect of underlayer on the via filling, the step coverage of CVD Al film was also examined on various underlayers.

EXPERIMENT

Three types of multilayer thin-film structures were manufactured: PVD Al (400 nm)/CVD Al (50 nm)/MOCVD TiN (5 nm)/I-PVD Ti (10 nm)/oxide/Si, PVD Al (400 nm)/CVD Al (50 nm)/I-PVD Ti (10 nm)/oxide/Si, and PVD Al (350 nm)/CVD Al (50 nm)/PVD Al (50 nm)/I-PVD Ti (10 nm)/oxide/Si. Oxide films were deposited from tetraethyl orthosilicate (TEOS) by the plasma-enhanced CVD method. Metal films were deposited in a cluster-type metal deposition system (Applied Materials, ENDURA 5500). The details of the system was reported elsewhere [5]. CVD Al thin films were deposited at 220°C under 25 Torr using DMAH as the precursor. After CVD Al, the sputter deposition of Al followed for CVD-PVD integration. PVD Al films were deposited at the wafer temperature of 380°C with the dc power of 2 kW.

The characteristics of CVD Al and CVD-PVD Al films were analyzed using various techniques: step coverage by scanning electron microscopy (SEM), surface roughness by atomic force microscopy (AFM), and crystallographic texture by X-ray diffraction (XRD). In order to investigate the via filling characteristics and electrical properties of CVD-PVD Al films, Si wafers were processed through a double-level-metal 0.18-μm device process flow. Vias were patterned using KrF lithography and etched using a high-density plasma oxide etcher. Via resistance was measured using a via chain structure with various via sizes ranging from 0.22 to 0.30 μm. The height of vias used in the electrical characterization was fixed at 850 nm.

RESULTS

Figure 1 shows the via resistance distributions of the Al plug processes and the conventional W plug process. The CVD-PVD Al plug process on the MOCVD TiN/I-PVD Ti showed via

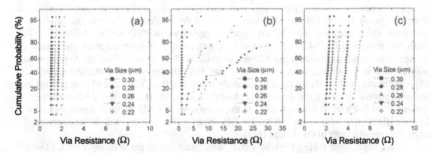

Figure 1. Via resistance distributions of the Al plug processes: (a) MOCVD TiN/I-PVD Ti underlayer, (b) I-PVD Ti underlayer; and (c) conventional W plug process (control).

resistance lower than that of the W plug process at various via sizes ranging from 0.22 to 0.30 µm. This is because the resistivity of Al is superior to that of W. When the I-PVD Ti film was used as an underlayer for CVD-PVD Al plug process, however, the via resistance and its deviation abruptly increased with decreasing via size, which shows the complete Al filling was not achieved at small vias (< 0.30 µm).

In order to investigate the effect of underlayer on the via filling in CVD-PVD Al plug process, we examined the step coverage of the CVD Al films deposited on various underlayers. Figure 2 (a) shows that the CVD Al film deposited on the MOCVD TiN/I-PVD Ti has excellent step coverage and that the top of all vias is open after the CVD Al deposition. However, the CVD Al film deposited on the I-PVD Ti underlayer sealed the top of vias, and the protruded Al grains, "bumps", were observed at the via entrance, as shown in figure 2 (b). If the CVD Al film seals the top of vias, the complete via filling by PVD Al is difficult to achieve, and this results in the poor distribution of via resistance shown in figure 1 (b).   Thinner CVD Al film (30 nm) was also tested on the I-PVD Ti underlayer, however, the complete via filling was not yet achieved.

The Al bump was considered to originate from the rapid abnormal growth of CVD Al on I-PVD Ti at the via entrance.   Similar phenomenon was reported in the previous work [5].   In an effort to suppress abnormal growth of CVD Al at the via entrance and maintain via open, a thin PVD Al film was deposited on the I-PVD Ti film prior to CVD Al at a low wafer temperature with the rf power of 12 kW.   Figure 2 (c) and (d) show the PVD Al film deposited on the I-PVD

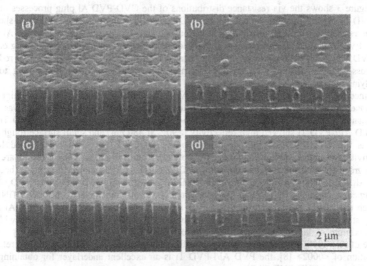

Figure 2.   Bird's-eye views of CVD Al film deposited on hole patterns (tilt angle, 40°): CVD Al film on (a) MOCVD TiN/I-PVD Ti and (b) I-PVD Ti; (c) PVD Al film on I-PVD Ti; (d) CVD Al film on PVD Al/I-PVD Ti.

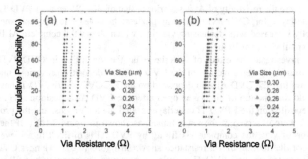

Figure 3. Via resistance distributions of Al plug processes: (a) MOCVD TiN/I-PVD Ti underlayer and (b) PVD Al/I-PVD Ti underlayer.

Ti film and the CVD Al film on the PVD Al/I-PVD Ti, respectively. The CVD Al film deposited on the PVD Al/I-PVD Ti showed excellent step coverage over hole pattern and no Al bump was observed. Moreover, the surface morphology of the CVD Al film was much better than that of the CVD Al film deposited on the MOCVD TiN/I-PVD Ti.

Figure 3 shows the via resistance distributions of the CVD-PVD Al plug processes on the MOCVD TiN/IMP Ti underlayer and on the PVD Al/I-PVD Ti underlayer. Both splits showed low via resistance with tight distribution at various via sizes. The via resistance of the Al plug on the PVD Al/I-PVD Ti underlayer is slightly lower compared with that of the Al plug on the MOCVD TiN/I-PVD Ti underlayer. This is because high-resistivity MOCVD TiN film reduces the cross-sectional area of Al plug and increases the contact resistance of Al plug to the underlying metal.

Since the microstructure of the CVD-PVD Al film is directly related to the electromigration resistance of metal wire [7], the microstructure of blanket CVD-PVD Al film was also investigated on various underlayers. The CVD-PVD Al films deposited on the I-PVD Ti and the PVD Al/I-PVD Ti showed high reflectivity (>90% at 480 nm) and small surface roughness, whereas the CVD-PVD Al film deposited on the MOCVD TiN/I-PVD Ti showed lower reflectivity and higher surface roughness, as shown in figures 4 and 5. These results are in a good agreement with the surface morphology of the CVD Al films shown in figure 2, because CVD Al film functions as the seed layer for PVD Al growth. Figure 6 shows the XRD <111> rocking curves of the CVD-PVD Al films deposited on various underlayers. MOCVD TiN film between the I-PVD Ti film and the CVD-PVD Al film deteriorated the texture of the Al film. Strongest texture of Al <111> was obtained on the PVD Al/I-PVD Ti underlayer. Full-width at half-maximum of the rocking curve was approximately 0.7° in this case. Since very strong texture of Al <111> is obtained in the PVD Al film deposited on the Ti film with the preferred orientation of <0002> [8], the PVD Al/I-PVD Ti is an excellent underlayer for obtaining the <111> texture of CVD-PVD Al film.

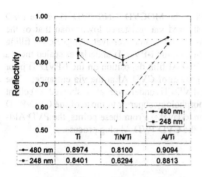

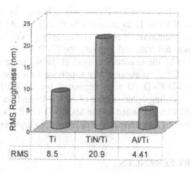

	Ti	TiN/Ti	Al/Ti
480 nm	0.8974	0.8100	0.9094
248 nm	0.8401	0.6294	0.8813

Figure 4. Reflectivity of CVD-PVD Al films deposited on various underlayers

	Ti	TiN/Ti	Al/Ti
RMS	8.5	20.9	4.41

Figure 5. Root-mean-square (RMS) roughness of CVD-PVD Al films deposited on various underlayers

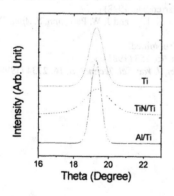

Figure 6. XRD Al <111> rocking curves of CVD-PVD Al films deposited on various underlayers

CONCLUSIONS

Excellent via filling was achieved by employing the MOCVD TiN/I-PVD Ti or the PVD Al/I-PVD Ti as an underlayer, and the Al plugs showed via resistance lower than that of the conventional W plug. When only I-PVD Ti film was used as an underlayer, complete via filling was not obtained, because the CVD Al film deposited on the I-PVD Ti underlayer sealed the via top by the abnormal growth at the via entrance. PVD Al film deposition on the I-PVD Ti prior to CVD Al successfully suppressed the abnormal growth of CVD Al at the via entrance. The CVD-PVD Al film deposited on the PVD Al/I-PVD Ti underlayer also showed excellent crystallographic texture of Al <111> and surface morphology superior to those of the CVD-PVD Al film deposited on the MOCVD TiN/I-PVD Ti underlayer. From these points, the PVD Al/I-PVD Ti structure is a very promising underlayer for the CVD-PVD Al plug process.

REFERENCES

1. J. K. Kim, *U. S. Pat.*, 5,804,501 (1998).
2. I. Beinglass and M. Naik, *Thin Solid Films*, **320**, 35 (1998).
3. A. Konecni, G. Dixit, N. M. Russell, J. D. Luttmer and R. H. Havemann, *Thin Solid Films*, **320**, 52 (1998).
4. L. M. Gignac, K. P. Rodbell, L. A. Clevenger, R. C. Iggulden, R. F. Schnabel, S. J. Weber, C. Lavoie, C. Cabral, Jr., P. W. DeHaven, Y.-Y. Wang and S. H. Boettcher, *Advanced Metallization Conference in 1997* (MRS, Warrendale, 1998) p. 79.
5. W.-J. Lee, B. Y. Kim, S. Y. Han, J. W. Lee, J. K. Kim and J. W. Park, *Jpn. J. Appl. Phys.*, **39**, (2000) (in press).
6. W.-J. Lee and J. W. Park, *Thin Solid Films*, submitted.
7. S. Vaidya and A. K. Sinha, *Thin Solid Films*, **75**, 253 (1981).
8. D. B. Knorr, S. M. Merchant, M. A. Biberger, *J. Vac. Sci. Technol. B*, **16**, 2734 (1998).

Mat. Res. Soc.Symp. Proc. Vol. 612 © 2000 Materials Research Society

Backside Copper Contamination Issues in CMOS Process Integration –
A Case Study

K. Prasad, K.C. Tee, L. Chan[1], and A. K. See[1]
School of Electrical & Electronic Engineering
Nanyang Technological University
Nanyang Avenue, Singapore 639798
[1]R&D Department
Chartered Semiconductor Manufacturing Limited
60 Woodlands Industrial Park Street II, Singapore 738406

ABSTRACT

NMOS and PMOS transistors of various (W/L) ratios, down to 0.24μm channel length, have been used to investigate the effects of copper diffusion (from the backside) on their electrical parameters. A thin layer of copper film was deposited on the back surface of the wafer. Over 10 hours of annealing at 400°C was carried out. Electrical parameters such as the threshold voltage (V_{T0}), the drain saturation current (I_{Dsat}) and the off-current (I_{off}), for transistors, and the leakage current for large diodes were measured. Secondary Ion Mass Spectroscopy (SIMS) was used to monitor the copper diffusion. Even after 10 hours of annealing at 400°C, electrical parameters of both NMOS and PMOS devices and leakage currents of diodes showed no significant degradation.

INTRODUCTION

There are a number of reported results on the use of diffusion barriers to prevent Cu diffusion in Si and SiO_2 for any sub-micron integrated circuit (IC) device technology using Cu metallization [1]. However, there is no specific information on the effects of copper contamination on the degradation of the electrical parameters of MOS devices. Copper may influence the diffusion length and lifetime of the minority carriers in silicon [2,3], but since MOS devices function by majority carrier injection, the impact has not really been scrutinized for sub-micron devices. Nevertheless, the concerns about copper contamination in the manufacturing industry are serious. In this paper, we report results on the effects of Cu diffusion on various electrical parameters of both NMOS and PMOS devices in CMOS ICs. Specifically, the effects of deliberate back surface Cu contamination on the performance of MOS devices will be investigated.

EXPERIMENT

NMOS and PMOS devices were fabricated on 8-inch p-type Si wafers, using twin well process. Conventional self-aligned silicide (SALICIDE) process was implemented. It consisted of pre-Ti deposition cleaning, Ti deposition, first step rapid thermal annealing (RTA), silicide-etchback and second step RTA [4]. NMOS and PMOS devices of various aspect ratios (W/L) were fabricated with an Inter Layer Dielectric, single Al (with 0.5% Cu) metal stack and silicon dioxide and silicon nitride for passivation.

The wafers were thinned down to about 300μm by back grinding. Three types of samples were prepared, viz., control sample, Al sample and Cu sample. The control sample was used as a reference for the experiment and had no metal deposited on the polished back surface. For the Al sample, the back surface of the sample was deposited with 300 nm Al. Lastly, the back surface of the Cu sample was deposited with 200 nm Cu.

Subsequently, all the samples were subjected to thermal annealing at 400^0 C in nitrogen ambient. Temperatures higher than 400^0C were deliberately avoided to minimize any possible failure of the Al metal lines for the interconnect. The devices were annealed for a total time of 10 hours.

The electrical parameters of devices were measured after each annealing step, using Hewlett-Packard HP4156 parameter analyzer. Three sets of NMOS and PMOS transistors were selected, viz., aspect ratios (W/L) of 20μm/20μm, 20μm/0.5μm and 20μm/0.24μm. Parameters monitored for transistors were threshold voltage (V_{T0}), drain saturation current (I_{Dsat}) and off-current (I_{off}). I_{Dsat} was measured at $|V_{GS}|=|V_{DS}|=2.5V$, and I_{off} was measured at $|V_{DS}|=2.75V$ with the source, the gate, and the substrate grounded. Both I_{Dsat} and I_{off} were normalized with respect to gate width (W).

For the diode leakage studies, both large n^+/p and p^+/n diodes were used. Two types of structures were used, viz., contact intensive and area intensive diode structures. The diodes were 460μm × 246μm in size, with SALICIDE block. The reason for blocking silicide formed on the diode was to ensure that the leakage current was caused by the p-n junction, and not by the titanium silicide and silicon interface. The contacts to the junction were 0.32 μm in diameter. Contact intensive structure comprised of 60,000 contacts, while the area intensive structure comprised of 600 contacts only.

Secondary ion mass spectroscopy (SIMS) was used to monitor the copper depth profile on both sides of the Cu sample. Copper at the back surface of the wafer was stripped using dilute sulfuric acid and hydrogen peroxide mixture, prior to SIMS analysis, to increase sensitivity. On the front side, the passivation oxide and nitride, and the metal stack were removed by wet etching. Hence, only 255 nm of poly-silicon gate and 4.5 nm of silicon dioxide were left above the silicon substrate for SIMS analysis. The depth profile was carried out on a large poly-silicon pad, where the pad size was about 0.5mm^2.

RESULTS & DISCUSSION

Electrical Measurements for Transistors

The initial values of electrical parameters of both NMOS and PMOS transistors for all the three types of samples were similar. Furthermore, after 10 hours of annealing at 400^0C, there was little change in the electrical parameters of all the three types of samples. This suggests that the presence of Cu, during annealing, on the backside of Si wafer has little impact on the electrical parameters of MOS devices.

In Fig. 1 and Fig. 2, results of 400^0C annealing on V_{T0}, and I_{off} for both NMOS and PMOS devices are shown, respectively, for the copper sample. All the parameters appeared to be quite stable after 10 hours of annealing. For NMOS transistors, V_{T0} was found to vary from 0.55 V to 0.65 V, while it varied between –0.55 V and –0.65 V for PMOS transistors. I_{Dsat} was stable for all the three different sizes of transistors. I_{off} was found to be in the range of pA/µm, although it fluctuated a little due to sensitivity of the measurements. Similar results were also observed for the control sample and the Al sample. They are omitted for the sake of brevity.

Electrical measurement for junction leakage

Both contact intensive and area intensive diode structures on the Cu sample exhibited no degradation in the junction leakage after annealing. For the n^+/p diodes, the reverse bias currents at 2.75V and 3.6 were 14.5±2.3 pA and 59.3±7.3 pA, respectively. For the p^+/n diode, the leakage currents were –7.6±0.9 pA at –2.75V and –17.2±1.6 pA at –3.6V. The diode leakage currents for the copper sample are shown in Fig. 3 both before and after 10 hours of 400^0C annealing. These results suggest that the p-n junctions were not affected despite the 10 hours of annealing at 400^0C.

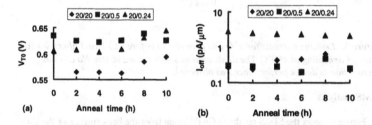

Figure 1. Effect of 400^0C annealing on (a) the threshold voltage V_{T0} and (b) the off-current I_{off} of NMOS transistors. The (W/L) values shown in the legend are in (µm/µm).

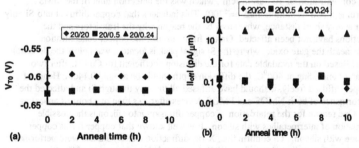

Figure 2. Effect of 400^0C annealing on (a) the threshold voltage V_{T0} and (b) the off-current I_{off} of PMOS transistors. The (W/L) values shown in the legend are in (µm/µm).

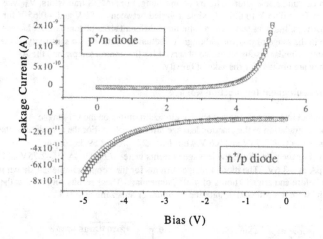

Figure 3. Leakage currents for n^+/p and p^+/n diodes before (□) and after (◊) 10 hours of annealing at 400°C. The diode area was 460 µm × 246 µm. No obvious degradation in device performance was observed.

SIMS analysis

Figure 4 shows the SIMS results of Cu diffusion from the back surface of the Cu sample after 10 hours of annealing at 400°C. The silicon signal was monitored as a matrix indicator and is not drawn to scale. Cu signal started to tail off at around 0.3µm at a concentration of 10^{16} atoms/cm^3, which was the detection limit of the SIMS instrument (Cu resolution: m/Δm=3700). This indicates that copper diffused into Si only over a very short distance, which may be the reason why the transistors and the junctions have not been affected. On the front surface, the copper concentration underneath the gate oxide, where the Si signal peak is located, was below the detection limit. Based on the available data for the diffusion coefficient for Cu [5], after two hours of annealing at 400°C, the diffusion depth should be around 0.14cm. Hence, if copper diffused freely, it should have diffused all the way through Si and affected the performance of both NMOS and PMOS devices after less than one hour of annealing.

One reason for the retardation of copper diffusion into silicon is the possible formation of intermetallics with silicon. It is well known that copper forms copper silicide with silicon. To confirm this, X-ray diffraction (XRD) studies were performed on copper contaminated samples. XRD data revealed the formation of Cu_3Si, consistent with other published results [6]. We believe that copper is either dissolved in copper silicide or precipitated in the first few microns of the back surface of the wafer. Thus,

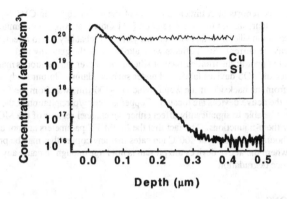

Figure 4. SIMS depth profile plot from the back surface of the Cu sample. Cu level drops below the detection limit after ~0.3 μm.

we would expect very little (below the SIMS detection limit) or no copper at either the front surface of the wafer or at the junctions. Even with copper contamination, the copper concentration at the front surface or the junctions may not be sufficient enough to degrade either the MOSFETs or the p-n junctions. Our experimental results appear to confirm this possibility.

In our experimental approach, the silicon wafer is thinned down to ~300 μm by back grinding. Such a step would possibly introduce defects into silicon, which might act as gettering centres for copper. To investigate this possibility, independent experiments were carried out on both thinned and unthinned silicon wafers, where copper was deliberately introduced from the backside through the use of Physical Vapour Deposition (PVD). Subsequently, thermal annealing was carried out at 400°C for times up to 10 hours. SIMS data on these samples showed the copper signal consistently levelling off at around 0.3 μm with a concentration of 10^{16} cm^{-3} (SIMS detection limit) and the XRD studies revealed the formation of copper silicide. These results clearly show that the effect of thinning, in our studies, is insignificant in hampering the diffusion of copper from the backside of the wafer.

Experimental observations by Rotondaro et al. [7] have clearly shown that 10^{12} atoms/cm^2 of copper contamination does not introduce any major electrically active defects during the device fabrication process. Also, Hozawa et al. [8] have demonstrated that copper will saturate at a concentration of 1.9×10^{14} cm^{-3}, when gettered by boron of 10^{17} cm^{-3} concentration in the p-well region. Furthermore, phosphorous is known to getter copper, gold and other undesirable contaminants [9]. Hence, the implanted species, especially for the n-well implant and the threshold adjustment implant, would be able to slow down the copper diffusion. Thus, the MOS parameters and the leakage currents for n$^+$/p and p$^+$/n diodes are unlikely to be affected. This agrees quite well with our experimental results.

There are some reports of detrimental effects of the use of copper in CMOS process [10,11]. In particular, studies by Vermiere *et al.* [11] show that copper contamination on pre-gate oxidation silicon surfaces affects the yield and reliability. Also affected is the junction leakage current. Such effects were attributed to the gettering and precipitation of copper at the front surface of silicon. However, no mention is made on the performance of MOS devices fabricated under such conditions. In our study, copper is introduced from the backside of the wafer, which is ~300 μm away from the front surface where the active devices are located. Copper gettering/precipitation at the back surface may not be able to significantly affect either the channel regions of the MOS devices and/or the p-n junctions. The fact that the MOSFET parameters are not affected even after 10 hours of annealing at 400^{0}C indicates that any copper that may be present in the region where the active devices are present is not large enough to cause any significant device degradation.

CONCLUSIONS

A systematic study of Cu diffusion from the back surface of silicon wafer was carried out to investigate its effects on the performance of NMOS and PMOS devices. Preliminary results of annealing at 400^{0}C show that no adequate copper species could diffuse from the back surface of the sample and reach either the channel region of the transistors or the p-n junctions to cause any device degradation. We believe that the copper cross contamination from the backside of wafer may not be a critical issue for copper integration in CMOS ICs.

REFERENCES

1. Shi-Qing Wang, *MRS bulletin,* 30, (1994).
2. A. A. Istratov, C. Flink, H. Hieslmair, T. Heiser, And E. R. Weber, *Appl. Phys. Lett.* **71**, 2121, (1997).
3. A. A. Istratov, H. Hieslmair, C. Flink, T. Heiser, And E. R. Weber, *Appl. Phys. Lett.* **71**, 2349, (1997).
4. C.S. Ho, K.L. Pey, H. Wong, K.H. Lee, R.P.G. Karunasiri, S.J. Chua, Y. Tang, S.M. Wong, L.H. Chan, *Proc. 14th International VLSI Multilevel Interconnection Conference*, Santa Clara, USA, 396, (1997).
5. R.D. Thompson, and K.N. Tu, *Appl. Phys. Lett,* **41**, 440, (1982).
6. L. Stolt, F.M. Dheurie and J.M.E. Harper, *Thin Solid Films,* **200**, 147 (1991).
7. A.L.P. Rotondaro, E.Vandamme, J.Vanhellemont, E.Simoen, M.M. Heyns and C. Claeys, *Solid State Phenomena*, **47-48**, 397, (1996).
8. K. Hozawa, T. Itogo, S. Isomae, and M. Ohkura, *1999 Symposium on VLSI Technology*, Tokyo, (1999).
9. S.M.Sze, "VLSI Technology", 2nd Edition, McGraw-Hill (USA), (1988).
10. G. Harsanyi, *IEEE Electron Device Lett.,* **20**, 5, (1999)
11. B. Vermiere, L. Lee, and H.G. Parks, *IEEE Trans. Semicond. Manufacuturing*, **11**, 232, (1998).

Mat. Res. Soc. Symp. Proc. Vol. 612 ⎣ 2000 Materials Research Society

Localized Measurement of Strains in Damascene Copper Interconnects by Convergent-Beam Electron Diffraction

Julie A. Nucci[1], Robert R. Keller[2], Stephan Krämer[1], Cynthia A. Volkert[1], and Mihal E. Gross[3]

[1]Max-Planck-Institut für Metallforschung, Seestraße 92, D-70174 Stuttgart, GERMANY
[2]N.I.S.T. Materials Reliability Division, 325 Broadway, Boulder, CO 80303, U.S.A.
[3]Bell Labs, Lucent Technologies, Murray Hill, NJ 07974, U.S.A.

ABSTRACT

Convergent beam electron diffraction (CBED) was used to measure localized lattice strains in damascene copper interconnects. This method provides data from areas of approximate diameter 20 nm, enabling evaluation of strain states within individual grains. Lattice parameters were determined by measuring the deficient higher order Laue zone (HOLZ) line positions in experimental zone axis patterns and subsequently comparing them to kinematical and dynamical simulations. Quantitative comparison was accomplished using a least squares analysis of distances between line intersections. Deposition-induced strains between 0.06% and 0.14% were measured in 2.0 µm wide lines. The uncertainty in strain determination was approximately 0.02%, as limited by the precision in HOLZ line detection. In addition to enabling localized analysis of strain states, another advantage of using CBED is that the microstructure can be fully evaluated. Used in conjunction with global methods such as X-ray diffraction, CBED may provide unique insight into localized failure phenomena such as electromigration void formation in damascene copper.

INTRODUCTION

Several methods are used to measure strains, and hence to infer stresses, in thin films. X-ray diffraction and wafer curvature, which are the most commonly used methods, provide averaged information and result in the determination of a global strain state. Data from such measurements is often correlated to average microstructure, such as texture strength or median grain size. As a result of their global nature, insight into localized phenomena that occur within interconnects is difficult to realize using these approaches. Examples of such localized behavior include electromigration or stress-induced void formation, grain to grain variation in thermal stresses due to elastic anisotropy, and dislocation clustering during plastic deformation in interconnects.

A method being developed for local strain measurement in interconnects is X-ray microdiffraction. Focussed beams of approximate diameter 1 µm [1] were recently produced using this technique. This is sufficient for obtaining strain information from larger individual grains. The main advantage of this technique is that no special specimen preparation techniques are needed; the measured sample strain state represents that of the as-fabricated structure. A disadvantage of this technique is that synchrotron radiation is required to produce such a finely focussed X-ray beam. While it is possible to extract local texture information using this technique, it is more difficult to select the precise location from which a measurement is taken since the microstructure cannot be imaged with the precision attained by electron microscopy.

Strain can be measured with spatial resolution of approximately 10 nm using convergent-beam electron diffraction (CBED) in a transmission electron microscope (TEM). Since the electron beam can also be used to image the sample, the location from which diffracted information is collected can be precisely selected. This technique was recently applied to measure local strain distributions in free-standing aluminum interconnects. Strain within an individual grain was measured during thermal cycling[2]. Grain-to-grain strain variations were measured in the vicinity of hillocks that formed as a result of electromigration testing [3].

The study of local strain variations is especially relevant to understanding the mechanical properties of copper lines, since strain gradients may produce large stress gradients due to elastic anisotropy. Besides anisotropy, analyzing copper samples requires more careful consideration of dynamical scattering and sample geometry effects, as compared to the analysis of free-standing aluminum lines. We address these issues in this paper and demonstrate the feasibility of measuring strain in sub-micrometer grains from damascene copper lines using CBED.

EXPERIMENTAL

The samples studied were damascene-processed copper lines. A 580 nm thermal oxide was grown on a silicon substrate, followed by a 120 nm Si_3N_4 barrier. Trenches of width 2.0 μm, spacing 2.0 μm, depth 0.5 μm, and length 0.5 mm were formed by reactive ion etching of a subsequent SiO_2 layer. Following deposition of a 50 nm tantalum barrier layer and a 100 nm copper seed layer by physical vapor deposition (as measured in the field between trenches), a 1 μm copper film was electroplated over the structure using a $CuSO_4/H_2SO_4$ bath containing organic additives for good damascene fill. The samples were chemical mechanically polished after room temperature recrystallization, which was deemed to be complete by focussed ion beam imaging.

Plan-view TEM specimens were prepared by the tripod-polishing method. A very brief (approximately 15-20 s) front-side polish was performed using 0.05 μm colloidal silica to remove any copper oxide. Following the setting of a 3° wedge angle, the silicon was polished parallel to the copper line length using wet diamond-embedded paper with grit sizes ranging from 30 μm to 0.1 μm. A schematic of the TEM sample and a micrograph of the copper microstructure are shown in figure 1. Note the extremely fine structure, including a high density of twins.

CBED was performed in a TEM operating at 120 kV. This voltage produces useful configurations of higher order Laue zone (HOLZ) lines, and it allows for penetration through 150 nm of copper, which was the approximate specimen thickness. The TEM was operated in scanning mode, so that a large, continuously variable convergence angle could be formed [4].

Convergence angles of approximately 1° produced Kossel-Möllenstedt conditions. Since the microscope was not equipped with an energy filter, the specimen was cooled to −180°C to improve HOLZ line visibility through suppression of thermal vibrations. The strain associated with this cooling was considered in the final strain analysis.

Figure 1. Schematic of specimen geometry and bright-field TEM micrograph showing as-deposited structure.

STRAIN MEASUREMENT BY CBED

CBED patterns are obtained by converging the electron beam into a cone shape. By illuminating the sample with such a beam, a continuous distribution of incident electron directions is produced within the bounds of the cone surface. The diffraction spots formed by parallel illumination become discs when the beam is converged. Under these conditions, each disc is formed since the incident beam directions within the cone simultaneously satisfy a single Bragg condition. The central disc is used for strain analysis.

Dark deficient lines within the central disc result from HOLZ reflections. Since these reflections often have large g-vectors, the correspondingly high Bragg angle leads to a relatively large shift of the HOLZ lines. In addition, lines become sharper with increasing |g|, so their positions in the pattern become better defined. As a result, lattice distortions can be measured with great sensitivity. Limits to the resolution attainable by CBED depend on the combination of which reflections are present, how sensitive those reflections are to the individual strain components, and how precisely the line positions can be measured. Figure 2 shows how the individual HOLZ line positions shift as a result of a strain applied in the plane normal to the incident beam. Figure 2a and 2b show the HOLZ line shifts induced by applying strain along the x and y directions. Figure 2c was produced by application of a shear strain in the xy plane In this figure, the x-direction runs horizontally, the y- vertically. Since the HOLZ lines respond differently to the applied strain, multiple strain components can be evaluated by considering a set of HOLZ lines. A single diffraction pattern can provide information about the triaxial strain state since there is also a measurable z-component of the reciprocal lattice vectors.

A recently developed methodology for quantitatively evaluating CBED patterns for strain analysis is detailed in reference 3, and briefly summarized here. Line positions in the experimental pattern are determined using a Hough transform[5], which is well-suited to precisely locating narrow lines of even faint contrast, with sub-pixel precision. The experimental patterns are then compared with simulated kinematic patterns that vary systematically with strain. However, the dynamical interaction between the incident electrons and the crystal potential of copper can result in HOLZ line shifts equivalent to shifts induced by 0.3% strain. Since the dynamical shift can be larger than shifts induced by the experimental strain, this effect

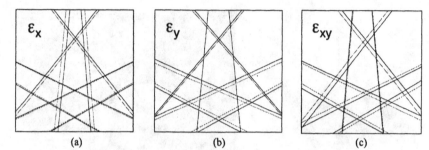

Figure 2. Effect of strain components on HOLZ line positions. x-direction runs horizontally, y-direction runs vertically. Thinner lines represent unstrained state.

must be properly considered. The goodness-of-fit between the theoretical and experimental data is determined by performing a nonlinear least-squares fit analysis using normalized distances between HOLZ line intersections. Standard optimization routines are employed to seek out the best fitting solution.

Multiple combinations of strain components can lead to the same solution. This multiplicity arises since the zone axis patterns display only the deficient lines; an absolute measure of Bragg angles cannot be made. For example, it is possible for a tensile in-plane strain to give rise to the same deficient line position as a compressive out-of-plane strain. A reasonable assumption about the geometry and mechanics of the specimen must therefore be made to eliminate physically non-realistic strain states.

RESULTS AND DISCUSSION

Figure 3 shows a [233] zone axis pattern obtained from a single grain in a copper interconnect, with the specimen held at $-180°C$. A kinematical simulation is overlaid onto this pattern in the adjacent image. The best-fit simulation was created using lattice constants corresponding to a longitudinal strain $\varepsilon_x = 0.16\%$, a transverse strain $\varepsilon_y = 0.19\%$, and a normal strain $\varepsilon_z = -0.13\%$. These values include the thermal strain (assumed to be elastic) at the low temperature.

Extrapolation of the result back to room temperature was accomplished with the aid of a two-dimensional finite element analysis, in which the thermal strain associated with a 200°C temperature drop from room temperature was modeled in addition to the loss of silicon-induced constraint due to thinning. The plan view TEM geometry was modeled as 5 μm long copper and oxide stripes fixed together and rigidly bound to silicon along their widths at one end of the sample. Only half of the copper and oxide linewidths were meshed, due to symmetry. A plane stress model was used to best approximate the strain state in the TEM sample since relaxation of macroscopic stresses in the z direction is possible. Strains varied greatly within approximately 1 μm of either edge of the line, and were reasonably uniform for the middle 3 μm. Within this uniform region, the copper exhibited a longitudinal tensile strain of approximately 0.10% and a transverse tensile strain of approximately 0.05%, due to the 200°C decrease in temperature. One might be concerned that the copper could undergo plastic deformation upon cooling to –180°C,

Figure 3. [233] zone axis pattern and same pattern with kinematical simulation overlaid. Pattern obtained from a copper grain, with specimen held at –180°C. Simulation corresponds to ε_x = 0.16%, ε_y = 0.19% at the low temperature.

since the thermal stress in the thinned specimen is of the order of 190 MPa. The lack of a uniformly high density of dislocations as shown in Figure 1 suggests that this did not occur, perhaps as a result of the extremely fine grain size. Upon subtracting these thermal strains from the strains measured by CBED, we can estimate the in-plane strain state of the specimen prior to cooling, assuming no plastic deformation occurred. The room temperature result is $\varepsilon_x = 0.06\%$ and $\varepsilon_y = 0.14\%$. We are aware of no other localized strain measurements at this scale, to which we can compare this result.

The uncertainty in the measurement was approximately 0.02% strain. This was limited by the goodness of fit between the experimental line positions and the simulated line positions. Use of an automated Hough transform and polynomial peak-fitting routine resulted in locating HOLZ line positions with an accuracy of approximately 0.2 pixel [2]. Subsequent use of a grid search optimization scheme led to the strain state depicted in figure 3.

Note that specimen thinning affects only a load applied to the original sample macroscopically, such as the strain due to thermal expansion mismatch. Localized sources of strain such as those described in the introduction are not expected to change dramatically upon removal of the silicon substrate. In electroplated copper lines, localized strains are also expected, due to either residual plating compounds in grain boundaries and other defects [6] or to fine scale dislocation substructure formation [7]. We have not specifically identified the localized source in this study.

Work in progress involves use of an energy-filtering TEM, which allows for the acquisition of good quality HOLZ lines at room temperature and above. Room temperature measurement simplifies the analysis since there is no need to consider a thermal strain correction. Ongoing work also involves the preparation of cross sectional TEM specimens using focussed ion beam methods in order to better preserve the bulk strain state. For example, when a very small region of the sample (approximately 20 μm long) is thinned along the length of the line to approximately 120 nm, the small electron transparent region is embedded in a much thicker, more mechanically stable piece of silicon. Analytical modeling of this sample geometry revealed that the strain along the line length is consistent with that along the line length in the bulk material. Relevant comparisons of the longitudinal strain among adjacent grains can then be made. Such comparisons should provide insight into the nature of the localized stress and strain distributions in damascene lines.

CONCLUSIONS

We have demonstrated the feasibility of using convergent-beam electron diffraction to locally measure strain states in damascene copper lines. Tensile strains between 0.06 and 0.14% were measured, with an uncertainty of 0.02% due to the line detection procedure. This paper addressed some of the issues associated with making measurements of strain using TEM samples. With proper consideration of relaxation due to thinning, reasonable conclusions about the original strain state can be inferred. By suitable combination of X-ray methods with CBED, the global and localized strain states in narrow interconnects can be more quantitatively characterized.

ACKNOWLEDGMENTS

RRK thanks the NIST Office of Microelectronics Programs.

REFERENCES

1 . N. Tamura, J. –S. Chung, G. E. Ice, and B. C. Larson in *Materials Reliability in Microelectronics IX*, edited by D. D. Brown, A. H. Verbruggen, and C. A. Volkert (Mater. Res. Soc. Proc. **563**, Pittsburgh, PA, 1999) pp. 175-180.

2 . S. Krämer, J. Mayer, *Proc. Fifth Int'l Workshop on Stress-Induced Phenomena in Metallization*, edited by O. Kraft, E. Arzt, C. A. Volkert, P. S. Ho, and H. Okabayashi (AIP Conf. Proc. 491, New York, 1999), pp. 289-297.

3 . S. Krämer, J. Mayer, C. Witt, A. Weickenmeier and M. Rühle, "Analysis of local strain in aluminium interconnects by energy-filtered CBED", *Ultramicroscopy*. **81**, in press (2000).

4 . D. B. Williams, *Practical Analytical Electron Microscopy in Materials Science* (Philips Electronic Instruments, Deerfield Beach, FL, 1984), p. 125.

5 . Hough, P. V. C., U.S. Patent 3,069,654 (1962).

6 . S. H. Brongersma, E. Richard, I. Vervoort, and K. Maex, in *Proc. Fifth Int'l Workshop on Stress-Induced Phenomena in Metallization*, edited by O. Kraft, E. Arzt, C. A. Volkert, P. S. Ho, and H. Okabayashi (AIP Conf. Proc. 491, New York, 1999), pp. 249-254.

7 . C. Cabral, Jr., P. C. Andricacos, L. Gignac, I. C. Noyan, K. P. Rodbell, T. M. Shaw, R. Rosenberg, J. M. E. Harper, P. W. DeHaven, P. S. Locke, S. Malhotra, C. Uzoh, and S. J. Klepeis, in *Proc. Advanced Metallization Conference in 1998*, edited by G. S. Sandhu, H. Koerner, M. Murakami, U. Yasuda, and N. Kobayashi (Mater. Res. Soc., Pittsburgh, PA, 1999), pp. 81-87.

Mat. Res. Soc. Symp. Proc. Vol. 612 © 2000 Materials Research Society

Strain Measurements From Single Grains In Passivated Aluminum Conductor Lines By X-Ray Microdiffraction During Electromigration

K. J. Hwang, G. S. Cargill III, and T. Marieb[†]
Materials Research Center, Lehigh University, Bethlehem, PA 18015, kjh2@lehigh.edu
[†]Components Research, Intel Corp., Hillsboro, OR 97124

ABSTRACT

We describe a method for determining the local strain state of passivated aluminum metal lines from single grains within 2.6 μm x 7.0 μm x 0.75 μm sized regions along the line. X-ray microbeam diffraction is used to obtain localized measurements of thermal and electromigration-induced strain during 37 hours of electromigration in a passivated 2.6 μm-wide, 300 μm-long pure Al conductor line at a current density of 4.2×10^5 A/cm^2 and temperature of 270°C. Diffraction from single grains is used to measure both the in-plane and normal components of strain and their evolution during electromigration at several positions along the line.

INTRODUCTION

Mechanical strain in thin film interconnect structures has long been a reliability concern in VLSI devices. Thermally induced hydrostatic strain in narrow lines can result in cavitation and ultimately line failure [1,2]. In passivated narrow lines, hydrostatic strains in excess of 2×10^{-3} have been observed by bending beam [3] and predicted in finite-element calculations [4-5]. Most previous investigations have used conventional x-ray methods with millimeter-size x-ray beams to measure the macroscopic strain in arrays of aluminum lines encapsulated by passivation [6-10]. At present, obtaining information about local strain in interconnects is difficult, and few techniques are available. Wang et al. [11-13] and Chung et al. [14] have demonstrated the feasibility of x-ray microbeam diffraction for measuring local strain in Al lines. In the present studies, x-ray microbeams have been used to measure microscopic strain distributions from single grains at multiple regions along a 2.6 μm-wide, 300 μm-long, and 0.75 μm-thick passivated pure Al conductor line and for a 120 μm x 120 μm pure Al contact pad. The sampling volume for regions along the line was 2.6 μm x 7.0 μm x 0.75 μm. Initial results from these measurements are described in this paper.

EXPERIMENT AND PROCEDURES

X-ray microbeam diffraction measurements were made using NSLS beamline X6A, a bending magnet white (7keV - 30keV) x-ray synchrotron beamline. The instrumentation used for these measurements is shown in Fig. 1. The novel feature of the instrument is that rotation of the sample is not required in measuring different strain components, since the irradiated sample volume remains fixed while measuring various diffracting planes. The spatial resolution of the measurements in this configuration is limited solely by the x-ray beam size or by the sample dimensions, whichever are smaller. The instrument consists of the following parts: a customized Huber diffractometer, a pinhole collimator providing a 7 μm x 10 μm x-ray beam, a high precision x-y-z sample positioning mechanism, a sample heating stage, a liquid-nitrogen cooled, Ge-based, energy dispersive x-ray detector (SSD), and an optical microscope for positioning and observing the sample.

For the conductor line sample used in these experiments, aluminum was sputter deposited on oxidized Si (100) substrates to a thickness of 0.75 μm and annealed at 400°C after SiO$_2$ deposition (0.7 μm thick) for 4 hours. The Al contact pads were 120 μm wide x 120 μm long. The lines and pads had a 100 Å Ti overlayer and a 450 Å Ti underlayer. Four 0.5 μm diameter Ti vias connected each end of the line to the contact pads. The Al had a strong (111) fiber texture.

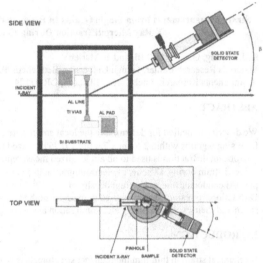

Fig. 1 Schematic of the instrumentation used in the microbeam x-ray experiments.

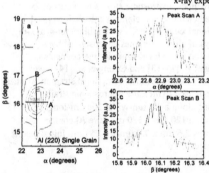

Fig. 2 Reciprocal space optimization of an Al (220) grain.

Each single grain diffraction measurement was optimized in both real space and reciprocal space. Fig. 2 shows an example of reciprocal space optimization for an Al (220) reflection. Angular scans of the solid state detector were performed in both α and β directions for each reflection, and gaussian functions were fitted to the resultant peaks, as shown in Fig. 2b and 2c, to determine the center of the diffracted x-ray beam. A similar procedure with position scans of the sample was followed for optimizing the position of the x-ray microbeam on the center of the diffracting grain in real space.

The lattice spacings of single grains with (111) fiber axis orientations in the Al line and in one Al contact pad were measured. Al (111), (200), (220), and (113) planes from several single grains in the Al pad were measured, and a white-beam $\sin^2(\Psi)$ technique was used to determine the in-plane strain of the Al pad and the unstrained-lattice parameter. The strain state at several locations along the Al line was determined using the technique of Dölle and Hauk [15]. The strains ε_x, ε_y, and ε_z along the principal axes at a given location can be determined from:

$$\varepsilon_{\Phi\psi} = (d_{\Phi\psi} - d_{hkl}^0)/d_{hkl}^0 = \varepsilon_x \cos^2\Phi \sin^2\psi + \varepsilon_y \sin^2\Phi \sin^2\psi + \varepsilon_z \cos^2\psi$$

where $\varepsilon_{\Phi\psi}$ is the measured strain from lattice spacing (hkl), $d_{\Phi\psi}$ is the measured (hkl) lattice spacing, d_{hkl}^0 is the unstrained (hkl) lattice spacing, ε_x is the principal strain in the direction along the width of the Al line, ε_y is the principal strain in the direction along length of the Al line, ε_z is the principal strain in the direction along the thickness of the Al line, ψ is the angle made by the out-of-plane component of the (hkl) scattering vector with the normal to the sample surface, and Φ is the angle made by the in-plane component of the (hkl) scattering vector with the direction along the length of the Al line. The (111) and (200) lattice spacings from single grains were measured to determine the strain state of the line. We assumed $\varepsilon_x=\varepsilon_y$ due to line width (2.6 um) being much larger than the thickness (0.75 um) [5]. Although we sought to measure planes from the same grain, it is the possible that the (111) and (200) measurements were from different grains, since the x-ray beam size was larger than the average grain size. However, it is certain that these measurements are within the irradiated volume of 2.6 μm x 7.0 μm x 0.75 μm, containing approximately 30 grains. The Si substrate was used for *in-situ* calibration and correction of the mechanically determined scattering angles.

RESULTS

Thermal Strain at 25°C and at 270°C

Thermal strain values for the Al pad determined from measured lattice spacings are shown in Table I. At 25°C the Al pad was in biaxial tension ($\varepsilon_{biaxial}= 2.69x10^{-3}$) corresponding to a biaxial stress of 305 MPa, and at 270°C the Al pad was in biaxial compression ($\varepsilon_{biaxial}= -0.99x10^{-3}$) corresponding to a biaxial stress of -100 MPa. The strain-free temperature of the pad was determined to be 192°C. Using the temperature dependent single crystal elastic constants for Al from Gerlich and Fischer [16] adapted for (111) fiber texture [6], the unstressed lattice parameter was determined from the Al pad measurements to be 4.046 Å at 25°C and 4.065 Å at 270°C.

Table I. Al Pad--Average Thermal Strain and Stress

Temp (°C)	$\varepsilon_{biaxial}x10^{-3}$	$\varepsilon_z x10^{-3}$	$\sigma_{biaxial}$ (MPa)
25	2.69	-2.69	305
270	-0.99	1.53	-100

Table II. Al Line--Average Thermal Strain and Stress

Temp (°C)	$\varepsilon_z x 10^{-3}$	$(\varepsilon_x = \varepsilon_y) x 10^{-3}$	$\varepsilon_{Hydrostatic} x 10^{-3}$	σ_z (MPa)	$\sigma_x = \sigma_y$ M Pa	$\sigma_{Hydrostatic}$ (MPa)
25	-0.83	3.00	1.73	248	464	392
270	2.33	0.48	1.10	120	204	176

Table III. Al Line--Localized Thermal Strain and Stress at 25°C

Location	$\varepsilon_z x 10^{-3}$	$(\varepsilon_x = \varepsilon_y) x 10^{-3}$	$\varepsilon_{Hydrostatic} x 10^{-3}$	σ_z (MPa)	$\sigma_x = \sigma_y$ M Pa	$\sigma_{Hydrostatic}$ (MPa)
a	-0.69	2.52	1.45	209	390	330
b	-0.96	3.49	2.00	287	538	455

Table IV. Al Line--Localized Thermal Strain and Stress at 270°C

Location	$\varepsilon_z x 10^{-3}$	$(\varepsilon_x = \varepsilon_y) x 10^{-3}$	$\varepsilon_{Hydrostatic} x 10^{-3}$	σ_z (MPa)	$\sigma_x = \sigma_y$ M Pa	$\sigma_{Hydrostatic}$ (MPa)
A	2.29	0.29	0.96	104	171	149
B	2.42	0.43	1.09	119	200	173
C	2.26	0.59	1.15	126	218	187
D	2.33	0.62	1.19	130	225	194

Thermal strain was measured at two regions (a and b) along the Al line at 25°C and at four regions (A, B, C, and D) at 270°C. The average principal strains and stresses at these locations are given in Table II. Sums of the principal strains were positive, indicating a hydrostatic tensile strain at both measured temperatures (1.73×10^{-3} at 25°C and 1.10×10^{-3} at 270°C). Furthermore, measurements from two locations along the line at 25°C, given in Table III, show more local variation in hydrostatic stress than the measurements at 270°C, given in Table IV.

Electromigration-Induced Strain at 270°C

Electromigration of the Al line was performed at 270°C using a current of 5.5 mA (current density of 4.2×10^5 A/cm^2). At four locations (A, B, C, and D) along the line, the (111) and (200) planes from single grains were periodically measured to determine strain evolution. Measurements were made prior to electromigration, during 36.5 hours of electromigration, and during the subsequent 49 hours with no current. Fig. 3 shows the locations of the measurements along the line and the principal strains measured at the four locations. Current flow began at zero hours. The measured principal strains at negative time are the initial thermally-induced strains and are the same as those in Table III. During electromigration, positions A, B, and C show reductions in the normal strain ε_z and in-plane strain $\varepsilon_x = \varepsilon_y$, which result in general increases in hydrostatic strain at these locations. At Position D, the hydrostatic strain appears to generally decrease, although with more erratic changes in principal strains as electromigration progresses. During the period after electromigration, strains at positions A and B remain largely

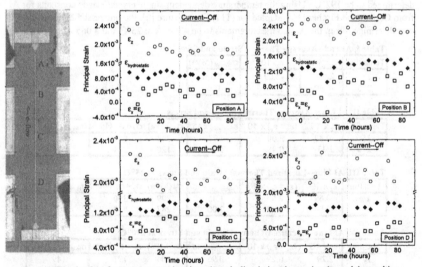

Fig. 3 Schematic of the four measurement locations along the line during electromigration and the resulting principal strains before, during, and after electromigration at four locations A, B, C, and D.

unchanged, but strains at position C appear to decrease in magnitude ($\varepsilon_{hydrostatic}$ and $\varepsilon_x = \varepsilon_y$) or remain constant ($\varepsilon_z$). At position D, $\varepsilon_{hydrostatic}$ and $\varepsilon_x = \varepsilon_y$ increase, but ε_z changes erratically.

Grain Rotations at 270°C

From our diffraction measurements, we could monitor small rotations of grains at the four measurement locations. Fig. 4 shows the change in angular position of the (200) reflections for each of the four locations during electromigration (closed triangles) and following electromigration (open circles) at the four locations (A, B, C, D) along the line. The data labels above each symbol denote the time of the measurement. The dotted lines are simulated

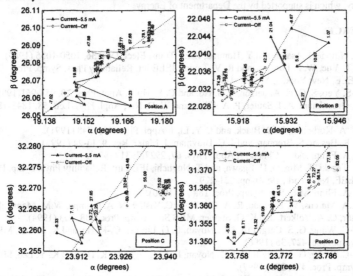

Fig. 4 Al (200) grain rotations due to electromigration and subsequent relaxation at the four locations along the Al line.

rotational paths of the (200) angular positions caused by grain rotation about the (111) plane normal (in-plane rotation). Positions A, C, and D show counterclockwise in-plane grain micro-rotations of approximately 0.03 degrees while Position B shows clockwise micro-rotation of approximately 0.01 degrees. It is interesting to note that the grain rotation directions do not reverse during the period following electromigration (current-off).

CONCLUSIONS

We have shown that local information on thermal and electromigration strain can be obtained by white beam microdiffraction from single grains. Thermal measurements of the Al pad were consistent with equibiaxial behavior while the Al line showed effects of hydrostatic tensile stress.

Localized strain evolution of the Al line during electromigration and subsequent relaxation was measured, and grain rotations associated with electromigration and relaxation were observed.

ACKNOWLEDGMENTS

The authors acknowledge H.-K. Kao and A. C. Ho for development of the microdiffraction instrument and other assistance, and C.-C. Kao for technical assistance at the NSLS. This work has been supported by NSF grants DMR-9796284 and DMR-9896002. The experiments were carried out at the National Synchrotron Light Source (NSLS) at Brookhaven National Laboratory, which is supported by the Department of Energy.

REFERENCES

[1]   K. Hinode, I. Asano, and Y. Homma, IEEE Trans. Elect. Dev. 36, 1050-1055 (1989).
[2]   J. T. Yue, W. P. Funsten, and R. V. Taylor, IEEE Int. Reliability Phys. Symp. Proc. (IEEE, New York, 1985), pp. 1-8.
[3]   I.-S. Yeo, S. G. H. Anderson, P. S. Ho, and C. K. Hu, J. Appl. Phys. 78, 953-961(1995).
[4]   B. Greenbaum, A. I. Sauter, P. A. Flinn, and W. D. Nix, Appl. Phys. Lett. 58, 1845-1847 (1991).
[5]   M. A. Korhonen, R. D. Black, and C-Y. Li, J. Appl. Phys. 69, 1748 (1991).
[6]   P. R. Besser, S. Brennan, and J. C. Bravman, J. Mater. Res. 9, 13-24 (1994).
[7]   P. A. Flinn, and C. Chiang, J. Appl. Phys. 67, 2927-2931 (1990).
[8]   A. Tezaki, T. Mineta, H. Egawa, and T. Noguchi, IEEE Int. Reliability Phys. Symp. Proc. (IEEE, New York, 1990), pp. 221-229.
[9]   P. A. Flinn, MRS Soc. Symp. Proc. 188 , 3 (1990).
[10] M.A. Marcus, W. F. Flood, R. A. Cirelli, R. C. Kistler, N. A. Ciampa, W.M. Mansfield, D. L. Barr, C. A. Volkert, and K. G. Steiner, MRS Soc. Symp. Proc. 338, 203 (1994).
[11] P.-C. Wang, G. S. Cargill III, I. C. Noyan, E. G. Liniger, C.-K. Hu, and K. Y. Lee, MRS Symp. Proc. 427, 35 (1996).
[12] P.-C. Wang, G. S. Cargill III, I. C. Noyan, E. G. Liniger, C.-K. Hu, and K. Y. Lee, MRS Symp. Proc. 473, 273 (1997).
[13] P.-C. Wang, G. S. Cargill III, I. C. Noyan, and C.-K. Hu, Appl. Phys. Lett. 72, 1296 (1998).
[14] J.-S. Chung, N. Tamura, G. E. Ice, B. C. Larson, and J. D. Budai, MRS Soc. Symp. Proc. 563, 169 (1999).
[15] D. Gerlich, and E. S. Fisher, J. Phys. Chem. Solids 30, 1197-1205 (1969).
[16] H. Dolle, and V. Hauk, Z. Metallk. 69, 410 (1978).

Mat. Res. Soc. Symp. Proc. Vol. 612 © 2000 Materials Research Society

Experimental Studies of the Reliability of Interconnect Trees

S.P. Hau-Riege and C.V. Thompson
Department of Materials Science and Engineering, M.I.T., Cambridge, MA

ABSTRACT

The electromigration resistance of simple straight-line interconnects is usually used to estimate the reliability of complex integrated circuits. This is generally inaccurate, and overly conservative at best. The shapes and connectedness of interconnects is not accounted for in standard reliability assessments. We have identified the interconnect tree as the fundamental reliability unit. An interconnect tree consists of connected conducting line segments lying within a single layer of metallization, and terminating at two or more nodes at which there is a diffusion barrier such as a W-filled via. We performed electromigration experiments on the simplest tree structures, such as 'L'- and 'T'-shaped interconnects, as well as straight lines with an additional via in the middle of the line, passing currents of different magnitudes and directions through the limbs of the trees. We found that metal limbs ending in other limbs can act as reservoirs for electromigrating metal atoms. Passive reservoirs, which are limbs that do not carry electrical current, are generally beneficial for reliability, whereas limbs that do carry electrical current, called active reservoirs, can be beneficial or detrimental, depending on the direction and magnitude of the current in the reservoir. However, our experiments show that bends in interconnects do not affect their reliability significantly. We also found that the reliability of an interconnect tree can be conservatively estimated by considering void-growth and void-nucleation-limited failures at the most heavily stressed junction in the tree, which can be found by analyzing the geometry and current configuration. Our experimentally verified model for tree reliability can be used with layout tools for reliability-driven computer-aided design (RCAD), through ranking of the reliabilities of trees in order to identify areas at risk from electromigration damage.

INTRODUCTION

Electromigration-induced failure continues to be an important reliability issue for integrated circuits. Electromigration is electronic-current-induced atomic diffusion due to momentum transfer from flowing electrons to host atoms [1-2]. As device dimensions shrink with the introduction of each new generation of technology, the cross sectional area of interconnects becomes smaller, and interconnects carry ever-higher current densities. Current design rules and design practices developed to prevent electromigration-induced failure tend to be overly conservative [3]. To optimize performance for a given technology, while maintaining a high overall reliability, it is necessary to develop a design practice that more accurately, and less conservatively, accounts for the effects of circuit layout on the risk of electromigration-induced failure.

To date, modeling, simulation, and experimental analyses of interconnect reliability have primarily focussed on the study of electromigration in simple straight-line interconnects. It has been found that in interconnects terminating at diffusion barriers, such as W-filled vias, the hydrostatic stress becomes tensile near the cathode and compressive near the anode [4]. The evolution of the hydrostatic stress has been described successfully using the Korhonen model for electromigration-induced stress evolution [5]. If the tensile stress at the anode-end of the line exceeds the critical stress necessary for void nucleation, a void nucleates. If conductive refractory-metal under-layers, over-layers, or liner materials are present, voiding does not lead to open-circuit failure because the electron current can shunt around a void through the electromigration-resistant refractory metal layers. In this case, when a void nucleates at the cathode end of the line, the hydrostatic stress is relieved around the void, and the void grows due to electromigration, leading to an increase in the electrical resistance of the interconnect segment. If the compressive stress at the cathode-end of the line exceeds the critical stress for cracking of the encapsulating insulator, metallic extrusions form, potentially leading to short-circuiting of interconnects and Cu diffusing into the silicon leading to device degradation.

While straight, junction-free interconnects are typically tested, laid-out integrated circuits often have interconnects with more complex structures that includes junctions. These complex shapes will be referred to as interconnect trees. A tree is defined as a unit of continuously connected high-conductivity metal lying within one layer of metallization, and terminating at diffusion barriers such as vias or contacts with refractory metal liners. In the general case, trees have junctions and more than one terminating branch. In this paper, we discuss the results of electromigration experiments performed on simple tree structures, such as 'L'- and 'T'-shaped interconnects, as well as straight lines with an additional via in the middle of the line, passing currents of different magnitudes and directions through the limbs of the trees. We compare the results of the lifetime experiments with predictions made using analytical models and numerical simulations of electromigration based on the Korhonen model.

SIMULATION OF ELECTROMIGRATION IN TREES

We simulated the effect of electromigration in interconnect trees using our electromigration simulator MIT/EmSim [6-7]. MIT/EmSim is a 1D-electromigration simulation based on the Korhonen model [5], which has been modified to account for a wide range of effects, including

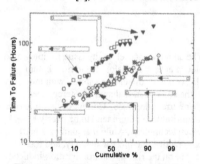

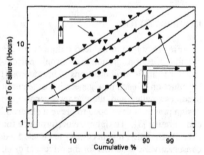

Figure 1: (a) Times to failure, defined as a 30% increase in electrical resistance, for 0.27μm-wide lines tested at T = 250°C and j = 2x10⁶ A/cm² for tree geometries and electron current configurations shown in (b). The line length, ℓ, is 500μm.

Figure 2: Comparison of lifetimes for populations of "L"-shaped interconect trees with only one or both limbs carrying current. The lines were stressed with a current density of 2x10⁶ A/cm² in the long limb, and the current density in the active reservoirs was 5x10⁵ A/cm². The temperature was 350°C.

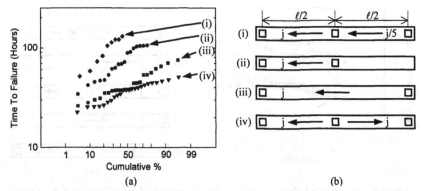

Figure 3: (a) Times to failure, defined as a 30% increase in electrical resistance, for 0.27μm-wide lines tested at $T = 250°C$ and $j = 2x10^6$ A/cm² for electron current configurations shown in (b). The line length, ℓ, is 500μm.

the effects of junctions [8]. The set of input parameters for the simulation was taken from Hau-Riege et al. [8]. Through use of this simulation, we can model stress evolution, void nucleation, and void growth. We can also model the associated increase of the resistance of the limbs in the trees.

EXPERIMENTS

We performed electromigration experiments on simple, passivated Al(0.5wt% Cu) tree structures in the first level of metallization, such as 'L'- and 'T'-shaped interconnects, as well as straight lines with an additional via in the middle of the line. The trees were electrically connected to the second layer of metallization through W-filled vias. This allowed independent application of currents of different magnitudes and directions in the limbs of the trees. Two sets of structures, one with linewidths of 0.27μm and one with linewidths of 3.0μm, were tested. The metallization stack consisted of 100Å Ti / 200Å TiN / 3500Å Al(0.5 wt % Cu) / 40Å Ti / 375Å TiN. The samples were stressed in a QualiTau MIRA electromigration test system, at a temperature of 250°C by forcing a constant current density. The voltage drops over the limbs of the tree were measured as a function of time. Processing of the electromigration samples and electromigration testing were done in collaboration with National Semiconductor.

RESULTS

Figure 1 shows times to failure on a cumulative lognormal probability plot for 0.27 μm-wide 'I'-, 'dotted I'-, 'L'-, and 'T'-structures stressed with a current density of $2x10^6$ A/cm² at 250°C. On a lognormal plot, times that are lognormally distributed fall on a straight line. The line length, ℓ, is 500 μm. Failure is defined by a 30% increase in resistance. Figure 2 compares results from 'L'-structures with one or both limbs active, and with different current configurations. The time to failure was defined as the time at which the resistance of the

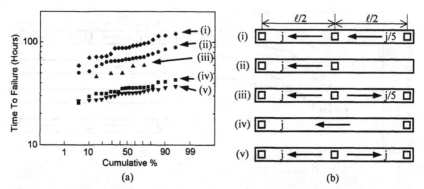

Figure 4: (a) Times to failure, defined as a 30% increase in electrical resistance, for 3.0μm-wide lines tested at $T = 250°C$ and $j = 1x10^6$ A/cm^2 for electron current configurations shown in (b). The line length, ℓ, is 500μm.

structure had increased by 5%. When the shorter limbs were active, they carried one quarter of the current in the longer limbs.

We have varied the current configuration near vias to investigate the effects on nodal reliability. Figure 3 shows times to failure of 0.27μm-wide and 500 μm-long 'dotted I'-structures stressed with a constant current density $j_1 = 2x10^6$ A/cm^2 in the left limb, and a current density with varying direction and magnitude, j_2, in the right limb at 250°C. We fit a lognormal distribution to the times to failure, and the median times to failure as a function of j_2 are shown in figure 5 (b). Figure 4 shows times to failure of 3.0μm-wide and 500 μm-long 'dotted I'-structures stressed with $j_1=1x10^6$A/cm^2 at 250°C, and the median times to failure as a function of j_2 are shown in figure 5 (c). The median times to failure as a function of j_2 of 0.27μm-wide and 500 μm-long 'dotted I'-structures stressed with $j_1=5x10^5$A/cm^2 at 350°C are shown in figure 5 (d).

DISCUSSION

We tested the effects of the presence and orientation of inactive metal limbs ('passive reservoirs') as shown in figure 1, as well as the effect of line bends on the reliability of interconnects. We found that metal limbs ending in other limbs can act as reservoirs for electromigrating metal atoms, thereby improving reliability. The orientation of the reservoir does not affect the time to failure. The lifetimes of interconnects with bends are similar to the lifetimes of straight-line interconnects, so that lines with bends can be treated in a similar way as straight lines with the same line width and length. Atomic diffusion in passive reservoirs is stress-induced. An electric current alters the effectiveness of a reservoir ('active reservoir') because in this case, electromigration provides an additional driving force for atomic diffusion in the reservoir. As shown in figure 2, an electron flow toward the junction increases the lifetime, whereas an electron flow away from the junction decreases the lifetime.

The results of these experiments on passive and active reservoirs demonstrate that reliability estimates of interconnect trees have to consider the current configuration as well as the connectedness of the limbs in a tree. To further quantify these effects, we assessed the reliability of the center node of the 'dotted-I' structure shown in figure 5 (a) through experiments, as shown in figures 5 (b), (c), and (d). Overlaid on this figure are the simulation results and the predictions from an analytic 'default model' discussed in references [8] and [9]. The default model estimates the times for void nucleation, t_{nucl}, and the times for void growth, t_{growth}. The estimated times to failure for a node, taken as the maximum of t_{nucl} and t_{growth}, are generally lower than the experimentally obtained times, demonstrating that the default model gives a conservative lifetime prediction. A comparison of the simulations with experiments shows very good general agreement.

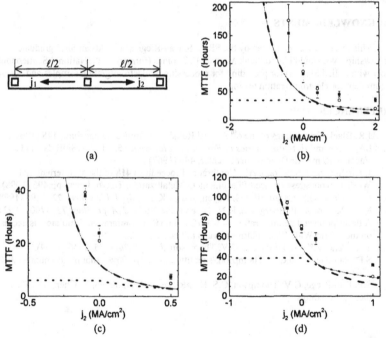

(a)　　　　　　　　(b)

(c)　　　　　　　　(d)

Figure 5: *Comparison of experimental data with the default model for the structure shown in (a) with $\ell = 500$ µm, and (b) $w = 0.27$ µm, $T = 250°C$, $j_1=2x10^6$ A/cm^2, (c) $w = 0.27$ µm, $T=350°C$, $j_1 = 5x10^5$ A/cm^2, and (d) $w = 3.0$µm, $T = 250°C$, $j_1 = 1x10^6$ A/cm^2. The dashed lines show the calculated times for void nucleation, t_{nucl}, the dotted lines show the times for void growth, t_{growth} and the continuous lines show the estimated times to failure taken as the maximum of t_{nucl} and t_{growth}. Overlaid are experimentally obtained median times to failure represented by solid square symbols and error bars indicating a 95% confidence interval. Also overlaid are times to failure obtained through simulations, represented by open circles.*

CONCLUSIONS

We have carried out a wide variety of experiments on electromigration in simple interconnect trees, including 'L'- and 'T'-shaped trees. With these experiments, we have investigated the effects of active and passive reservoirs on electromigration (examples of which are shown in figure 1 and 2), and we have demonstrated that these effects can be accurately accounted for in simulations. These experiments also allowed investigation of the effects of relative current directions and magnitudes on the reliability of lines meeting at a junction or via (as illustrated in figure 3 and 4). These results have led to the development and validation of a junction-based default model for assessment of the reliability of mortal interconnect trees (as illustrated in figure 5).

ACKNOWLEDGMENTS

This research was supported by the SRC. S. Hau-Riege also holds an SRC graduate fellowship. We would like to thank Michael E. Thomas (formerly with National Semiconductor, now with Allied Signals) for providing for processing of electromigration samples and access to equipment for electromigration testing.

REFERENCES

1. J.R. Black, Proceedings of the 6th Annual Reliability Physics Symposium, 148 (1967).
2. I.A. Blech and H. Sello, *Physics of Failures in Electronics* **5**, ed. T.S. Shilliday and J. Vacarro, Rome Air Development Center, 496 (1967).
3. J. Kitchin and T.S. Sriram, AIP Conference Proceedings **418** of the 4th International Workshop on Stress Induced Phenomena in Metallization, Tokyo, Japan, pp.495 (1998).
4. P.-C. Wang, G.S. Cargill III, I.C. Noyan, and C.-K. Hu, *Appl. Phys. Lett.* **72**, 1296 (1998).
5. M. A. Korhonen, P. Boergesen, K.N. Tu, and Che-Yu Li, *J. Appl. Phys.* 73, 3790 (1993).
6. A description and demonstration of MIT/EmSim (Electromigration Simulator) is accessible on the World Wide Web at http://nirvana.mit.edu/emsim.
7. Y.-J. Park, V.K. Andleigh, and C.V. Thompson, *J. Appl. Phys.* **85**, 3546 (1999).
8. S.P. Hau-Riege and C.V. Thompson, submitted to *J.Appl.Phys.*, submission number JR00-0361.
9. S.P. Hau-Riege, C.V. Thompson, C.S. Hau-Riege, V.K. Andleigh, Y. Chery, and D. Troxel, these proceedings.

Mat. Res. Soc. Symp. Proc. Vol. 612 © 2000 Materials Research Society

Grain Orientation and Strain Measurements in Sub-Micron wide Passivated Individual Aluminum Test Structures

N. Tamura[1], B. C. Valek[2], R. Spolenak[3], A. A. MacDowell[1], R. S. Celestre[1], H.A.Padmore[1], W. L. Brown[3], T. Marieb[5], J. C. Bravman[2], B. W. Batterman[1] and J. R. Patel [1,4]
[1] ALS/LBNL, 1 Cyclotron Road, Berkeley CA 94720, USA
[2] Dept. of Mat. Sci. & Eng., Stanford University, Stanford, CA 94305, USA
[3] Bell Laboratories, Lucent Technologies, Murray Hill NJ 07974, USA
[4] SSRL/SLAC, Stanford University, Stanford, CA 94309, USA
[5] Intel Corp., Portland, OR 97124, USA

ABSTRACT

An X-ray microdiffraction dedicated beamline, combining white and monochromatic beam capabilities, has been built at the Advanced Light Source. The purpose of this beamline is to address the myriad problems in Materials Science and Physics that require submicron x-ray beams for structural characterization. Many such problems are found in the general area of thin films and nano-materials. For instance, the ability to characterize the orientation and strain state in individual grains of thin films allows us to measure structural changes at a very local level. These microstructural changes are influenced heavily by such parameters as deposition conditions and subsequent treatment. The accurate measurement of strain gradients at the micron and sub-micron level finds many applications ranging from the strain state under nano-indenters to gradients at crack tips. Undoubtedly many other applications will unfold in the future as we gain experience with the capabilities and limitations of this instrument. We have applied this technique to measure grain orientation and residual stress in single grains of pure Al interconnect lines and preliminary results on post-electromigration test experiments are presented. It is shown that measurements with this instrument can be used to resolve the complete stress tensor (6 components) in a submicron volume inside a single grain of Al under a passivation layer with an overall precision of about 20 MPa. The microstructure of passivated lines appears to be complex, with grains divided into identifiable subgrains and noticeable local variations of both tensile/compressive and shear stresses within single grains.

INTRODUCTION

The reliability of integrated circuits depends on the time to failure of the polycrystalline Al or Cu based lines used as device interconnections. Understanding interconnect failure mechanisms in order to predict and improve their lifetime is therefore a major concern in the semiconductor industry. Electromigration is the material transport resulting from the momentum transfer of the electron wind on the constitutive atoms under high current density. Damage by void formation, growth, and migration, ultimately causing open circuit failure, was very early identified as the main mechanism responsible for failure [1]. Voids form to overcome high stresses, which develop in the material during both fabrication and electromigration. The elevated residual stress, which builds up in the lines during the fabrication process, is essentially due to the thermal mismatch between the line, substrate, and surrounding oxide (passivation layer). Electromigration leads to the development of a compressive stress at the anode with possible hillock formation and of a tensile stress at the cathode with void formation. In the ideal case, the resulting stress gradient would counterbalance the atomic flux due to electromigration, leading to a steady state regime with no further damage ("immortal line"). Accurate measurements of this stress gradient as a function of such parameters as microstructure, initial residual stress state, line aspect ratio, chemical composition, and nature of passivation layer, constitute important inputs for electromigration damage modeling. On the other hand, void dynamics have been shown to be rather complex, with behavior such as drifting and coalescence observed [2,3]. Voids preferentially appear at grain boundaries or other imperfections on the lines and their locations depend on the microstructure and on the local stress state of the individual grains. Though progress has been recently made on the simulation of void dynamics [4-6], a full understanding of void formation is limited due to the lack of experimental information on microstructure and stress at the submicron level.

EXPERIMENTAL PROCEDURE

The beamline has a unity magnification toroidal mirror that produces a 50 by 200 micron focus just inside an x-ray hutch at the position of an x-y slit. The beam path in the hutch consists of source defining slits, a four bounce Ge or Si monochromator, followed by elliptically bent Kirkpatrick-Baez mirror pair which focuses the beam from the slits to sub micron dimensions (0.8 x 0.8 microns). An important feature of this arrangement is the ability to switch between white and monochromatic beams that are essential for characterizing crystal grains in the sub-micron range. The sample stage rests on a state of the art six-circle diffractometer equipped with encoders in the main rotation stages calibrated to a second of arc. The detector is a 4K x 4K CCD (Bruker) with a 9x9 cm view area mounted on a detector arm that can be positioned around the sample. The detector itself can also be positioned within 1 micron over 40 centimeters along the detector arm.

Diffraction experiments at the sub-micron level raise certain issues not encountered or ignored in conventional diffraction. One overriding concern is, except for translations within the focal plane under the small beam, the sample has to remain fixed (i.e. no rotation can be allowed). Since the sphere of confusion of modern, well-constructed diffractometers is more than 10 microns, any rotation would move the sub-micron region of interest out of the beam. We have facilities for precision translation of the specimen to allow us to scan different crystal regions or grains. The equipment described above enables us to keep the sample fixed and rotate the CCD to the desired position around the sample. A similar technique is used at the APS [7-9] and we apply it to both Al interconnects (this paper) and damascene Cu interconnects [10].

The samples investigated are pure Al two level test structures 10 or 30 μm long, 0.7 μm wide and 0.75 μm thick. Shunt layers of Ti cover the bottom and the top of the lines. The lines are passivated under 0.7 μm of oxide (PETEOS). The test lines are connected to the Al pads by Al vias of thickness 2250 Å.

The samples were scanned with white beam using a 0.5 μm interval between images to obtain a grain orientation map of the entire structure. Next both the white and monochromatic beam features of this instrument were used to determine the stress state in individual grains under electromigration conditions at 225°C.

RESULTS AND DISCUSSION

The texture scan on the 10x0.7x0.75 microns passivated Al interconnect reveals a peculiar microstructure with overlapping grains of dimensions ranging from less than a micron to 3-4 microns in length. All of the 66 Laue patterns display at least two Al grains. The texture map itself does not immediately reveal the actual grain size because of the inability of the analysis software to adequately discriminate between top and bottom overlapping grains (the depth sensitivity of the technique is too weak to resolve the stacking sequence in a 0.75 thick sample), or between two neighboring grains giving similar intensity ratios. On the other hand, it is able to effectively find and index several grains in composite Laue patterns, so that the actual position and size of each single grain can be inferred. The analysis of this particular line reveals 4 different grains over 3 microns in size and several other minor micron size grains (the total number of detected grains is 17). Strictly speaking, the line microstructure is not of bamboo type. Some of the grains clearly overlapped and this seems to be a feature of this particular line since the investigation of other lines does not show systematic composite Laue patterns. This test structure does demonstrate the ability of the technique to probe buried grains. It also shows that even if a line seems to be bamboo-like when surface features are examined, it may contain boundaries which are far from being perpendicular to the surface and could constitute fast diffusion paths during electromigration. The analysis of the orientation of these grains reveals a rather poor fiber texture with out-of-plane (111) reflections ranging from 2 to 10° from the surface normal.

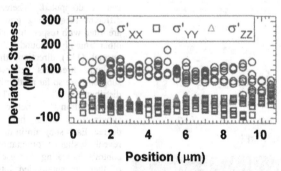

Fig 1.- Cumulative data on the deviatoric residual stress in the 10 µm long wire

The deviatoric residual stress tensor [11] has been computed for each grain on each image and the results are cumulated in Fig 1. The deviatoric stress tensor is calculated from the deviatoric strain tensor by using the anisotropic stiffness constants of the material recalculated in the sample reference frame (we follow the usual convention labeling the X, Y and Z directions along the line, across the line, and along the surface normal respectively).

The plot shows a decrease of the stress values at the two ends of the lines. This indicates that grains from the pads, expected to be only slightly strained, are also hit by the microbeam in these areas. In the wire, the grains are tensile along the direction of the wire (X) and compressed along the two other directions (Y and Z) by approximately the same amount [the terms tensile and compressive refers to the deviatoric component of the stress only and not the complete stress].

This result is in agreement with the observation of Hosoda et al. [12] who studied the stress values as a function of the aspect ratio (width/thickness) of the line. According to their work, an aspect ratio of one gives equal Y and Z stress components. The absolute differences between the X component and the Y,Z components is about 120-130 MPa. These values are comparable to those obtained by other authors [13] even if they cannot be directly compared because of differences in aspect ratio, fabrication process, and passivation layers.

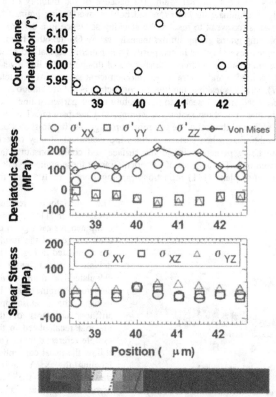

A study of texture and strain/stress variations within individual grains is rich in detail (Fig. 2). The largest grains are made of subgrains about a micron in size stacked mainly in the direction of the line, as revealed by the shifts of the Laue spots as we scan along the grain. Since the indexation of the Laue patterns provides the full orientation matrices of each subgrain, the tilt axis together with the corresponding tilt angles between the subgrains can be computed. Whereas different neighboring grains are tilted with respect to each other mainly around the normal (111) direction, the tilt axis directions between subgrains are so far randomly distributed. The misorientation between subgrains is about half a degree. Each subgrain in turn reveals slight orientation contrast indicating that each of them is subdivided into finer submicron subgrains with smaller misorientations, giving to the whole microstructure a mosaic aspect.

Fig. 2.- Texture and deviatoric stresses in a single grain of a pure Al interconnect. The bottom picture is a texture map of the grain. Lines have been added as guidelines in delimiting subgrains. The top plot shows orientation variations (tilts of the out-of-plane (111) with respect to the surface normal). The middle plots represent deviatoric stresses along X,Y,Z. and variations of the shear components of the stress.

The deviatoric stress profiles of individual grains show slight variation along them (Fig 2.). Larger bumps, which sometimes occurred at the boundaries between two subgrains, could be artifacts due to the fitting process of overlapping peaks coming from these two subgrains. The Von Mises stress, which is a measure of the distortion of the material, is often used as a reliability criterion in the industry. If the Von Mises stress locally exceeds the yield stress of the material, plastic deformation will occur. The observed tensile stress along the direction of the line and the observed subgrain microstructure are clear indications that plastic deformation occurred during or after the fabrication process of the line to partially relieve the resulting high stresses.

Stress in MPa	Initial 225°C	Final 225°C	150°C	75°C	25°C
σ'_{XX}	37.3	-2.8	39.9	81.2	118.4
σ'_{YY}	-10.3	28	22.2	-6.1	-20
σ'_{ZZ}	-26.9	-25.2	-61.9	-74.9	-97.8

Table 1.- Evolution of the deviatoric stress in a single grain before and after an electromigration test at 225 °C and during cooling to room temperature

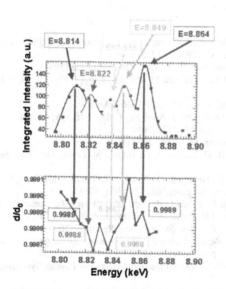

Fig. 3.- Energy scan of the (4 2 2) reflection at 225 °C

The deviatoric stress changes in the 30x0.7x0.75 µm line after electromigration failure and during cooling is shown in Table 1 for a grain near the anode end. The results indicate a clear change in the deviatoric stress during electromigration. The Z component didn't change but the deviatoric stress along the X direction becomes compressive. When cooling down the deviatoric component of the stress in the grain becomes tensile along the X direction and compressive along the Z direction. The total change in stress (hydrostatic + deviatoric) between 225 °C and 25 °C is 279.5 MPa, 110.3 MPa and 86 MPa along the X, Y and Z directions respectively.

The Laue patterns give the angular positions of the diffracted beams and their approximate energy can be deduced (assuming zero hydrostatic strain).

The small energy range around the peak can therefore be scanned with the monochromator and the peak position can be recorded with the CCD detector in function of the incoming beam energy. Fig. 3 shows an example of such an energy scan. It was taken with a broad monochromatic beam of 3 by 3 µm in order to illuminate the whole grain and see its microstructure. The energy peak is very irregular in shape with subpeaks corresponding to individual subgrains. Each subpeak can be fit to a gaussian function and the absolute value of the strain/stress is calculated for each of the subgrains. The observation indicates only a slight variation in strain. On the other hand, each subpeak has a certain width (5-15 eV wide) compared to an energy peak of a perfect Si sample (3-4 eV). These subpeak widths reflect the texture variation within the subgrains, confirming the mosaic nature of the microstructure in passivated Al interconnects.

CONCLUSION

We have demonstrated a new experimental X-ray microdiffraction technique, which is able to resolve the complete strain and stress tensors at submicron level in buried grains. The technique does not necessitate any sample preparation so that the true stress and texture state are measured. It was shown that the microstructure of thin Al lines under passivation is extremely complex and rich with both stress and texture variations within single grains. The largest grains have a mosaic substructure resulting from plastic deformation during fabrication.

REFERENCES

[1] I.A. Blech and H. Sello, Physics of failure in Electronics Series proceedings (USAF Rome Air Development center Reliability, Rome, NY, 1967), Vol. 5, p. 496
[2] E. Arzt, O. Kraft, W.D. Nix, and J.E. Sanchez, Jr., J. Appl. Phys., 76, 1563 (1994)
[3] T. Marieb, P. Flinn, J. C. Bravman, D. Gardner, and M. Madden, J. Appl. Physics, 78, 1026-1032 (1995)
[4] M. R. Gungor and D. Maroudas, J. Appl. Phys., 85, 2233-2246 (1999)
[5] Y.-J. Park, V. K. Andleigh, and C. V. Thompson, J. Appl. Phys., 85, 3546-3555 (1999)
[6] R.-J. Gleixner and W. D. Nix, J. Appl. Phys., 86, 1932-1944
[7] J.-S. Chung, and G. E. Ice, J. Appl. Physics, 86, 5249-5255 (1999)
[8] J.-S. Chung, N. Tamura, G. E. Ice, B. C. Larson, J. D. Budai, W. Lowe, Mat. Res. Soc. Symp. proc, 563, 169-174 (1999)
[9] N. Tamura, J.-S. Chung, G.E. Ice, B. C. Larson, J. D. Budai, J. Z. Tischler, M. Yoon, E. L. Williams, and W. P. Lowe, Mat. Res. Soc. Symp. proc, 563, 175-180 (1999)
[10] R. Spolenak, D.L. Barr, M.E. Gross, K. Evans-Lutherodt, W.L. Brown, N. Tamura, A.A. MacDowell, R.S. Celestre, H.A.Padmore, J.R. Patel, B.C. Valek, J.C. Bravman, P. Flinn, T. Marieb, R.R. Keller, B.W. Batterman, Mat. Res. Soc. Symp. Proc., submitted (2000)
[11] I. C. Noyan and J. B. Cohen, *Residual Stress: Measurement by Diffraction and Interpretation* (Springer, New York, 1987), p. 33
[12] T. Hosoda, H. Yagi, and H. Tsuchikawa, 1989 International reliability Physics Symposium Proceedings, IEEE, p. 202-206 (1989).
[13] P. R. Besser, *X-ray determination of thermal strains and stresses in thin aluminum films and lines*, PhD, Stanford University (1993).

Mat. Res. Soc. Symp. Proc. Vol. 612 © 2000 Materials Research Society

The Influence of Stress-Induced Voiding on the Electromigration Behavior of AlCu Interconnects

A.E. Zitzelsberger and A.H. Fischer
Infineon Technologies
Reliability Methodology
Otto Hahn Ring 6
D-81739 Munich, Germany

ABSTRACT

The influence of stress-induced voiding on the electromigration (EM) behavior has been investigated on narrow via-line structures. For this purpose the EM performance of metal lines containing stress-induced voids already before the electrical operation has been compared with samples without stress-induced voids. We found, that pre-existing stress voids in the metal lines do not affect the activation energy E_a and lead only to a small decrease of the EM median time to failure, but cause a relevant reduction of the current density exponent down to $n = 1$. As a consequence, a tremendous decrease of the electromigration limited life time is obtained.

INTRODUCTION

Stress-induced voiding in metal interconnects is an important aspect in reliability methodology. The phenomenon is driven by the relaxation of thermally induced and intrinsic stress. The thermally induced tensile stress is obtained in the interconnect due to the thermal mismatch between metal and surrounding $SiO2$-dielectric during cooldown from high deposition temperatures. As a result, stress-induced voids are observed in the metal line before electrical operation. In our particular case the occurance of these voids was found to be related to a high density plasma process (HDP) during the deposition of the intrametal dielectric. The stress voids could completely be suppressed by introducing an anneal process after the metal patterning combined with a reduced HDP-temperature. Other methods to avoid initial voids are described in [1,2].

In most cases the discussion of stressmigration related reliability risks is limited to the resistance increase which correlates to the growth of stress-induced voids and affects the functionality of the circuit. This primary risk can be assessed by providing a model which allows, similar to Black's equation for electromigration, the transformation of the failure distribution from highly accelerated to operation conditions [3,4].

However, the influence of stress-induced voids on the electromigration performance has to be considered as a secondary risk. As proposed by Lloyd [5,6], the current density exponent n in Black's equation will be close to $n = 1$ if the electromigration failure in the metal line is due to the growth of pre-existing voids. Systematic investigations supporting this behavior will be presented in this paper.

EXPERIMENTAL

Electromigration tests have been performed on metal layer M1 of a 3 level AlCu-metallization. Two different types of via-line teststructures were investigated, one with a short 30µm line and one with a long 500µm line (Fig.1). Both teststructures were stressed with an electron flow direction from metal layer M0 (W) into metal layer M1 (AlCu). In order to determine the activation energy E_a and current density exponent n the EM tests were performed at various temperatures (170 - 250°C) and current densities (5 – 15mA/µm2), respectively.

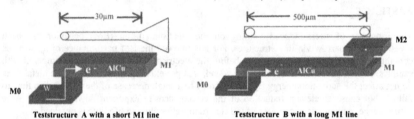

Teststructure A with a short M1 line   Teststructure B with a long M1 line

Figure 1:
Schematic drawing of the teststructures, the arrow indicates the direction of electron flow

RESULTS

Table 1 summarizes the results of the electromigration test. One sample was stored for 2000h at 225°C (high temperature storage HTS) before starting the EM test.

Teststructure			Failure distribution @ j_{str} and 250°C			activation energy and current density exponent	
Type	Stress-voids before EM ?	HTS before EM ?	j_{str} [mA/µm²]	t_{50} [h]	σ	E_a [eV]	n
A	No	no	12.0	12.4	0.33	0.88	1.7
A	Yes	no	12.0	11.8	0.34	0.89	1.4
B	No	no	10.0	52.0	0.22	0.88	1.7
B	Yes	no	10.0	45.0	0.20	0.88	1.0
B	Yes	yes	10.0	38.0	0.20	0.88	1.0

Teststructure A with short line:
No significant differences in t_{50} and σ between lines with and without pre-existing stress voids were obtained. Additionally no difference in activation energy was found (Ea = 0.89eV for lines with pre-exing voids and Ea = 0.88eV for lines without voids (Fig. 2)). A lower current density exponent n = 1.4 for structures with stress voids was observed (Fig. 3).

Teststructure B with long line:
Median time to failure (t_{50}) for lines with pre-existing voids is slightly smaller in comparison with lines without stress voids. Even lower values are obtained for samples which have been

stored at 225°C for 2000h before EM test. Sigma values are nearly identical for samples with and without stress voids. A current density exponent of n = 1 was found for lines with pre-existing voids and n = 1.7 for lines without voids (Fig.4)

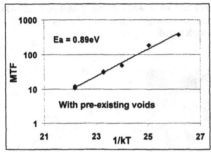
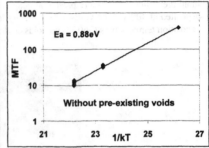

Figure 2:
Activation energy E_a for teststructure A.

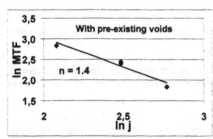
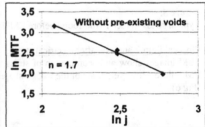

Figure 3:
Current density exponent n for teststructure A.

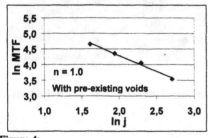
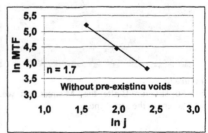

Figure 4:
Current density exponent n for teststructure B.

For both types of teststructures physical failure analysis was performed, before as well as after electrical stress.

Physical failure analysis before electrical stress:
In all metal lines voids were present already *before* EM stress when the encapsulating oxide was deposited at 400°C. The voids were typically found to be wegde shaped (Fig.5). For lower HDP deposition temperatures (360°C) no voids could be detected.

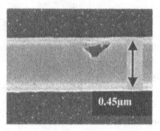

Figure 5:
Typical stress induced void found before EM Test. The HDP deposition temperature was 400°C)

Physical failure analysis after electrical stress:
SEM images show an accumulation of stress voids to one larger void which finally caused the EM failure. However, some of the wedge-shaped stress voids remain unchanged in the same line (Fig 6).

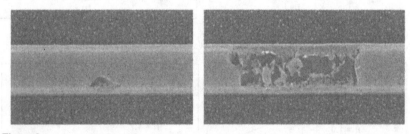

Figure 6:
SEM of structure B after the EM stress, left: wedge-shaped stress void, right: void which leads to EM failure in the same line.

DISCUSSION

For metal lines with pre-existing voids a smaller current density exponent n was found in comparison to lines without voids. In the case of long lines (teststructure B) a decrease from n = 1.7 to 1.0 was observed, and in the case of short lines (teststructure A) from n = 1.7 to 1.4. This can possibly be explain by the electromigration-induced growth of pre-existing stress voids. In contrast, in intact lines the voids must be nucleated first. The fact that the current density exponent in the long lines is even more reduced can probably be explained with the void spacing. Here, the average spacing between pre-existing stress voids was found to be in the range 20 - 50μm [7]. Hence the electromigration fail due to the growth of pre-existing voids becomes more likely in lines with lengths considerable larger than the average void spacing.

Despite of the significant changes of the current density exponents, the activation energy was observed to be unaffected by the presence of stress-induced voids. This implies the same diffusion mechanism for samples with and without pre-existing stress voids.

The observation of lower current density exponents but unchanged activation energies coincides with the failure scenario described by Lloyd [5,6]. He associates n = 1.0 with a failure mechanism that is based on void growth rather then void nucleation.

To get an idea how strong the calculated EM-life time is changed by smaller n values, an example is given below:

Black's equation is used to transform failure times gathered from the stress tests (j_{str}, T_{str}) to operation conditions (j_{target} and T_{op}). Considering the maximum allowed failure frequency. The life time at operation conditions can be expressed by

$$t_{LT} = t_{50} \left(\frac{j_{str}}{j_{target}} \right)^n \exp(\Phi^{-1}(P) \cdot \sigma) \cdot \exp\left(\frac{E_a}{k} \left(\frac{1}{T_{op}} - \frac{1}{T_{str}} \right) \right)$$

where t_{LT} is the life time at operation condition, T_{op} and T_{str} are the temperatures at operation and EM-stress, t_{op} is the operation time, t_{50} is the median time to failure at j_{str} and T_{str}, j_{str} and j_{target} are the current densities at stress and operation condition and $\Phi^{-1}(P)$ is the quantile of the standard normal distribution of the allowed failure frequency.

With an exemplary data set (t_{50str} = 10h, j_{str} = 12mA/μm^2, T_{str} = 250°C, σ = 0.3, E_a = 0.88eV), the life time is reduced from **30a** to **5a** using n = 1.7 and n = 1.0 respectively.

CONCLUSION

Besides the resistance increase due to stressvoiding, a lower current density exponent for lines with stress-induced voids was found to be an additional reliability risk. In the presented example, a very strong decrease of the electromigration limited life time from 30a to 5a is obtained as consequence of pre-existing voids.

REFERENCES

[1] Lee, S.G. et al., to be published in proceedings of the IITC meeting (1999).
[2] Besser, P. et al.,Proc. AMC, 699-704 (1998).
[3] Sullivan, T.D., proposed as JEDEC standard, JC-14.2-98-189 (1998).
[4] Fischer A.H. et al., „Stressmigration behavior of multilevel ULSI AlCu Metallization", to be published in this proc.
[5] Lloyd, J.R., „Electromigration in integrated circuit conductors", J. Phys. D: Appl. Phys.32, R109-118 (1999).
[6] Lloyd, J.R., „Electromigration and mechanical stress", Microelectronic Engineering 49, 51-64 (1999).
[7] Zitzelsberger A.E. and Lehr M., IRW final report, 10-13 (1999).

Mat. Res. Soc. Symp. Proc. Vol. 612 © 2000 Materials Research Society

Electromigration-Induced Stress Interaction between Via and Polygranular Cluster

Young-Joon Park, In-Suk Choi* and Young-Chang Joo*
Thin Film Technology Research Center, Korea Institute of Science and Technology,
Seoul 130-650, Korea
*Seoul National University, School of Materials Science and Engineering,
Seoul 151-742, Korea

ABSTRACT

We have investigated the stress interaction between via and polygranular cluster in the pure Al line using 1-dimensional computer simulation. The conventional belief was that the fastest stress evolution at the via occurs when the polygranular cluster is just below (or above) the via. However, the electromigration induced stress at the via would be faster when a cluster is apart from via because the stress interaction between via and clusters may assist electromigration. We simulated the time that the via reaches a certain stress value as a function of the distance of the cluster. It gives a specific distance where the time was minimum (*i.e* the fastest stress evolution). We named the position as the *Fa*stest *S*tress *E*nhancing *P*olygranular cluster *P*osition (FaSEPP). As a function of the current density, the FaSEPP decreases.

INTRODUCTION

Electromigration is atomic diffusion driven by a momentum transfer from conducting electrons. With every new generation of integrated circuits, as interconnect linewidths have been reduced, electromigration from high current densities leads to concern about the interconnect reliability [1]. Electromigration-induced failures occur at the site of flux divergences. For example, void is nucleated where incoming flux is smaller than outgoing flux and hillock is formed in the other way. Accumulation or depletion due to these flux divergences generates compressive or tensile stress respectively [2]. When the maximum stress reaches a critical value, electromigration-induced damages occur. In order to make highly reliable integrated circuit, the detail understanding of the mechanism is necessary. However considering the technical difficulties in measuring electromigration-induced stress directly, computer simulations may be an alternative to understand electromigration effect inside interconnects. In this respect, several attempts have been made for computer simulations of electromigration [3-6].

There are two important sites of flux divergence, the line-ends and the cluster-ends. The line-ends are connected with the via made of tungsten or blocked by a diffusion barrier between the line-end and the via. Therefore the flux from the via to the cathode end of the line or from the

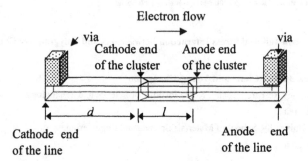

Figure 1 *The schematic structure of the pure Al line used in this simulation. The total length of line was 100μm and the cluster length, l, was 20μm. The distance of cluster from the via, d, varied from 0 to 70 um.*

anode end of the line to the via, is zero because of materials discontinuity. Local variation of the grain structures causes another important flux divergence. A grain boundary is a faster diffusion path than bulk or interfaces. If a grain boundary is discontinuous along the line like near-bamboo lines, flux divergence occurs where bamboo grain meets polygranular cluster [7].

Usually, the effect of via and polygranular clusters has been investigated separately, and there is little study on the interaction of these two important flux divergence sites. It has been commonly believed that the via has the worst lifetime when a cluster is just under (or above) the via because electromigration diffusion is faster in polygranular clusters than bamboo grains. However, stress field effect should be taken account of. Generally, the negative stress gradient (with respect to the electron flow) such as in the cluster is against electromigration. On the other hand, the positive stress gradients assist electromigration. In this study we investigated the stress interaction between via and cluster with various distances by computer simulation.

SIMULATION OUTLINE

Fig. 1 shows the schematic diagram of the line structure used in this simulation. Using one-dimensional Finite Element Method, an interconnect was broken into segments of length Δx and the atomic flux in each segment was tracked after time increments Δt over the length of the line. The flux of atoms was calculated as detailed by Park *et al.* [5]. The initial simulation condition was as follows: the total length of line was 100μm and the cluster length, *l*, was 20μm. The distance of cluster from the via, *d*, varied from 0 to 70 um. The current density was $1 \times 10^6 \text{A/cm}^2$ and the temperature was 473K. The atomic diffusivity in a polygranular cluster was assumed 200 times larger than that in bamboo grains [8]. Other details of the simulation conditions are described elsewhere [5].

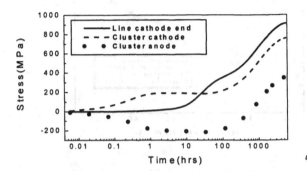

Figure 2 *The stress evolution with simulation time at the cathode end of the line, the cathode end of the cluster, and the anode end of the cluster. The current density is 1MA/cm² and the temperature is 473K. The total length of line is 100μm and the cluster length, l, is 20μm*

RESULTS AND DISCUSSION

First, we take one specific case in which the distance between the via and the cluster is 10μm. Fig. 2 illustrates the stress evolution with the simulation time at the cathode end of the line, the cathode end of the cluster, and the anode end of the cluster. At the beginning of the test, stress piles up faster at the cathode end of the cluster than at the cathode end of the line. The flux divergence at the cathode end of the line is $\Delta J_{via}=J_{bulk}-0$ and that at the cathode end of the cluster is $\Delta J_{cluster}=J_{gb}-J_{bulk}$. In most cases, the grain boundary diffusivity is over an order of magnitude larger than that of bulk or interface diffusion, so that the flux divergence is bigger at the cathode end of the cluster than the cathode end of the line [5-6]. But as the test goes on, the stress in the cluster maintains local-steady-state and the tensile stress at the cathode end of the cluster reaches the local plateau because back stress force is balanced with electromigration wind force in the cluster. At the local steady state, the stress at the cathode end of the line starts to increase rapidly and then exceeds the stress at the cathode end of the cluster (Fig. 2). These features can be explained that the stress from the cathode of the cluster travels, interacts with the tensile stress from the cathode of the line and forms a positive gradient which assists electromigration.

If there is a stress interaction between via and cluster, the stress at the cathode end of the line is affected by the distance of the cluster. A clear overview for the effect of the distance can be gained from Fig. 3. The time that the stress at the cathode end of the line reaches a certain value, for example, 400MPa, is plotted with the distance from the via. When the cluster is located around 14μm apart from the via, the stress at the cathode end of the line achieves 400MPa earlier

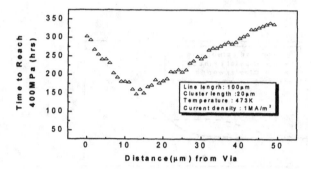

Figure 3 *The time that the stress at the cathode end of the line reaches 400MPa, is plotted with the distance from the via. The lowest is around 14μm, where the stress at the cathode end of the line achieves 400MPa earlier than any other. We named this cluster position as the Fastest Stress Enhancing Polygranular cluster Position (FaSEPP).*

than any other. From now on, we call this cluster position as the Fastest Stress Enhancing Polygranular cluster Position (FaSEPP). The existence of the FaSEPP is against the conventional belief that the stress is the highest when cluster located just below via.

The fundamental idea to explain the FaSEPP can be obtained using Fig. 4. The local steady state is maintained in the cluster, because the negative stress gradient in the cluster works against the electromigration. Simultaneously at the line end, the stresses from the via and the cluster travel toward each other and form the bell- shaped stress distribution, as seen in Fig. 4 (a) and (b).

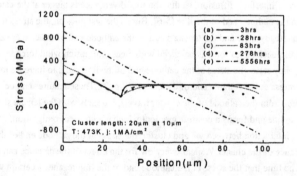

Figure 4 *The stress profiles along a 100-μm–long line with a 20-μm-long cluster located at 10μm apart from the via at five different simulation times. The current density is 1MA/cm² and temperature is 473K.*

Since its net stress gradient is positive, the bell-shape moves upward at a fast rate until the local steady state develops to the quasi-local-steady-state which is the same as the extrapolation of the local-steady-state (Fig. 4-(c)). After the quasi-local-steady-state, the stress evolves much slower than before as indicated from the time differences between Fig. 4 (b), (c) and (d). This feature is also observed in other conditions.

   Now we apply this fundamental idea to account for the FaSEPP. If a cluster is farther from the via compared to the FaSEPP, it takes longer for the tensile stresses from the cluster and the via to meet each other. In addition, since the stress gradient at the local steady state has little deviation with the cluster location, it can be assumed that the stress gradient of local-steady-state is constant for the same cluster size and the tensile stress of the cluster end is same as all. Therefore the net stress gradient outside of the cluster is less than that at the FaSEPP. In case that a cluster is closer than the FaSEPP, the stress at the cathode end of the line initially increases very fast because of shorter time to interact and high net stress gradient. However after being the quasi-local-steady- state, the stress growth rate at the cathode end of the line becomes much slower than before, so that the total time to reach 400MPa is longer than that at the FaSEPP.

   Based on the idea, we also explain the change of the FaSEPP with variation of current densities. As a current density decreases, the FaSEPP increases, as shown in the Fig. 5. As we assumed, the stress gradient at the local steady state is constant regardless of the cluster location. If a current density decreases, the tensile stress at the cluster end at the local steady state becomes lower and the slope of the local-steady-state is less steep. Therefore to reach higher

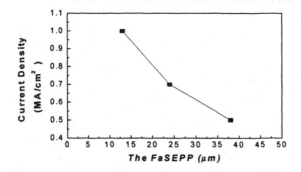

Figure 5 *The change of the FaSEPP (Fastest Stress Enhancing Polygranular cluster Position) with current density when the stress at the line end reached 400MPa. The total length of line was 100μm and the cluster length, l, was 20μm. The temperature is 473K. As a current density decreases, the FaSEPP increases.*

extrapolated value, the cluster has to move far from the via.

SUMMARY

We investigated the stress interaction between via and cluster with various distances by computer simulation. When the cluster is located apart at specific position, the stress at the cathode end of the line achieves a certain value fastest. We named this cluster position as the Fastest Stress Enhancing Polygranular cluster Position (FaSEPP). The existence of the FaSEPP is against the conventional belief that the stress at the via reach fastest when cluster is located just below via. The FaSEPP is associated with the value of the local steady state and obtained approximately by the extrapolation of the local steady state. As a function of the current density, the FaSEPP decreases. With more theoretical study of this idea, we will try to clearly explain the stress interaction in the line between cluster and via.

REFERENCE

1.   Y. C. Joo and C. V. Thompson, J, Appl. Phys **76**, 7339 (1994).
2.   M.A. Korhonen, P. Borgesen, K. N. Tu, and C. -Y. Li, J. Appl. Phys. **73**, 3790 (1993).
3.   D Brown, J. E. Sanchez, M. A. Korhonen, and C. Y. Li, Appl. Phys. Lett. **67**, 439 (1995).
4.   B. D Knownlton, J. J. Clement, and C. V. Thompson, J. Appl. Phys. **81**, 6073 (1997).
5.   Y. J. Park and C. V. Thompson, J. Appl. Phys. **82**, 4277 (1997).
6.   Y. J. Park, V. K. Andleigh and C. V. Thompson, J. Appl. Phys. **85**, 3546 (1999).
7.   E. Kinsbron, Appl. Phys. Lett. **36**, 968 (1980).
8.   J. R. Lloyd and J. J. Clement, Appl. Phys. Lett. **69**, 2486 (1996).

Poster Session
Interconnects: Texture

Mat. Res. Soc. Symp. Proc. Vol. 612 © 2000 Materials Research Society

TRENCH AND VIA FILLING WITH ELECTROPLATED COPPER: EFFECT OF CURRENT DENSITY AND PULSE WAVEFORM

[1]C. H. Seah, [2]S. Mridha, [2]Y. K. Siew, [2]G. Sarkar and [3]L. H. Chan
[1]Thin Film Dept., Chartered Silicon Partners Pte Ltd.,
60 Woodlands Industrial Park D, Street 2, Singapore 738406, Singapore
Phone/Fax: (65)-3946348/3946516, Email: seahch@charteredsemi.com
[2]School of Applied Science, Nanyang Technological University,
Nanyang Avenue, Singapore 639798, Singapore
[3]Research & Development Dept., Chartered Semiconductor Manufacturing Ltd.,
60 Woodlands Industrial Park D, Street 2, Singapore 738406, Singapore

ABSTRACT

A study was carried out to investigate the effect of current density and pulse waveform on the filling of line trenches and contact vias with aspect ratio of 2:1 for sub-0.25 μm device manufacturing using normal pulse plating of copper. The growth pattern of the copper films deposited using 0.05 and 0.10 A/cm^2 current density gave no significant difference. Small grains were seen to have nucleated uniformly across the line trenches and via holes after 1 second of electroplating. With increasing the deposition time to 2 seconds, a slight buildup of the film thickness was observed in both trenches and vias without significant increase in the size of the copper grains. Grain growth involving the coalescence of small grains occurred after 5 seconds of plating and a further buildup in thickness and fill up of the trenches and vias occurred after 10 seconds of deposition.

When the patterned wafers were plated with a pulse waveform of 3 ms on and 0.5 ms off, the filling of trenches could not be complete after 30 seconds of electroplating. A complete filling of the trenches was achieved within 30 seconds of deposition using a pulse waveform of 6 to 8 ms on and 1 to 2 ms off. When the on-period was increased above this range to 9.9 ms, voids were observed at the centre of the via holes.

INTRODUCTION

Electroplating has become the technology of choice for the first generation of copper dual damascene interconnects used for fabricating advanced CMOS devices [1-2] because of its relatively low cost, high deposition rates and ease of filling high aspect ratio features. Successful electrochemical deposition of copper into trenches lined with copper seed and diffusion barrier, produced film resistivity of <2.0 μΩ-cm, has been demonstrated and show good via-filling capabilities [3-7].

Many authors have reported the success of filling via holes and line trenches with copper using the pulse electroplating technique. Pulse plating on Cu/Ta/Si is reported to fill the vias and trenches successfully in sub-0.5 μm devices with the formation of the fine and dense copper grains [4,8]. The pulse plating allows the electrolyte to eliminate concentration gradient in the via/trench and that helps uniform copper plating within the trench without any voids [9]. Lopatin et al. [10] reported that the cleft depth of the via decreased when applied current density was reduced from 20 to 10 mA/cm^2 using reversed pulse plating. However, no much information is available about the effect of the pulse waveform and current density on via/trench filling in silicon devices. This paper describes the effect of current density and pulse waveform on the filling capability of line trenches and contact vias using normal pulse copper plating in sub-0.25 μm devices.

PROCEDURES

The substrate used was (100)-oriented p-type patterned Si wafers, on which trenches and vias at 0.25 μm width of aspect ratio 2:1 were etched into the SiO$_2$ dielectric pattern. Ionized metal plasma (IMP) technology was used to sputter a layer of 350 Å TaN which acted as the diffusion barrier and adhesion promoter, followed by 1200 Å Cu seed layer.

The electrochemical deposition was done at 30°C with current densities of 0.05 and 0.10 A/cm^2 and deposition times up to 10 seconds to understand the effect of growth pattern of the copper films. In the case of the pulse waveform study, a two step electroplating process was employed in order to achieve a high deposition rate. The first step involved a short deposition time of 2 seconds using a low current density of 0.05 A/cm^2. The purpose of this step was to nucleate small copper crystallites uniformly across the surfaces and to prevent pinch-off at the neck of the trench [11]. The second step involved the use of a deposition time of 30 seconds with a higher current density of 0.15 A/cm^2 in order to have a faster copper deposition. Previous study has shown that a current density of 0.15 A/cm^2 has been found suitable to produce copper film with 0.2 μm grains which has a more uniform grain distribution i.e. less degree of bimodal distribution [12].

The electrolyte used was standard copper plating electrolyte composed of CuSO$_4$.5H$_2$O and H$_2$SO$_4$ solution without additives. The anode used was a piece of electronic grade copper sheet while the cathode was the substrate to be plated. Field emission scanning electron microscopy (FESEM) was used to characterize the growth pattern of these copper films.

RESULTS AND DISCUSSION

Effect of Current Density

The sequence of via and trench filling using different deposition times at current densities of 0.05 A/cm^2 at 30°C are shown in the SEM micrographs of Fig. 1. The micrographs show that small copper grains nucleated uniformly across the sub-0.25 μm trench/via after 1 second of electroplating, giving an average grain size of 1000 Å (Fig. 1a). When deposited for two seconds, a slight buildup of the film thickness was observed in both vias and trenches without significant increase in the size of the copper grains (Fig. 1b). Grain growth occurred after plating for 5 seconds involving the coalescence of small grains, producing larger grains of 2500 Å (Fig. 1c). After 10 seconds of deposition, the grains grew in size to as large as 4000 Å and a further buildup in thickness and fill up of the trench and via occurred (Fig. 1d).

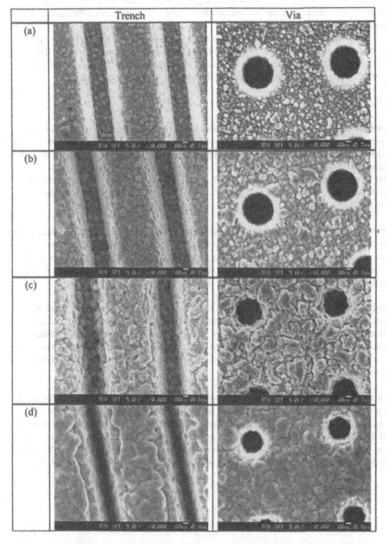

Fig. 1: SEM micrographs of electroplated copper filling into line trenches and via holes using a pulse waveform of 6 ms on + 2 ms off, current density of 0.05 A/cm^2 and deposition times of (a) 1 sec, (b) 2 sec, (c) 5 sec and (d) 10 sec.

A similar observation in term of growth pattern and grain size was also made for the copper specimens deposited with 0.10 A/cm² current density. There is no significant difference in the grain feature observed from the specimens produced using current densities of 0.05 and 0.10 A/cm² from a deposition time of 2 seconds onwards. A small degree of bimodal distribution could be seen from the morphology of both films after plating for 10 seconds (Fig. 1d). The results of this work suggest that there is no significant effect of the current density on the grain size and growth pattern of the copper film during the trench and via filling process.

Effect of Pulse Waveforms

Fig. 2 shows the SEM micrographs of the trench and via filled with electroplated copper using different pulse waveforms of (a) 3 ms on + 0.5 ms off, (b) 6 ms on + 2 ms off, (c) 8 ms on + 1 ms off and (d) 9.9 ms on + 2.5 ms off, after 30 seconds of copper deposition. A complete filling of the trenches was achieved within 30 seconds of electroplating using an on-period of 6 ms or greater. However when plated with 3 ms on + 0.5 ms off, the via filling was not completed as evident from the micrograph in Fig. 2a. These results suggest that the filling rate of the line trenches and contact vias depend strongly on the on-period of the total pulse cycle.

Cross-sectional SEM of the trench and via was performed on the completely filled specimens and the micrographs are presented in Fig. 2 which show that complete filling of vias and trenches could be achieved using the two step electroplating process, although a seam line is seen in the line trenches, see Figs. 2b and 2c. Fig. 2d shows that voids are present within the centre of the via while the copper film built up in thickness on top of the trench.

An important observation may be noted from the above micrograph that for the line trenches, the thickness of copper deposits at both sides and at the bottom of the trenches are similar. This suggests that the reduction of Cu^{2+} ion to neutral Cu, followed by adsorption onto the surface, was a surface reaction-controlled process [8]. The presence of the seam line may be unavoidable since the growth of copper film from both sides of the trench would meet at the trench centre, forming a seam line.

It is seen from Fig. 2d that the thickness of the copper film on top of the via built up to about 1.5 μm when plated with an on-period of 9.9 ms and an off-period of 2.5 ms, and at the same time, a void was seen within the trench. This suggests that too high an on-period during the pulse cycle may result in too fast a deposition rate and hence greater growth of copper at the top corner of the via, leading to pinch-off of the via. This caused premature closure of the via which prevented further copper ions from diffusing into the trench and therefore resulted in the formation of voids.

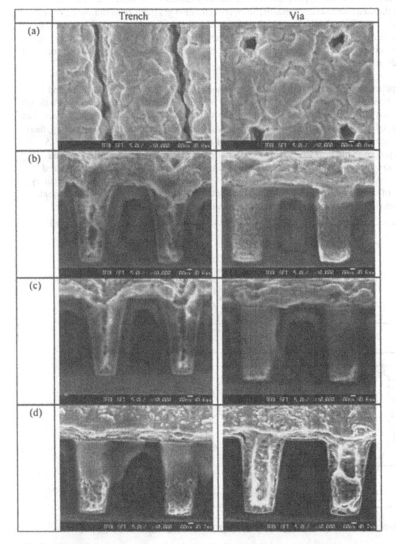

Fig. 2: SEM micrographs of line trenches and via holes filled with electroplated copper using different pulse waveforms of (a) 3 ms on + 0.5 ms off, (b) 6 ms on + 2 ms off, (c) 8 ms on + 1 ms off and (d) 9.9 ms on + 2.5 ms off.

It has been shown [9] that application of the deposition current in the form of a pulse helps diminish the concentration overpotential since concentration gradients relax during the off-period of the pulse. Additionally, if the on-period of the pulse is less than the diffusion time constant (l^2/D, where l is a characteristic length such as trench depth and D is the diffusivity of the cupric ion), concentration gradients do not develop to the extent they were for a current density of equal magnitude applied as D. C.

Normal pulse plating of copper involved a continuous repeated sequence of nucleation and growth processes, in which grain growth occurred with increasing deposition time but at a slower rate than that during D. C. plating. Because of the slower rate, it produced fine grains of size less than or equal to 0.1 μm and hence filled the vias and trenches. In the case of D. C. plating, there was only one nucleation step and growth of copper grains occurred with deposition time. A model is proposed in Fig. 3 to explain how normal pulse plating filled the sub-0.25 μm line trenches and via holes with finer copper grains. Fig. 3a shows the fine grains of copper seed deposited by IMP process. During electroplating, the nucleation of copper grains will occur by diffusion of copper ions from the electrolyte preferably at the high energy sites i.e. at the grain boundaries (Fig. 3b). The growth of copper grains thus continues at these sites with time.

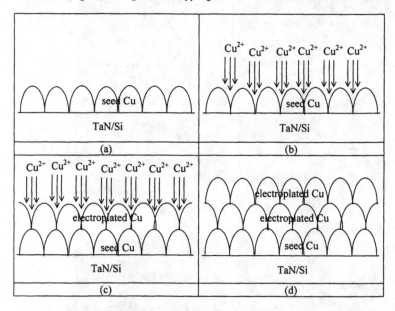

Fig. 3: Proposed model for smaller copper grain growth during normal pulse plating.

Because the current was alternatively turned on and off in the total pulse cycle, new nucleations will occur at the high energy sites during the on-period of every pulse cycle and thus new grains will grow continually along the new grain boundaries of the electroplated copper, see Fig. 3c. At the same time, cathodic reduction reactions of the copper ions depleted the concentration of the dissolved species being reduced adjacent to the cathode surface. Therefore, during the off-period of the pulse cycle, any concentration gradient adjacent to the surface would be eliminated and the copper concentration homogeneity at the surface would be restored back to the normal concentration of the bulk solution. During the electroplating process, any cathodic reaction will proceed to the formation of hydrogen gas. The cycle repeated and as a result, the grain size of the copper film remained smaller and thus allowed the sub-0.25 μm trenches and vias to be filled up (Fig. 3d).

CONCLUSION

There is no significant difference in the growth pattern of the copper films deposited using 0.05 and 0.10 A/cm^2 current density. Small grains of 1000 Å were seen to have nucleated uniformly across the line trenches and via holes after 1 second of electroplating. With increasing the deposition time to 2 seconds, a slight buildup of the film thickness was observed in both trenches and vias without significant increase in the size of the copper grains. Grain growth involving the coalescence of small grains into larger grains of 2500 Å occurred after 5 seconds of plating and a further buildup in thickness and fill up of the trenches and vias occurred after 10 seconds of deposition.

When the patterned wafers were plated with a pulse waveform of 3 ms on and 0.5 ms off, the filling of trenches could not be complete after 30 seconds of electroplating. A complete filling of the trenches was achieved within 30 seconds of electroplating using a pulse waveform of 6 to 8 ms on and 1 to 2 ms off. When the on-period was increased above this range to 9.9 ms, voids were observed at the centre of the via holes.

REFERENCES

1.   S. Venkatesan et al., Proc. IEDM Conf., (1997) p. 769
2.   D. Edelstein et al., Proc. IEDM Conf., (1997) p. 773
3.   R.J. Contolini, L. Tarte, R.T. Graff, L.B. Evans, J.N. Cox, M.R. Puich, J.D. Gee, X.-C. Mu and C. Chiang, VMIC 12 (1995) p. 322
4.   V. M. Dubin, C. H. Ting and R. Cheung, VMIC 14 (1997) p. 69
5.   J. Heidenreich et al., IITC 1 (1998) p. 151
6.   V. Blaschke et al., IITC 1 (1998) p. 154
7.   T. Ritzdorf, L. Graham, S. Jin, C. Mu and D. Fraser, IITC 1 (1998) p. 166
8.   Principles of Electronics Packaging, by D. P. Seraphim, R. Lasky and C. Y. Li, McGraw-Hill (1989)
9.   P.C. Andricacos, C. Uzoh, J.O. Dukovic, J. Horkans and H. Deligianni, Adv. Metallization Conf. in 1998, (MRS 1999) p. 29
10.   S. Lopatin et al., Adv. Metallization Conf. in 1998, (MRS 1999) p. 35
11.   J. Reid et al., IITC 2 (1999) p. 284
12.   C.H. Seah, S. Mridha and L.H. Chan, IITC 1 (1998) p. 157

Mat. Res. Soc. Symp. Proc. Vol. 612 © 2000 Materials Research Society

Deposition of Smooth Thin Cu Films in Deep Submicron Trench by Plasma CVD Reactor with H Atom Source

Masaharu Shiratani, Hong Jie Jin, Yasuhiro Nakatake, Kazunori Koga, Toshio Kinoshita and Yukio Watanabe
Department of Electronics, Graduate School of Information Science and Electrical Engineering, Kyushu University, Hakozaki, Fukuoka 812-8581, Japan

ABSTRACT

Effects of H irradiation on purifying Cu films and improving their surface roughness as well as size and orientation of Cu grains in the films have been examined using a newly developed plasma CVD reactor equipped with an H atom source, in which $Cu(hfac)_2$ is supplied as the source material. The H irradiation is effective in purifying the Cu films, increasing the grain size, and reducing the surface roughness, while it has no effect on the grain orientation. The decrease in dissociation degree of material gas leads to reduction of the surface reaction probability of Cu-containing radicals, which is important to realize conformal deposition in fine trenches. Using the control of dissociation degree of material gas independent of H irradiation, we have demonstrated conformal deposition of smooth Cu films in the trench using the developed plasma CVD reactor.

INTRODUCTION

Because of its lower resistivity and better electromigration properties, copper metallurgy is a desirable alternative to aluminum for ULSI interconnects. While the electroplating method is currently employed for formation of Cu interconnects in industry, it needs thin Cu seed layers deposited by other means. However, at present, there is no method to obtain such thin layers in small via and contact holes, of a width below 0.13 μm and a depth above 0.8 μm [1, 2]. For the thermal and plasma CVD, conformal deposition of such holes can be achieved when the surface reaction probability β of Cu-precursors is less than 0.02 [3]. For the thermal CVD, its processes are essentially characterized by heterogeneous decomposition of material gas on an activated substrate-surface and hence the β value is controlled by varying substrate temperature (T_s) and/or flow rate of material gas. However, such control also influences the deposition rate and film properties such as purity and resistivity. The plasma CVD has a significant advantage over the thermal CVD in β control, because β can be changed by varying the dissociation degree of material gas under a constant T_s [4, 5]. Previously, we showed the useful effects of H atoms on removing impurities in Cu thin films deposited using the plasma CVD [6, 7]. Based on this result, we have newly developed a plasma CVD reactor equipped with an H atom source (hereafter referred ot as HAPCVD reactor, namely, H assisted plasma CVD reactor).

In this paper, we will show that this HAPCVD reactor is useful for improving properties of Cu films, controlling β of Cu-containing radicals and depositing smooth thin Cu films in a small trench.

EXPERIMENTAL

The source material for Cu deposition used in this study was Cu(hfac)$_2$, bis(hexafluoro-acetylacetonato) copper (II), dissolved in ethanol (C$_2$H$_5$OH) at a concentration of 0.1 mol/l. The material was vaporized at 155°C and then transported to reactors with H$_2$ carrier gas.

Experiments were performed using two capacitively coupled parallel plate reactors, A and B, both of which had an additional H atom source. These reactors have been described in detail elsewhere [4-7].

The reactor A was employed to study the effects of H irradiation on Cu film properties such as purity, morphology, grain size, and crystal orientaion. The H atom source was composed of coaxial stainless steel tubes. The outer tube, 130 mm in inner diameter and 38 mm in length, was used as a grounded electrode. The inner one, 8 mm in outer diameter and 22 mm in length, was used as a powered electrode. The source was operated at a frequency of 400 kHz and a power P$_{as}$ below 40 W. For the main discharge, powered mesh (14 mesh/inch) and grounded plane electrodes of 50 mm in diameter were placed at a distance of 30-60 mm. The main discharge was driven at a frequency of 13.56 MHz and a power P$_m$ of 15 W. These two sets of electrodes were installed in a reactor of 160 mm in diameter and 240 mm in height. Relative concentrations of impurities in films were measured by *in situ* FT-IR spectroscopy.

The reactor B was employed to show that the degree of dissociation of Cu(hfac)$_2$ can be controlled independently of the concentration of H atoms and conformal deposition in a small trench can be realized. The H atom source was composed of coaxial stainless steel tubes. The outer tube, 48 mm in inner diameter and 145 mm in length, was used as a grounded electrode. The inner one, 8 mm in outer diameter and 110 mm in length, was used as a powered electrode. The source was operated at a frequency of 28 MHz and a power P$_{as}$ below 50 W. For the main discharge, a powered mesh (50 mesh/inch) electrode of 55 mm in diameter and a grounded plane electrode of 85 mm in diameter were placed at a distance of 42 mm. The main discharge was driven at a frequency of 13.56 MHz and a power P$_m \leq 45$ W. These two sets of electrodes were installed in a reactor of 250 mm in diameter and 315 mm in height.

RESULTS AND DISCUSSION

Firstly, in order to study effects of the H irradiation on purifying Cu films containing impurities, they were deposited at room substrate temperature using the reactor A. Time evolution of impurity concentration in them during irradiation of H atoms supplied from the H atom source without using the main discharge was studied by *in situ* FT-IR spectroscopy. Since absorption intensities at wavenumbers related to the impurities have quite similar time evolution, we use the absorbance at 1200 cm^{-1}, which is the strongest among them, as a measure of impurity concentration in the Cu films [7]. Figure 1 shows the results at T$_s$ = 120°C as a parameter of the distance between the H atom source and Cu films. The absorbance monotonously decreases with time t after the discharge initiation and becomes almost zero at t=8, 13, and 29 min for the distance of 3, 4, and 6 cm, respectively, indicating that the H irradiation is effective in reducing impurities in the Cu films. Further results concerning this impurity removal effect have been reported elsewhere [6, 7].

Figure 1. Time evolution of absorbance at 1200 cm⁻¹ due to C-CF₃ stretch in Cu films during H irradiation as parameter of distance between H atom source and Cu films obtained with reactor A. Cu films were deposited under following conditions: material gases H_2 (50%) and C_2H_5OH [Cu(hfac)₂], total flow rate 72 sccm, pressure 133 Pa, P_m = 15 W, and T_s = RT. H_2 flow rates at vaporizer and H atom source were 18 and 18 sccm, respectively. Film thickness was 15 nm. H irradiation was carried out under conditions of H_2(100%), flow rate 36 sccm, pressure 133 Pa, P_{as} = 25 W, and T_s = 120 ℃.

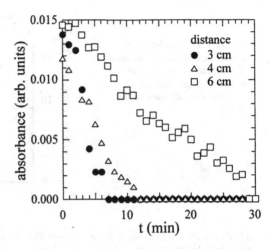

Next, to study effects of the H irradiation on size and orientaion of Cu grains, high purity Cu films (≈100% Cu) were deposited onto Si(100) substrates using the reactor A and then they were irradiated by H atoms supplied from the H atom source without using the main discharge. Grain size of the films after H irradiation was studied by transmission electron spectroscopy as a parameter of the irradiation time. Orientation of Cu grains observed by X-ray diffraction was random, suggesting that the H irradiation had no effect on the orientaion. The results concerning

Figure 2. Time evolution of Cu grain size during H irradiation as parameter of film thickness obtained with reactor A. Cu films were deposited under following conditions: material gases H_2 (83%) and C_2H_5OH [Cu(hfac)₂], total flow rate 216 sccm, pressure 133 Pa, P_m = 15 W, and T_s = 170 ℃. H irradiation was carried out under conditions of H_2(100%), flow rate 36 sccm, pressure 133 Pa, P_{as} = 40 W, and T_s = 250 ℃. H_2 flow rates at vaporizer and H atom source were 90 and 90 sccm, respectively.

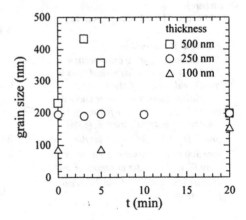

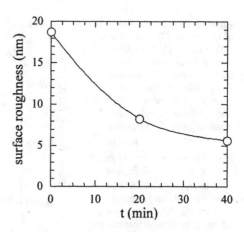

Figure 3. Time evolution of surface roughness of films during H irradiation obtained with reactor A. Experimental conditions are the same as those in Fig. 2.

the grain size are shown in Fig. 2. For the films of 500 nm in thickness, the grain size increases rapidly from 230 nm at t=0 min to 435 nm at t=5 min, being close to the film thickness, then decreases to 200 nm at t=20 min. Therefore, there is an optimum H irradiation time to obtain the largest size, while the mechanism for grain-size variation due to H irradiation is unclear at present. It should be noted that both the increasing rate and the largest value of grain size obtained by the H irradiation at $T_s = 250°C$ are considerably high compared to those due to one hour annealing in H_2 gas at $T_s = 400°C$ [8]. For the as-deposited films of 250 nm in thickness, the grain size of which is almost the same as the film thickness, the irradiation do not change the size. For the films of 100 nm in thickness, the grain size increases gradually from 80 nm at t=0 min to 150 nm at t=20 min. The largest size for 100 nm thickness can surpass the films thickness, probably because some voids exist in the as deposited films as observed in the SEM photograph (not shown here).

Effects of H irradiation on surface

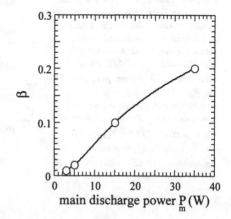

Figure 4. P_m dependence of β deduced comparing cross-section SEM photographs of deposited films in trench with results of Monte Carlo simulation. Experiments were carried out with reactor B under following conditions: material gases H_2 (83%) and $C_2H_5OH[Cu(hfac)_2]$, total flow rate 300 sccm, pressure 73 Pa, and P_{as} = 60 W. H_2 flow rates at vaporizer and H atom source were 10 and 240 sccm, respectively.

Figure 5. Cross section SEM photograph of Cu film deposited in trecnh 0.75 μm wide and 3 μm deep obtained with reactor B. Cu films was deposited under following conditions: material gases H_2 (67%) and $C_2H_5OH[Cu(hfac)_2]$, total flow rate 150 sccm, pressure 73 Pa, $P_m = 45$ W, $P_{as} = 60$ W, and $T_s = 170°C$.

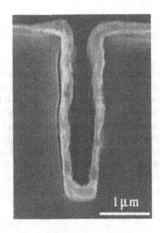

roughness is evaluated using the cross-section SEM. The samples of 100 nm in thickness are prepared by the same method as those in Fig. 2. Time evolution of surface roughness deduced from the standard deviation of film thickness is shown in Fig. 3. The roughness decreases considerably from 18 nm at t=0 min to 6 nm at t=40 min, indicating that the H irradiation contributes significantly to reducing the surface roughness of Cu films. This feature is important to obtain smooth thin Cu seed layers for the Cu electroplating.

In order to deposit high purity films in small trenches and holes, it is important to control dissociation degrees of H_2 and $Cu(hfac)_2$ independently. For this purpose, we employed the plasma CVD reactor B. The coverage shape of Cu film in a trench of 0.3 μm in width and 0.9 μm in depth was examined under deposition conditions of high-purity (≈100%) Cu films. Filling property of Cu films essentially depends on the β of Cu-containing radicals, which is the sum of the probabilities for the radicals to contribute to film and to form volatile molecules, s and γ respectively. To deduce the β from the trench coverage profiles obtained experimentally, the cross-section profile of film in the trench was simulated by Monte-Carlo method as a parameter of β [5]. The β was obtained comparing the profile by the simulation with the experimental one by cross-section SEM photographs. Figure 4 shows its P_m dependence. The β decreases significantly with decreasing P_m, that is, decreasing dissociation degree of material gas and a low β value of 0.01 is realized for P_m=3 W.

Finally, we tried conformal deposition of smooth Cu films in a trench using the HAPCVD reactor. Figure 5 shows the cross-section SEM. As shown in this figure, conformal deposition of smooth thin Cu films in the trench is realized by using the reactor.

CONCLUSIONS

Effects of H irradiation on removal of impurities in Cu films, surface roughness, and size and orientation of Cu grains have been examined using the plasma CVD reactor with the H atom source. The H irradiation is effective in purifying the Cu films, increasing the grain size, and reducing the surface roughness, while it has no effect on the grain orientation. We have evaluated the Cu-filling property in the trench using the HAPCVD reactor under deposition conditions of high-purity ($\approx 100\%$) Cu films. Comparison of the profile of film in the trench between the results by the Monte Carlo simulation and experiments reveals that the decrease in dissociation degree of material gas leads to the reduction in β of Cu-containing radicals, which is essential for the conformal deposition in fine trenches. Using the control of dissociation degree of material gas independent of H irradiation, we have demonstrated conformal deposition of smooth thin Cu films in the trench using the HAPCVD reactor.

ACKNOWLEDGMENTS

This research was partly supported by a grant from the Casio Science Promotion Foundation. We are indebted to Dr. N. Sonoda, Fukuryo Semi-con Engineering, for SEM measurements and to Dr. S. Samukawa, NEC and Dr. G. Chung, Tokyo Electron LTD., for supplying Si substrates with trenches. We would like to acknowledge the assistance of Mr. H. Matsuzaki who contributed greatly to the preparation of the experimental set-up.

REFERENCES

1. Copper Metallization in Industry, Mater. Res. Soc. Bull., **19**, 15 (1994).
2. W. W. Lee and P. S. Locke, Thin Solid Films, **262**, 39 (1995).
3. A. Burke, G. Braeckeimann, D. Manger, E. Eisenbraun, A. E. Kaloyeros, J. P. McVittie, J. Han, D. Bang, J. F. Loan and J. J. Sullivan, J. Appl. Phys., **82**, 4651 (1997).
4. H. J. Jin, M. Shiratani, T. Kawasaki, H. Kawasaki, M. Toyofuku, T. Fukuzawa, T. Kinoshita and Y. Watanabe, Jpn. J. Appl. Phys., **38**, 4492 (1999).
5. H. J. Jin, M. Shiratani, Y. Nakatake, K. Koga, T. Kinoshita and Y. Watanabe, Res. Rep. Information Science and Electrical Engineering Kyushu Univ., **5**, 57 (2000).
6. M. Shiratani, H. Kawasaki, T. Fukuzawa, T. Kinoshita and Y. Watanabe, J. Phys. D, **29**, 2754 (1996).
7. H. J. Jin, M. Shiratani, T. Kawasaki, H. Kawasaki, M. Toyofuku, T. Fukuzawa, T. Kinoshita and Y. Watanabe, J. Vac. Sci. & Technol. A, **17**, 726 (1999).
8. C. R. Kee-Won, A. L. S. L oke, H. Lee, T. Nogami, V. M. Dubin, R. A. Kavari, G . W. Ray and S . S. Wong, IEEE Trans. Electron Devices., **46**, 1113 (1999).

Mat. Res. Soc. Symp. Proc. Vol. 612 © 2000 Materials Research Society

Formation of Al$_x$O$_y$N$_z$ barriers for advanced silver metallization

Y. Wang and T.L. Alford
Department of Chemical, Bio, and Materials Engineering
NSF Center for Low Power Electronics
Arizona State University, Tempe, AZ 85287-6006, USA

Abstract

Silver has been explored as a potential candidate for future advanced interconnects due to its lowest electrical resistivity, when compared with Al and Cu. As in the case of Cu metallization, an additional layer between the Ag film and underneath dielectric is necessary in order to improve adhesion and to block the diffusion of Ag atoms. In this study, thin aluminum oxynitride (Al$_x$O$_y$N$_z$) diffusion barriers have been formed in the temperature range of 400-725 °C by annealing Ag/Al bilayers on oxidized Si substrates in ammonia ambient. Rutherford backscattering spectrometry showed that the out-diffused Al reacted with both the ammonia and oxygen in the ambient and encapsulated the Ag films. Higher process temperatures and thinner original Al layers showed to improve the resistivity of the encapsulated Ag layers. The resulting Ag resistivity values are ~1.75 ± 0.35 µΩ-cm. The thermal stability test of these diffusion barriers showed that these barriers sustained the interdiffusion between Cu and Ag up to 620 °C at least for 30 min in either vacuum or flowing He-0.5% H$_2$. This temperature is a 200°C improvement over previously reported values for the self-encapsulated Cu and Ag films. X-ray diffraction spectra showed no formation of any high resistive intermetallic compounds, *i.e.*, Ag$_3$Al, Ag$_2$Al, and AlAg$_3$.

I.   INTRODUCTION

Due to more stringent demands on interconnects for the advanced ultra-large-scale-integration (ULSI) technologies, advanced metallization schemes, other than Al- and Cu -based ones, has been explored to achieve higher current densities and faster switching speeds[1-2]. Compared with Al and Cu, silver is more attractive due to its lowest bulk electrical resistivity (1.47 µΩ-cm)[3] since RC delay (inherent with the metallization schemes) will become more significant with the development of ULSI technology. But advanced silver metallization has serious shortcomings, such as poor adhesion to dielectrics and the potential degradation of Si based semiconductor devices[4]. To address these issues, the addition of another metal layer (*e.g.*, Ti or Cr with Ag[1-2], Ti, Al, or Mg with Cu metallization[1-2, 5-8]) has been explored. Using the bilayer configuration and annealing in an appropriate ambient, a metal-nitride or metal-oxide encapsulation layer was formed on top of the Ag or Cu layers. The encapsulation layer reduced the rate of Ag sulfidation[1-2] or Cu oxidation[1-2, 5-8] in the corrosive environments and hence ensured the structure integrity of the metal lines. Addition of Al has shown good results for Cu metallization by forming Al-oxide passivation layers on Cu layers during anneals in a low pressure of O$_2$ ambient[6]. However, the use of Al underlayer has not been extensively investigated for Ag metallization. Compared with Al-oxide, Al-nitride has higher thermal conductivity (200Wm^{-1}K^{-1}) which allows for excellent thermal management for multichip modules and faster switching times[9]. It also has been explored as the diffusion barriers for metal-insulator-semiconductor (MIS) devices on silicon carbide substrates[10]. This article reports the self-encapsulation of Ag/Al bilayers in ammonia (NH₃) ambient to form diffusion barriers.

II. EXPERIMENTS

Bilayer structures of 100 or 200 nm Ag on 8 nm Al were deposited sequentially by electron-beam evaporation onto thermally oxidized (100) Si substrates without breaking vacuum. The base and operation pressures were 10^{-7} and ~10^{-6} Torr, respectively. Self-encapsulation was performed in a Lindberg single-zone quartz tube furnace with flowing electronic grade NH₃ (99.99%) for 30 min at temperatures ranging from 400 to 725 °C. Thermal stability of these diffusion barriers was evaluated by deposition of 50 nm Cu films onto the encapsulated samples in the same system described above. The samples were then annealed in either vacuum or flowing He-0.5%H₂. Rutherford backscattering spectrometry (RBS) and x-ray diffraction (XRD) were used to characterize the diffusion barriers and the underlying Ag layers. The sheet resistance of the encapsulated Ag layers was obtained with a van der Pauw four-point probe[11].

III. RESULTS AND DISCUSSION
A. Barrier formation and thermal stability

A backscattering spectrum of the sample nitridized at 725 °C is shown in figure 1 and is compared with that of the as-deposited sample. The schematic diagrams of the sample configuration before and after encapsulation are shown as the inset of figure 1. During the NH₃ anneal, aluminum diffused through the silver layer; there it reacted with NH₃ or residual O₂ and formed a thin layer of Al-oxynitride on the surface. It is worth to mention that although the surface segregation of Al was apparent for all the annealed samples (as denoted by "Al" peak in figure 1); only those samples annealed over 675 °C showed significant nitrogen peaks as well as the accompanying strong surface oxygen signals. The RUMP simulation software[12] was used to determine the thickness and composition of the surface Al-oxynitride films formed at different

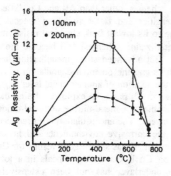

FIG. 3. The Ag resistivity of Ag(100nm)/Al(80nm) (open circle) and Ag(200nm)/Al(80nm) (solid circle) bilayers versus annealing temperature.

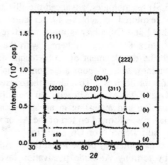

FIG. 4. θ-2θ X-ray diffraction spectra from several Ag(100nm)/Al(30nm) bilayers annealed at: (a) 725 °C (b) 675 °C (c) 625 °C and (d) as-dep. All the Ag peaks are indexed, except the one denoted as (004) is due to Si substrate.

temperatures. Results show that the Al-oxynitride layers contained ~35 atomic % of oxygen and the typical thicknesses were 10 nm. The thermodynamic data supported these results since the Gibbs free energy for consumption of one mole of Al to form Al_2O_3 at 725 °C is much lower, -606.079 KJ/mol (-6.292 eV/Al atom), than that to form AlN, -217.923 KJ/mol (-2.262 eV/Al atom)[13]. It is hence expected that oxygen-rich, Al-oxynitride films form at this temperature range instead of pure Al-nitride layers.

Copper films about 50 nm were deposited on the encapsulated sample and the the thermal stability evaluation of the $Al_xO_yN_z$ diffusion barrier were performed by a series of high temperature anneals of the above samples. The test results of the as-formed diffusion barrier after 30 min anneal at 620 °C are shown in the RBS spectra of figure 2. Inspection of this data revealed that no interdiffusion occurred between Cu and Ag. These results were the same for anneals in either vacuum or the forming gas ambient. Comparison showed that the $Ag/Al_xO_yN_z$ barrier has a much higher thermal stability (up to 620 °C), and is at least 200 °C higher than that of TiN barriers between Ag and Cu layers fabricated by the same method[14].

B. Resistivity of Ag films

Achievement of a low Ag resistivity is the most pertinent concern for the encapsulation process. Figure 3 shows the resistivities of two sets of encapsulated Ag layers of different thicknesses versus annealing temperatures. The error bars depict the standard deviationes of the eight measurements from each sample. It is assumed that the resistivity obtained with the van der Pauw four-point probe analysis is equal to that of the underlying Ag layer since the thin Al-oxynitride layer will have a much higher resistance than the underlying Ag layer. The thickness of the underlying Ag layer is obtained from the simulation of RBS data. It is observed that the resistivity initially increased dramatically after lower temperature anneals due to the alloy effects of Al on the resistivity of the Ag according to the dilute alloy theory[3]. Then it dropps to the lower value after the higher temperature anneals. The lowest resistivity values, 1.75 ± 0.35 $\mu\Omega$-cm, are obtained after the 725 °C anneal. This value is much lower than the ~4 $\mu\Omega$-cm obtained by the

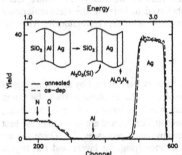

FIG. 1 RBS spectra (3.7 Mev He^{++}, 70° tilt) obtained from a Ag (200nm)/Al (8nm) bilayer on oxidized Si substrate before (dashed) and after (solid) a 0.5 hr anneal at 725 °C in flowing NH_3.

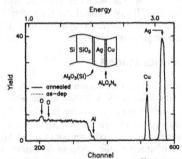

FIG. 2. RBS spectra (3.7 Mev He^{++}, 7° tilt) of the diffusion barriers before (dot) and after (solid) testing at 620 °C for 0.5 hr in flowing atmospheric N_2-5%H_2 ambient.

preliminary work on the self-encapsulation of Ag/Ti bilayers in ammonia ambient[2], and is also lower than the ~2.4 $\mu\Omega$-cm obtained by annealing Cu/Al bilayers in air or low pressure O_2[6-8].

Figure 4 shows the XRD data for the encapsulated Ag layers annealed at different temperatures. A strong $\langle 111 \rangle$ fiber texture is observed for the as-deposited sample; and is enhanced after high temperature (725 °C) anneal. This advantageously textured structure will enhance the electron migration resistance of the Ag layers. It is evident that none of the intermetallic compounds (e.g. Ag_3Al, Ag_2Al, and $AlAg_3$) appears in XRD spectra. Lower residual Al concentration in silver layer is obtained after higher temperature anneals and correlated with the peak shifts in Fig. 4.

However, the Ag-Al phase diagram[15] shows that the solid solubility of Al in Ag is relatively high in the temperature range employed by this study. According to the experimental data for dilute Ag-Al alloy[3], the increase of resistivity of Ag with the increase of atomic % Al $(\Delta\rho/\Delta c)$ is relatively high (i.e. ~ 1.95 $\mu\Omega$-cm/atomic % Al). Therefore the resistivity of Ag layers should increase dramatically. This obviously contradicts the results obtained here. The explanation of this phenomenon is related to the kinetics of Al diffusion and intermetallic compound formation. The activation energy for Al diffusion in Ag is low, about 1.28 eV/atom[16]. Using the relation of $x = 2(Dt)^{1/2}$, the diffusion distance at 725 °C for half an hour is 110 μm, which is much larger than the thickness of the Ag layers applied here. In addition, the reaction of Al with NH_3 or O_2 allows the free surface to serve as a sink for Al atoms and accelerates the Al atoms to diffuse through the Ag layers. Hence a lower Al concentration is obtained after the high temperature anneals. A similar mechanism was reported for Cu_xTi_y dealloying during self-encapsulation in ammonia ambient[17]. Therefore, it is not surprising that Al diffuses rapidly though the Ag layers to the AlN formation front instead of accumulating in Ag layer during high temperature anneals.

The above results and analysis show that the high temperature (725 °C) anneal is an optimal process for the Ag encapsulation process. It leads to the formation of a more effective $Al_xO_yN_z$ diffusion barrier that has a higher thermal stability than those previously formed. The lowest Ag resistivity so far was achieved and this will provide reduced RC delays for future ULSI technologies. Highly textured Ag films obtained is believed to improve the electromigration resistance. Further more, the application of Al is compatible with the conventionally Al-based metallization process currently employed by industry.

IV.   CONCLUSIONS

In this article, thin oxygen-rich $Al_xO_yN_z$ diffusion barriers were formed by annealing Ag/Al bilayers on oxidized Si substrate in the ammonia ambient. The resulting barriers had high thermal stability (i.e., 620 °C for 30 min). The lowest resistivity values, 1.75 ± 0.35 $\mu\Omega$-cm, were obtained from the underlying Ag layers encapsulated at 725 °C, and are lower than the best values previously obtained from both Cu and Ag self-encapsulation processes[2, 6-8]. There was no evidence of intermetallic compounds formation in the underlying Ag layers and the higher annealing temperatures assisted in the removal of Al from the Ag layers. A strong $\langle 111 \rangle$ fiber texture existed in the as-deposited sample and was enhanced after high temperature (725 °C) anneal.

Acknowledgments

The work was partially supported by The National Science Foundation, (L. Hess, Grant No. DMR-9624493), to whom the authors are greatly indebted. This work was carried out at the National Science Foundation's State/Industry/University Cooperative Research Centers' (NSF-S/I/UCRC) Center for Low Power Electronics (CLPE). CLPE is supported by the NSF (Grant #EEC-9523338), the State of Arizona, and the following companies and foundations: Burr-Brown, Inc., Conexant, Gain Technology, Intel Corporation, Medtronic Microelectronics Center, Microchip Technology, Motorola, Inc., The Motorola Foundation, Raytheon, Texas Instruments and Western Design Center. We are grateful to J. W. Mayer for his support of this project.

Reference:

1. T. L. Alford, J. Li, J.W. Mayer, and S.-Q. Wang, Thin Solid Films **262**, (1995)
2. T. L. Alford, D. Adams, T. Laursen, and B. Manfred Ullrich, Appl. Phys. Lett. **68**, 23 (1996).
3. *CRC Handbook of Electrical Resistivities of Binary Metallic Alloys*, edited by Klaus Shröder (CRC, Boca Raton, FL, 1983), p.44.
4. R. J. Gutmann, A. E. Kaloueros, and W. A. Lanford, Thin Solid Films **236**, 257 (1993).
5. W. A. Lanford, P. J. Ding, W. Wang, S. Hymes, and S. P. Murarka, Thin Solid Films **262**, 234 (1995).
6. W. Wang, W. L. Lanford ,and S. P. Murarka, Appl. Phys. Lett. **68**, 12 (1996).
7. P. J. Ding, W. Wang, W. A. Lanford, S. Hymes, and S. P. Murarka, Appl. Phys. Lett. **65**, 14 (1994).
8. P. J. Ding, W. Wang, W. A. Lanford, S. Hymes, and S. P. Murarka, J. Appl. Phys. **75**, 7 (1994).
9. I. Shalish, S. M. Gasser, E. Kolawa, M.-A. Nicolet, and R. P. Ruiz, Thin Solid Films **289**, 166 (1996).
10. C.-M. Zetterling, M. Östling, K. Wongchotigul, M. G. Spencer, X. Tang, C. I. Harris, N. Nordell, and S. S. Wong, J. Appl. Phys. **82**, 2990 (1997).
11. D. K. Schroder, *Semiconductor Material and Device Characterization*, (Wiley, New York, 1990), P9.
12. L. R. Doolittle, Nucl. Inst. Meth. Res. **B9**, 344 (1985).
13. *Lange's Handbook of Chemistry*, No. 14, 11th ed, edited by J. A. Dean (McGraw-Hill, New York, 1992) P6-69.
14. D. Adams, Ph. D. Dissertation, Arizona State University, 1996
15. *Landolt Börnstein* New Serie IV/5a, (Springer-Verlag, Berlin, New York, 1961) p5.
16. K.-N. Tu, James W. Mayer, and L. C. Feldman, *Electronic Thin Film Science for Electrical Engineers and Materials Scientists*, (Macmillan Publishing Company, New York, 1992).
17. S. W. Russell, T. L. Alford, and J. W. Mayer, J. Electrochem. Soc. 142, 1308 (1995).

Mat. Res. Soc. Symp. Proc. Vol. 612 © 2000 Materials Research Society

Kinetics model for the self-encapsulation of Ag/Al bilayers

Y. Wang, T. L. Alford, and J. W. Mayer
Department of Chemical, Bio, and Materials Engineering
NSF Center for Low Power Electronics
Arizona State University, Tempe, AZ 85287-6006, USA

Abstract

A model is proposed to describe the temperature dependence of the aluminum oxynitride ($Al_xO_yN_z$) diffusion barrier formation during a silver self-encapsulation process. These barrier layers form in the temperature range of 500-725 °C during anneals of the Ag/Al bilayers on oxidized Si substrates in an ammonia ambient. Experimental results show that temperature has a significant effect on the kinetics of this process. In this investigation, the diffusion of Al atoms through the Ag layers during self-encapsulation process is modeled using an analytical solution to a modified diffusion equation. This model shows that higher anneal temperatures will minimize the retardation effect by *i*) reducing the chemical affinity between Al and Ag atoms, and *ii*) allowing more Al atoms to surmount the interfacial energy barrier between the metal layer (Ag) and the newly formed $Al_xO_yN_z$ diffusion barriers. The theoretical predictions on the amount of segregated Al atom correlate well with experimental results from Rutherford backscattering spectrometry. This model in addition confirms the self-passivation characteristics of $Al_xO_yN_z$ diffusion barriers formed by Ag/Al bilayers annealed between 500~725 °C.

I.   INTRODUCTION

In the future, ULSI technologies will employ thinner and narrower multiple metal layers in order to achieve higher device speeds and higher component density[1]. This tendency will make the *RC* delay (inherent with the metallization schemes) much more significant than before. This is why Ag with the lowest resistivity among all the metals has been explored as an alternative for future metallization schemes[2-4]. At the same time, this scaling tendency will also require more robust diffusion barriers and better thermal management.   The previous study on self-encapsulation of Ag/Al bilayer provided a good approach for the above questions since this process has achieved the lower resistivity (~ 1.75 μΩ-cm) of the as-processed Ag layer and formed a thin $Al_xO_yN_z$ diffusion barrier with higher thermal stability than TiN barrier[4]. Our experimental results showed that higher temperatures increased the speed of Al diffusion through the Ag layer and helped to deplete Al atoms from the Ag layer by accelerating both the diffusion and the reaction of Al with NH_3 or O_2 at the surface. Since most of the research on the encapsulation process[2-5] was only focused on the experimental issues; it is hence worthy to investigate this process theoretically and to focus especially on the kinetics aspect in order to optimize this self-encapsulation process further. In this article, we quantitatively confirmed our previous explanation and obtained a better understanding of the self-encapsulation process.

II.   EXPERIMENTS

Ag(100nm)/Al(8nm) and Ag(200nm)/Al(8nm) bilayers were deposited sequentially by electron-beam evaporation without breaking the vacuum. The base and operation pressures were ~10^{-7} and ~10^{-6} Torr, respectively. Self-encapsulation was performed in an AST rapid thermal annealer (RTA) with a flowing gas mixture of N_2 and 1% NH_3 for different times at 500 and 725 °C, respectively. Rutherford backscattering spectrometry (RBS) and X-ray diffraction (XRD) was used to monitor the diffusion of Al atoms.

III.   EXPERIMENTAL RESULTS AND MODEL
A.   Experimental results

It has been shown that during the high temperature anneals (>500 °C) of Ag/Al bilayers, Al atoms diffuse through the silver layer; there they reacts with either NH_3 or residual O_2 and form a thin layer of Al-oxynitride on the surface[4]. A series of RBS measurements was done to obtain the time evolution of the surface Al peak after annealed at different temperatures. Table I lists the normalized backscattering yield of the surface Al atoms as a function of anneal times. The results were obtained from both the Ag(100nm) and the Ag(200nm) bilayer samples. Inspection of Table I reveals that at 725 °C, the height of the Al peaks from both sets of samples increased with the anneal time; but the Al peaks remained constant for anneal times longer than 10~15 min. While the peak heights after the 500 °C anneals increased much slower than in the previous case. The amount of surface Al atoms after the above anneals was determined by using the RUMP simulation software[8]. These values were used to validation our proposed encapsulation model.

Our previous analysis showed that the shift of Ag diffraction peak correlates with the concentration of Al atoms in silver layer after the encapsulation process. A lower residual Al concentration is also obtained after higher temperature anneals, and is indicated by the peak shifts dependence on anneal temperature[4]. Figure 1 shows the Ag (111) peaks versus anneal time respectively at 725 and 500 °C obtained from the sample described above. It is observed that at 725 °C, the Ag peaks gradually shifts back towards the original direction, while at 500 °C, no apparent movement is observed. This comparison confirms our previous expectation that the high anneal temperature accelerates Al diffusion through Ag films. Also it shows that high temperature (725 °C) anneal enhances the ⟨111⟩ fiber texture of Ag films as observed previously[4].

Table I.   Normalized surface Al backscattering yield of Ag(100nm) and Ag(200nm) bilayer samples and the corresponding anneal times at 725 and 500 °C, respectively.

Normalized Backscattering Yield

Thick	Temp	Time (min)							
(nm)	(°C)	0	3	5	10	15	20	25	30
100	500	0		0.410	0.431	0.468	0.498	0.536	0.578
100	725	0		0.915	1.048	1.106	1.097	1.097	1.097
200	725	0	0.856	1.061	1.086		1.099		1.099

B. Retardation of encapsulation

The key process during the encapsulation is the diffusion of Al atoms. But the boundary conditions for the standard solution of the diffusion equation are not valid for our case; therefore a new model was developed to address the finite diffusion conditions.

The schematic depiction of this diffusion process is shown in Fig. 2. There are two possible mechanisms retarding the encapsulation process. Calculation of the Gibbs energy to form a Ag-6% Al alloy[9] at 450 °C is about -3 eV/atom (Al). With the high temperatures employed here, this value becomes more negative and the mixing process is definitely favorable in view of thermodynamics. Therefore as the diffusion proceeds, Al atoms tend to be trapped by the Ag atoms(which is depicted by the trapped Al atom, solid circle at the bottom of the valley in Fig. 2). According to the chemical kinetics, the trapped Al concentration (S) is assumed to be directly proportional to the concentration (C) of the free-diffusing Al atoms:

$$S = RC \qquad (1)$$

where R is a constant[10].

While another possible retardation mechanism is related with the interfacial energy between the newly formed Al-oxynitride barrier and the underneath Ag layer. At this interface, the Al atoms accumulate before the formation of the Al-oxynitride barriers. Other works showed that the interfacial energy between solid Ag and solid Al_2O_3 is about 0.7 eV/atom (1630 erg/cm^2 at 700 °C)[11]. Therefore, 0.7 eV/ atom (Al) is chosen as an activation energy (ΔE) in the following discussion; and hence each Al atom must acquire enough thermal energy to overcome the interfacial energy barrier before it can be extracted from the Al-Ag solid solution.

C. Kinetics of the encapsulation process

The model of this encapsulation process is based on Fick's second law. The following assumptions are made based on the above discussion:

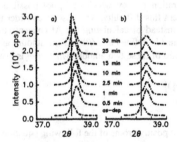

FIG. 1 The shift of Ag(111) x-ray diffraction peaks with anneal time at different temperature (a) 725°C and (b) 500°C.

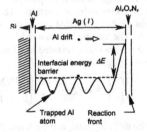

FIG. 2 Schematic depiction of Ag/Al bilayer encapsulation process. The retardation mechanisms are represented by the trapped Al atoms (solid circle at the bottom of the valley) and the interfacial energy barrier ΔE at the reaction front

1. Due to the presence of trapped Al atoms, the customary diffusion equation is modified as follows[21]:

$$\frac{\partial C}{\partial t} = \frac{D}{(R+1)} \frac{\partial^2 C}{\partial x^2} \qquad (2)$$

which is seen to be the usual form of the diffusion equation governed by a diffusion coefficient multiplied by a factor of $1/(R+1)$. Here R accounts for the chemical effects between Al and Ag on the diffusion process and is chosen as 500 to fit the experimental data. Considering the high annealing temperature ($T/T_{m,Ag} = 0.6\sim0.8$) employed here, the effect such as the high diffusivity path through the grain boundary of Ag can be neglected[12]; and hence a constant self-diffusivity of Al (D) instead of a grain boundary diffusivity is chosen here.

2. Since experimental results confirmed that the encapsulation starts instantly as the anneal starts, it is reasonable to assume an initial uniform Al concentration (C_2) within the Ag layer.

3. At the interface between the Ag layer and the newly formed Al-oxynitride barrier, there is an Al atom flux to the surface which corresponds to the depletion of the diffusive substance. This introduces another boundary condition represented by the flux across the reaction front as:

$$-D\frac{\partial C}{\partial x} = \alpha \left(C_0 - C_s \right) e^{-\Delta E/kT} \qquad (3)$$

Here the exponential term denotes the effect of the interfacial energy barrier on the depletion speed. This is similar to the approach used by Lea and Seah[13] on their study of surface segregation where they used the free energy difference as an activation factor to describe the evaporation speed. The C_0 is the equilibrium concentration of Al in the $Al_xO_yN_z$ layer, and C_s is the actual Al concentration within the Ag layer during diffusion. The ratio α is a proportional constant with units of velocity (cm/sec). An average α value of 0.098 is obtained from the RBS analysis and corresponds to the ratio of the area density of Al atoms in the $Al_xO_yN_z$ layer to that of the Ag atoms in the Ag layer. T and k are temperature (K) and the Boltzman's constant, respectively. Based on the discussions above, the solution of the modified diffusion equation (2) is a function of the unknown equilibrium concentration C_0; by which, it is not possible to compare with the experimental results. To circumvent this difficulty, we chose to consider the total amount of Al transported across the Ag layer instead of calculating the Al concentration profile $C(x)$. The transported amount of Al after time t is defined as $M(t)$ and the amount after infinite time ($t\to\infty$) is represented by $M(\infty)$. The ratio of these two amount $M(t)/M(\infty)$ is given by the following equation:

$$\frac{M(t)}{M(\infty)} = 1 - \sum_{n=1}^{\infty} \frac{2L^2 \exp\{-\beta_n^2 Dt/l^2(R+1)\}}{\beta_n^2 \left(\beta_n^2 + L^2 + L \right)} \qquad (4)$$

where t is the diffusing time, and the L and β_n's are the positive roots of the following equations:

$$\beta \tan \beta = L \qquad (5)$$

$$L = \frac{l\alpha}{D} e^{-\Delta E/kT} \qquad (6)$$

The curves in Fig. 3 show the calculated ratio of the transported Al atoms through Ag layer (100 nm) as a function of diffusing time; the upper one is for 725 °C and the lower one for 500 °C. Experimental data obtained from the RBS measurements are shown on the graph as well. It is obvious that all of the calculated curves match with the experimental results very well, especially for the long anneals.

The creditability of this model is verified by varying the film thickness at constant anneal temperature (725 °C). Figure 4 shows both the calculated curves and the experimental results; the upper one is for the 100 nm Ag film and the lower one is for the 200 nm Ag film. The fact that the experimental data overlapped well with these two curves supports our model on the time dependence of the segregation process, especially for long periods. A slight discrepancy between this model and the experimental data for the short periods may be due to the assumption that the initial Al concentration is uniform within the Ag layer. Regardless of this discrepancy, all the calculated curves and the experimental data show that the ratio increased more rapidly for the 725 °C anneals than that for the 500 °C anneals. This model therefore demonstrates the temperature enhancement on the encapsulation process and confirms our previous explanation that the higher temperature anneal extracts Al more readily and lowers the probability for intermetallic Al-Ag compound formation. These results give raise to the low resistivity in the underlying Ag layer.

IV. CONCLUSIONS

In this paper, the kinetics of the self-encapsulation process were investigated and the theoretical model quantitatively agreed with the available experimental data. This model described a competitive behavior between the diffusion and the trapping of Al atoms. It is believed that the trapping was related with both the chemical effects between Al and Ag and the

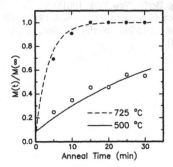

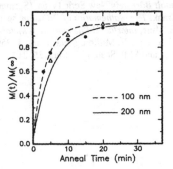

FIG. 3. Transportation ratio of Al atoms through Ag layer annealed at 725°C (solid circle) and 500°C (open circle) in the flowing gas mixture of N_2 and 1% NH_3. The overlapped curves are theoretical values from Eq. 4.

FIG.4 Transportation ratio of Al atoms through Ag layers of different thickness: 100 nm (solid circle) and 200 nm (open triangle) as a function of annealing time at 725°C. The overlapped curves are calculated with Eq. 4.

interfacial energy barrier at the $Ag/Al_xO_yN_z$ interface. It was also shown, both experimentally and theoretically, that higher temperature anneal are helpful to minimize the effects of the trapping, and to achieve a low resistive underlying Ag layer.

Acknowledgement

The work was partially supported by The National Science Foundation, (L. Hess, Grant No. DMR-9624493), to whom the authors are greatly indebted. This work was carried out at the National Science Foundation's State/Industry/University Cooperative Research Centers' (NSF-S/I/UCRC) Center for Low Power Electronics (CLPE). CLPE is supported by the NSF (Grant #EEC-9523338), the State of Arizona, and the following companies and foundations: Burr-Brown, Inc., Conexant, Gain Technology, Intel Corporation, Medtronic Microelectronics Center, Microchip Technology, Motorola, Inc., The Motorola Foundation, Raytheon, Texas Instruments and Western Design Center.

Reference:

1. T. Seidel and B. Zhao, Mat. Res. Soc. Symp. Proc.,**427**, 3 (1996).
2. T. L. Alford, D. Adams, T. Laursen, and B. Manfred Ullrich, Appl. Phys. Lett. **68**, 23 (1996).
3. T. L. Alford, J. Li, J.W. Mayer, and S.-Q. Wang, Thin Solid Films **262**, (1995).
4. Y. Wang and T. L. Alford, Appl. Phys. Lett. **74**, 52 (1999).
5. W. Wang, W. I. Lanford, and S. P. Murarka, Appl. Phys. Lett. **68**, 12 (1996).
6. I. Shalish, S. M. Gasser, E. Kolawa, M.-A. Nicolet, and R. P. Ruiz, Thin Solid Films **289**, 166 (1996).
7. *Interfacial Segregation*, edited by W. C. Johnson and J. M. Blakely (American Society for Metals, Metal Park, Ohio, 1977), p39.
8. L. R. Doolittle, Nucl. Inst. Meth. Res. **B9**, 344 (1985).
9. *Landolt Börnstein* New Serie IV/5a, (Springer-Verlag, Berlin, New York, 1961) p5.
10. J. Crank, *The Mathematics of Diffusion*, second edition, (Claredon Press, Oxford, 1975) p.60
11. L. E. Murr, *Interfacial Phenomena in Metals and Alloys*, (Addison-Wesley, London, 1975).
12. J. Kucera and B. Million, Meta. Trans. **1**, 2599 (1970).
13. C. Lea and M.P. Seah, Phil. Mag, **35**, 213 (1977).

Mat. Res. Soc. Symp. Proc. Vol. 612 © 2000 Materials Research Society

The Integration of Low-k Dielectric Material Hydrogen Silsesquioxane (HSQ) with Nitride Thin Films as Barriers

Yuxiao Zeng[1], Linghui Chen, and T. L. Alford

Department of Chemical, Bio and Materials Engineering,
NSF Center for Low Power Electronics,
Arizona State University, Tempe, Arizona 85287-6006, USA
[1]Current address: Epitronics Corporation, an ATMI company, Mesa, AZ 85210,USA

ABSTRACT

   HSQ (hydrogen silsesquioxane) is one of the promising low-k materials used in VLSI technology as an intra-metal dielectric to reduce capacitance-related issues. Like any other dielectrics, the integration of HSQ in multilevel interconnect schemes has been of considerable importance. In this study, the compatibility of HSQ with different nitride barrier layers, such as PVD and CVD TiN, PVD TaN, and CVD W_2N, has been investigated by using a variety of techniques. The refractory metal barriers, Ti and Ta, are also included for a comparison. The degradation of HSQ films indicates a strong underlying barrier layer dependence. With CVD nitrides or refractory metals as barrier, HSQ exhibits a better structural and property stability than that with PVD nitrides. The possible mechanisms have been discussed to account for these observations.

INTRODUCTION

   HSQ (hydrogen silsesquioxane), an inorganic spin-on material, is receiving attention as one of the promising low-*k* dielectrics due to its low dielectric constant, excellent gapfill and planarization performance, and capability of using standard spin-on production technique [1-3]. HSQ has a general formula $(HSiO_{1.5})_{2n}$, where n = 3 to 8. It has a cagelike structure with each silicon atom connected to one hydrogen atom and 1.5 oxygen atoms.

   In real device structures, the potential interaction between HSQ and metal via is a reliability issue deleterious to device properties. A common failure mode is via poisoning [4-5] Furthermore, for the HSQ films, it is preferred to avoid any undesired interaction in order to maintain the dielectric properties of HSQ. The alleviation of these problems requires a proper diffusion barrier to be used. Due to their high melting points and high thermal stability, refractory metals and their nitrides, such as Ti, Ta, TiN, TaN, and W_2N, have been intensely investigated as diffusion barriers in interconnect schemes [6-8] Specifically, Ti and TiN are mainly aimed at Al metallizations, and the other three are primarily intended for Cu metallizations. In this paper, we choose these barriers to study their potential interactions with the HSQ and evaluate their effectiveness as the barrier between the HSQ and the metal via. The conventionally used intermetal dielectric PETEOS - silicon dioxide deposited from tetraethylorthosilicate (TEOS) by plasma-enhanced chemical vapor deposition process is also selected for the purpose of comparison.

EXPERIMENTAL

HSQ films (350 nm thick) were deposited on (100) Si wafers by spin-on technique, followed by sequential bakes at different temperatures (100-300 °C) and a final cure process (400 °C). PETEOS films with a thickness of 1 μm were deposited by using TEOS and oxygen plasma. Barrier thin films were sequentially deposited on these two dielectrics, respectively, by using two techniques: physical vapor depostion (PVD) and chemical vapor deposition (CVD). The thicknesses of all these barrier layers were determined to be approximately 50 nm. All the anneals on different types of samples were conducted for 60 min in a vacuum furnace with a base pressure lower than 4×10^{-8} Torr at various temperatures.

XRD analysis was used for structural characterization of the barrier layers. The microstructure of the barrier films was observed by using plane-view and cross-section Transmission Electron Microscopy (TEM). A series of analysis was carried out in a General Ionex tandetron accelerator to characterize interdiffusions of barrier/dielectric structures upon annealing. The conventional Rutherford backscattering spectrometry (RBS) measurement was primarily utilized to profile heavy- and medium-Z elements in these structures. Forward REcoil Spectrometry (FRES) (also called ERD, Elastic Recoil Detection) is suitable for analyzing hydrogen. In this technique, the H areal density (the product of H concentration and HSQ film thickness) is obtained by using a calibration standard. The detailed description of this technique was give elsewhere [9]. Resonance scattering technique was used to analyze O and C elements by employing $^{16}O(\alpha,\alpha)^{16}O$ and $^{12}C(\alpha,\alpha)^{12}C$ elastic resonance.

RESULTS

The experimental results indicate that the most significant compositional change in HSQ films upon annealing is the loss of H. The normalized areal densities of H in HSQ for the Al-based barriers and Cu-based barriers are plotted in Figs. 1(a)-(b) respectively, with the data for bare HSQ samples included as a reference. For the bare HSQ film, the H concentration in HSQ progressively decreases upon annealing. It is interesting to note that for the barrier-capped HSQ films, within the experimental error, the H concentration variation upon annealing exhibits an apparent barrier layer dependence within the temperature range up to 500 °C, which is of practical interest. Compared to the bare HSQ film, the HSQ films with Ti, Ta and CVD TiN overlying barriers exhibit a slower H concentration change, in contrast to a faster H concentration decrease for the HSQ films with PVD TaN and PVD TiN as top layers. There is essentially no difference for W$_2$N-capped HSQ and bare HSQ samples.

Figure 2 presents the normalized areal densities of H in PETEOS as a function of annealing temperature for barriers Ta, PVD and CVD TiN, and PVD TaN. Quite unlike the case of HSQ, there is little difference in the H compositional change of PETEOS with various overlying barrier films.

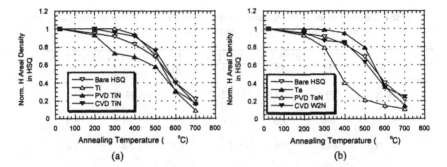

Figure 1 *Normalized H areal density of HSQ as a function of annealing temperature for bare HSQ and (a) Al-based barriers/HSQ, and (b) Cu-based barriers/HSQ samples, as obtained from FRES measurements.*

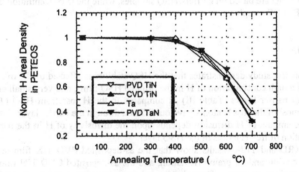

Figure 2 *Normalized H areal density of PETEOS as a function of annealing temperature for PVD and CVD TiN, Ta and PVD TaN/PETEOS samples, as obtained from FRES measurements.*

Table I summarizes the experimental results based on the elastic resonance and FRES measurements. Elastic resonance techniques were used to detect O and C impurities in the as-deposited and annealed barrier films. The estimated detection limit is ~1 at.%. The H impurity content in these barriers is estimated from FRES. Due to the detection limits of these techniques, these data are only sufficient for the purpose of qualitative comparison. From Table I, it is obvious that the as-deposited CVD barrier films, especially the CVD TiN, have a much higher H content than the PVD films. The CVD TiN also has a higher residual C content than other nitride films.

Table I Summary of the concentrations of impurities H, O and C in barrier films before and after annealing (units: $\times 10^{15}$ atoms/cm^2 for H; at.% for O and C).

		Ti	PVD TiN	CVD TiN	Ta	PVD TaN	CVD W$_2$N
H*	As-dep	22	17	50	20	15	25
	700 °C	9	17	22	11	9	12
O	As-dep	6.0	3.6	2.0	8.0	3.4	3.4
	700 °C	45.0	4.0	3.0	8.0	3.4	3.4
C	As-dep	2.4	2.0	2.6	1.4	1.0	1.0
	700 °C	9.0	2.1	2.6	2.2	1.0	1.0

* The H and O data are based on barriers/HSQ samples, while the C concentration data are based on barriers/PETEOS samples.

DISCUSSION

The present study demonstrates that the HSQ stability is affected by the overlying barrier. Since there is a high concentration of H in the original HSQ films, and very small amount of H is detected in all barrier layers (Table III) as compared to the H loss from HSQ (Table II), it is obvious that most of the lost H atoms have outdiffused through the barrier layers into the vacuum ambient. The amount of H diffusion depends upon the diffusivity of H in the barrier layer and the concentration of free H atoms in the HSQ layer.

The XRD and TEM results show that the as-deposited PVD TiN film has a columnar microstructure with random grain orientation, and the as-deposited CVD TiN film has a (200)-textured fine-grained structure. The columnar microstructure can provide a rapid diffusion path for H atoms, while the fine-grained microstructure means a long diffusion path for H atoms. Furthermore, it is indicated in the present study that CVD TiN films contain more impurities such as H and C, since an organic precursor has been used for CVD TiN deposition. Thus it is speculated that the C and H impurities may also stuff the grain boundaries of the fine-grained CVD TiN barrier and block atom diffusion, at least at the beginning of the outdiffusion of H atoms from HSQ. The CVD W$_2$N film has lower H and C contents than the CVD TiN, therefore it is not as effective as the CVD TiN in preventing H outdiffusion.

Generally, the atom diffusion through grain boundaries is much faster than that *via* crystal lattice. However, for the H atom diffusion, due to its small atomic size and extremely high mobility [10], the diffusion of H atoms through the lattice is fast as well. In a nitride barrier, such as TiN, N atoms can block the lattice diffusion paths for H. As a result, it is assumed that the lattice diffusion coefficients of H in metal barriers (Ti and Ta) are higher than those in nitride barriers. Nitride barriers are thereby expected to be more resistant to the H outdiffusion than metal barriers. The experimental results are however quite different.

Moreover, the H concentration decrease is even faster when HSQ is capped with PVD TaN or PVD TiN than that for the bare HSQ. A "nitride effect" mechanism is proposed to attempt to account for this discrepancy. It is noted that in HSQ, H atoms are normally bonded to Si atoms to form Si-H bonds. It is also noted that the bonding energy of a Si-H bond is 3.12×10^5 J/mol (all bonding energy data from reference 11), which is lower than that of a N-H bond (3.89×10^5 J/mol). It implies that the N-H bond is more chemically stable than the Si-H bond. In other words, N atoms in nitride barriers have an energy advantage to assist in breaking some Si-H bonds in HSQ at elevated temperatures. Taking this effect into account, we anticipate a higher concentration of free H atoms for the nitride barrier/HSQ system than that for the metal barrier/HSQ system. It can explain why PVD TaN and PVD TiN induce the HSQ degradation.

Although this mechanism remains to be confirmed and further investigated, it does give a reasonable interpretation to what have been observed. In the meantime, our finding that there is essentially no difference in the H concentration decrease for all the barrier/PETEOS samples (see Fig. 2) is a piece of evidence to support this mechanism. The H in PETEOS exists mainly in the form of C-H and/or O-H [12] instead of Si-H. The bonding energies of the C-H bond (4.14×10^5 J/mol) and O-H bond (4.64×10^5 J/mol) are higher that of the N-H bond (3.89×10^5 J/mol). Therefore, such a driving force does not exist any more for nitride barriers to induce H outdiffusion from PETEOS.

CONCLUSIONS

The HSQ degradation during annealing occurred in all barrier/HSQ systems and it had an obvious overlying barrier dependence. Compared to the case of bare HSQ, Ti, Ta, and CVD TiN barriers could resist HSQ degradation to some degree, while PVD TaN and PVD TiN could induce the degradation of HSQ. The interpretations were attempted from the microstructural and structural points of view, and also by a presented "nitride effect" mechanism.

ACKNOWLEGMENTS

The great help from Dr. Stephen W. Russell (Micron Technology) and Dr. Andrew J. McKerrow (Texas Instruments) is highly appreciated. The work was partially supported by the National Science Foundation (L. Hess, DMR-96-24493) to whom the authors are greatly indebted. Support for the Center for Low Power Electronics is partially provided by NSF, State of Arizona, Analog Devices, Analogy, Burr-Brown, Hughes Aircraft, Intel, Microchip, Motorola, National Semiconductor, Rockwell, Scientific Monitoring, Texas Instruments, and Western Design Center.

REFERENCES:

1. C.H. Ting and T.E. Seidel, Mat. Res. Soc. Sym. Proc., **381**, San Francisco, CA, April 17-19, 1995.
2. S.-P. Jeng, K.J. Taylor, T. Seha, M.-C. Chang, J. Fattaruso, R.H. Havemann, VLSI Tech. Symp., Kyoto, Japan, 1995.
3. J.N. Bremmer, Y. Liu, K.G. Gruszynski, F.C. Dall, Mat. Res. Soc. Sym. Proc., San Francisco, CA, April 1-4, 1997.

4. H.G. Tompkins and C. Tray, J. Electrochem. Soc. **136**, 2331 (1989).
5. J.D. Romero, M. Khan, H. Fatemi, and J. Turlo, J. Mater. Res. **6**, 1996 (1991).
6. S.-Q. Wang, I. Raijmakers, B.J. Burrow, S. Suthar, S. Redkar, and K.-B. Kim, J. Appl. Phys. **70**, 5176 (1990).[12]C.Y. Ting and M. Wittmer, Thin Solid Films, **96**, 327 (1982).
7. G. Gagnon, J.F. Currie, J.L. Brebner, and T. Darwall, J. Appl. Phys. **79**, 7612 (1996).
8. P.J. Pokela, C.-K. Knok, E. Kolawa, S. Raud, and M.-A. Nicolet, Appl. Surf. Sci. **53**, 364 (1991).
9. Y. Zeng, S. W. Russell, A. J. McKerrow, L.-H. Chen, T. L. Alford, J. Vac. Sci. Technol B **18** (2000) 221.
10. H. Wipf, *Hydrogen in Metals III*, Springer, Berlin, 1997.
11. R.T. Sanderson, *Chemical Bonds and Bond Energy*, Academic Press, New York, 1976.
12. S.K. Ray, C.K. Maiti, S.K. Lahiri, and N.B. Chakrabarti, J. Vac. Sci. Technol. B **10**, 1139 (1992).

Mat. Res. Soc. Symp. Proc. Vol. 612 © 2000 Materials Research Society

VOLATILE LIQUID PRECURSORS FOR THE CHEMICAL VAPOR DEPOSITION (CVD) OF THIN FILMS CONTAINING TUNGSTEN

Roy G. Gordon, Seán Barry, Randy N. R. Broomhall-Dillard, Valerie A. Wagner and Ying Wang
Department of Chemistry and Chemical Biology, Harvard University, Cambridge, MA 02138

ABSTRACT

A new CVD process is described for depositing conformal layers containing tungsten, tungsten nitride or tungsten oxide. A film of tungsten metal is deposited by vaporizing liquid tungsten(0) pentacarbonyl 1-methylbutylisonitrile and passing the vapors over a surface heated to 400 to 500 °C. This process can be used to form gate electrodes compatible with ultrathin dielectric layers. Tungsten nitride films are deposited by combining ammonia gas with this tungsten-containing vapor and using substrates at temperatures of 250 to 400 °C. Tungsten nitride can act as a barrier to diffusion of copper in microelectronic circuits. Tungsten oxide films are deposited by adding oxygen gas to the tungsten-containing vapor and using substrates at temperatures of 200 to 300 °C. These tungsten oxide films can be used as part of electrochromic windows, mirrors or displays. Physical properties of several related liquid tungsten compounds are described. These low-viscosity liquids are stable to air and water. These new compounds have a number of advantages over tungsten-containing CVD precursors used previously.

INTRODUCTION

Chemical vapor deposition (CVD) is a widely-used process for forming solid materials, such as coatings, from reactants in the vapor phase. CVD processes can have many advantages: good step coverage, dense films, high deposition rates, scalability to larger areas and low cost.

CVD of tungsten is usually carried out using tungsten hexafluoride, WF_6, as the tungsten source.[1] However, the byproduct hydrogen fluoride is highly corrosive and can damage substrates. Residual fluoride contamination in the films can cause problems such as loss of adhesion, or diffusion of fluorine into gate oxides causing threshold voltage shifts.[2]

Tungsten hexacarbonyl, $W(CO)_6$, is another source for CVD which can avoid the problems with fluorine.[3] Gate electrodes of tungsten have been formed by CVD from $W(CO)_6$ on ultra-thin dielectrics needed for high speed/high density MOS and CMOS devices.[4] $W(CO)_6$ has also been used to deposit tungsten nitride, W_2N, with properties suitable for barriers to diffusion of copper in microelectronics.[5] CVD using both $W(CO)_6$ vapor and oxygen gas, O_2, has produced electrochromic films of tungsten oxide.[6]

However, tungsten hexacarbonyl does have some practical disadvantages. Sublimation from solid $W(CO)_6$ is not a reproducible source of vapor, since the surface area of a solid changes as the solid evaporates. $W(CO)_6$ is highly toxic and has sufficient vapor pressure at room temperature that toxic concentrations of vapor can be emitted from any spilled material. Also, the carbon monoxide byproduct from its CVD reactions is highly toxic and lacks any warning odor.

We have synthesized some new tungsten compounds that overcome the disadvantages of tungsten hexacarbonyl while maintaining similar capabilities for CVD of tungsten, tungsten nitride and tungsten oxide. The new compounds are formed by replacing one of the carbonyl ligands in tungsten hexacarbonyl by an alkyl isonitrile, RNC, where R is a hydrocarbon. With suitable choices of the alkyl group, these compounds are low-viscosity liquids at room temperature and can be vaporized and distilled at higher temperatures (typically 50 to 80 °C under vacuum). These liquids have negligible vapor pressure at room temperature, so they are safer to handle than tungsten hexacarbonyl. They are stable to air and water. Their CVD byproducts have a pungent odor even at low concentrations, so any failure of the exhaust scrubbing system can easily be identified before toxic concentrations of carbon monoxide are released.

SYNTHESIS OF VOLATILE LIQUID TUNGSTEN COMPOUNDS

The isonitrile ligands were synthesized in two steps from commercially available amines. First the amine is formylated by refluxing an equimolar mixture of the amine and ethyl formate:[7]

$$RNH_2 + HC(O)OC_2H_5 \longrightarrow RNHC(O)H + C_2H_5OH \qquad (1)$$

The resulting formamide is then dehydrated by reaction with p-toluenesulfonyl chloride and quinoline to form an alkyl isonitrile:[8]

The new tungsten compounds are readily synthesized by reacting the alkyl isonitrile with tungsten hexacarbonyl:

$$RNC + W(CO)_6 \longrightarrow RNCW(CO)_6 + CO \qquad (3)$$

The tungsten hexacarbonyl is dissolved in tetrahydrofuran, THF, and the liquid alkyl isonitrile is added dropwise with stirring at room temperature. A small amount of palladium oxide (0.1% by weight of the tungsten hexacarbonyl) is used to catalyze the reaction.[9] Carbon monoxide gas is evolved, and the reaction is complete within about 15 minutes. Slightly longer reaction times are needed for the isonitriles with larger alkyl groups. The solution is filtered to remove the palladium oxide, and the THF is removed under vacuum. The crude product is generally over 95% pure. Purification is then done by falling film molecular distillation[10] under high vacuum (typically 0.01 Torr).

This method is a considerable improvement on previous synthetic techniques for this type of compound. Previously, replacement of a carbonyl ligand with an isonitrile was done in refluxing toluene.[11] These conditions generally lead to considerable amounts of unwanted byproducts, such as bis- and tris-substituted isonitrile compounds, as well as cluster compounds containing two or more tungsten atoms.

During the reaction replacing carbon monoxide by isonitrile, the temperature of the solution stayed constant to within ± 0.1 °C. Thus the bond strengths of W-CO and W-CNR are practically identical. The reaction is driven to completion by the entropy generated by the release of the carbon monoxide gas.

Some physical properties of the new tungsten compounds are given in Table I.

Table I. Physical properties of some tungsten(0) pentacarbonyl alkylisonitrile compounds

Alkyl on Isonitrile	No. of carbons in alkyl group	M.P. of LW(CO)$_5$ °C	Viscosity at 40 °C centiPoise	Liquid Density g/cc	Molecular Com-plexity	B.P. of LW(CO)$_5$ °C/Torr
isopropyl	3	58-61	solid		1.07	80/0.002
propyl	3	89-90	solid		1.07	
tert-butyl[11]	4	131-132	solid			
sec-butyl	4	43-44	solid		1.11	
isobutyl	4	47-49	solid		1.09	
n-butyl	4	34-36	6.85		1.09	115/1.2
2,2-dimethylpropyl	5	88-90	solid		0.93	
1,1-dimethylpropyl	5	74-76	solid		1.06	
1-ethylpropyl	5	33-34	7.68		1.04	
1,2-dimethylpropyl	5	31-33	10.13	2.02[12]	1.05	
isopentyl (isoamyl)	5	30-33	9.51		1.02	
2-methylbutyl	5	21-23	8.95	1.77	1.11	
1-methylbutyl	5	9	7.03	1.76	1.11	
n-pentyl (n-amyl)	5	-9 to -8	8.63	1.75	1.08	
1,3-dimethylbutyl	6	12-14	7.39	1.64	1.15	
n-hexyl	6	-10 to -9	9.91	1.64	1.00	
1-methylhexyl	7	2-4	9.49	1.62	0.95	
n-octyl	8	<20	10.35	1.62	0.88	

The last six compounds in Table I are liquids at room temperature, having melting points below 20 °C. All of the compounds with alkyl groups having 4 or fewer carbons are solids at room temperature. All of these compounds with alkyl groups having 6 or more carbons are liquids at room temperature. Of the 8 isomeric compounds with 5-carbon alkyl groups, 2 are liquid and the other 6 are solids with low melting points.

The molecular masses of these new compounds were determined by cryoscopy in p-xylene solution. Their "molecular complexities," defined as the ratio of the cryoscopic molecular mass to the theoretical monomeric value, generally fall between about 0.9 and 1.1. Thus the compounds are monomeric in solution.

The crystal structure (shown in Figure 1) verifies the monomeric nature of tungsten(0) pentacarbonyl 1,2-dimethylpropyl-isonitrile. The tungsten atoms have nearly perfect octahedral coordination by the six bonded carbon atoms. The tungsten-carbon bond to the isonitrile is 2.12 Angstroms long and the tungsten-carbon bonds to the carbonyls range from 2.03 to 2.05 Angstroms. These lengths are within the range of values previously observed for similar types of bonds.[13] The two enantiomers of the isonitrile ligand, whose chiral center is at carbon C2, appear to be disordered in the centrosymmetric crystal.

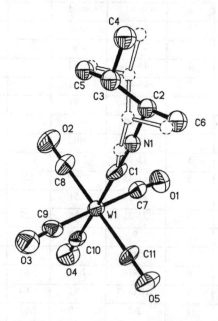

Figure 1. Molecular structure of solid tungsten(0) pentacarbonyl 1,2-dimethylpropylisonitrile.

These tungsten compounds are stable in contact with air and water at room temperature. Exposure to light, however, does cause gradual disproportionation into tungsten hexacarbonyl and tungsten tetracarbonyl bis-isonitrile and precipitation of dark tungsten cluster compounds.

CVD EXPERIMENTS

Tungsten(0) pentacarbonyl 1-methylbutylisonitrile was evaporated from a reservoir at 80 °C into a stainless steel vacuum chamber initially at 10^{-7} Torr. Tungsten metal films were deposited on silicon wafers, glass and glassy carbon substrates placed on a substrate holder heated to 500 °C. Rutherford Backscattering Spectroscopy (RBS) showed that the films were tungsten metal with about 1% molybdenum impurity derived from the starting tungsten hexacarbonyl. X-ray Photoelectron Spectroscopy detected some carbon and oxygen impurities. X-ray diffraction shows that the films are polycrystalline.

The precursors were also vaporized by flash evaporation. For this purpose, liquid tungsten(0) pentacarbonyl 1-methylbutylisonitrile was mixed with liquid mesitylene to lower its viscosity below about 3 centipoise. At this viscosity the solution could be nebulized easily into tiny droplets (less than about 20 microns in diameter) by a high-frequency (1.4 MHz) ultrasonic system.[14] The resulting fog was entrained by a flow of nitrogen and ammonia carrier gas at atmospheric pressure into a tube furnace with substrates placed on a heated aluminum substrate holder in the tube. The solid precursors could also be vaporized in this manner by dissolving them in mesitylene. Tungsten nitride films were deposited on substrates at 250 to 400 °C. RBS analysis showed that the films have a composition near to W_2N. X-ray diffraction showed that these films are amorphous.

Similar experiments with oxygen gas in place of the ammonia gave films of amorphous tungsten oxide. RBS analysis gave a composition $WO_{3.5}C_{0.025}$.

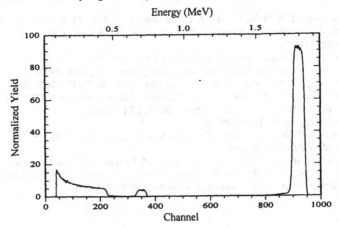

Figure 2. Rutherford Backscattering Spectrum of a CVD tungsten oxide film deposited on glassy carbon.

CONCLUSIONS

Six new liquid tungsten compounds have been synthesized and distilled under vacuum. These monomeric compounds have five carbonyl ligands and one alkyl isonitrile ligand octahedrally coordinated to the tungsten atom. The liquid with the lowest viscosity and highest volatility is tungsten(0) pentacarbonyl 1-methylbutylisonitrile. Eleven new solid compounds with similar structures were synthesized and sublimed under vacuum. Pure tungsten metal, tungsten nitride, or tungsten oxide can be deposited from their vapors.

ACKNOWLEDGMENTS

This work was supported in part by the National Science Foundation grants ECS-9975504 and CTS-9974412. We thank Dr. Richard Staples for his determination of the X-ray structure. The CCD-based X-ray diffractometer at Harvard University was purchased through NIH grant 1S10RR11937-01. The Rutherford Backscattering Spectra were taken using the Cambridge Accelerator for Materials Science operated by the Materials Research Science and Engineering Center at Harvard University and supported by the National Science Foundation.

REFERENCES

1. J. E. J. Schmitz, *Chemical Vapor Deposition of Tungsten and Tungsten Silicides for VLSI/ULSI Applications*, Noyes Publications, Park Ridge, New Jersey, USA, 1992.
2. Reference 1, pp. 199-203.
3 . See, for example, J. Haigh, G. Burkhardt and K. Blake, J. Crystal Growth, **155**, 266-271 (1995).
4 . D. A. Buchanan, F. R. McFeely and J. J. Yurkas, Appl. Phys. Lett., **73**, 1676-1678 (1998).
5. J. E. Kelsey, C. Goldberg, G. Nuesca, G. Peterson and A. E. Kaloyeros, J. Vac. Sci. Technol., B **17**, 1101-1104 (1999).
6. D. Davazoglou, Chimica Chronica, New Series, **23**, 423-428 (1994).
7. J. Moffat, M. V. Newton and G. J. Papenmeier, J. Org. Chem., **27**, 4058 (1962).
8. R. E. Schuster, J. E. Scott and J. Casanova, Jr., J. Org. Synth., **46**, 75-77 (1966); R. E. Schuster, J. E. Scott and J. Casanova, Jr., Org. Syntheses Collective Volume **5**, 772-774 (1973).
9. N. J. Coville and M. O. Albers, Inorg. Chim. Acta **65**, L7-L8 (1982).
10. Kontes Catalog number 285600-0000.
11. M. O. Albers, E. Singleton and N. J. Coville, J. Chem. Ed., **63**, 444 (1986).
12. Density of solid from X-ray data.
13. C.-L. Chen, H.-H. Lee, T.-Y. Hsieh, G.-H. Lee, S.-M. Peng and S.-T. Liu, Organometallics, **17**, 1937-1940 (1998); F. E. Hahn and M. Tamm, J. Organomet. Chem., **410**, C9-C12 (1991).
14. R. G. Gordon, International Patent Application WO 99/28532 (1999).

Mat. Res. Soc. Symp. Proc. Vol. 612 © 2000 Materials Research Society

A Comparative Study of Ti/Low-k HSQ (Hydrogen Silsesquioxane) and Ti/TEOS (Tetraethylorthosilicate) Structures at Elevated Temperatures

Yuxiao Zeng[1], Linghui Chen, and T. L. Alford

Department of Chemical, Bio and Materials Engineering,
NSF Center for Low Power Electronics,
Arizona State University, Tempe, Arizona 85287-6006, USA
[1]Current address: Epitronics Corporation, an ATMI company, Mesa, AZ, USA

ABSTRACT

For the benefit of reducing capacitance in multilevel interconnect technology, low-k dielectric HSQ (hydrogen silsesquioxane) has been used as a gapfill material in Al-metallization-based non-etchback embedded scheme. The vias are consequently fabricated through the HSQ layer followed by W plug deposition. In order to reduce the extent of via poisoning and achieve good W/Al contact, thin Ti/TiN stack films are typically deposited before via plug deposition. In this case, HSQ makes direct contact with the Ti layer. The reliability of the Ti/HSQ structures at elevated temperatures has been systematically studied in this work by using a variety of techniques. These results are also compared with those from Ti/TEOS (Tetraethylorthosilicate) structure, where TEOS is a conventional intra-metal dielectric. When the temperature is below 550 °C, a significant number of oxygen atoms are observed to diffuse into the titanium layer. The primary source of oxygen is believed to come from the HSQ film. When the temperature is above 550 °C, HSQ starts to react with Ti. At 700 °C, a $TiO/Ti_5Si_3/HSQ$ stack structure forms. The Ti/HSQ system exhibits a higher reactivity than that of the Ti/TEOS system.

INTRODUCTION

A number of low-k dielectric materials are being actively studied [1-3] to replace conventional interlayer dielectric SiO_2 due to the following benefits: to increase device speed by lowering interconnect RC delay, to diminish crosstalk by reducing the capacitance between parallel running lines, and to reduce power dissipation because the heat given off is directly proportional to the capacitance. Hydrogen silsesquioxane (HSQ) is one of the promising candidates because of its low dielectric constant (<3), excellent gapfill and planarization capability and low electrical leakage [4-5]. The general formula for HSQ is $(HSiO_{1.5})_{2n}$ where n = 3 to 8. Each Si atom is connected to one H atom and 1.5 O atoms.

In typical Al-metallization-based non-etchback embedded architecture which utilizes dielectric materials SOG (spin-on-glass), SOG usually has to be exposed to the sidewalls of the vias. This may lead to via poisoning due to SOG outgassing during the thermal cycle of the subsequent deposition processes [6-7]. As a result, a barrier is often used between the SOG and the via. A Ti/TiN stack structure deposited before via fill is such a choice. Many earlier researchers [8-10] have reported that Ti reacts with thermally grown SiO_2 at high temperatures. In this paper, we are interested in the possible interaction between Ti and HSQ films, since this is associated with the integrity of the HSQ film, the adhesion, and also the via poisoning issue. The results are also compared with the studies on the conventional intermetal dielectrics SiO_2 and plasma-enhanced tetraethylorthosilicate (PETEOS).

EXPERIMENTAL

The HSQ films with a thickness of 350 nm were spin-coated onto (100) Si wafers. The PETEOS films with a thickness of ~350 nm were prepared on (100) Si wafers by the use of TEOS and oxygen plasma. The 50-nm-thick Ti films were subsequently sputter-deposited on these dielectric layers. All the anneals of these Ti/dielectric samples were performed in a vacuum quartz tube furnace with a base pressure better than 4×10^{-8} Torr at various temperatures (300-800 °C) for 30 or 60 min.

Sheet resistance measurements were carried out at room temperature on both as-deposited and annealed Ti/dielectric samples using a four-point-probe apparatus. XRD analysis was used for structural characterization of the Ti layers. Conventional Rutherford backscattering spectrometry (RBS) analysis was performed to analyze medium and high Z elements in a General Ionex tandetron accelerator. The spectra were analyzed with the aid of an analysis software RUMP [11]. Elastic Recoil Detection (ERD) technique was employed in this study to profile H element. The detailed description of this technique was given elsewhere [12]. Nuclear resonance analysis (NRA) technique was used to analyze O element by employing $^{16}O(\alpha,\alpha)^{16}O$ elastic resonance at an energy of approximately 3.054 MeV. Auger electron spectrometry (AES) and secondary ion mass spectrometry (SIMS) were also used to profile element depth distribution of the samples. Pressure-resolved thermal desorption spectroscopy (TDS) was used to analyze the outgassing behavior of HSQ at elevated temperature.

RESULTS

The sheet resistance of 50 nm Ti films as a function of annealing is displayed in Fig. 1(a) for both Ti/HSQ and Ti/PETEOS structures. It is noted that the sheet resistance of the Ti films starts to increase remarkably at 300 °C, and then undergoes an abrupt fall at 550 °C for Ti on HSQ, and 600 °C for Ti on PETEOS. After that, the sheet resistance rises again to some degree and then undergoes a slight decrease with increasing temperature.

XRD analysis under a standard θ-θ configuration shows that the as-deposited Ti on HSQ or PETEOS exhibits a very strong (002) texture. As the anneal proceeds (T > 300 °C), the Ti (002) peak shifts towards lower θ angle. The peak shift is almost completed at 600 °C for Ti on HSQ, and at 650 °C for Ti on PETEOS. The shift of the Ti (002) peak is also reported by Russell et al. [10] in their studies on the Ti/SiO_2 system. Their studies show that for the Ti/SiO_2 system reacting at high temperatures, liberated oxygen atoms interstitially diffuse into the Ti lattice to form initially a solid solution of Ti and thus lead to the expansion of the Ti lattice and corresponding shift of Ti diffraction peaks. This suggests that the Ti (002) peak shift in the present study may arise from the O incorporation as well. Figure 1(b) is a plot of the O concentration in 50 nm Ti films (O/Ti ratio) as a function of annealing temperature (60 min annealing time), which shows an increasing O content in the Ti films as the anneal proceeds (T > 300 °C).

The correlation between the lattice constant C_0 of Ti, as measured from (002) peak in x-ray diffraction pattern, and the O/Ti ratio of the Ti layer is presented in Fig. 2. A straight line fits to those data from both Ti/HSQ and Ti/PETEOS quite well. A good correlation also exists between the Ti sheet resistance and O/Ti ratio in the Ti layer if one compares Figs. 1(a) and 1(b). The XRD spectra under a 2° glancing angle scan mode for the Ti/HSQ samples were given in Fig. 3. For 600 °C annealing, there is a shift of Ti peaks but no new phase forms. Both Ti_5Si_3 and TiO phases are observed in the 800 °C annealed sample. Similar observations apply in the Ti/PETEOS system. Figure 4 compares RBS spectra for Ti/HSQ samples, as-deposited and annealed at 600 and 700 °C for 60 min, respectively. Upon 600 °C annealing, a significant height decrease of the Ti signal and the appearance of a small step at the front edge of the Si signal are observed. As annealing temperature increases to 700 °C, the height of the Ti signal further decreases and the Si step becomes more apparent. RUMP simulation reveals that the Si step corresponds to the formation of Ti_5Si_3, and the final stable Ti peak consists of Ti from the surface TiO layer and Ti from the underlying Ti_5Si_3 layer. The above observations are analogous to the previous reports on the Ti/SiO_2 system [13].

Both AES and SIMS depth profiling results also confirm the formation of TiO and Ti_5Si_3 layers in the 700 °C, 60 min annealed Ti/HSQ sample. Based on SIMS data, the relative Ti_5Si_3 thickness as a function of annealing temperature for both Ti/HSQ and Ti/PETEOS structures is plotted as shown in Fig. 5. It can be seen that at 550 °C, where Ti_5Si_3 seems to just begin to form, there is little difference between Ti/HSQ and Ti/PETEOS; at 600 °C, there is more Ti_5Si_3 formation in Ti/HSQ; and at 650 °C, the difference becomes smaller. Similar results are also obtained from RBS measurements by using RUMP simulation to determine the thickness of Ti_5Si_3 formed in these annealed samples.

DISCUSSION

The Ti/HSQ system behaves differently in different annealing temperature regimes. According to XRD, RBS, SIMS, and AES results, Ti reacts with HSQ at high temperatures to form a final $TiO/Ti_5Si_3/HSQ$ stack structure. Prior to the formation of TiO (~700 °C), liberated O from HSQ diffuses through the silicide into the Ti lattice to initially form a Ti(O) solid solution. The liberated O atoms resulting from the Ti/HSQ reaction account primarily for the dramatic increase of the O composition in Ti in the high temperature regime.

In the low temperature regime, there is also a considerable amount of O dissolved in the Ti layer for the Ti/HSQ system [Fig. 1(a)]. Furthermore, it is apparent that at 300-400 °C, O incorporates more significantly than does it at 400-550 °C. In order to explain why there is such a significant amount of O diffusing into Ti in the low temperature regime, the Ti/PETEOS case is examined first. Figure 1(b) also shows a gradual O content increase in the low temperature regime for the Ti/PETEOS. Two sources are most possible to cause this. One is from the vacuum ambient. The other potential source of O is a slight Ti/PETEOS reaction in the low temperature regime. Barbour et al. [14] report that approximately 38 at. % oxygen, which is roughly equal to the solubility limit

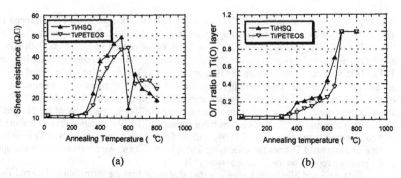

Figure 1 *(a) Sheet resistance variations and (b) w (O/Ti ratio) values in Ti(O) vs. annealing temperature (60 min) for 50 nm Ti films on HSQ and PETEOS.*

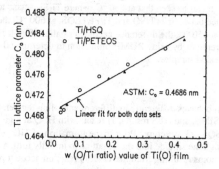

Figure 2 *A plot showing the correlation between the lattice constant C_0 of Ti, and O/Ti ratio in Ti.*

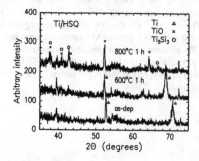

Figure 3 *XRD patterns under a glancing angle geometry (glancing angle 2°) for 50nm Ti/HSQ samples as-deposited and after annealing at 600 and 800 °C for 60 min.*

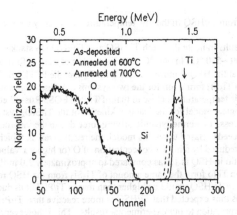

Figure 4 *RBS spectra (2.0 MeV He⁺⁺, tilt 60°) for 50nm Ti/HSQ samples as-deposited and after annealing for 60 min at 600 and 700 °C.*

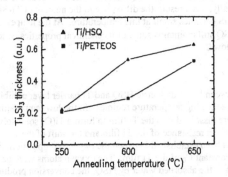

Figure 5 *The normalized Ti_5Si_3 thickness as a function of annealing temperature for 60 min annealed Ti/HSQ and Ti/PETEOS samples, as obtained from SIMS spectra.*

of O in Ti, is dissolved into a 0.7-nm-thick Ti layer after room temperature deposition of Ti in a vacuum chamber with a base pressure of 5×10^{-11} Torr, due to the high activity of Ti in dissociating SiO_2. This suggests that for the present case, even in the low temperature regime, Ti may still dissociate a limited amount of PETEOS, although significant Ti/PETEOS reaction can occur only at higher temperature. These two possible O sources certainly apply to the Ti/HSQ system as well. However, the difference in the O content for the Ti/HSQ and Ti/PETEOS at low temperatures as indicated in Fig. 1(b), implies that there should be some additional O sources for Ti/HSQ. Our TDS experiments on Ti/HSQ samples show that some water vapor and organic fragments evolve in the low temperature regime and significant H_2 evolution occurs at ~600 °C. The outgassing water vapor is assumed to be primarily from the initial adsorption of water by

HSQ. The evolved H_2O from Ti/HSQ in the low temperature regime may also come from the conversion reactions of HSQ.

SIMS and RBS results indicate that both Ti/HSQ and Ti/PETEOS stack structures begin to react and form silicide at ~550 °C. In 600 °C, there is much more Ti_5Si_3 formed in the Ti/HSQ, which implies that the HSQ is more reactive with Ti than the PETEOS. In 650 °C, however, the difference in Ti_5Si_3 formation for the two systems gets less, suggesting that the reactivity of Ti/HSQ at this temperature is close to that of Ti/PETEOS. Russell et al. [13] have developed a phenomenological model for the reaction kinetics of the Ti/SiO_2 system. This model takes into account the decrease in thermodynamic driving force due to the incorporation of O into Ti as the reaction proceeds. Based upon this model, the reaction rate at which SiO_x-type material reacts with Ti is reduced by a higher concentration of O (or higher x value) in SiO_x. Since the nominal O/Si ratio in HSQ is 1.5 as compared to approximately 2.0 in PETEOS, less O needs to be removed by the Ti to form the same amount of Ti_5Si_3 from the HSQ relative to the PETEOS. The Ti activity in Ti/HSQ would be higher than that in Ti/PETEOS due to less O incorporation into Ti. It is thus expected that Ti/HSQ is more reactive than Ti/PETEOS. This can give a reasonable interpretation to our experimental results. This is however complicated by the competing reaction in which HSQ is substantially converted into SiO_2 as the temperature gets much higher. ERD and TDS results show that a considerable amount of H loss from HSQ occurs at 550-650 °C. This implies that at 650 °C, HSQ has lost most of its H and is thus more equivalent to SiO_2 chemically. As a result, the difference in the amount of Ti_5Si_3 formed for Ti/HSQ and for Ti/PETEOS becomes small at this temperature. At the temperature below 650 °C, we can assume that HSQ still maintains some of its chemical properties, since the O/Si ratio of the HSQ film is still below 2.

CONCLUSION

The interaction between low-k dielectric HSQ and Ti barrier layer exhibited different characteristics in different annealing temperature regimes. In the low temperature regime (300-550 °C), oxygen atoms were dissolved into the Ti film to form a Ti(O) solid solution, which caused the increase of the sheet resistance of the Ti film and the shift of the Ti (002) diffraction peak. Good correlation has been established between the oxygen composition, the sheet resistance, and the lattice constant C_0 of the Ti film. The oxygen atoms were believed to come from various sources, such as the absorbed water by HSQ, the conversion products of HSQ, and the vacuum ambient. The gettering of O-containing sources by the Ti layer can reduce the risk of the via poisoning in device processing, since less amount of these outgassing materials will be absorbed by the via. In the meantime, the interaction between Ti and HSQ can promote the adhesion. In the high temperature regime (550-700 °C), a final TiO/Ti_5Si_3/HSQ stack structure was formed due to the reaction of HSQ with Ti. Due to the stoichiometry difference between HSQ and PETEOS (or SiO_2), HSQ was more reactive with Ti than PETEOS and SiO_2. In real-time processing, the maximum post-HSQ-deposition processing temperature should be controlled to avoid the reaction between Ti and HSQ, since this may impair the integrity and dielectric property of the HSQ material.

ACKNOWLEGMENTS

The great help from Dr. Stephen W. Russell (Micron Technology) and Dr. Andrew J. McKerrow (Texas Instruments) is highly appreciated. The work was partially supported by the National Science Foundation (L. Hess, DMR-96-24493) to whom the authors are greatly indebted. Support for the Center for Low Power Electronics is partially provided by NSF, State of Arizona, Analog Devices, Analogy, Burr-Brown, Hughes Aircraft, Intel, Microchip, Motorola, National Semiconductor, Rockwell, Scientific Monitoring, Texas Instruments, and Western Design Center.

REFERENCES:

1. S.-P. Jeng, M.-C. Chang, T. Kroger, P. McAnally, R. H. Havemann, VLSI Tech. Symp. Tech. Dig. 1994, p73.

2. K. J. Taylor, M. Eissa, J. Gaynor, H. Ngugen, Mat. Res. Soc. Sym. Proc. 476 (1997) 82.

3. W. W. Lee, P. S. Ho, MRS Bulletin 22 (1997) 19.

4. S.-P. Jeng, K.J. Taylor, T. Seha, M.-C. Chang, J. Fattaruso, R. H. Havemann, VLSI Tech. Symp. Kyoto, Japan, 1995, p61.

5. J. N. Bremmer, Y. Liu, K. G. Gruszynski, F. C. Dall, Mat. Res. Soc. Sym. Proc. 427 (1997) 103.

6. C. Chiang, N.V. Lam, J. K. Chu, N. Cox, J. Fraser, J. Bozarth, B. Mumford, V-MIC Conf. IEEE, 1987, p404.

7. H. G. Tompkins, C. Tracy, J. Vac. Sci. Technol. B 8 (1990) 558.

8. C. Y. Ting, M. Wittmer, S. S. Lyer, S. B. Brodsky, J. Electrochem. Soc., 131 (1984) 2934.

9. A. E. T. Kuiper, M. F. C. Willemsen, J. C. Barbour, Appl. Surf. Sci. 35 (1988-89) 186.

10. S. W. Russell, J. W. Strane, J. W. Mayer, S.-Q. Wang, J Appl. Phys. 76 (1994) 257.

11. L. R. Doolittle, Nucl. Instrum. Meth. B 9 (1985) 344.

12. Y. Zeng, S. W. Russell, A. J. McKerrow, L.-H. Chen, T. L. Alford, J. Vac. Sci. Technol B 18 (2000) 221.

13. S. W. Russell, J. W. Strane, J.W. Mayer, S.-Q. Wang, J Appl. Phys. 76 (1994) 264.

14. B. K. Patnaik, C. V. Barros Leite, G. B. Baptista, E. A. Schweikert, D. L. Cocke, L. Quinones, N. Magussen, Nucl. Instrum. Meth. B 35 (1988) 159.

Mat. Res. Soc. Symp. Proc. Vol. 612 © 2000 Materials Research Society

HIGH DENSITY PLASMA SILICON CARBIDE AS A BARRIER/ETCH STOP FILM FOR COPPER DAMASCENE INTERCONNECTS

Hichem M'Saad, Seon-Mee Cho, Manoj Vellaikal, Zhuang Li

Applied Materials, Inc.
Dielectric Systems and Modules
Santa Clara, CA 95054

ABSTRACT

A low κ dielectric barrier/etch stop has been developed for use in copper damascene application. The film is deposited using methane, silane and argon as precursors in a HDP-CVD reactor. The film has a dielectric constant of 4.2 which is lower than the dielectric constant of conventional SiC or plasma silicon nitride (>7). Film characterization including physical, electrical, adhesion to ILD films, etch selectivity, and copper diffusion barrier properties show that this film is a better barrier than silicon nitride for low κ copper damascene interconnects. This film consists of a refractive index in the range of 1.7 to 1.8, a compressive stress of 1.0-1.5×10^9 dynes/cm^2, and a leakage current of 5.0×10^{-10} A/cm^2 at 1 MV/cm. When integrated in-situ with HDP-FSG, an effective dielectric constant of 3.5 can be achieved.

INTRODUCTION

In copper dual damascene interconnects, the dielectric material is embedded with a copper diffusion barrier and an etch stop. In first generation copper damascene devices, the barrier is usually silicon nitride because of its good barrier properties to copper diffusion, its selectivity to oxides, its good dielectric strength, mechanical and chemical stability, and electrical insulation. However, silicon nitride suffers from a high dielectric constant, around 7.2. This high dielectric value increases the overall ILD time delay (Figure 1). Therefore, there is a need to substitute silicon nitride with a lower dielectric constant barrier, especially for the lower dielectric constant materials. Indeed, Table 1 shows that substituting SiN with a lower κ barrier (κ=4.2) reduces the effective κ by 17% in conjunction with black diamond but by only 12% in conjunction with silicon oxide. Therefore, we have embarked on developing a lower dielectric constant barrier which can be used in combination with low κ materials in Cu interconnects.

Table 1: *Effective Dielectric Constant for different IMD schemes in copper damascene interconnects*

Barrier K Value	ILD K Value		
	USG K=4.1	FSG K=3.3	BD K=2.7
1,000Å SiN (K = 7.2)	4.7	4.2	3.6
1,000Å HDP-SiC (K = 4.2)	4.1	3.5	3.0

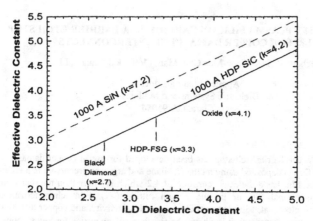

Figure 1: Barrier impact on effective dielectric constant for a 400nm ILD and a 100nm barrier. Data was generated by assuming that the barrier and low k material form a parallel plate capacitor. Black Diamond is a carbon-doped oxide low κ material [1-4] while HDP-FSG is a fluorine-doped oxide [5].

THEORY

Silicon carbide is known to exhibit excellent etch resistance to fluorine chemistry. Hence, SiC would be an ideal etch stop because it would provide good selectivity to oxide-based dielectrics. However, conventional silicon carbide is electrically leaky and exhibit a high dielectric constant, similar to silicon nitride. Leakage in conventional SiC is attributed to conduction through silicon. Because of its smaller size, carbon is less polarizable than silicon. Si has an electronic polarizability that is 20 times that of carbon. Therefore, by substituting the less polarizable carbon atom for silicon in SiC, we should obtain a lower dielectric constant SiC. Furthermore, this material will be electrically insulating because of the reduction in the silicon content.

Plasma CVD processes are ideal in depositing films with the appropriate stoichiometry. Low κ SiC film deposited using PE-CVD technique has been reported in the literature [6-7]. When deposition is performed in a high density plasma environment, we have the added benefit of depositing a dense and compressive film. Hence, Ultima HDP-CVD™ reactor was used to deposit carbon-rich silicon carbide. Silane, methane and argon were the deposition precursors.

RESULTS AND DISCUSSION

A plot of the dielectric constant measured at 1MHz versus refractive index for different HDP-CVD SiC stoichiometries is indicated in Figure 2. Films were deposited at different temperatures, ranging from 250°C to 400°C, which account for the scatter in the data. This figure shows that dielectric constant decreases with refractive index as would be expected since the refractive index is related to the optical dielectric constant by the relationship of $k=n^2$. It also demonstrates that the dielectric constant for the HDP-CVD

SiC film is tunable and can be much lower than that of conventional SiC films (κ>7). It also shows that κ in the range of 4.0 to 4.5 can be obtained by targeting a refractive index between 1.7 to 1.8. These films were compressive in nature with stress values in the 1.0-1.5×10^9 dynes/cm^2 range.

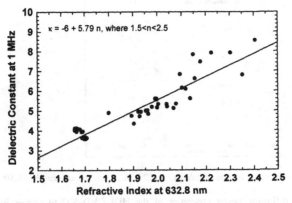

Figure 2: Dielectric Constant versus refractive index. Conventional SiC has a κ value of 7.2 which corresponds to a refractive index of 2.3.

FT-IR spectra for HDP silicon carbide films show that carbon-rich HDP-SiC contains considerably more $Si-CH_3$ and $Si(CH_2)_n$ bonds than conventional SiC (Fig. 3).

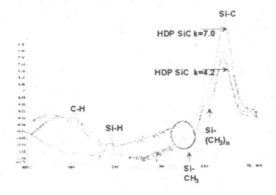

Figure 3: FT-IR of HDP-SiC at different κ values.

Conventional silicon carbide is known to be a leaky film. We have measured a current leakage density greater than 10^{-6} A/cm^2 for this film. However, HDP-SiC has a leakage density of 5×10^{-10} A/cm^2. As shown in Figure 4, the leakage current of low κ HDP SiC is comparable to HDP-CVD silicon nitride at 1 MV/cm.

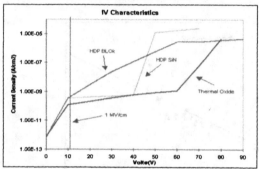

Figure 4: Leakage Current as a Function of Voltage for HDP-SiC, Thermal oxide and HDP- SiN.

A study of the diffusion barrier properties of the HDP-CVD SiC to copper has provided interesting findings; namely that HDP-CVD SiC is a better diffusion barrier to Cu than SiN. If we define the Cu diffusion length as being the distance from the Cu/dielectric interface at which the copper concentration decreases by three orders of magnitude, then the diffusion length of Cu in HDP-SiC is 100Å. Meanwhile, diffusion length of Cu in silicon nitride is ~300Å (Figure 5).

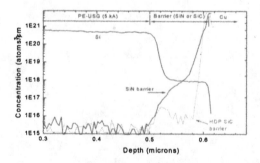

Figure 5: Cu diffusion profile for SiN and HDP SiC barrier films in a sandwich structure consisting of 500nm PE-USG/ 100nm barrier/ 400nm Cu/ Si. Structure was subjected to a 3 hours anneal at 400°C (6 anneal cycles of 30 minutes each).

The film has been successfully integrated with HDP-FSG. A multi-layered stack with six alternating layers of 1000Å HDP SiC and 1.0 μm HDP-FSG was deposited on a silicon wafer and was subsequently furnace annealed for 2 hours at 500°C in a nitrogen ambient. Post anneal results are shown in Figure 6. From this figure, it can be deduced that HDP-SiC adheres well to HDP-FSG. Figure 7 shows SIMS data for this multi-layered stack after annealing for 2 hours at 500°C. No fluorine diffusion into the SiC film was observed.

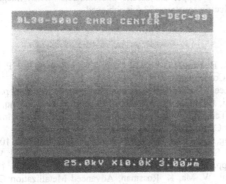

Figure 6: Post anneal results for the six layered multi-layered HDP SiC/FSG structure.

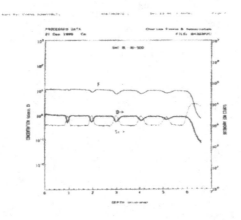

Figure 7: SIMS data for the stack above showing no F diffusion from FSG after anneal.

CONCLUSIONS

In summary, we have developed an HDP-CVD based SiC film as a barrier/etch stop in Cu damascene applications. Film characterization including physical, electrical, adhesion to ILD films, and copper diffusion barrier properties show that this film is a better barrier than silicon nitride for low κ copper damascene interconnects. When integrated with HDP-FSG, an effective dielectric constant of 3.5 is achieved.

REFERENCES

1. T. Poon, J. Ma, M. Naik, B. Tang, M. Yu, C-I Lang, I. Lou, D. Sugiarto, P. Lee, W-F. Yau, D. Cheung, VMIC Conference, p. 190 (1999).
2. B. Pang, W. F. Yau, P. Lee, M. Naik, Semiconductor Fabtech, 10th Edition, p. 285.
3. M. Naik, S. Parikh, P. Li, J. Educato, D. Cheung, I. Hashim, P. hey, S. Jenq, T. Pan, F. Redeker, V. Rana, B. Tang, D. Yost, Proceedings of the IEEE 1999 International Interconnect Technology Conference p.181-3 (1999).
4. R. P. Mandal, D. Cheung, W.F. Yau; B. Cohen, S. Rengarajan, E. Chou, 10th Annual IEEE/SEMI. Advanced Semiconductor Manufacturing Conference and Workshop. ASMC 99 Proceedings, pp. 299-303 (1999).
5. H. M'Saad, M. Vellaikal, W. Ma, K. Rossman, Advanced Metallization Conference Proceedings, Sept. 1999, p. 22.
6. M. J. Loboda, J.A. Seifferly, F.C. Dall, J. Vac. Sci. Technol. A, 12(1), p. 90(1994).
7. P. Xu, K. Huang, A. Patel, S. Rathi, B. Tang, J. Ferguson, J. Huang, C. Ngai, IITC Conference, 1999.

Mat. Res. Soc. Symp. Proc. Vol. 612 © 2000 Materials Research Society

Evolution of Surface Morphology During Cu(TMVS)(hfac) Sourced Copper CVD

Daewon Yang, Jongwon Hong, and Timothy S. Cale
Focus Center – New York, Rensselaer: Interconnections for Gigascale Integration,
Rensselaer Polytechnic Institute, Troy, NY 12180

ABSTRACT

In this paper, we describe an experimental study of the nucleation and growth stages during Cu(TMVS)(hfac) sourced Cu CVD on TaN substrates. In particular, we have investigated the effects of water vapor as a co-reactant on evolving surface morphology. The results of short (less than 10 s) depositions without/with water vapor indicate that water vapor helps to reduce the incubation time and to enhance the nuclei formation, uniformity, and adhesion (based on AFM analysis). Introducing water vapor during only the initial stage of deposition results in lower roughnesses, larger grain sizes, and lower short-range roughnesses as compared to the films deposited without water vapor. From this study, we conclude that water vapor enhances Cu nucleation and that a relatively small amount of water vapor before or during the initial stage of deposition improves surface morphology in terms of roughness and grain size.

INTRODUCTION

As the device densities of integrated circuits (ICs) increases, overall circuit performance becomes increasingly dependent on the properties of the metal films used to interconnect the devices [1,2]. Copper based metallization has been introduced into leading edge ICs because of its promising reliability performance, as well as the potential cost savings associated with damascene processing [3-5]. The cost savings and IC performance increases due to Cu introduction will be more fully realized if deep sub-quarter micron, high aspect ratio, features can be filled inexpensively with barrier material and copper. This is one of the most important issues for multilevel metallization (MLM) process flows today [1].

Over the last several years, chemical vapor deposition (CVD) of Cu has been heavily investigated as a way of depositing Cu into tight contact holes [4-7]. Cu(TMVS)(hfac) (TMVS is trimethylvinylsilane and hfac is 1,1,1,5,5,5-hexafluoroacetylacetonate) is the most widely used Cu(I) precursor. The overall reaction [3,6] has been proposed as:

$$2Cu^{I}(TMVS)(hfac)_{(g)} \rightarrow Cu^{0}_{(s)} + Cu^{II}(hfac)_{2\,(g)} + 2TMVS_{(g)} \qquad (1)$$

$Cu^{II}(hfac)_2$ and TMVS are thermally stable up to at least 523 K; and leave the surface without decomposing. Complete TMVS desorption from the surface occurs around 398 to 423 K.

The addition of water vapor during CVD Cu deposition has been studied by several groups [4,5,7]. Kim et al. [7] reported that shorter incubation time and enhanced growth rate could be achieved using water vapor. Based on analyses of final thick film properties, the addition of water vapor into the reaction chamber enhanced nucleation [5,7] and increased deposition rate [4,5,7]. However, using excess water vapor resulted in higher resistivity [5,8]. Mermet et al. [5] reported that deposited film resistivity decreased upon annealing at 723 K for 30 min.

Even though CVD of Cu has been studied for several years [5,6], many important physical processes associated with film growth are still not well understood. Many properties of thin films are directly controlled by surface morphology, which in turn is largely determined by the initial stages of deposition. Since most studies have focused on the growth stage of deposition, relatively little has been reported about the early stages of deposition. Therefore, our experiments include studies of both the early stages and the growth stage of Cu deposition in a LPCVD reactor. The study of the early stage of CVD processes will help to us understand the roles of nucleation and grain growth during continuous film formation.

EXPERIMENT

In this study, an LPCVD microreactor was used to investigate CVD Cu nucleation and growth using Cu(TMVS)(hfac) as the precursor. Cu was deposited on a heated TaN substrate. TaN substrates were chosen for this study because it is a commonly investigated diffusion barrier for Cu. The substrate temperature was estimated by measuring the heating plate temperature with two thermocouples. The temperature readings were calibrated periodically and the temperature deference between the substrate surface and the heating plate was measured at several different process conditions. The reactor pressure was sensed by a capacitance manometer, and adjusted by a butterfly valve via a closed-loop pressure controller. A base pressure of 1 mTorr inside reaction chamber was achieved. A load lock chamber was installed to avoid air contamination. A scrubber (Mystaire wet scrubbing system) was installed on the exhaust line to remove any unreacted and hazardous gases.

Cu(TMVS)(hfac) is a Cu compound containing an excess of the stabilization chemical (TMVS) and TMVS is highly volatile compared to Cu(TMVS)(hfac). Conventional bubbling of this precursor may result in inconsistent deposition properties, such as decreasing deposition rates, because the carrier gas possibly strips the stabilizing components from the precursor. Therefore, in our reactor system, Cu(TMVS)(hfac) was delivered to the reaction chamber using a direct liquid injection (DLI) system (Porter Instrument Company, Inc). In this DLI system, liquid precursor is pushed through a vaporizer by helium (He) gas at 25 psi and the liquid flow is controlled by a mass flow meter. Semiconductor grade He gas (purity higher than 99.9995%) was used as the carrier/diluent gas, which was mixed with liquid precursor at the vaporizer, then injected into the reaction chamber in the vapor phase. A process line was installed just before showerhead in order to introduce co-reactants using a bubbler system and the flow was controlled using a mass flow controller. He gas was used to carry the co-reactant into the reaction chamber. Process gas delivery lines were maintained at 333 K in order to avoid condensation of the precursor during delivery. For the purposes of this discussion, the process line only includes the parts from the vaporizer to the showerhead.

Experiments were performed to investigate the evolution of CVD Cu films. Those experiments included studies of the early stage and the growth stage of deposition with and without water vapor. For the study of the early stages of deposition, both with and without water vapor, Cu was deposited for 5 or 10 s on TaN substrates. For the study of the growth stage of Cu CVD, depositions were conducted for 10 min.

A surface profilometer (α-step) was used to estimate the average deposited film thickness after making steps in the Cu deposited film near the center of the sample by photolithographic and wet etching methods [8]. Atomic force microscopy (AFM) was used in the contact mode to analyze surface morphologies of the deposited Cu films. Surface roughnesses reported are root-mean-square (RMS) values. The properties and dimensions of the cantilever play an important role in determining the sensitivity and resolution of the AFM. Specifications of the probes used in this study are listed in Table I.

RESULTS AND DISCUSSION

Figure 1(a) shows surface morphology of a representative TaN substrate used in this study. According to the AFM analysis over the $1\mu m \times 1\mu m$ area shown in the Figure 1(a), the mean z height was 4.12 nm and RMS roughness was 0.78 nm. Lu *et al.* [10] suggested three parameters to describe rough surfaces: interface width, w; lateral correlation length, ξ; and roughness exponent, α. Figure 1(c) shows the three parameters describing surface morphologies as well as how to estimate the parameters using the height-height correlation function H(r). The height-height correlation function, H(r), is defined as $<[h(r)-h(0)]^2>$, where h(r) is the surface height at position r. The interface width (w, vertical correlation) is the amplitude of surface fluctuations, which is given by the saturation value, $H(r) = 2w^2$ as $r \to \infty$. The lateral correlation length (ξ) is the wavelength of surface fluctuations, which is given quantitatively by the 'knee' in H(r) from the graph in Figure 1(c). The roughness exponent (α) describes how "wiggly" the short-range (local) surface is. The roughness exponent can be extracted from the slope at small values of r (see Figure 1(c)) and has a value between 0 and 1. A small α value implies a rougher surface. Figure 1(b) shows log-log plots of the height-height correlation function of the TaN surface (shown in Figure 1(a)) as a function of position r. From Figure 1(b), the interface width (w) is estimated to be about 0.78 nm, the lateral correlation length (ξ) is about 23 nm, and the roughness exponent (α) is about 0.58.

Table I. Specifications of probes used [9].

Manufacturer	ThermoMicroscopes
Model name	Sharpened Microlever
Cantilever length (μm)	180
Cantilever width(μm)	18
Cantilever thickness(μm)	0.6
Force constant (N/m)	0.05
Resonant frequency (kHz)	22
Typical radius of tip (nm)	< 20
Tip half angle (degrees)	18
Tip shape	pyramidal
Reflective coating	gold

Nucleation stage study

Experiments for the study of CVD Cu nucleation with and without water vapor were performed on TaN substrates. Water vapor was introduced using a bubbler type evaporator system. Depositions were conducted for 5 s periods using different process protocols at 473 K substrate temperature, 20 mg/min precursor flow, 100 sccm carrier gas flow, and 0.5 Torr total pressure. According to AFM micrographs of Cu nuclei deposited for 5 s without and with water vapor (shown in Figure 2), the nuclei sizes and densities vary significantly in different deposition

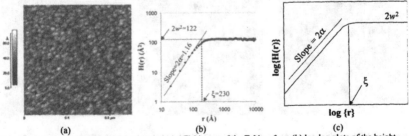

(a) (b) (c)

Figure 1. Surface morphology of the substrate: (a) AFM image of the TaN surface, (b) log-log plots of the height-height correlation function measured from the surface of TaN substrate, and (c) three parameters describing surface morphologies.

processes. The 5 s depositions without water vapor (Figure 2(a)) result in very few sparsely distributed nuclei. This is probably due to the low deposition temperature (not enough energy) at the given process conditions. However, the nuclei densities increase significantly by introducing 50 sccm of water vapor flow before/during the deposition, as shown in Figures 2(b) and (c). It seems that the incubation time is reduced by introducing water vapor during the initial stage of deposition. Apparently, the largest number of small-size nuclei results from introducing water vapor during the deposition. These results indicate that nucleation rate could be enhanced by proper process conditions.

Hydroxyl groups on the surface are thought to be as the primary sites for adsorption of Cu precursor [11]. According to Jain *et al.* [4], introducing water vapor during deposition results in formation of new hydroxyl groups on the surface, which can act as sites for nucleation. Comparing the results of the depositions with water vapor introduced before deposition to those during deposition, nuclei density is lower and they are larger. These results indicate that water vapor may also improve gas phase reactions that accelerate Cu nucleation. According to Jain *et al.* [12], the Cu-TMVS bond could be weakened in the presence of water by hydrogen bridging between the oxygen in the 'hfac' ring and the hydrogen in the water molecule and/or by the electron donation by the oxygen in the water molecule to the Cu center. Dissociation of TMVS from Cu(TMVS)(hfac) has been reported to be the rate limiting step for CVD Cu using this precursor. The presence of water might accelerate this step, and result in higher deposition rates.

During AFM analysis, image scans in the contact mode failed several times for the films deposited for a short time (less than 10 s) without water vapor (refer to poor images shown in Figure 3(a)), which is most likely because the AFM tip moves the nuclei along substrate surface. However, the surface scans in the contact mode for the nuclei deposited with water vapor (refer to Figure 3(b)) were performed without difficulty and scanned images were clear. This result indicates that the adhesion of Cu nuclei on TaN films is apparently stronger when water vapor is introduced during the initial deposition, compared to those deposited without water vapor.

Growth stage study

The upper figures in Figure 4 show topographies of CVD Cu films deposited for 10 min at 448 K without and with 10 sccm water vapor introduced for 2 min during the initial stage of

(a)   (b)   (c)

Figure 2. AFMs resulting from 5 s CVD Cu on TaN at 473 K substrate temperature, 0.5 Torr total pressure, 20 mg/min precursor flow, and 100 sccm carrier gas flow: (a) without water vapor, (b) with 50 sccm water vapor flow, and (c) water vapor introduced for 30 s before the deposition.

deposition. Thicknesses of the films in Figures 4(a) and (b) are about 0.64 µm and 0.84 µm, respectively. This result indicates that deposition rate increases by introducing water during deposition. Based on scratch and tape tests, adhesion of the film deposited with water vapor to TaN is much stronger than the film deposited without water vapor.

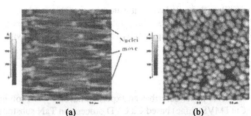

(a)   (b)

Figure 3. AFMs of CVD Cu nuclei: (a) 10 s deposition without water vapor and (b) 2 s deposition with 20 sccm water vapor. Deposition conditions were 448 K substrate temperature, 1 Torr total pressure, 20 mg/min precursor flow, and 50 sccm He flow.

The surfaces of these films are analyzed using the height-height correlation function, H(r), (refer to Figure 1). The lower figures in Figure 4 show log-log plots of the height-height correlation functions as functions of position. The H(r) analysis of the surface of the film deposited without water vapor (Figure 4(a)) results in about 64 nm interface width (w), about 138 nm lateral correlation length (ξ), and about 0.37 roughness exponent (α). The H(r) analysis of the surface of the film deposited using water vapor (Figure 4(b)) results in about 45 nm interface width, about 187 nm lateral correlation length, and about 0.67 roughness exponent. These results show that the surface of films deposited with water vapor during the initial stage results in lower roughnesses, larger grain sizes, and lower short-range roughnesses as compared to the surface of

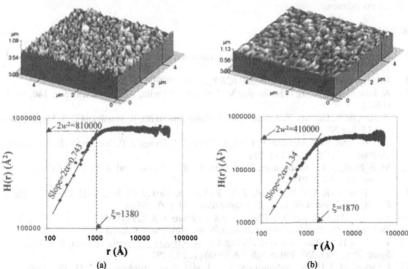

(a)   (b)

Figure 4. Topography (upper) and log-log plots of the height-height correlation function (lower) of CVD Cu films deposited at 473 K for 10 min: (a) without water vapor and (b) with 10 sccm water vapor introduced for 120 s during the initial stage of deposition.

the films deposited without water vapor. This can be explained with the results from the early stage study that introducing water vapor lowers the kinetic barrier of nucleation on TaN substrates and that apparently forms more Cu nuclei with uniform distribution, which may result in smoother films.

CONCLUSIONS

This paper describes an experimental study of evolving surface morphology in Cu(TMVS)(hfac) based Cu CVD process on TaN substrates. The effects of water vapor as a co-reactant during the nucleation and growth stages of deposition are studied. Compared to those without water vapor, depositions with water vapor result in higher nuclei density, uniform distributions, and apparently stronger adhesion (based on AFM analysis) on TaN substrates. From this study, it can be concluded that water vapor enhances nucleation and that introducing water vapor during the initial stage of deposition improves growth rate, surface roughness, and adhesion. The best process protocol for Cu CVD in the reactor system used in this study is introducing less than 10 sccm of water vapor before or during the initial stage of deposition at 473 K substrate temperature, 20 mg/min precursor flow, 50 sccm carrier gas flow, and 1 Torr total pressure.

ACKNOWLEDGMENTS

The authors gratefully acknowledge support from Tokyo Electron Arizona and the National Science Foundation.

REFERENCES

1. International Technology Roadmap for Semiconductors, 1999 edition, (http://www.itrs.net/ 1999_SIA_Roadmap/Home.htm).
2. R. Singh and R. K. Ulrich, *INTERFACE* **8(2)**, 26 (1999).
3. A. Jain, K.-M. Chi, T. T. Kodas and M. J. Hampden-Smith, *J. Electrochem. Soc.* **140**, 1434 (1993).
4. A. Jain, A. V. Gelatos, T. T. Kodas, M. J. Hampden-Smith, R. Marsh and C. J. Mogab, *Thin Solid Films* **262**, 52 (1995).
5. J.-L. Mermet, M.-J. Mouche, F. Pires, E. Richard, J. Torres, J. Palleau and F. Braud, *Journal De Physique IV* **5**, C5-517 (1995).
7. M. B. Naik, S, K. Lakshmanan, R. H. Wentorf, R. R. Reeves and W. N. Gill, *J. Cryst. Growth* **19**, 133 (1998).
8. J.-Y. Kim, Y.-K. Lee, H.-S. Park, J.-W. Park, D.-K. Park, J.-H. Joo, W.-H. Lee, Y.-K. Ko, P. J. Reucroft and B.-R. Cho, *Thin Solid films* **330**, 190 (1998).
10. S. Kim, J.-M. Park and D.-J. Choi, *Thin Solid Films* **315**, 229 (1998).
11. Website of ThermoMicroscopes (at http://www.park.com/products).
12. T.-M. Lu, H.-N. Yang and G.-C. Wang, Fractal Aspects of Materials (Mater. Res. Soc. Symp. Proc. Vol. 367, Pittsburgh, PA 1995) pp. 283-292.
13. J. Farkas, M. J. Hampden-Smith and T. T. Kodas, *J. Electrochem. Soc.* **141**, 3539 (1994).
14. A. Jain, T. T. Kodas, T. S. Corbitt and M. J. Hampden-Smith, *Chem. Mater.* **8**, 1119 (1996).

Mat. Res. Soc. Symp. Proc. Vol. 612 © 2000 Materials Research Society

Studies of Copper Surfaces modified by Thermal and Plasma Treatments

G.P. Beyer, M. Baklanov, T. Conard, and K. Maex
Imec
Kapeldreef 75
3001 Leuven
BELGIUM

ABSTRACT

It was found that copper surfaces, which had been exposed to a clean room atmosphere, were covered by a layer, whose chemical composition can be described by $Cu(OH)_2 \cdot CuCO_3$. This layer can effectively be removed by either a short thermal treatment in vacuum at 350°C, a hydrogen plasma treatment, or a combination of both. Ex-situ photoelectron spectroscopy measurements show little difference of the chemical composition of the surface after the respective treatments. The thermal treatment, however, gives rise to re-crystallisation of the copper film due to the difference in temperature of deposition and the anneal. Ex-situ ellipsometry measurements indicate that the hydrogen plasma not only removes $Cu(OH)_2 \cdot CuCO_3$ but also passivates the copper surface.

INTRODUCTION

Advanced interconnect structures of integrated circuits incorporate copper for the transmission of the electrical signal. Although copper has a lower resistivity than aluminium it has been introduced into integrated circuit production only lately because of concerns of copper diffusion into the active region of the electronic devices. In order to suppress the copper diffusion the copper interconnect is usually encapsulated by metallic and dielectric diffusion barriers. This encapsulation scheme raises a number of issues such as interface diffusion in electromigration [1] and interface reactions during the deposition of the dielectric diffusion barrier on top of copper [2]. In the latter study it was demonstrated that the reaction between the silicon nitride and the copper was retarded by the presence of a native passivation layer on the copper surface. The disadvantage of this passivation layer, however, is that it is not self-limiting and stable as in the case of aluminium. The exposure of the copper surface to wet chemicals has been described by Apen et al. [3]. In this study the influence of a thermal treatment or hydrogen plasma on the copper surface has been studied.

EXPERIMENTAL DETAILS

Copper layers were deposited by sputtering on substrates consisting of a silicon wafer, covered by oxide and tantalum nitride. The temperature during deposition was approximately 50°C. The thickness of the layer was 150 nm. The wafers were stored in a clean room ambient for about 6 months. Then the copper layers were subjected to a thermal treatment, a hydrogen plasma treatment, or a combination of both. The experimental details of the various treatments are as follows. The wafers were heated in a vacuum chamber to a temperature of 350°C for two minutes. The back ground pressure of the chamber is 10^{-7} Torr. During processing the pressure rises to about 1 mTorr due to the presence of argon, which couples the wafer thermally to the heated chuck. The plasma treatment occurs in a mixture consisting of 5% hydrogen and 95% helium at pressures below 100 mTorr and lasts for one minute.

DISCUSSION

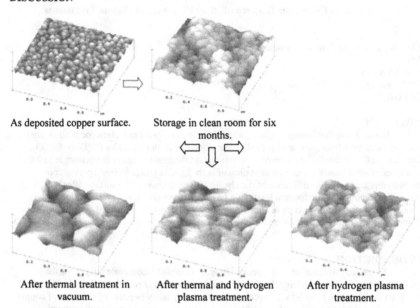

As deposited copper surface.　　Storage in clean room for six months.

After thermal treatment in vacuum.　　After thermal and hydrogen plasma treatment.　　After hydrogen plasma treatment.

Fig. 1 AFM pictures of the copper surface after various treatments. The height scale is 50 nm per division.

The Atomic Force Microscopy (AFM) pictures in Fig. 1 show the evolution of the copper surface as a result of the various treatments. As deposited layers have a clearly recognisable grain structure with a diameter of about 80 nm. After storage for six months the grain structure is still preserved though less defined. Upon thermal treatment in vacuum the mean grain size increases by approximately a factor of five. The re-crystallisation of the copper film is due to the difference of the anneal and deposition temperature. When the thermal treatment is followed by an exposure to the hydrogen plasma the large grains are observed, too. A single hydrogen plasma treatment, however, does not seem to alter the grain structure compared with the structure of the stored sample.

The composition of the residual gas in the anneal chambers was monitored in-situ by a quadrupole mass spectrometer. This made it possible to study the desorption of gas molecules from the copper surface in-situ. Fig. 2 displays the evolution of the main desorption products during the thermal treatment. The main desorption species are mass 44, 28, 16, and 18 amu. The latter is due to the removal of water from the copper surface. Mass 44 is most likely carbon dioxide, which fragments into carbon monoxide (mass 28), and oxygen (mass 16). Mass 18 is water. The signal at mass 38 is an isotope from the argon gas, which transfers the heat from the chuck to the wafer.

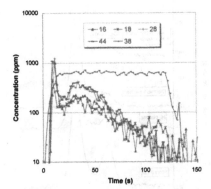

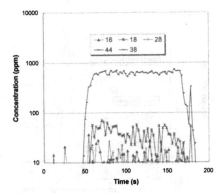

Fig. 2 Desorption products from a copper surface, which had been stored in the clean room. Wafer temperature 350°C, pressure 1 mTorr argon, duration of thermal treatment two minutes.

Fig. 3 Desorption analysis of a copper surface, which had been cleaned by a hydrogen plasma. The signals do not rise above the background level.

The desorption appears to occur in 2 steps. In the pre-equilibrium of the thermal treatment, i.e. before the argon signal at 38 amu reaches steady state, all the molecules record the highest concentration in form a sharp peak within the first ten seconds. At this stage the argon pressure at the back of the wafer is still increasing which indicates that the wafer temperature is rising. Within 30-35 seconds after the start of the thermal treatment– at which point the argon signal and the wafer backside pressure are stable - the signal of the desorbing molecules goes through a second, albeit smaller maximum. The signals then decays monotonically until it levels out at the background noise of the measurement after about 100 seconds after the start of the heat treatment. This double peak structure, which is not observed in the desorption signal of e.g. water from silicon oxide layers, is interpreted as being due to the desorption of first loosely bonded species on the surface of copper followed by the desorption of chemisorbed species.

In another experiment a wafer with a copper layer was first exposed to a hydrogen plasma and then heated in the anneal chamber to see if desorption would still occur after the plasma treatment (Fig. 3). No signals above the noise level were recorded.

These results show that volatile compounds can efficiently be removed from the copper surface by a thermal treatment or hydrogen plasma. The appearance of carbon di and monoxide not only in the low temperature but also in the high temperature part of the desorption spectrum indicates the presence of chemically bonded carbon oxide on the copper surface.

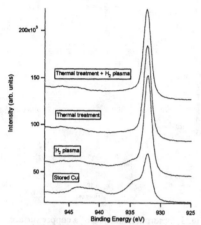

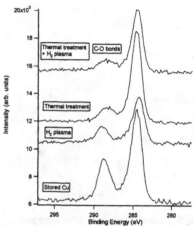

Fig. 4 Cu$2p$ spectra after storage, after hydrogen plasma, after a thermal treatment in vacuum, and after a combination of both thermal treatment and hydrogen plasma.

Fig. 5 C$1s$ spectra after storage, after hydrogen plasma, after a thermal treatment in vacuum, and after a combination of both thermal treatment and hydrogen plasma.

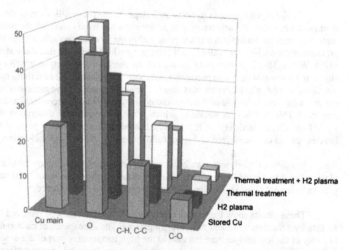

Fig. 6 Chemical make-up of the copper surface after storage, after hydrogen plasma, after a thermal treatment in vacuum, and after a combination of both thermal treatment and hydrogen plasma.

The chemical make-up of the copper surface was studied by ex-situ X-ray photoelectron · spectroscopy (XPS). Fig. 4 and 5 show the $Cu2p_{3/2}$ and $C1s$ spectra from the different samples. The $Cu2p$ reference spectrum in Fig. 4 contains two structure at ~932 eV and ~934 eV, which are attributed to metallic Cu and Cu^{++} (CuO or Cu hydroxide). Between 938 and 945 eV a shake up structure associated to the Cu^{++} is observed. After the hydrogen plasma, the Cu^{++} peaks are strongly reduced. The reduction is even more pronounced after the thermal treatment. No differences are observed in the $Cu2p_{3/2}$ photoelectron spectrum between the thermal treatment and thermal treatment + hydrogen plasma. The main peak of the carbon spectrum (Fig. 5) is attributed to C-H and C-C bonds while the structure at the higher binding energy is associated to C-O bonds. It is clearly observed that the relative amount of C-O bonds decreases after hydrogen plasma, thermal treatment, or a combination of both. Fig. 6 represents the concentration of each element assuming that the signals are derived from a homogeneous layer. After treatment, regardless if an anneal, a hydrogen plasma or a combination of both is given, the copper concentration increases. A drop in the oxygen and in the carbon concentration, linked with oxygen, accompanies this increase.

The optical characteristics of the copper surface were studied by ex-situ ellipsometry. The main parameters of an ellipsometry measurementsare the angles delta and psi. A decrease of delta is related to film growth. Therefore, a large delta value is associated with a clean metal surface. A decrease in psi is associated with the formation of a non transparent layer or roughening of the surface. The storage of as deposited copper at atmosphere reduces both the delta and the psi value (Fig. 7). This is in agreement with both the XPS and the AFM measurements, which indicate a contamination and roughening of the copper surface by exposure atmosphere. A thermal treatment in the vacuum brings the delta value back to close the original value. This indicates a cleaning effect of the anneal on the copper surface, which complies with the results derived from the XPS measurements.

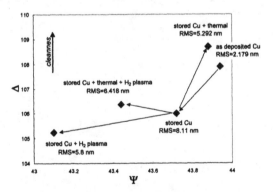

Fig. 7 Ellipsometry measurement of the copper surface after deposition, after storage in the clean room, and after a thermal treatment in vacuum, a hydrogen plasma treatment or a combination of both. The RMS values indicate the surface roughness of the various samples.

If the copper surface undergoes a hydrogen plasma treatment both the delta and the psi values decrease. Although the surface topography of the hydrogen plasma treated sample resembles the one of the stored sample the RMS values indicate a smoother surface after the treatment, which shows that the decrease of psi is not related to surface roughening. What is more the delta value, which is lower after the plasma treatment, cannot be reconciled with the XPS measurement that clearly demonstrated the cleaning effect of the hydrogen plasma.

If the hydrogen plasma is preceded by a thermal treatment the delta value remains the same, but the psi decreases. This points to the formation of a non-transparent layer as a by product of the hydrogen plasma. We postulate that the hydrogen plasma gives rise to the passivation of the copper surface by a hydride [4], which is stable in atmosphere for at least a few hours.

CONCLUSIONS

The storage of as deposited copper gives rise to a coverage of the surface by a layer of $Cu(OH)_2 \cdot CuCO_3$. The surface can be cleaned by a thermal treatment in vacuum, a hydrogen plasma, or a combination of both. As a by-product of the thermal treatment re-crystallisation of the copper layer is observed. There is evidence that the hydrogen plasma treatment leads to surface passivation.

REFERENCES

[1] L. Vanasupa, Y.-C. Joo, P.R. Besser, and S. Pramanick, *Journal Applied Physics*, 85(5), p. 2583 (1999)
[2] K.S. Low, W. Pamler, M. Schwerd, H.J. Barth, and Koerner, to be published in *Proceedings Advanced Metallisation Conference*, 1999, Florida
[3] E. Apen, B.R. Rogers, and J. Sellers, *J. Vac. Sci. Technol*, A16(3), (1998) p. 1227
[4] Y.L. Chan, P. Chuang, and T.J. Chuang, *J. Vac. Sci. Technol*, A16(3), p. 1023(1998)

Mat. Res. Soc. Symp. Proc. Vol. 612 © 2000 Materials Research Society

Study of Ta as a Diffusion Barrier in Cu/SiO$_2$ Structure

J. S. Pan[1], A. T. S. Wee[2], C. H. A. Huan[2], J. W. Chai[1], J. H. Zhang[3]
[1]Institute of Materials Research & Engineering, 3 Research Link, Singapore 117602
[2]Department of Physics, National University of Singapore, 10 Kent Ridge Crescent, Singapore 119260
[3]School of Mechanical and Production Engineering, Nanyang Technological University, Nanyang Avenue, Singapore 639798

Abstract

Tantalum (Ta) thin films of 35 nm thickness were investigated as diffusion barriers as well as adhesion-promoting layers between Cu and SiO$_2$ using X-ray diffractometry (XRD), Scanning electron microscopy (SEM), Auger electron spectroscopy (AES) and X-ray photoelectron spectroscopy (XPS). After annealing at 600°C for 1h in vacuum, no evidence of interdiffusion was observed. However, XPS depth profiling indicates that elemental Si appears at the Ta/SiO$_2$ interface after annealing. In-situ XPS studies show that the Ta/SiO$_2$ interface was stable until 500°C, but about 32% of the interfacial SiO$_2$ was reduced to elemental Si at 600°C. Upon cooling to room temperature, some elemental Si recombined to form SiO$_2$ again, leaving only 6.5% elemental Si. Comparative studies on the interface chemical states of Cu/SiO$_2$ and Ta/SiO$_2$ indicate that the stability of the Cu/Ta/SiO$_2$/Si system may be ascribed to the strong bonding of Ta and SiO$_2$, due to the reduction of SiO$_2$ through Ta oxide formation.

1. Introduction

Copper is an attractive material for ULSI metallization due to its lower resistivity (1.7 µΩcm) and higher electromigration resistance compared to Al and Al-based alloys [1-4]. However, the implementation of Cu metallization requires the use of a barrier layer since Cu diffuses easily into Si and SiO$_2$ even at quite modest temperatures, which leads to degradation of device reliability. Also, copper lacks the ability to adhere to SiO$_2$ and most insulating substrates [4]. This is due to the inability of copper to reduce SiO$_2$, which results in a purely mechanical bond at the interface. Thus thin film layers should be used between Cu and SiO$_2$ to improve the adhesion of Cu to the SiO$_2$ layer [5-7]. The characteristics of the adhesion promoter must have good adhesion with copper and the oxide layer. This means that it should have some limited interaction with both of these layers. A good diffusion barrier should have minimal interaction with copper so that it does not affect the resistance of the copper interconnect. Another problem with metallization is the stress that is developed in the thin metal lines during annealing. Excessive stresses can lead to failure of interconnect lines. Bilayer metallization is especially susceptible to the stress problem because any interaction between the layers tends to alter the stress state of the metallization as a whole.

There has been considerable effort to identify a suitable diffusion barrier for Cu metallization. Ta and Ta(N) materials are among the most promising barrier materials for Cu metallization [8-13], since Ta has relatively high melting (2996 °C) and silicidation (~650 °C) temperatures, and very low solubility in Cu [9, 10]. In addition, Ta does not react with Cu and the diffusion of Cu through Ta deposited on oxidized Si substrates does not occur up to 700 °C [9, 10]. However, it has been reported that the failure temperature of the Ta diffusion barrier is still as low as 600 °C [10]. The diffusion of Cu through microstructural defects such as grain boundaries, protrusions, and voids in Ta thin film is responsible for the low failure temperature [9, 14]. Therefore, current research of Ta as a diffusion barrier for Cu metallization has focused on improving barrier properties of Ta thin films using different deposition methods. To our knowledge, no study on the interdiffusion and/or interfacial reactions taking place in the Cu/Ta/SiO$_2$/Si metallization scheme has been reported. It is important to understand the basic properties of materials of potential use as the Cu diffusion barrier on SiO$_2$, As this knowledge is applicable to metallization technology in ULSI.

In the present study, we investigate the thermal stability of Ta barrier layers in a Cu metallization system. The interdiffusion and/or reactions in the stacked structure of Cu/Ta/SiO$_2$/Si are elucidated.

2. Experimental

Cu (~100 nm)/Ta (35 nm) bilayer films were sequentially sputtered on thermally grown SiO$_2$ layers on p-type Si(100) substrates using a dc magnetron sputtering system without breaking vacuum. The base pressure of the sputtering chamber was lower than 1×10^{-6} Torr. Deposition of all films was performed under the conditions of Ar pressure of 8 mTorr and target power of 120W without intentional substrate heating. The sputtering Ta and Cu targets used have purities of 99.95% and 99.997% respectively. The Ar gas used was of electronic grade. The wafers were subsequently diced into about 1×1 cm^2 pieces for various thermal treatments. Some samples were then annealed at 600°C for 1h in a vacuum of 1×10^{-7} Torr. For the purposes of comparison, single layer ~30nm films of Ta and Cu were also deposited on SiO$_2$/Si for interface analysis.

The crystallographic structures of the phase formed by the solid phase reactions after annealing were characterized by X-ray diffractometry (XRD), using a Schimadzu XRD-6000 diffractometer with Cu Kα radiation. To enhance the sensitivity for X-rays, the measurements were performed in parallel beam geometry (glancing mode) at a constant incident angle of α = 1°. The surface morphology of the samples before and after annealing was investigated in a Hitachi S-3500N scanning electron microscope (SEM). The degree of interdiffusion and/or reactions taking place in the Cu/Ta/SiO$_2$/Si structure were deduced from the variation in the compositional depth profiles obtained by Auger electron spectroscopy (AES) analysis with Ar ion sputtering. The AES analysis was performed in VG ESCALAB system using a 10 keV primary electron beam and a concentric hemispherical analyzer (CHA) set at a constant retard ratio (CRR) of 4. In the AES depth profiles, atomic percentages were calculated by summing the observed Auger peak-to-peak intensities in differential spectra, corrected using standard sensitivity factors. The chemical states of the elements at the interface were characterized by combining the X-ray photoelectron spectroscopy (XPS) measurement with Ar ion sputtering. The XPS experiments were performed in a VG ESCALAB 220i-XL instrument using a monochromatic Al Kα x-ray source (1486.6 eV) and a concentric hemispherical analyzer (CHA) set at a constant analyzer energy of 20 eV. All XPS spectra were taken at a take-off angle of 90°, measured with respect to sample surface plane. To compensate for the surface charging effect, all XPS binding energies were referenced to the adventitious C 1s peak at the binding energy (BE) of 284.8 eV.

3. Results and discussion

The phase of the specimens before and after annealing was analyzed with glancing XRD. As shown in Fig. 1, no phases due to the reaction within the structure such as Ta-Si and Cu-Si compounds were observed after annealing at 600 °C. The main difference in the XRD patterns before and after annealing was the intensity of the Cu(111) peak. After annealing, the intensity of this peak increases while the full width at half maximum (FWHM) decreases from around 0.6° to 0.3°. This is ascribed to Cu grain growth after annealing. The surface morphology of the Cu/Ta/SiO$_2$/Si structure before and after annealing was observed using SEM. As expected from the XRD spectra, no noticeable change was found after annealing at 600 °C, indicating that no reaction between Cu and Ta or Si occurred during annealing at 600°C. The Cu/Ta/Si structure was on the other hand completely degraded by the reactions involving Cu, Ta and Si around 550°C [9, 13]. This implies that the interface between Ta and SiO$_2$ is more stable than that between Ta and Si. It is noted that high magnification images, which are not shown here, show voids on the surface of the annealed samples. The voids formed during annealing can be related to thermal stress. Film stress may be created due to accommodation of surface tension of the Cu thin film during the thermal heating and cooling cycle and the different grain growth rates of the Cu and underlying Ta films [15, 16]. The voids are formed to relieve this thermal stress.

The AES depth profiles of the Cu/Ta/SiO$_2$/Si structure before and after annealing is shown in Fig. 2. AES depth profile shows that Si and Cu interdiffusion are clearly prevented by the Ta barrier layer. No interdiffusion between Cu/Ta and Ta/SiO$_2$ interfaces was also observed. Throughout the barrier layer, there is a significant amount of oxygen, especially at the two interfaces. The AES and XRD results suggest that the Ta thin film is a good barrier in preventing Cu diffusing into SiO$_2$ and interfacial reaction up to 600°C. There are two possible mechanisms for the barrier failure. One is the diffusion of Cu into the SiO$_2$ layer and the subsequent reaction with Si to form Cu silicides, and another is the interfacial reaction between the barrier layer and SiO$_2$ layer. The AES depth profile clearly shows that Ta layer effectively prevents Cu diffusion to the SiO$_2$ layer. However, AES cannot give detailed interfacial reaction information which can be obtained by XPS measurements. The XPS results are shown below.

Since Cu is thermodynamically stable with Ta at high temperature, only the interface of Ta/SiO$_2$ was studied by XPS, while the Cu/SiO$_2$ interface was analyzed for comparison. Fig. 3 shows XPS spectra of Ta 4f and Si 2p across the interface of Ta/SiO$_2$ after 600 °C annealing. The Ar ion sputter time of the estimated starting point of the interface region is denoted as 0 min. Clearly, the interface Ta 4f spectra exhibit two pairs of peaks, the doublets being due to spin-orbit coupling. The higher binding energy pair (Ta 4f$_{7/2}$ at 27.4 eV) corresponds to Ta oxide and the pair at low binding energy (Ta 4f$_{7/2}$ at 22.0 eV) is assigned to metallic Ta. Si 2p spectra show that the main peak at binding energy of 103.8 eV corresponds to SiO$_2$ and a peak with binding energy of 98.9 eV is assigned to elemental Si, indicating elemental Si is presented at interface. Moreover, the O 1s spectra, which are not shown here, show a clear binding energy shift from around

531.3 eV to 533.2 eV, suggesting the transmission from Ta oxide complex to Si oxide as sputtering progressed [17]. These results suggest that a strong bonding was formed between Ta and SiO_2, along with a slight reduction of SiO_2 by the Ta thin film. The XPS spectra of Cu 2p and Si 2p across the interface of Cu/SiO_2 are shown in the Fig.4. Unlike the Ta/SiO_2 interface, the spectra of Si 2p in the adjacent SiO_2 do not exhibit discernible elemental Si. The XPS spectra of O1s, which are not shown here, only exhibit a small shift from 533.0eV to 533.7eV. This suggests the Cu atoms do not reduce the adjacent SiO_2 substantially and consequently, may not have strong bonding with the SiO_2. However, it is interesting to note that Cu 2p spectra exhibit a change from the state of metallic Cu to Cu-oxide, suggesting the chemical environment of interface Cu atoms is modified by the adjacent SiO_2. The reduction of SiO_2 by Ti was observed by Takeyama et al. [5] in the $Cu/Ti/SiO_2/Si$ system. In their system, the reduction of SiO_2 through the formation of Ti oxides at the interface occurs even in the as-deposited specimen, indicating that the reaction at Ti/SiO_2 interface can occur at room temperature. Indeed, Taubenblatt and Helms [18] examined the reaction of Ti deposited on SiO_2 and reported that Ti reacts strongly with oxygen atoms in SiO_2 at room temperature and reduces the interface layer of SiO_2, leading to the release of Si atoms from their oxidized site. However, in our $Cu/Ta/SiO_2/Si$ system, we did not observe reduction of SiO_2 layer in the as-deposited sample, indicating that Ta cannot react with SiO_2 at room temperature. In order to understand the reaction temperature of Ta and SiO_2, *in-situ* XPS studies of the effect of thermal annealing at the Ta/SiO_2 interface was done. In this experiment, the Ta layer thickness was reduced using Ar ion sputtering so that XPS can detect the Ta/SiO_2 interface. The sample was heated in vacuum from room temperature to 600 °C. At the same time XPS was used to monitor the change in interface between Ta and SiO_2. XPS Si 2p spectra (Fig. 5) obtained at different temperatures show that the Ta/SiO_2 interface is stable until 500 °C. A dramatic change occurred at 600 °C, with 32% SiO_2 being reduced to elemental Si. This means that Ta reacts strongly with oxygen atoms in SiO_2 at 600 °C and reduces the interface layer of SiO_2, leading to the release of Si atoms from their oxidized site. However, after the sample was cooled down to room temperature, we found that some elemental Si recombined to form SiO_2 again, and only 6.5% elemental Si remained. Further work is needed to clearly explain this reversible phenomenon.

The XPS results of Ta/SiO_2 suggest the presence of a stable interface zone (Ta-Si-oxide complex) after high temperature annealing. A stable Ta/SiO_2 interface has the following contribution to the stability of the $Cu/Ta/SiO_2/Si$ structure. First, the driving force for Cu inward diffusion (Cu-Si compound formation) is reduced significantly. Secondly, no additional free state Si may be released from the interface and consequently, outdiffuse to the Cu surface thin film. For the Cu/SiO_2 interface, the lack of reaction and bonding of Cu with SiO_2 results in no stable interface zone being formed in $Cu/SiO_2/Si$, which may result in the easy diffusion at high temperatures.

4. Conclusion

We have used the Ta thin film layer as a diffusion barrier as well as an adhesion-promoting layer between Cu and SiO_2 to examine the possible diffusion and/or reaction taking place in $Cu/Ta/SiO_2/Si$ structure. Through characterization of the annealed structure by XRD, SEM, AES and XPS, we found that a thin Ta film (~35nm) effectively suppresses Cu diffusion to the SiO_2 layer, with the $Cu/Ta/SiO_2/Si$ structure remaining thermally stable up to 600 °C. The reduction of SiO_2 through the formation of Ta oxides at the Ta/SiO_2 interface occurs at 600 °C. At this temperature, 32% SiO_2 was reduced to elemental Si. This means that Ta reacts strongly with oxygen atoms in SiO_2 at 600 °C and reduces the interfacial layer of SiO_2, leading to the release of Si atoms from their oxidized site. However, after the sample was cooled down to room temperature, some elemental Si recombined to form SiO_2 again, and only 6.5% elemental Si remained. No reduction of SiO_2 was observed at the Cu/SiO_2 interface. This stable Ta/Si-oxide interface range is crucial for the stability of the $Cu/Ta/SiO_2/Si$ structure. Based on this understanding, Ta is therefore a promising candidate for use as a diffusion barrier as well as an adhesion promoter for Cu interconnection lines on SiO_2.

Reference:

1. P. L. Pai and C. H. Ting, IEEE Electron Device Lett. 10 (1989)423.
2. V. M. Donnelly and M. E. Gross, J. Vac. Sci. Technol. A 11(1993)66.
3. J. C. Chiou, K. C. Juang, M. C. Chen, J. Electrochem. Soc. 142(1995)177.
4. S. P. Murarka, Mater. Sci. and Eng. R 19(1997)87.
5. M. Takeyama, A. Noya, K. Sakanishi, H. Seki and K. Sasaki, Jpn. J. Appl. Phys. 35(1996)4027.
6. M. Y. Kwak, D. H. Shin, T. W. Kang, K. N. Kim, Jpn. J. Appl. Phys. 38(1999)5792.
7. M. Y. Kwak, D. H. Shin, T. W. Kang, K. N. Kim, Thin Solid Films 339(1999)290.
8. E. Kolawa, J. S. Chen, J. S. Reid, P. J. Pokela and M. A. Nicolet, J. Appl. Phys. 70(1991)1369.
9. K. Holloway, P. M. Fryer, C. Cabral, Jr., J. M. E. Harper, P. J. Bailey and K. H. Kelleher, J. Appl, Phys. 71(1992)5433.
10. S. Y. Jang, S. M. Lee and H. K. Baik, J. Mater. Sci.: Mater. Electron. 7(1996)271.
11. M. Stavrev, C. Wenzel, A. Möller, and K. Drescher, Appl. Surf. Sci. 91(1995)257.
12. K. H. Min, K. C. Chun, and K. B. Kim, J. Vac. Sci. Technol. B 14(1996)3263.
13. D. S. Yoon, H. K. Baik, and S. M. Lee, J. Vac. Sci. Technol. B 17(1999)174.
14. S. Q. Wang, S. Suther, C. Hoeflich and B. J. Burrow, J. Appl. Phys. 73(1993)2301.
15. J. C. Chang and M. C. Chen, Thin Solid Films 322(1998)213.
16. S. P. Murarka, *VLSI Technology*, 2nd edn., ,S. M. Sze (Ed.), McGraw-Hill, Singapore, 1988, p 375.
17. J. F. Moulder, W. F. Stickle, P. E. Sobol, and K. D. Bomben, *Handbook of X-ray Photoelectron Spectroscopy*, Physical Electronics, Inc., Eden Prairie, MN, 1995.
18. M. A. Taubenblatt and C. R. Helms, J. Appl. Phys. 53 (1982)6308.

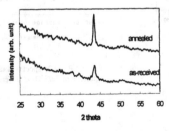

Fig. 1 Glancing angle XRD patterns of the as-received and 600°C-annealed $Cu/Ta/SiO_2/Si$ specimen.

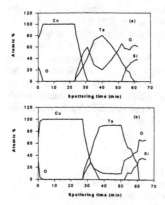

Fig. 2 AES depth profiles of the Cu/Ta/SiO₂/Si system; (a) as-deposited and (b) after annealing at 600 °C.

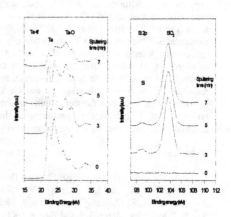

Fig. 3 XPS Ta 4f and Si 2p spectra obtained from the interface of Ta/SiO₂ after 600 °C annealing.

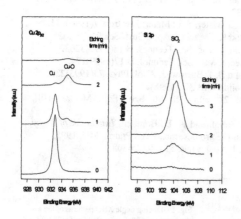

Fig. 4 XPS Cu 2p₃/₂ and Si 2p spectra obtained from the interface of Cu/SiO₂ after 600 °C annealing.

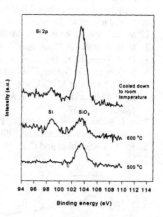

Fig. 5 XPS Si 2p spectra obtained from the interface of Ta/SiO₂ at different

Mat. Res. Soc. Symp. Proc. Vol. 612 © 2000 Materials Research Society

Organic Solution Deposition of Copper Seed Layers onto Barrier Metals

H. Gu, R. Fang, T. J. O'Keefe, M. J. O'Keefe, W.-S. Shih[1], J. A. M. Snook[1], K. D. Leedy[2] and R. Cortez[2]
University of Missouri-Rolla, Dept. of Metallurgical Engineering and Materials Research Center, Rolla, MO 65409, U.S.A.
[1]Brewer Science, Inc., Rolla, MO 65401, U.S.A.
[2]Air Force Research Laboratory, Sensors Directorate, Wright-Patterson AFB, OH 45433, U.S.A

ABSTRACT

Spontaneous deposition of copper seed layers from metal bearing organic based solutions onto sputter deposited titanium, titanium nitride, and tantalum diffusion barrier thin films has been demonstrated. Based on electrochemically driven cementation exchange reactions, the process was used to produce adherent, selectively deposited copper metal particulate films on blanket and patterned barrier metal thin films on silicon substrates. The organic solution deposited copper films were capable of acting as seed layers for subsequent electrolytic and electroless copper deposition processes using standard plating baths. Electroless and electrolytic copper films from 0.1μm to 1.0μm thick were produced on a variety of samples on which the organic solution copper acted as the initial catalytic seed layer. The feasibility of using organic solution deposited palladium as a seed layer followed by electroless copper deposition has also been demonstrated. In addition, experiments conducted on patterned barrier metal samples with exposed areas of dielectric such as polyimide indicated that no organic solution copper or palladium deposition occurred on the insulating materials.

INTRODUCTION

The incorporation of low-k dielectrics and copper interconnects in the fabrication of high speed silicon integrated circuits (ICs) requires significant changes to the current method of using blanket deposition and etch back processes. While physical vapor deposition (PVD) techniques, such as sputtering, appear to be viable for the formation of thin barrier layers between the low-k dielectric and copper interconnect, technical and economic benefits of using chemical vapor deposition (CVD) [1], electrolytic plating [2], and electroless plating [3] processes for build up of the copper interconnects make these approaches attractive alternatives. Electrochemical deposition of copper is relatively inexpensive compared to vapor deposition methods but suffers from the fact that the barrier layers, typically titanium or tantalum based metals or metal nitrides, are difficult to electrochemically activate and plate with adherent copper. In fact, in the primary electrolytic copper metals industry, titanium is employed as a re-usable cathode material because the copper is easily removed from the titanium surface after plating. Therefore, it is necessary to deposit adherent, thin seed layers of copper onto the barrier layer prior to deposition of thicker copper films by electrochemical methods. Although PVD and CVD copper seed layers can be used to fabricate electroplated copper interconnects, an electrochemical process for depositing copper seed layers and interconnects has many benefits, including lower cost of ownership, simplified processing parameters, and the ability to scale with increasing wafer size [4].

Traditional electrochemical deposition of metal films and coatings utilizes aqueous electrolytes. A novel process for depositing metals from organic solutions based on cementation exchange reactions has been demonstrated for a number of metal/metal ion systems, including Au on Zn and Pb on Fe [5,6]. Most of the previous work on electrochemical deposition of metals from organic solutions has focused on copper deposition onto sputtered aluminum thin films [7]. In general, the cementation exchange reaction is based on the dissolution of a less noble metal substrate material (M_2) into the organic solution ($R-M_{x\,(org)}$) while simultaneously depositing more noble metal atoms (M_1) onto the surface of the substrate [5]:

$$\overline{R - M^+_{1\,(org)}} + M_{2(s)} = \overline{R - M^-_{2(org)}} + M_{1(s)} \tag{1}$$

An inherent characteristic of the process is high selectivity of the depositing atoms as the cementation exchange reaction occurs only at the surface of electrochemically active, less noble cathodic sites and not on insulating or dielectric materials. In contrast, standard palladium activation from aqueous stannous chloride/palladium solutions that catalyze electroless copper deposition can be used on both conducting (metal) and non-conducting (polymer) surfaces since it is not selective to the underlying substrate material.

In this study, we report for the first time the use of organic solution deposition of copper onto barrier metals such as Ti, TiN and Ta, for example

$$\overline{R - Cu^{2+}_{\,(org)}} + Ti_{(s)} = \overline{R - Ti^{2+}_{\,(org)}} + Cu_{(s)} \tag{2}$$

in which adherent Cu seed layers are deposited onto sputter deposited barrier metal films. To demonstrate the feasibility of the seed layers with subsequent electrochemical copper deposition, standard electroless and electrolytic copper plating solutions were used to build up 0.1μm to 1.0μm thick copper layers on the barrier films after deposition of the catalytic seed layers.

EXPERIMENT

Deposition of Ti, TiN and Ta barrier films onto Si or SiO_2 wafer substrates was done using a Denton Vacuum Discovery 18 dc magnetron sputtering system with a base vacuum of 1.4×10^{-6} Pa. The pressure in the chamber was held constant at ~0.53 Pa during deposition and the forward power was fixed at 300W. Titanium and tantalum films were deposited using argon gas while the TiN was reactively deposited using an Ar/N_2 mixture. A water cooled substrate holder was used without external heating for all of the depositions. Film thickness was varied from 50nm to 1000nm. Patterning of 100nm Ti films was accomplished by contact lithography. Polyimide dielectric which was spin coated, exposed and cured over the patterned Ti lines and bond pads was left on the substrates during subsequent electrochemical plating operations.

Copper seed layer deposition onto the barrier metal films was done for less than one minute using organic solutions in which the copper ion concentration was < 1 g/L. Subsequent electroless copper deposition was accomplished using a formaldehyde based aqueous formulation of 10-15 g/L copper sulfate ($CuSO_4$-$5H_2O$), 40 g/L Na2EDTA (ethylenediamaine tetra-acetate), 10-15 g/L formaldehyde, 12 g/L NaOH to pH of 12.5, 65°C bath temperature for 10 minutes. A copper plating bath consisting of 180 g/L copper sulfate ($CuSO_4$-$5H_2O$), 65 g/L H_2SO_4, 60 parts per million chloride, and brightener and leveler additives at 35°C was used to deposit electrolytic films at a current density of 25 mA/cm^2 for 5 minutes.

Analysis of the deposited films was done using two different scanning electron microscopes. A Hitachi S4700 field emission microscope and a FEI Dual Beam 620 focused ion beam (FIB) microscope were used to image and conduct chemical analysis of broadface and cross-sectional samples after deposition.

RESULTS

Attempts to directly deposit copper onto the different barrier metals using only the electroless and electrolytic copper plating baths were unsuccessful. In all cases, either no copper plated on the surface or the copper film that did plate on the surface was not adherent. This result was expected since Ti and Ta surfaces are normally not receptive to deposition of adherent copper films by electrochemical plating operations without special surface preparation. Results from organic solution treated barrier metal films will be presented by the type of barrier film.

Ti Films

In this study preliminary investigations on organic solution deposited copper onto barrier films has focused on sputter deposited titanium. Although not normally used as a barrier film, Ti is often used as an adhesion and/or protection layer for metallic films such as aluminum and TiN. For the purposes of this study, it served the role of a reproducible, model surface for conducting organic solution deposition experiments.

Depicted in Fig. 1a and 1b are broadface scanning electron microscope (SEM) micrographs of organic solution deposited copper particles on the surface of a 1000nm thick Ti film. The light or white particles in Fig. 1 are the deposited copper. As can be seen in Fig. 1a, the distribution of sub-micrometer size copper particles on the surface of the Ti is fairly uniform. Fig. 1b is a higher magnification of the same sample in which the largest copper particle is approximately 400nm in diameter with a typical copper particle size of <100nm. Adherence tape testing of the seed layers after organic solution deposition indicated that the copper particles could not be removed from the titanium. Chemical characterization of the large particles by energy dispersive x-ray analysis indicated only a copper signal.

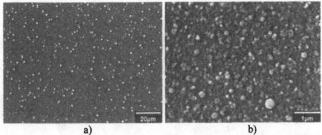

a)   b)

Figure 1. Broadface micrographs of organic solution copper particles on titanium: a) low magnification, b) high magnification. The white or bright areas are the copper particles.

After deposition of the copper seed layers from organic solutions, the Ti barrier film samples were subjected to electroless and electrolytic plating. Depicted in Fig. 2a is a broadface SEM micrograph after organic seed layer and electroless copper deposition while Fig. 2b is after organic seed layer and electrolytic copper plating. Both organic solution seed layer/aqueous electrochemical copper samples passed qualitative adherence testing of the samples using standard scotch tape.

As presented in equation (2), an inherent part of the process is dissolution of the barrier film into the organic solution. Lack of uniformity or preferential attack of the barrier film is undesirable from a reliability standpoint. Figs. 3a and 3b are cross-sectional micrographs from a 1000nm thick Ti film sample after a three step copper deposition process: first, organic copper seed layer, then electroless copper deposition, and finally electrolytic copper deposition. Fig. 3a is the sample after cleaving the wafer while Fig. 3b is a similar sample that was ion milled in the FIB microscope. Both figures demonstrate that a continuous, void free copper film was electrochemically deposited on the sputter deposited Ti film. Examination of similar samples prepared on 100nm thick Ti films gave the same results in that adherent, void free copper was deposited on the barrier film without appreciable or non-uniform removal of the Ti layer.

a)   b)

Figure 2. *Broadface micrographs of copper films by a) electroless and b) electrolytic plating on top of organic solution copper seeded titanium.*

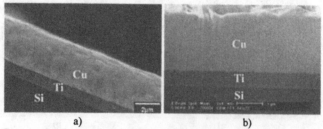

a)   b)

Figure 3. *Cross sectional micrographs of a) cleaved and b) focus ion beam milled, electrochemically deposited copper on sputter deposited titanium.*

The previous Ti barrier film results were on unpatterned wafers. As mentioned in the introduction, the process is inherently area selective in that deposition should only occur on electrochemically active, less noble metal surfaces. Patterned Ti films on Si/SiO$_2$ wafers with polyimide as the dielectric were subjected to a copper bearing organic solution and then an electroless copper plating bath. Shown in Fig. 4a is a section of the wafer before exposure to the electrochemical deposition solutions while Fig. 4b is the same area of the wafer after copper deposition. Although difficult to show without a color micrograph, copper was deposited on top of all of the exposed Ti in Fig. 4b. No copper was deposited on the polyimide. Similar experiments that used dielectrics such as Si$_3$N$_4$ and benzocyclobutene (BCB) had comparable results: copper deposited only on the exposed metal and not on the dielectric.

Preliminary results using Pd loaded organic solutions, in place of the copper loaded organic solutions, have paralleled the copper seed layer experiments on titanium. After organic Pd seed layer processing the Ti surface could be selectively plated with electroless copper.

TiN Films

Attempts to produce copper seed layers on the surface of 50nm to 100nm thick TiN layers have been dependent on the structure, phase and nitrogen content of the TiN. Adjustments to the chemical composition of the organic solution have been required in order to use Cu or Pd to activate the TiN surface for subsequent electroless or electrolytic copper plating. Preliminary analysis of the surface of the TiN after organic solution copper deposition indicates the surface is devoid of any distinguishing features showing copper nuclei (Fig. 5a) but that after electroless copper deposition the film surface appears similar to the Ti samples (Fig. 5b). The electroless copper is essentially void free and adherent to the underlying substrate.

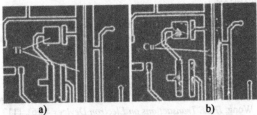

a)   b)

Figure 4. *Optical micrographs of patterned titanium/polyimide samples a) as processed and b) after organic solution plus electroless copper deposition.*

a)   b)

Figure 5. *Broadface micrographs of sputter deposited TiN after a) copper bearing organic solution processing and b) after subsequent electroless copper deposition.*

Ta Films

Work has only recently been started using metal bearing organic solutions to deposit seed layers on Ta and Ta(N) barrier films. Initial screening tests indicate that the process is feasible but the passivity of the tantalum films is much more tenacious than titanium films, requiring a more aggressive media to remove the surface oxide layer. However, non-uniform seed layers have been deposited and subsequently plated with electroless copper in localized areas. Studies to improve deposition uniformity and adherence have been initiated.

CONCLUSIONS

A spontaneous, electrochemically driven process for activating the surface of Ti, TiN and Ta barrier films for subsequent electroless and electrolytic copper plating processes has been demonstrated. Organic based solutions with dissolved copper were used to activate and catalyze the surface of the barrier metals which resulted in adherent copper seed layers compatible for use with standard electroless and electrolytic plating baths. Void free copper films with a final thickness of up to $1\mu m$ were deposited onto the barrier films. Selective area deposition of organic solution and then electroless copper onto the exposed surface of patterned titanium/polyimide samples indicated that deposition occurred only on the metal and not the dielectric. Preliminary results using palladium loaded organic solutions indicated the feasibility of using alternative metal systems to activate the surface of barrier metals for subsequent electroless copper deposition.

ACKNOWLEDGMENTS

This work was funded by the Small Business Innovation Research (SBIR) program through the Sensors Directorate of the Air Force Research Laboratory, contract # F33615-97-C-1074.

REFERENCES

1. C. Ryu, K.-W. Kwon, A. L. S. Loke, H. Lee, T. Nogami, V. M. Dubin, R. A. Kavari, G. W. Way, and S. S. Wong, *IEEE Transactions on Electron Devices* **46**(6), 1113 (1999).
2. R. D. Mikkola, Q.-T. Jiang, and B. Carpenter, *Plating & Surface Finishing*, March, 81 (2000).
3. V. Dubin and S.-D. Yosi, *J. Electrochem. Soc.* **44**(3), 898 (1997).
4. T. J. O'Keefe, M. Stroder, M. J. O'Keefe in *Advance Interconnects and Contact Materials and Processes for Future Integrated Circuits*, ed. S. P. Murarka, D. B. Fraser, M. Eizenberg, R. Tung, R. Madar, Mater. Res. Soc. Proc. **514**, 473, Warrendale, PA (1998).
5. T. J. O'Keefe, Method for Stripping Metals in Solvent Extraction, U. S. Patent #5,228,903, July 20, 1993.
6. L. M. Chia, M. P. Niera, C. Flores, and T. J. O'Keefe in *Extraction and Processing for the Treatment and Minimization of Wastes*, ed. J. P. Hager et al, TMS Annual Meeting, 279, Warrendale, PA (1994).
7. M. J. O'Keefe, K. D. Leedy, J. T. Grant, M. Fang, H. Gu, and T. J. O'Keefe, *J. Vac. Sci. Tech. B* **17**(5), 2366 (1999).

Mat. Res. Soc. Symp. Proc. Vol. 612 © 2000 Materials Research Society

Stress, Microstructure and Temperature Stability of Reactive Sputter Deposited WN$_x$ Thin Films

K. D. Leedy, M. J. O'Keefe[1], J. G. Wilson, R. Osterday[2] and J. T. Grant[3]

Air Force Research Laboratory, Sensors Directorate, Wright-Patterson AFB, OH 45433
[1]University of Missouri-Rolla, Dept. of Metallurgical Engineering, Rolla, MO 65401
[2]Southwestern Ohio Council for Higher Education, Dayton, OH 45420
[3]Research Institute, University of Dayton, Dayton, OH 45469

ABSTRACT

Tungsten nitride (WN$_x$) thin films can be used as Schottky barriers in high power, high temperature semiconductor devices or as diffusion barriers between Cu, low-k dielectric and silicon because each application requires a thermally stable film. Therefore, it is important to understand the thin film properties of WN$_x$ as a function of deposition conditions and elevated temperature exposure. In this investigation, the influence of nitrogen content and post deposition annealing on the stress and microstructure of reactive dc magnetron sputter deposited WN$_x$ films was analyzed. With an increasing N$_2$ to Ar flow ratio, the as-deposited crystal structure of the films changed from α-W to ß-W to amorphous WN$_x$ and finally to W$_2$N. Rapid thermal anneals up to 650°C resulted in large tensile stress increases and phase transformations to W$_2$N in the nitrogen-containing films. Grain growth during annealing decreased as the concentration of nitrogen in the film increased. The nitrogen content in the films was determined using x-ray photoelectron spectroscopy (XPS).

INTRODUCTION

Tungsten nitride thin films have been studied extensively for use as thermally stable Schottky contacts to GaAs [1-4]. Another potential application of WN$_x$ is as a barrier to Cu diffusion in Si-based integrated circuits. An optimal barrier layer should have a dense, amorphous microstructure, a smooth surface morphology, thermal stability with Cu and Si, low film stress and minimal thickness (10 to 20 nm) [5, 6]. Amorphous thin films are considered more effective diffusion barriers than polycrystalline thin films because of the lack of grain boundaries which function as fast diffusion paths [7]. Several transition metals and their nitrides are good candidate materials, including tungsten nitride [8]. Because of high temperature anneals used in the fabrication of semiconductor devices, the thermal stability of WN$_x$ films is an important consideration with respect to crystallinity, grain growth and stress.

Fabrication methods for WN$_x$ films include metal-organic chemical vapor deposition [9], plasma enhanced chemical vapor deposition (PECVD) [10, 11] and reactive sputtering [12-14]. Lee [15] investigated PECVD WN$_x$ as a diffusion barrier between Al and Si while So [13] studied reactive sputtered WN$_x$. The Cu barrier performance of reactive sputtered WN$_x$ was examined by Uekubo [14] with 25 nm thick WN$_x$ films and Suh [12] with 5 to 100 nm WN$_x$ thick films. Alternatively, a large grained reactive sputtered WN$_x$ film fabricated by Yongjun [16] exhibited low stress, high surface smoothness and high thermal stability. It was found that a low nitrogen concentration in a W seed layer followed by a rapid thermal anneal in N$_2$ formed a large grained WN$_x$ barrier.

The intrinsic stress of WN_x thin films has been addressed in only limited studies. Yu [1] measured a compressive stress range of ~ 3 to 1.75 GPa in 300 nm thick WN_x films deposited at 5 mTorr with N_2 partial flows from 0 to 40 %. Lee [10] studied the influence of PECVD W/WN_x bilayers on the intrinsic film stress.

In this paper the influence of film thickness and nitrogen content on the microstructure and stress of reactive sputtered WN_x films in the as-deposited state and after 400 and 650°C anneals was studied.

EXPERIMENT

Tungsten nitride thin films were deposited from a 99.95% pure W target using a Denton Vacuum Discovery 18 dc magnetron sputtering system with a base vacuum of 1.4×10^{-6} Pa. A mass flow regulated Ar-N_2 sputtering pressure of ~ 0.53 Pa and 250 W forward power were held constant resulting in a nominal deposition rate of 0.3 nm/s. The N_2 partial flow rate (the ratio of the nitrogen flow rate to the total flow rate of nitrogen and argon) ranged from 0 to 50 %. Thermally oxidized (100) silicon wafers of 75 mm diameter and holey carbon support films on 3 mm grids were used as substrate materials during the study. Films deposited onto the holey carbon grids were 20 nm thick while films deposited onto the Si wafers were either 20 or 200 nm thick. A water-cooled substrate holder was used without external heating for all of the depositions.

The intrinsic stress of the WN_x films was calculated using the wafer curvature technique and Stoney's equation. A laser reflectometry Flexus 2900 thin film stress system was used to measure the intrinsic stress of the as-deposited WN_x films on the Si substrates and to measure film stress *in-situ* during annealing. The nitrogen purged Flexus had a 5°C/min heating rate from 25 to 650°C, a 5 min. hold at 650°C and then a 5°C/min cooling rate to 25°C. Rapid thermal anneals of films on holey carbon grids were performed in an AG Associates Heatpulse 210 at 400 and 650°C for 120 sec in flowing Ar-5%H_2 forming gas. X-ray diffraction (XRD) of 200 nm thick samples was done using a Rigaku D-MAX III thin film diffractometer with a fixed, low angle (5°) Cu Kα radiation source and a rotating sample holder. Particle size and crystal structure of 20 nm thick films were measured by transmission electron microscopy (TEM) using a Philips CM 200 at 200 keV. Particle sizes were determined by a statistical analysis of grain diameters in bright and dark field broadface images. Film composition was determined with a Physical Electronics model 5700 x-ray photoelectron spectroscopy (XPS) system utilizing monochromatic Al Kα radiation.

RESULTS

Composition

The atomic concentration of nitrogen in the films was determined from XPS measurements. These measurements were made several weeks after the films had been grown, so the surfaces were affected by the atmospheric environment in which they had been stored. XPS analysis showed that the surfaces were oxidized and contaminated with carbon. The W $4p_{3/2}$ and N 1s photoelectron signals obtained from these surfaces are shown in Fig. 1, and were used for the quantitative analysis. (Note that the spectra have been offset for clarity). Since the W $4p_{3/2}$ and N 1s photoelectrons have similar kinetic energies, about 1060 and 1090 eV

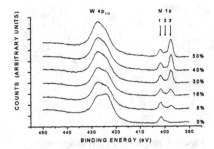

Figure 1 The W 4p$_{3/2}$ and N 1s XPS spectra obtained from the films grown with N$_2$ partial flows of 0, 8, 16, 30, 40, and 50%.

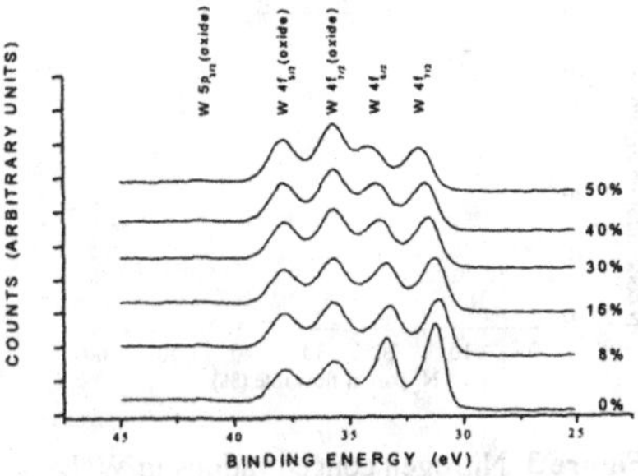

Figure 2 The W 4f and 5p$_{3/2}$ XPS spectra obtained from the films grown with N$_2$ partial flows of 0, 8, 16, 30, 40, and 50%.

respectively, their intensities will be similarly attenuated by surface contamination. The N 1s signals are comprised of three peaks (1, 2, and 3). The intensities of peaks 1 and 2 from each sample are fairly constant, whereas the intensities of peak 3 increase with N$_2$ flow rate. Peak 3 corresponds to nitrogen in the film, whereas peaks 1 and 2 are due to surface reactions where the nitrogen is in a higher oxidation state. (Carbon and the nitrogen peaks 1 and 2 all decreased very rapidly with inert gas sputtering of the surface). The W 4p$_{3/2}$ lineshape also varies between samples and is due to different degrees of surface oxidation of the films. The degree of surface oxidation was determined from the W 4f and 5p$_{3/2}$ photoelectron spectra, as these peaks are relatively sharp and shifted to higher binding energy due to surface oxidation. These W spectra are shown in Fig. 2. The fractions of the total non-oxide tungsten signals obtained from the W 4f and 5p$_{3/2}$ (barely visible in the figure) spectra were then applied to the total W 4p$_{3/2}$ intensity to remove the amount due to surface oxidation. The nitrogen concentrations in the films were then obtained by calculation using Scofield cross-sections, and asymmetry parameter, analyzer transmission and atom size corrections [17], and are plotted in Fig. 3. The increase in film nitrogen content is also reflected by an increase in resistivity from 16 to 244 μΩcm for 0 and 43 at. % nitrogen films, respectively.

Microstructure

The crystal structure of as-deposited WN$_x$ films with no nitrogen was bcc α-W and remained α-W after rapid thermal anneals of 400 and 650°C. Films with 12 at. % nitrogen were β-W (A-15) and transformed completely to W$_2$N after 650°C anneals. Films with 29 and 35 at. % nitrogen had a predominantly amorphous structure in the as-deposited condition and after 400°C anneals. The W$_2$N phase existed in the as-deposited and annealed 43 at. % nitrogen films. All films with nitrogen contained W$_2$N after 650°C anneals and films with > 29 at. % nitrogen exhibited some conversion of W$_2$N to α-W after 650°C anneals. The conversion of WN$_x$ to α-W due to annealing was studied extensively by Lin [11, 18]. A phase map by Suh [12] shows the same general trend in WN$_x$ microstructure evolution with temperature using 450 to 850°C furnace anneals for 1 hr. As-deposited and furnace annealed WN$_x$ data from Yu [1] at 700°C/30 min. also exhibited the same type of crystal structure development. Table I shows the WN$_x$

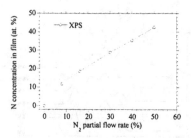

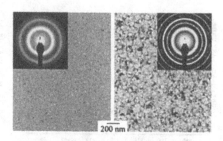

Figure 3 Nitrogen concentrations in WN$_x$ films as a function of N$_2$ partial flows during growth.

Figure 4 Bright field TEM images with inset diffraction patterns of WN$_x$ film with 19 at. % nitrogen. As-deposited (L) and after 650°C rapid thermal anneal (R).

crystal phases of the 20 nm thick as-deposited and rapid thermal annealed films. For films with multiple phases present, the first species listed was most prevalent. XRD of 200 nm thick WN$_x$ films showed the same structures.

WN$_x$ particles were 5 to 10 nm in diameter in the as-deposited state. Figures 4 and 5 show that rapid thermal anneals caused grain growth in films with lower nitrogen contents, associated primarily with transformations to W$_2$N. The diffraction patterns on the left and right in Fig. 4 are of amorphous WN$_x$ and W$_2$N, respectively. The highest nitrogen content films showed no measurable grain growth after anneals, attributed to the stability of W$_2$N. Accurate grain size measurements in films with no nitrogen were complicated by a nonuniform morphology. Although some of the WN$_x$ films are listed as amorphous, ~ 10 nm diameter particles were observed and measured.

Table I Crystal structure of WN$_x$ films as-deposited and after 400 and 650°C rapid thermal anneals for 2 min. (a-WN$_x$, amorphous WN$_x$).

Nitrogen Content (at. %)	As-deposited	400°C	650°C
0	α-W	α-W	α-W
12	β-W	β-W + α-W	W$_2$N
19	a-WN$_x$ + α-W	β-W + α-W	W$_2$N
29	a-WN$_x$	a-WN$_x$	α-W + W$_2$N
35	a-WN$_x$ + W$_2$N	a-WN$_x$ + W$_2$N	W$_2$N + α-W
43	W$_2$N	W$_2$N	W$_2$N + α-W

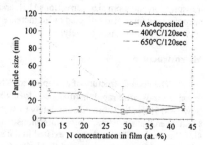

Figure 5 Particle size of WN$_x$ films vs. film nitrogen content, as-deposited and after 400 and 650°C rapid thermal anneals.

Film Stress

As-deposited 20 nm thick WN_x films exhibited compressive intrinsic stresses that became more compressive with increasing film nitrogen content. Figure 6 shows stress as a function of nitrogen content for as-deposited films and films subjected to rapid thermal anneals at 400 and 650°C. The rapid thermal anneals resulted in tensile stress increases in all films with a maximum increase of > 2000 MPa compared to the as-deposited films. Factors contributing to the generation of tensile stress include grain growth [19] and differential thermal expansion between the film and substrate.

The difference in intrinsic stress between 20 and 200 nm thick WN_x films is shown in Fig. 7. As-deposited stresses of both thickness films were similar except in lower nitrogen content films where 200 nm films exhibited more tensile stresses. Furnace anneals at 650°C resulted in large tensile stress increases in 200 nm thick WN_x films while the only significant stress change in 20 nm thick WN_x was a tensile increase in 43 at. % nitrogen films. Furnace anneals were performed in flowing N_2, so the propensity for film oxidation is higher than in the forming gas used in rapid thermal anneals.

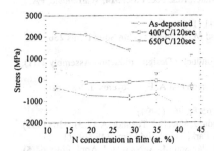

Figure 6 Film stress vs. nitrogen concentration in 20 nm thick WN_x films, as-deposited and after 400 and 650°C rapid thermal anneals.

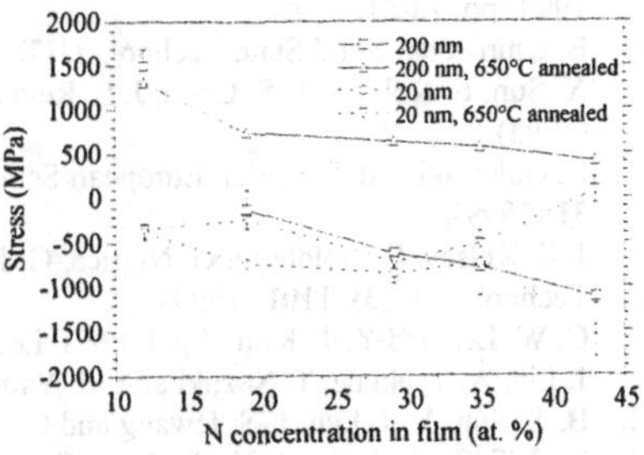

Figure 7 Film stress vs. nitrogen concentration in 20 and 200 nm thick WN_x films, as-deposited and after 650°C furnace anneals.

CONCLUSIONS

Reactive sputter deposited tungsten nitride thin films displayed a range of stable and metastable crystal phases depending on film nitrogen concentration, from α-W with no nitrogen concentration to W_2N at high nitrogen concentrations. XPS was used to calculate the film nitrogen contents. As-deposited particle sizes were nominally 10 nm for all nitrogen concentrations. Maximum grain growth up to 90 nm after 400 and 650°C rapid thermal anneals occurred in films with the lowest nitrogen contents. Rapid thermal anneals and furnace anneals at 650°C caused substantial tensile stress increases in most films, decreasing grain growth with increasing nitrogen concentration, and a predominantly W_2N crystal structure.

ACKNOWLEDGMENTS

The authors gratefully acknowledge support provided by the Materials and Manufacturing Directorate (J.T.G.).

REFERENCES

1. K. M. Yu, J. M. Jaklevic, E. E. Haller, S. K. Cheung and S. P. Kwok, J. Appl. Phys. **64(3)**, 1284 (1988).
2. Y. T. Kim and C. W. Lee, J. Appl. Phys. **76(1)**, 542 (1994).
3. A. E. Geissberger, R. A. Sadler, F. A. Leyenaar and M. L. Balzan, J. Vac. Sci. Technol. A **4(6)**, 3091 (1986).
4. J. S. Lee, C. S. Park, J. W. Yang, J. Y. Kang and D. S. Ma, J. Appl. Phys. **67(2)**, 1134 (1990).
5. S. S. Wong, C. Ryu, H. Lee and K.-W. Kwon in *Advanced Interconnects and Contact Materials and Processes for Future Integrated Circuits*, edited by S. P. Murarka, D. B. Fraser, M. Eizenberg, R. Tung, R. Madar (Mater. Res. Soc. Proc. **514**, Warrendale, PA, 1998) pp. 75-81.
6. B. Chin et al., Solid State Technol. **41(7)**, 141 (1998).
7. X. Sun, E. Kolawa, J.-S. Chen, J. S. Reid and M.-A. Nicolet, Thin Solid Films **236**, 347 (1993).
8. C. Galewski and T. Seidel, European Semiconductor Design Production Assembly **21(1)**, 31 (1999).
9. J. E. Kelsey, C. Goldberg, G. Nuesca, G. Peterson and A. E. Kaloyeros, J. Vac. Sci. Technol. B **17(3)**, 1101 (1999).
10. C. W. Lee and Y. T. Kim, Appl. Phys. Lett. **65(8)**, 965 (1994).
11. J. Lin, A. Tsukune, T. Suzuki and M. Yamada, J. Vac. Sci. Technol. A **17(3)**, 936 (1999).
12. B.-S. Suh, Y.-J. Lee, J.-S. Hwang and C.-O. Park, Thin Solid Films **348**, 299 (1999).
13. F. C. T. So, E. Kolawa, X.-A. Zhao, E. T.-S. Pan and M.-A. Nicolet, J. Appl. Phys. **64(5)**, 2787 (1988).
14. M. Uekubo, T. Oku, K. Nii, M. Murakami, K. Takahiro, S. Yamaguchi, T. Nakano and T. Ohta, Thin Solid Films **286**, 170 (1996).
15. C. W. Lee, Y. T. Kim, C. Lee, J. Y. Lee, S.-K. Min and Y. W. Park, J. Vac. Sci. Technol. B **12(1)**, 69 (1994).
16. H. Yongjun, U. S. Patent No. 5633200, 1997.
17. M.P. Seah in *Practical Surface Analysis, 2nd Ed., Vol.1, Auger and X-ray Photoelectron Spectroscopy*, ed. by D. Briggs and M.P. Seah, (John Wiley, Chichester, 1990), Chapter 5.
18. J. Lin, A. Tsukune, T. Suzuki and M. Yamada, J. Vac. Sci. Technol. A **16(2)**, 611 (1998).
19. F. Spaepen, Acta Mater. **48**, 31 (2000).

Joint Session:
Grain Evolution
of Metals

Mat. Res. Soc. Symp. Proc. Vol. 612 © 2000 Materials Research Society

Room Temperature Recrystallization of Electroplated Copper Thin Films: Methods and Mechanisms

D. Walther[1], M. E. Gross[1*], K. Evans-Lutterodt[1], W. L. Brown[1], M. Oh[2], S. Merchant[2], P. Naresh[1]
[1]Bell Labs, Lucent Technologies, Murray Hill, NJ 07974
[2]Bell Labs, Lucent Technologies, Orlando, FL 32819

Abstract

We report a comparison of the room temperature recrystallization of electroplated (EP) copper in blanket films as a function of thickness measured by focused ion beam (FIB) microscope images and sheet resistance measurements. Both sets of data show an increase in rate with film thickness from 0.75 μm up to 5 μm, while little recrystallization is observed in films thinner than 0.75 μm. Interestingly, the recrystallization rates from FIB analysis are consistently faster than those from the sheet resistance measurements. These data suggest that the recrystallization is initiated close to the top surface of the EP Cu film, but that in thinner films a high surface-to-volume ratio allows surface inhibition or pinning to retard the transformation. A Johnson-Mehl-Avrami-Kolmogorov (JMAK) analysis of the two data sets yields unusually high values for the Avrami exponent α of up to 7 for the FIB data, while lower values of around 4 are obtained for the sheet resistance data. X-ray diffraction pole figures of the films have also been collected and correlations between the crystallographic texture, film thickness and recrystallization are discussed.

I. Introduction

Cu is increasingly replacing Al as the interconnect metal in high performance integrated circuits for increased speed due to higher conductivity and improved reliability due to higher electromigration resistance.[1,2] However, understanding the influence of the microstructure and texture of Cu on its electrical performance and reliability is still at an early stage.[2,3] Cu electroplating involves a Cu sulfate/sulfuric acid plating bath to which HCl and organic species have been added to achieve the desirable bottom-up fill by suppression of plating at feature corners and regions between the trenches and acceleration within the trenches.[4] This major alteration of the Cu surface chemistry and the exchange current density during plating result in fine-grained (0.1-0.2μm) Cu deposits incorporating a fraction of the organic and inorganic additives and a high density of defects and twins.[5-7]

The ramifications of the use of additives in the plating bath extends beyond the as-plated microstructure to the subsequent secondary recrystallization that proceeds at room temperature to give a final grain size of up to several microns (Figure 1). This remarkable transformation is related to the influence of additives in the plating bath on the microstructure, composition, defect density and/or strain of the plated films.[8-17] The influence of topography, radial position, plating bath chemistry, barrier layers, and low temperature cycling on the room temperature recrystallization rate of EP Cu have been reported recently as indirect means of learning something about the mechanisms. As we investigate ways to obtain direct measurements of the different factors that may be influencing the recrystallization process (e.g., impurities, dislocations, strain, twins), we

---

* author to whom correspondence should be addressed: mihal@lucent.com

Time after plating:

| 5.4 hours | 7.6 hours | 10.8 hours | annealed at 300° C |

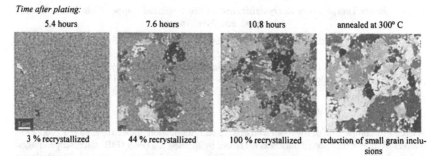

| 3 % recrystallized | 44 % recrystallized | 100 % recrystallized | reduction of small grain inclusions |

Figure 1. Plan-view FIB secondary electron images showing recrystallization with increasing time after plating. (SiO$_2$ / 300 Å PVD Ta / 1000 Å PVD Cu // 2 μm EP Cu)

are also continuing our investigations to gain insights indirectly. Sheet resistance measurements have been used most widely to monitor the recrystallization process, where the decrease in sheet resistance is associated with a decrease in the grain boundary volume. A more laborious task is the image analysis of focused ion beam (FIB) secondary electron micrographs. The two methods of extracting recrystallization rates are complementary and can provide additional insight into the mechanisms of the room temperature recrystallization. In this paper, we compare the two methods to examine the thickness dependence of the recrystallization rate for EP Cu films ranging from 0.25 to 5 μm.

II. Experimental

Duplicate sets of samples were prepared for FIB imaging and four point probe sheet resistance measurements to monitor the room temperature recrystallization of the EP Cu as a function of time and film thickness. Samples consisted of 200 mm Si wafers onto which were deposited sequentially 300 Å Ta diffusion barrier and 1000 Å Cu seed layers using Novellus Inova hollow cathode magnetron (HCM) sputter deposition cluster tool. EP Cu films of varying thickness (0.25, 0.50, 0.75, 1.0, 2.0, 3.0, 4.0, 5.0 μm) were then deposited using a Novellus Sabre electroplating system and Enthone CuBath M chemistry. Monitoring of the room temperature recrystallization was begun as soon as possible after plating.

Four-point probe sheet resistance measurements were recorded using a KLA-Tencor Prometrix RS75 system. 121 points were measured on each wafer at the center and in five concentric rings. To reduce complications from a possible radial dependence in the recrystallization rate[16,17] and for the purposes of direct comparison with the FIB measurements, the sheet resistance data discussed in this paper correspond to the mean value of the 16 measurements made at a radial position 38 mm from the wafer center. Once the room temperature recrystallization was deemed to be complete as indicated by little or no further change in sheet resistance over several days, the wafers used for the sheet resistance measurements were annealed to 300°C for 30 min. in a N$_2$ ambient.

FIB secondary electron plan view images were collected using a Micrion 9000 FIB system and rastering the focused Ga$^+$ beam across a 40 μm x 40 μm area until the surface oxide was removed and the grains of Cu were clearly delineated. Each image was recorded in a previously untouched area of the sample, as the implantation of Ga in the film during imaging retards the recrystallization process. The region analyzed on each sample was approximately 38 mm from the wafer center.

X-ray diffraction pole figure data were collected on a Bruker-AXS GADDS diffractometer equipped with an area detector. Orientation distribution functions and complete pole figures were generated using software from HyperNex.[18]

III. Measurement Methods

Sheet Resistance

The sheet resistance decrease of approximately 20% in EP Cu films after plating has been associated with the decrease in grain boundary volume during the room temperature recrystallization process.[8,10,11,14] The fraction of recrystallized material is obtained from a linear interpolation of the sheet resistance, according to equation (1), where the resistance measured immediately after plating, R_0, and the resistance after thermal annealing (300° C, 30 min), R_{anneal}, serve as starting and end point, respectively.

$$\text{Fraction Recrystallized}(t) = \frac{R_0 - R(t)}{R_0 - R_{anneal}} \tag{1}$$

Focused Ion Beam Imaging

A series of 40 μm x 40 μm FIB plan view images was recorded for each sample at time intervals until we could not detect any further changes in grain size. These images were recorded digitally with a resolution of 1024 x 1024 pixels and 256 gray values. An image analysis program specifically written for this purpose was used to distinguish and quantify the small-grained (as-plated) and large-grained (recrystallized) regions. The program samples a fixed environment around each pixel and differentiates the small-grained and large-grained regions based on statistical properties of the gray value distribution in this environment. The program then generates a transparent overlay mask that colors those pixels of the FIB image that show large grains and yields a value for the integrated area that is equivalent to the fraction of recrystallized material according to:

$$\text{Fraction Recrystallized}(t) = \frac{\#\text{pixels (large grained)}}{\#\text{pixels (overall)}} \tag{2}$$

Comparison of the methods

The two methods differ in the emphasis on certain aspects of the recrystallization. This is shown in the following comparison

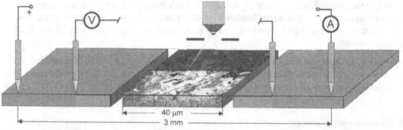

Figure 2. Schematic of the two methods of measuring the recrystallization rate. Note the different length scales of the two measurements.

FIB method	Sheet resistance method
• direct evidence for recrystallization	• indirect evidence for recrystallization
• measurement made at one radial position on the wafer	• mapping the entire wafer at 121 points each time in five equally spaced concentric rings
• provides 2-dimensional view of the abnormal grain growth that can be quantified through image analysis	• provides 3-dimensional information about the recrystallization process as it tracks the decrease in grain boundary area

The last point is particularly important for the comparison of the data obtained from the two methods. Figure 2 shows a graphical comparison of the methods.

IV. Kinetics

The samples have been monitored by both measurement procedures over several hundred hours. In Figures 3a and 3b, data for the recrystallized fraction are plotted for the first 50 hours after plating. The rate of recrystallization varies with EP Cu film thickness as well as between the two measurements.

FIB versus Sheet Resistance Measurements

The FIB images provide 2-dimensional information about the transformation occurring at or close to the surface of the Cu film. In contrast, the sheet resistance measurements sample the entire volume of Cu, EP film *plus* PVD seed layer, and, therefore, provide 3-dimensional information on the progression of the recrystallization through *both* layers. The recrystallization rates from the FIB data are consistently faster for all films that recrystallize, that is, for films ≥ 0.75 µm. At the time when the FIB images show a completely recrystallized film *surface*, the sheet resistance measurements typically indicate that less than 50% of the total volume of Cu has recrystallized. These observations support a scenario in which nucleation of recrystallization occurs close to the top surface. Subsequently, the transformation spreads more quickly laterally than vertically. Depending on the film thickness, the recrystallization can extend fully through the thickness of the film.

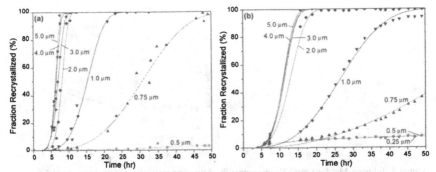

Figure 3. Fraction of Cu recrystallized as a function of time after electroplating for EP Cu films of different thickness, as derived from (a) FIB secondary electron images, and (b) sheet resistance measurements, at a radial distance approximately 38 mm from the center of the wafer.

Two earlier reports[11,16] that have compared FIB and sheet resistance changes during recrystallization of 1 μm EP Cu films indicate that the two types of measurements yield the same kinetics of recrystallization, in contrast to our observations. Differences in the electroplating process are important, as is the location on the wafer where measurements are made since radial variations in recrystallization rates have been reported.

<u>JMAK Analysis</u>

Transformations involving nucleation and growth processes, including recrystallization, have been modeled using the Johnson-Mehl-Avrami-Kolmogorov (JMAK) equation,

$$X(t) = 1 - \exp\left[-\left(\frac{t}{t_R}\right)^\alpha\right], \qquad (3)$$

where $X(t)$ is the fraction of material that has recrystallized at time t, t_R is a characteristic time constant corresponding to a approximately 63% recrystallization, and α is the Avrami exponent. Although some of the presumptions of this model (such as isotropic growth, randomly distributed nucleation) are not fulfilled in the case of room temperature recrystallization of EP Cu, the parameters obtained from this model can be useful for getting insights into the mechanisms of this phenomenon.

Equation 3 can also be written in the form,

$$\ln[-\ln(1 - X(t))] = \alpha \ln(t) - \alpha \ln(t_R), \qquad (4)$$

such that by plotting $\ln[-\ln(1 - X(t))]$ versus $\ln(t)$ we obtain the Avrami exponent α as the slope of the curve. This analysis is depicted in Figures 4a and 4b for the two different measurement methods. The FIB data are fit very well by straight lines. The sheet resistance data are also fit well between 5% and 95% recrystallization. The greater deviations from linearity observed in Figure 4b for the 0.75 and 1.0 μm films at shorter times suggest an incubation period that may be

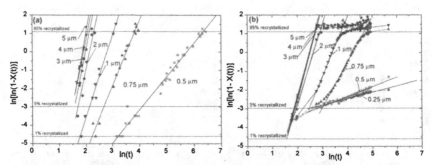

Figure 4. Johnson-Mehl-Avrami-Kolmogorov fit of the recrystallization data determined from (a) FIB image analysis and (b) sheet resistance measurements.

transitional between the thinner and thicker films. The glacial pace of recrystallization in the thin films with high surface-to-volume ratios may be due to surface inhibition caused by grain boundary pinning by impurities. An insufficient concentration of the species and/or defects that play a role in driving the recrystallization process may also be a factor. The 3-5 μm thick films exhibit a rapid "bulk volume" recrystallization rate and no incubation time within the resolution of our data.

Figure 5 compares the Avrami exponents extracted from the straight-line fits in Figure 4 for each of the two measurement procedures as a function of the EP Cu film thickness. There is a notable similarity in the trends of the α values from the two techniques; however, the values from the FIB data are consistently higher than those from the sheet resistance data. The values of α from the sheet resistance data increase with film thickness until reaching a plateau between 4 and 5 for films ≥ 1.5 μm. This is in agreement with the value of 4 that is expected for three dimensional recrystallization with a constant nucleation rate.[19] The Avrami exponents from the FIB data are consistently higher than the values from the sheet resistance data, reaching a plateau at a value of 7. Such a high value for alpha can not easily be explained by the JMAK theory. It would mean a nucleation rate $\propto t^3$, which was clearly not observed in the FIB images.

We conclude that the JMAK theory, taking the fraction recrystallized at a given time as its single input, is insufficient for handling the more complex mechanisms involved in the recrystallization of the EP Cu films. More detailed theories must be developed to take account of the specific features that occur in this recrystallization process such as twinning, the concentration and distribution of incorporated additives, defect densities, etc. Additionally, further developments are needed in our analytical capabilities to achieve both the spatial resolution and the sensitivity needed to provide quantitative inputs to the models.

Thickness Dependence

The increase in recrystallization rate with EP Cu thickness is consistent with previous reports of rates based on sheet resistance measurements.[10,13,15] By examining the variation over a wider range of thicknesses, we note a sharp drop in rate for films thinner than about 0.75 μm that suggests surface inhibition or pinning of grain boundary mobility. The 0.25 μm EP Cu film shows

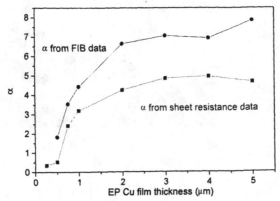

Figure 5. Avrami exponent α as a function of EP Cu film thickness from JMAK fits of data from FIB (circles) and sheet resistance measurements (squares).

no indication of recrystallization within the resolution of the FIB image, and only a slight decrease in sheet resistance, whereas some widely dispersed grain growth can be observed in the FIB images of the 0.5 μm sample after several hundred hours that shows a comparable change in sheet resistance.

The 0.75 and 1.0 μm films exhibit two distinct growth regimes, with an initial slow rate or incubation period, followed by a rapid growth regime, that is seen most clearly in the JMAK curves for the sheet resistance data that represent the transformation occurring in the entire volume of the Cu (Figure 4b). Additional factors that need to be considered beyond the impurities or defects that provide the driving force for recrystallization are the surface-to-volume ratio of the film and the ratio of the EP-to-PVD Cu film thickness.[20]

Note that the recrystallization of the EP Cu films also extends into the initially small-grained 1,000Å PVD Cu film in those samples where the EP Cu film itself fully recrystallizes. The two layers are finally indistinguishable in cross-section FIB or transmission electron micrographs.[17] A more detailed study of the recrystallization of PVD Cu layers driven by EP Cu will be reported separately.[20]

V. Crystallographic Texture

To complement the recrystallization rate studies, we carried out X-ray diffraction pole figure analyses of the EP Cu films after two months at room temperature following plating. The thicker films were fully recrystallized at this point, whereas the 0.25 μm film was still small-grained. The largest texture component in the Cu films is (111) and we focus here on the complete pole figures for the (111) orientation. Shown in Figure 6 are the FIB plan-view images and associated complete (111) pole figures for the 0.25, 1.0, and 5.0 μm EP Cu films.

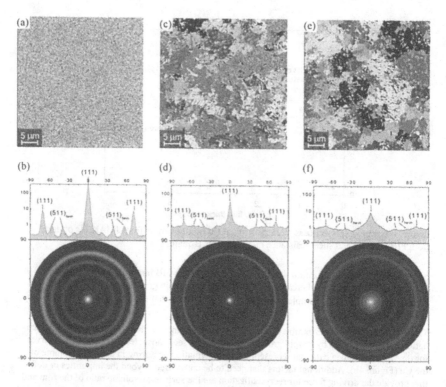

Figure 6. Plan-view FIB images and complete (111) pole figures of (a,b) 0.25 µm EP Cu; (c,d) 1µm EP Cu; (e,f) 5 µm EP Cu after 2 months at room temperature following plating. Note that the 0.25 µm film has not recrystallized in this time.

The 0.25 µm EP Cu film, which has not recrystallized even after several months (Figure 6a) exhibits a (111) pole figure (Figure 6b) that is essentially that of an as-plated film. The (111) peaks at 0 and 70° indicate a preferred (111) orientation. Twinning around one of these (111) directions produces grains with $(511)_{twin}$ orientations, whose (111) poles appear at 39, 56, and 70°. Second generation twins give rise to peaks at 22, 66, and 88°, although only the 22° peak is clearly distinguishable. This pole figure is similar to that of a reference PVD Cu film with no EP Cu. The high degree of twinning is typical for Cu and other fcc metals that have low stacking fault energies. Note, however, that despite the similarity of the (111) pole figures of the PVD and EP Cu films, the nature of the twin boundaries cannot be distinguished from these data. Incoherent twin boundaries, which have been associated with the recrystallization process, have significantly higher stored energy (498 mJ/m^2) than coherent twin boundaries (21 mJ/m^2).[21] A higher concentration of incoherent twin boundaries may be one of the contributing factors driving the room temperature recrystallization of the EP films.

The (111) pole figure of the fully recrystallized 1 μm EP Cu film exhibits (111) and (511)$_{twin}$ peaks (Figure 6d) that are broader and weaker than in the unrecrystallized 0.25 μm film. The higher background level corresponds to a higher random fraction that may in part result from multiple generations of twinning.[22] Numerous twins are distinguishable as parallel boundaries of alternating contrast in the large grains seen in Figure 6c. The loss of (111) texture upon room temperature recrystallization has previously been reported, however there are other reports where an increase in texture is observed.[11] We believe that these reports are not contradictory, but reflect the different influences of electroplating process, barrier and seed layers, composition, and microstructure, in different EP Cu films.

The (111) pole figure of the 5 μm EP Cu film, which has also completely recrystallized, exhibits (111) and (511)$_{twin}$ peaks that are further broadened to the point where the (511)$_{twin}$ peaks are almost completely subsumed in the higher random background. The increase in the random component of the (111) pole figures may arise from the formation of multiple generations of twins during recrystallization as well as from the absence of any preferred orientation in the nucleation of the grain growth. After multiple generations of twins the large number of resultant peaks may come to resemble, and no longer be distinguishable from, a film with a high random component.

VI. Conclusions

We have demonstrated that recrystallization rate data from FIB imaging and sheet resistance measurements can provide additional insights into the nucleation and growth of EP Cu films. The combination of 2- and 3-dimensional information from the two techniques leads us to conclude that nucleation occurs near the surface of the film. Grain growth then extends laterally more rapidly than vertically, as indicated by the continuing change in sheet resistance after the FIB images show no further change in grain size.

The thickness dependence of the recrystallization rate shows an extremely slow transformation for films thinner than 0.75 μm, consistent with surface inhibition or grain boundary pinning. The rate increases significantly for films thicker than 0.75 μm, as the surface-to-volume ratio decreases. A "bulk volume" recrystallization becomes dominant for films thicker than 2 μm.

Modeling the recrystallization data using the JMAK equation leads to unusually high values for the Avrami exponent α of 7 for the data obtained from FIB images. We conclude that the JMAK method is insufficient to describe the complex recrystallization process of EP Cu. A more detailed model is needed that accounts for specific features such as twinning, concentration and distribution of incorporated additives, defect densities, etc.

Texture analysis of the recrystallized films shows a preferred (111) texture that decreases with film thickness, as the texture becomes increasingly random. A high degree of twinning is indicated by the presence of (511)$_{twin}$ peaks, which also decrease in intensity with increasing film thickness. Multiple generations of twins, as well as an absence of a preferred orientation for nucleation of grain growth, can account for these changes. These data are consistent with a mechanism in which twinning plays an integral role in the recrystallization process.

References

1. D. Edelstein, J. Heidenreich, R. Goldblatt, W. Cote, C. Uzoh, N. Lustig, P. Roper, T. McDevitt, W. Motsiff, A. Simon, J. Dukovic, R. Wachnik, H. Rathore, R. Schulz, L. Su, S. Luce and J. Slattery, IEEE 1997 Intl. Electron Devices Meeting Digest, 773 (1997).
2. C.-K. Hu, R. Rosenberg, and K.Y. Lee, Appl. Phys. Lett. **74**, 2945 (1999).
3. C. Ryu, K.-W. Kwon, A.L.S. Loke, V.M. Dubin, R.A. Rahim, G.W. Ray, and S.S. Wong, Digest IEEE 1998 Symp. VLSI Tech., p. 156.
4. P.C. Andricacos, C. Uzoh, J.O. Dukovic, J. Horkans, and H. Deligianni, IBM J. Res. Devel. **42**, 567 (1998).
5. A. Gangulee, J. Appl. Phys. **43**, 867 (1972).
6. E.M. Hofer and H.E. Hintermann, J. Electrochem. Soc. **112**, 167 (1965).
7. I.V. Tomov, D.S. Stoychev, and I.B. Vitanova, J. Appl. Electrochem. **15**, 887 (1985).
8. T. Ritzdorf, L. Graham, S. Jin, C. Mu, and D. Fraser, in *Intl. Interconnect Technol. Conf. Abstracts*, pp. 166-168 (San Francisco, CA, June 1998).
9. C. Lingk, M.E. Gross, J. Appl. Phys. **84**, 5547 (1998).
10. Q.-T. Jiang and K. Smekalin, in *Adv. Metalliz. Conf. 1998*, edited by G.S. Sandhu, H. Koerner, M. Murakami, Y. Yasuda, and N. Kobayashi (Mater. Res. Soc., Pittsburgh, PA, 1999), pp. 209-215.
11. C. Cabral, Jr., P.C. Andricacos, L. Gignac, I.C. Noyan, K.P. Rodbell, T.M. Shaw, R. Rosenberg, J.M.E. Harper, P.W. DeHaven, P.S. Locke, S. Malhotra, C. Uzoh, and S.J. Klepeis, in *Adv. Metalliz. Conf. 1998*, edited by G.S. Sandhu, H. Koerner, M. Murakami, Y. Yasuda, and N. Kobayashi (Mater. Res. Soc., Pittsburgh, PA, 1999), pp. 81-87.
12. L.M. Gignac, K.P. Rodbell, C. Cabral, Jr., P.C. Andricacos, P.M. Rice, R.B. Beyers, P.S. Locke, and S.J. Klepeis, Mat. Res. Soc. Proc. **564**, 373 (1999).
13. M.E. Gross, R. Drese, C. Lingk, W.L. Brown, K. Evans-Lutterodt, D. Barr, D. Golovin, T. Ritzdorf, J. Turner, and L. Graham, Mat. Res. Soc. Symp. Proc. **564**, 379 (1999).
14. J.M.E. Harper, C. Cabral, Jr., P.C. Andricacos, L. Gignac, I.C. Noyan, K.P. Rodbell, and C.K. Hu, J. Applied Phys. **86**, 2516 (1999).
15. S.H. Brongersma, E. Richard, I. Vervoort, H. Bender, W. Vandervorst, S. Lagrange, G. Beyer, and K. Maex, J. Appl. Phys. **86**, 3642 (1999).
16. S.G. Malhotra, P.S. Locke, A.H. Simon, J. Fluegel, C. Parks, P. DeHaven, D.G. Hemmes, R. Jackson, and E. Patton, in *Adv. Metalliz. Conf. 1999*, edited by M.E. Gross, T. Gessner, N. Kobayashi, and Y. Yasuda (Mater. Res. Soc., Pittsburgh, PA, 2000), in press.
17. M.E. Gross, R. Drese, D. Golovin, W.L. Brown, C. Lingk, S. Merchant, and M. Oh, in *Adv. Metalliz. Conf. 1999*, edited by M.E. Gross, T. Gessner, N. Kobayashi, and Y. Yasuda (Mater. Res. Soc., Pittsburgh, PA, 2000), in press.
18. HyperNex, State College, PA 16801
19. F. J. Humphreys and M. Hatherly, *Recrystallization and Related Annealing Phenomena* (Pergamon, New York, 1996).
20. M.E. Gross, and D. Walther, unpublished results.
21. D.A. Porter and K.E. Easterling, *Phase Transformations in Metals and Alloys* (Chapman & Hall, London, 1992), p. 123.
22. C. Lingk, M.E. Gross, W.L. Brown, J. Appl. Phys. **87**, 2232 (2000).

Mat. Res. Soc. Symp. Proc. Vol. 612 © 2000 Materials Research Society

The Effects of the Mechanical Properties of the Confinement Material on Electromigration in Metallic Interconnects

Stefan P. Hau-Riege and Carl V. Thompson
Department of Materials Science and Engineering, M.I.T., Cambridge, Massachusetts 02139

ABSTRACT

New low-dielectric-constant inter-level dielectrics are being investigated as alternatives to SiO_2 for future integrated circuits. In general, these materials have very different mechanical properties from SiO_2. In the standard model, electromigration-induced stress evolution caused by changes in the number of available lattice sites in interconnects is described by an effective elastic modulus, B. Finite element calculations have been carried out to obtain B as a function of differences in the modulus, E, of interlevel dielectrics, for several stress-free homogeneous dilational strain configurations, for several line aspect ratios, and for different metallization schemes. In contradiction to earlier models, we find that for Cu-based metallization schemes with liners, a decrease in E by nearly two orders of magnitude has a relatively small effect on B, changing it by less than a factor of 2. However, B, and therefore the reliability of Cu interconnects *can* be strongly dependent on the modulus and thickness of the liner material.

INTRODUCTION

Electromigration continues to be one of the most important reliability issues for integrated circuit metallization systems [1]. Electromigration is electronic-current-induced atomic diffusion due to momentum transfer from flowing electrons to host atoms. As atoms electromigrate, volumes in which atoms accumulate develop more compressive stresses, while volumes from which atoms are depleted develop more tensile stresses. Electromigration-induced stress evolution in interconnects has been successfully described by the Korhonen model [2]. In this model, the relationship between a change in the number of available lattice sites per unit volume, dC, and a change in the hydrostatic stress, dσ, is described through

$$\frac{dC}{C} = -\frac{d\sigma}{B},\qquad(1)$$

where B is an effective elastic modulus. B has been calculated analytically for an elliptical aluminum interconnect embedded in an infinite SiO_2 or Si matrix [2], based on Eshelby's theory of inclusions [3].

The Korhonen model was originally developed for SiO_2-embedded Al-based metallization systems for which grain boundaries provide the fastest diffusion paths for electromigration [4]. However, the increase in the ratio of the wiring delay to the total intrinsic transistor delay has provided the motivation for the IC industry to move from aluminum-based interconnects embedded in SiO_2 to copper-based metallization systems with inter-level dielectrics (ILD) having lower dielectric constants, k, than SiO_2 [5]. Low-k ILD's are often polymer-based, and are mechanically much softer than SiO_2 [6], having Young's moduli that are more than an order of magnitude smaller than that of SiO_2. A change of the mechanical properties of the ILD alters the degree to which electromigration-induced stresses build up in

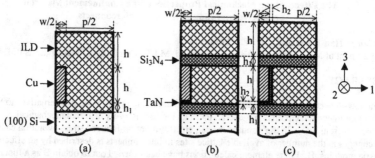

Figure 1: *Sketches of the geometry and coordinate system of the model. The finite element mesh is free to expand in the vertical direction. Mirror symmetry is applied horizontally, and translational symmetry is applied along the line in direction 2. The dimensions were taken to be $h = 0.4\mu m$, $h_1 = 0.1\mu m$, $h_2 = 35nm$, $h_{wafer} = 10\mu m$, and $p/w = 9.3$.*

interconnects. In the Korhonen model for electromigration, this is accounted for through the effective modulus, B. In this paper, we analyze how B changes when the mechanical properties of the ILD are changed, by performing finite element analyses (FEM) of stress changes due to dilational strains in realistic interconnect structures. We also investigate the effects of the line aspect ratio, as well as the presence of liner or barrier materials. Finally, we discuss the effects of changes in B on electromigration and electromigration-induced failure.

SIMULATION TECHNIQUE

Finite element calculations were performed using the commercial code Abaqus [7] with three-dimensional continuum-stress/displacement first-order-interpolation elements. The simulations are described in detail in reference [8]. The materials properties used in the simulations [8] were chosen for a typical electromigration test temperature of 200°C, and all materials were assumed to behave perfectly elastically. The geometries of the models representing three different metallization schemes are sketched in figures 1 (a), (b), and (c). The model sketched in figure 1 (a) is closest to the geometry studied analytically by Korhonen et al. [2]. We have modeled three different line widths, w = 0.24 μm, 0.40 μm, and 1.0 μm. We assumed perfect adhesion (traction) of Cu to the liner, ILD, and Si$_3$N$_4$. The orientation of the coordinate axes is shown in figure 1.

Divergences in the electromigration flux lead to an accumulation or depletion of atoms. With reasoning similar to that described by Korhonen et al. [2], we have modeled the deposition of atoms in a continuum model by assuming a set of stress-free homogeneous dilational strains of the interconnect, ε_1^T, ε_2^T, and ε_3^T, as detailed below. The hydrostatic stress for a given set of free strains was calculated using FEM and was related to the dilation by [9]

$$\Delta \equiv \varepsilon_1^T + \varepsilon_2^T + \varepsilon_3^T = \frac{dC}{C}. \tag{2}$$

With this result, the effective modulus was obtained using equation (1).

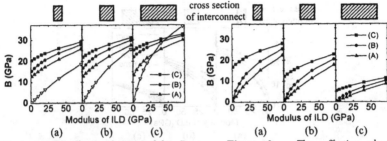

Figure 2: *The effective elastic modulus, B, as a function of the elastic modulus of the ILD for ε_1^T = ε_2^T, and ε_3^T = 0 for line widths: (a) w = 0.24 µm, (b) w = 0.40 µm, and (c) w = 1.00 µm. The results of the analytic model of Korhonen et al. [2] are shown by the curve marked with open triangles. (A), (B), and (C) correspond to the models sketched in figures 1 (a), (b), and (c).*

Figure 3: *The effective elastic modulus, B, as a function of the elastic modulus of the ILD for ε_1^T = ε_2^T = 0 for line widths: (a) w = 0.24 µm, (b) w = 0.40 µm, and (c) w = 1.00 µm. (A), (B), and (C) correspond to the models sketched in figures 1 (a), (b), and (c).*

Depending on the primary diffusion path, metallic atoms deposited in the interconnect can lead to different types of deformation. In our discussion, we focus on two different scenarios: (i) ε_1^T = ε_2^T, and ε_3^T = 0, which is the case for a polygranular interconnect with a columnar grain structure, and in which the grain boundaries are the primary diffusion paths; and (ii) ε_1^T = ε_2^T = 0, which is the case for atoms diffusing along and deposited on the top surface of the interconnect (which might apply to the case of Damascene Cu with Si_3N_4 as an interlayer diffusion barrier).

SIMULATION RESULTS

The effective elastic modulus, B, as a function of the Young's modulus of the ILD, E, for the different metallization schemes sketched in figure 1, and for three different line widths, is shown in figures 2 (a) to (c) for the case of ε_1^T = ε_2^T, and ε_3^T = 0. The Young's modulus ranges from 71.4 GPa, which is the elastic modulus of amorphous SiO_2, down to 1 GPa. Figures 3 (a) to (c) show B for the case of ε_1^T = ε_2^T = 0 for different aspect ratios, and figures 4 (a) to (c) show B for the case of ε_1^T = ε_2^T = 0 for different liner thicknesses. For comparison, the results of the analytic analysis described by Korhonen et al. [2] is overlaid on the curves in figure 2, assuming an elliptical copper interconnect inside an infinite, isotropic ILD matrix.

DISCUSSION OF SIMULATION RESULTS

We will first discuss the case of a metallization scheme without liner or barrier materials, and without Si_3N_4 layers, as sketched in figure 1 (a). σ_{11}/σ_{22} describes the degree of deviation from a state of biaxial stress. If σ_{11}/σ_{22} = 100%, the stress state is fully biaxial, which was

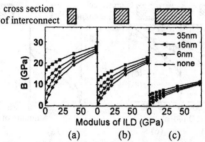

Figure 4: *The effective elastic modulus, B, as a function of the elastic modulus of the ILD for ε_1^T = ε_2^T = 0 for different barrier thicknesses and for line widths: (a) w = 0.24 μm, (b) w = 0.40 μm, and (c) w = 1.00 μm based on the model sketched in figure 1 (c).*

assumed for the analytic model described by Korhonen et al. [2]. Our calculations show that, in general, the lines do not exhibit a biaxial stress state. For $\varepsilon_1^T = \varepsilon_2^T$ and $\varepsilon_3^T = 0$, $\sigma_{11}/\sigma_{22} \approx 53\%$ for an ILD with a Young's modulus similar to that of SiO_2, and $\sigma_{11}/\sigma_{22} \approx 2\%$ if the Young's modulus of the ILD is 1 GPa. The softer the dielectric, the larger is the deviation from the biaxial stress state. This is the case because the interconnect can expand more easily into the ILD in direction 1, whereas the Si substrate prevents the interconnect from expanding in direction 2 along the interconnect. It can therefore be expected, and we verified this using FEM calculations, that the analytic model described by Korhonen et al. [2] does not predict valid effective moduli, B, for soft ILD's. For the case of $\varepsilon_1^T = \varepsilon_2^T = 0$ and $\varepsilon_3^T \neq 0$, the effective modulus decreases with decreasing aspect ratio h/w, because the material can expand more freely into direction 3 perpendicular to the wafer, and B becomes small for a small ILD Young's modulus, as can be seen in figure 3. Because $\varepsilon_2^T = 0$, the Si substrate does not prevent the interconnect from expanding in direction 2. This analysis shows that for the metallization scheme sketched in figure 1 (a) the effective elastic modulus strongly depends on the set of free dilational strains.

For the metallization scheme shown in figure 1 (b), the effective modulus shows a similar dependence on the Young's modulus of the ILD as for the case shown in figure 1 (a). B is slightly larger than in the case shown in figure 1 (a), because the Si_3N_4 as well as the liner material on the bottom of the trench somewhat restrain the interconnect from expanding into the ILD.

Finally, the metallization scheme shown in figure 1 (c) shows a different behavior for soft ILD's than the other two metallization schemes, because the interconnect is significantly constrained in all three directions. The liner material at the side wall of the trench prevents the interconnect from expanding into direction 3 or direction 1. As in other cases, the extension of the interconnect in direction 2 is prevented by the Si wafer, so that for a 35 nm-thick liner, B changes by less than a factor of 2 when the Young's modulus of the ILD decreases by nearly two orders of magnitudes. Only for very thin liners is B much smaller for soft ILD's than for SiO_2, as shown in figure 4.

DISCUSSION OF THE EFFECTS OF THE EFFECTIVE ELASTIC MODULUS ON ELECTROMIGRATION

The tensile stress that develops at the electron-source via eventually becomes large enough that the strain energy reduction associated with void formation is larger than the surface energy cost associated with voiding, and a void nucleates. In the Korhonen model, if it is assumed that the diffusivity is independent of stress, the stress evolution at the electron-source end of a semi-infinite line prior to voiding is given by [2]

$$\sigma(t) = \sigma_0 + \sqrt{\frac{4t}{\pi} \frac{\Omega D}{kT} \frac{\rho q^* j}{\Omega}} \cdot \sqrt{B},$$ (3)

so that the stress increase is proportional to $\sqrt{B}$. t is the time, D is the atomic self diffusivity, k is Boltzmann's constant, T is temperature, q^* is the effective charge, Ω is the atomic volume, ρ is the electrical resistivity of the high-conductivity metal, and j is the current density. The electromigration-induced atomic flux in the absence of back stress effects is independent of the effective elastic modulus and, once a void has nucleated, the void length is given by [10]

$$L_{void} = \frac{q^* \rho j}{kT} Dt.$$ (4)

If it is easy for voids to nucleate, interconnect failure is controlled by void growth. In the case of little or no back-stress, void growth is independent of the effective elastic modulus, so that the mechanical properties of the ILD do not affect interconnect reliability in this case. A likely scenario for this behavior is a copper-Damascene interconnect with a weakly adhering Si_3N_4 layer on top of the Cu, making it easy for Cu to delaminate from the Si_3N_4. On the other hand, if liner materials and over-layers adhere well to the metallic interconnect, voids are difficult to nucleate and interconnect failure is determined by the rate at which stresses build up, so that the time to failure is inversely proportional to the effective elastic modulus (assuming $\sigma_0 = 0$). In this case, once a void has nucleated the stress relief around the void will quickly lead to rapid void growth and to a resistance increase large enough for the line to fail [10].

SUMMARY AND CONCLUSION

The analytic model developed by Korhonen et al. [2] to describe the elastic response when lattice sites are added or removed from Al embedded in SiO_2 was applied to Cu embedded in a low-E ILD. The predicted response, as embodied in the effective modulus B, was found to be of the same order as the changes in the modulus of the dielectric E, so that the predicted changes in interconnect reliability will be very significant when low-E ILD's are used. However, using more accurate finite element analyses, we have found that for Cu-based metallization schemes with liners and with dimensions as shown in figure 1 (c), a decrease in the elastic modulus of the ILD has little effect on the effective modulus, B, which suggests that the rates of electromigration-induced stress change and of void growth are nearly independent of the mechanical properties of the ILD. Because the copper is not in direct contact with the ILD, but is surrounded by a liner material, the critical tensile stress above which void nucleation occurs is also not affected by a change in the dielectric. This indicates that the reliability of interconnects should not significantly change if the ILD material is changed from SiO_2 to a softer material, as long as a sufficiently thick liner is used.

Our calculations also show that B only weakly depends on the line cross section. However, in future process generations, thinner, and therefore weaker, liners will be used, so that the effective modulus will be lower. A lower effective modulus slows the build up of stress, which, in turn, reduces the rate at which electromigration-induced damage initiates. On the other hand, the modulus does not affect the electromigration-induced atomic flux, and therefore the growth of voids, once nucleated, is unaffected. In addition, the failure mode of interconnects with thin liners may be different from the case of interconnects with thick liners, because if the liner is very thin, electromigration-induced stresses might lead to failure of the liner material itself, so that copper can diffuse into the ILD and Si to cause short-circuit failures or device failures. The extent to which the use of softer low-k ILD's will affect Cu-based interconnect reliability will depend most strongly on their effects on the mechanical reliability of diffusion-barrier liners.

In summary, our results suggest that as long as the liner materials for Cu interconnects remain intact, the mechanical properties of alternative (e.g., low-k) dielectrics will have less of an effect on interconnect reliability than the standard model of Korhonen et al. [2] would indicate. While the Korhonen model overestimates the effects of alternative dielectrics, the choice of dielectric material, the choice of liner material, and interconnect dimensions still have a significant effect on the mechanical reliability of interconnects, in ways that can be predicted using finite-element mechanical modeling in conjunction with electromigration models and simulations.

ACKNOWLEDGEMENT

This research was supported by the SRC. S. Hau-Riege also holds an SRC graduate fellowship. We would like to thank Steve C. Seel for proofreading the manuscript and helpful discussions.

REFERENCES

[1] C.-K. Hu, R. Rosenberg, H.S. Rathore, D.B. Nguyen, and B. Agarwala, Proceedings of the IEEE 1999 International Interconnect Technology Conference, pp.267-9, IEEE, Piscataway, NJ (1999).

[2] M. A. Korhonen, P. Boergesen, K.N. Tu, and Che-Yu Li, *J. Appl. Phys.* **73**, 3790 (1993).

[3] J. D. Eshelby, *Proc. Roy. Soc.* **A 241**, 376 (1957).

[4] C.V. Thompson and H. Kahn, *J. Electr. Materials* **22**, 581 (1993).

[5] D. Edelstein, J. Heidenreich, R. Goldblatt, W. Cote, C. Uzoh, N. Lustig, P. Roper, T. McDevitt, W. Motsiff, A. Simon, J. Dukovic, R. Wachnik, H. Rathore, R. Schulz, L. Su, S. Luce, and J. Slattery, *IEEE Intl. Electron Devices Meeting Digest*, 773 (1997).

[6] J. Waeterloos, M. Simmonds, A. Achen, and M. Meier, *Europ. Semicond.* **21**, 26 (1999).

[7] Abaqus, Version 5.8, general purpose finite element program, Hibbit, Karlson, and Sorensen, Inc., Pawtucket, RI, 1998.

[8] S.P. Hau-Riege and C.V. Thompson, submitted to *J. Mat. Res.*

[9] I.C. Noyan and J.B. Cohen, *Residual Stress*, Springer Verlag, New York, NY, 1987.

[10] Y.-J. Park, V.K. Andleigh, and C.V. Thompson, *J. Appl. Phys.* **85**, 3546 (1999).

Mat. Res. Soc. Symp. Proc. Vol 612 © 2000 Materials Research Society

Microtexture and Strain in Electroplated Copper Interconnects

R.Spolenak*, D. L. Barr*, M. E. Gross*, K. Evans-Lutterodt*, W. L. Brown*, N. Tamura†,
A. A. Macdowell†, R. S. Celestre†, H. A. Padmore†, B. C. Valek%, J. C. Bravman%, P.
Flinn%, T. Marieb!, R. R. Keller$, B. W. Batterman†§ and, J. R. Patel†§
* Bell Labs/Lucent Technologies, Murray Hill, NJ
† Advanced Light Source, Lawrence Berkeley National Lab., Berkeley, CA
% Dept. of Materials Science and Engineering, Stanford University, Stanford, CA
! Intel Corporation, Portland, OR
$ National Institute of Standards and Technology, Materials Reliability Division, Boulder, CO
§ SSRL/SLAC, Stanford University, Stanford, CA

ABSTRACT

The microstructure of narrow metal conductors in the electrical interconnections on IC chips has often been identified as of major importance in the reliability of these devices. The stresses and stress gradients that develop in the conductors as a result of thermal expansion differences in the materials and of electromigration at high current densities are believed to be strongly dependent on the details of the grain structure. The present work discusses new techniques based on microbeam x-ray diffraction (MBXRD) that have enabled measurement not only of the microstructure of totally encapsulated conductors but also of the local stresses in them on a micron and submicron scale. White x-rays from the Advanced Light Source were focused to a micron spot size by Kirkpatrick-Baez mirrors. The sample was stepped under the micro-beam and Laue images obtained at each sample location using a CCD area detector. Microstructure and local strain were deduced from these images. Cu lines with widths ranging from 0.8 μm to 5 μm and thickness of 1 μm were investigated. Comparisons are made between the capabilities of MBXRD and the well established techniques of broad beam XRD, electron back scatter diffraction (EBSD) and focused ion beam imagining (FIB).

INTRODUCTION

Currently, many companies are in the transition of dramatically changing their materials and processes for the multilevel interconnect structure in IC manufacturing. Well-known materials like aluminum and silicon dioxide are being replaced by copper and low-k dielectrics and the process architecture is changing from subtractive metal to damascene. The reliability of the new materials and structures has to be assessed and basic research and novel techniques are needed to gain the knowledge required for understanding their properties. For aluminum, stress gradients and high local stresses have been considered to be responsible for failure modes such as stress voiding and electromigration[1,2]. The measurement of stresses on a local (submicron) length scale has been found to be extremely difficult[3-7]. Even the non-destructive determination of the crystallographic orientation of buried single grains has recently only been accomplished by MBXRD[3,6].

In this paper we will demonstrate how an x-ray microbeam can be used to determine the orientation and the strain in single Cu grains and to produce orientation and stress maps of entire line segments. This new technique will be compared to established techniques such as XRD, EBSD and FIB imaging.

EXPERIMENTAL

The x-ray source for this experiment was the advanced light source (ALS) at the Lawrence Berkeley National lab (LBNL). An intense white x-ray beam (6 – 14 keV) was generated by a bending magnet and focused to a spot size of 0.8x0.7 μm^2 by a pair of elliptically bent Kirkpatrick-Baez mirrors. Laue patterns were acquired using a CCD area detector positioned 30 mm above the sample, which was tilted at 45° relative to the x-ray beam. The indexing of the Laue patterns yields the crystallographic orientation of the grains. The deviation of the position of the Laue spots from the predicted position they would have for an unstrained cubic crystal, allows for the determination of the deviatoric component of the strain tensor. Typically 8 to 10 reflections from single grains were used to evaluate the deviation. In order to obtain the dilatational component of the strain tensor the exact energy of each reflection was determined by using a four-crystal Ge monochromator. The stress tensor is calculated from the complete strain tensor by using the tensor of the elastic constants. The geometry of the four-crystal monochromator allows for illuminating the sample at the same spot as with white light. The monochromator was calibrated to within 1 eV.

Custom-made software was used to determine the orientation, the deviatoric part of the strain tensor in every pixel and the energy of selected reflections. It was possible to index even multiple grains within the illuminated volume. Further details of the technique can be found in Tamura *et al*.[8,9]

The samples examined in this work were single level damascene Cu lines especially designed for *in situ* electromigration experiments. Single Cu line segments were connected electrically by Ta links. The line width ranged from 0.8 μm to 5 μm and the line thickness was 1 μm. The dielectric used was PE-TEOS and the samples were either uncapped or capped by 200 nm of SiN_x. The trenches etched in the dielectric were lined with a Ta barrier layer and a sputtered Cu seed layer and filled with electroplated Cu from an Enthone bath. All samples were annealed at 400 °C for 30 min prior to the experiment.

RESULTS

The first set of experiments studied uncapped Cu samples. Figure 1 shows maps of the orientation as well as of the deviatoric strain components in a 5 μm wide Cu line segment. The pixel size is 0.5 by 0.5 μm^2. The gray level of the orientation map corresponds to the angle between the (111) crystal orientation and the normal to the sample surface. Black area in general reflects the absence of data. The horizontal black line in the maps in Figure 1and the black area in the lower right corner are two such regions. Two orientation maps are shown, which were produced by stepping the sample under the x-ray beam. When the x-ray illuminated volume in the sample consists of more than one grain, composite Laue patterns are obtained. The analysis software is able to index more than ten grains in such a case, as long as each grain is represented by at least six reflections. In the left MBXRD map in Figure 1 ("big grains") only the grains with the strongest reflections within a pixel are shown. In the second map("small grains") the grains with the second strongest reflections are shown revealing a more detailed microstructure. When comparing the second map to the FIB image in the figure, areas of smaller grains, typically twins, can be identified.

As described above, the deviatoric part of the strain tensor can also be determined by the white beam Laue images. Here, the stress components out of plane and across the line are shown. As only the deviatoric part is shown, the component along the line can be omitted, as it has to add up with the other parts to make the trace of the tensor zero. The stresses are not constant throughout a single grain. The largest tensile and compressive deviatoric stresses are found in the regions close to the grain boundaries. This can be seen even better in the map of the Von Mises stresses (second from the right), which are a measure of the maximum shear stresses within the sample.

The furthest right strip in Figure 1 shows an FIB image visualizing the surface grain structure at a high resolution. The topography visible within a single grain is an artifact of the FIB imaging. The FIB image shows more detail, but the grain structure visible in the orientation maps and the FIB image are in quite good agreement. The comparison between the x-ray orientation map and the EBSD map is fairly good. The EBSD map was taken after FIB imaging and suffered from FIB produced ion damage and topography.

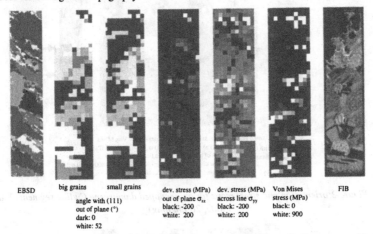

EBSD	big grains	small grains	dev. stress (MPa)	dev. stress (MPa)	Von Mises	FIB
		angle with (111) out of plane (°) dark: 0 white: 52	out of plane σ_{zz} black: -200 white: 200	across line σ_{yy} black: -200 white: 200	stress (MPa) black: 0 white: 900	

Figure 1:orientation and stress maps for an uncapped damascene Cu line segment (5 μm wide, 20 μm long)

Figure 2 shows results for a 2 micron wide uncapped damascene Cu line. The line is basically divided into a region with relatively big grains (in the lower half) and another with smaller grains. This can be seen in the orientation maps as well as the FIB image. In the "big grains" MBXRD map, which is based on only the strongest reflections in each pixel, the smaller grain with dark contrast in the FIB image (arrow) is not included. However, by discriminating against the most intense Laue reflections in each pixel, it shows up clearly in the "small grains" orientation map (second from the left). One has to bear in mind that MBXRD gives information about the entire volume of material, whereas an FIB image reveals only the structure closest to the surface.

In the deviatoric stress maps of Figure 2 only the strongest grains are shown. Here, the light (FIB) grain (see arrow) shows compressive deviatoric stresses (dark contrast) out of plane and tensile stresses across the line (light contrast).

Figure 3 shows results for a 2 micron wide line capped with 200 nm of SiN_x. This cap does not interfere with the MBXRD study of the line. The first map on the left shows again the angle between the (111) direction and the out of plane normal and thus also reveals the grain structure. In this case the grain structure is close to being bamboo. Here, the stresses measured not only originate from the post deposition thermal treatment, but also from an electromigration current of 0.5 MA/cm^2 that had been supplied for 20 hr at 300 °C. An interesting feature can be seen close to the grain boundary below a (111) oriented grain in the center of the segment. This can be seen in the compressive (dark) region in the deviatoric out of plane stress and in the tensile (white) region in the stress across the line. The differences in the two deviatoric components leads to a significant Von Mises stress (white) in the same region.

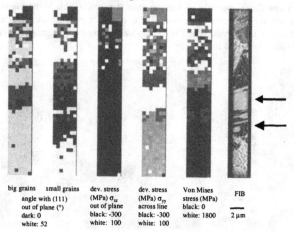

big grains	small grains	dev. stress (MPa) σ_{zz} out of plane	dev. stress (MPa) σ_{yy} across line	Von Mises stress (MPa)	FIB
	angle with (111) out of plane (°)	black: -300	black: -300	black: 0	
	dark: 0	white: 100	white: 100	white: 1800	2 μm
	white: 52				

Figure 2:orientation and stress maps for an uncapped damascene Cu line segment (2 μm wide, 20 μm long)

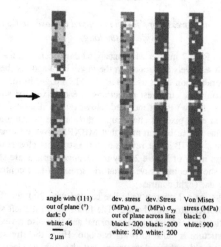

	angle with (111) out of plane (°) dark: 0 white: 46 2 μm	dev. stress (MPa) σ_{zz} out of plane black: -200 white: 200	dev. Stress (MPa) σ_{yy} across line black: -200 white: 200	Von Mises stress (MPa) black: 0 white: 900

Figure 3:orientation and stress maps for a capped damascene Cu line segment (2 μm wide, 30 μm long)

Figure 4 shows the orientation components for a 0.8 micron wide line that is also essentially bamboo type. In addition to the overall orientation map three more narrowly defined maps are shown emphasizing three different texture components. Three grains show a broad (111) texture, three to four seem to originate from a sidewall component or a second order twin and the majority belongs to a broad (115) texture. Results for the hydrostatic stress component for Al are shown in Valek *et al.* [Valek, 2000 #23]. The results for Cu will be shown in a future paper.

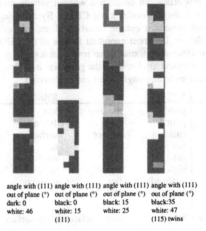

angle with (111)	angle with (111)	angle with (111)	angle with (111)
out of plane (°)	out of plane (°)	out of plane (°)	out of plane (°)
dark: 0	black: 0	black: 15	black:35
white: 46	white: 15	white: 25	white: 47
	(111)		(115) twins

Figure 4:orientation maps for a capped damascene Cu line segment (0.8 μm wide, 30 μm long)

DISCUSSION AND CONCLUSIONS

Considering the orientation maps of the different lines not many (111) oriented grains are observed. This observation is in contrast to the texture known from sputtered Cu films. However, it is agreement with observations made on annealed electroplated films[10]. The (115) twin texture had been found to be predominant. This is verified in the experimental results for the 0.8 micron wide line. Most of the grains are (115) oriented out of plane. Additionally, there is a (111) component and indications for a sidewall texture, a feature of electroplated Cu damascene that has been previously reported[11]. The manifold of texture components and the highly anisotropic elastic constants of Cu seem to be the origin of significant shear stresses, especially at grain boundaries as can be seen in Figures 1 to 3.

We have demonstrated the capabilities of a new method for orientation analysis and stress measurements on a submicron level. Table 1 shows the comparison between established techniques and this new x-ray technique. As in any local technique with a relatively long data acquisition time, the number of grains that can be analyzed is limited. For obtaining quantitative texture distributions conventional XRD is still the best approach. However, for the investigation of local phenomena, the local techniques are essential. At present FIB and EBSD have significantly better spatial resolution than the new MBXRD, however, they are limited to the analysis of volumes very close to the surface and cannot penetrate capping layers.

The strength of MBXRD is that it provides information from the entire volume of a thin film and that it in principle detects all grains that are present within the illuminated volume. The software mentioned above is able to identify more than ten grains within one Laue pattern. As long as the exposure time is long enough to provide enough data for indexing at least six Laue spots all the grains within the illuminated volume can be identified. The second advantage compared to the other local techniques is the property of x-rays to penetrate capping layers. In this way materials properties can be studied without sample preparation that may change the stress state, which is a major drawback for CBED. By acquiring images at different detector distances even buried structures can be investigated and the origin of the reflections can be identified by triangulation. The most important feature of the MBXRD technique is its ability to provide the complete three-dimensional strain tensor for a single grain without changing the stress state through sample preparation. Provided the grain size is comparable to or bigger than the spot size, even strain variations within single grains can be detected.

	XRD	FIB	EBSD	MBXRD	CBED
statistics	good	medium	low	low	very low
spatial resolution	poor	5 nm	100 nm	800 nm	10 nm
analyzed volume	entire	surface	surface	entire	entire
strain	yes	no	no	yes	yes
multiple grains	yes	no	no	yes	no
availability	high	high	high	scarce	scarce
measurement time	fast/medium	fast	medium	slow	slow
orientational accuracy	1°	10°	1°	0.1°	0.5°
sample preparation	none	clean surface	clean surface	none	TEM prep.

Table 1: Comparison of XRD (x-ray diffraction), FIB (focused ion beam), EBSD (electron back-scatter diffraction, MBXRD (microbeam x-ray diffration) and CBED (convergent beam electron diffraction)

Summarizing, this technique has the potential to become one of the key elements in the research of thin film phenomena on the micron and submicron scale. Reliability issues such as local plastic deformation caused by electromigration or thermal stresses are being investigated.

ACKNOWLEDGEMENTS
The authors would like to thank the members of the SFRL (Silicon Fabrication Research Laboratory) namely, M. Buonanno, M. Hoover, S. Fiorillo, M. D. Morris, K. Takahashi, J. Frackoviak, T. Craddock, J. F. Miner, G. R. Weber, W. Mansfield, W. W. Tai, F. Klemens, A. Kornblit, R. Keller, and E. Ferry for their contributions in preparation of the samples.

REFERENCES
1. I. A. Blech and K. L. Tai, Appl. Phys. Lett. 30 (8), 387-389 (1977).
2. P.-C. Wang, G. S. Cargill III, I. C. Noyan, and C.-K. Hu, Appl. Phys. Letters 72 (11), 1296-1298 (1998).
3. J.-S. Chung and G. E. Ice, J. Appl. Phys. 85, 3546-3555 (1999).
4. S. Kraemer, J. Mayer, C. Witt, A. Weickenmeier, and M. Ruehle, Ultramicroscopy 81, in press (2000).

5. H. J. Maier, R. R. Keller, H. Renner, H. Mughrabi, and A. Preston, Philosophical Magazine **A74**, 23-46 (1996).
6. N. Tamura, J.-S. Chung, G. E. Ice, B. C. Larson, J. D. Budai, and W. Lowe, Mat. Res. Soc. Symp. Proc. **563**, 175-180 (1999).
7. A. J. Wilkinson, Ultramicroscopy **62** (4), 237-247 (1996).
8. N. Tamura, B. C. Valek, R. Spolenak, A. A. MacDowell, R. S. Celestre, H. A. Padmore, W. L. Brown, J. C. Bravman, P. Flinn, T. Marieb, B. W. Batterman and, J. R. Patel, Mat. Soc. Rec. Proc. **submitted** (2000).
9. A. A. MacDowell, C. H. Chang, H. A. Padmore, J. R. Patel, and A. C. Thompson, Mat. Res. Soc. Proc. **524**, 55-58 (1998).
10. R. Spolenak, C. A. Volkert, K. M. Takahashi, S. A. Fiorillo, J. F. Miner, and W. L. Brown, Mat. Res. Soc. Proc. **in print** (1999).
11. C. Lingk, M. E. Gross, and W. L. Brown, Applied Physics Letters **74** (5), 682-684 (1999).

Mat. Res. Soc. Symp. Proc. Vol. 612 © 2000 Materials Research Society

Characterization of Cu-Al alloy/SiO₂ interface microstructure

Pei-I Wang[1], S. P. Murarka[1], G.-R. Yang[1], E. Barnat[1], T.-M. Lu[1], Y.-C. Chen[2], Xiang Li[2] , and K. Rajan[2]

[1]Center for Integrated Electronics, Electronics Manufacturing and Electronic Media, Rensselaer Polytechnic Institute, Troy, NY 12180

[2]Department of Materials Science & Engineering, Rensselaer Polytechnic Institute, Troy NY 12180

ABSTRACT

Cu-Al alloys have been recommended for application as the diffusion barriers/adhesion promoters for advanced copper based metallization schemes. This approach to barrier formation is to generate an ultra-thin interfacial layer through Cu alloying without significantly affecting the resistivity of Cu. In this paper the microstructure of the bilayers of Cu/Cu-5 at%Al and Cu-5 at%Al/Cu sputter deposited on SiO₂ before and after thermal annealing is investigated by transmission electron microscopy (TEM). Interfacial layer is observed in both cases. The variation of the resistance of the Cu-Al alloy film is consistent with its microstructure. The x-ray diffraction (XRD) spectra of Cu-5 at%Al on SiO₂ shows that the addition of Al into Cu intends to favor the Cu (111) texture. These results will be presented and discussed showing that films of Cu doped with Al appear to act as a suitable barrier and adhesion promoter between SiO₂ and Cu.

INTRODUCTION

Copper interconnect-technology has been moving rapidly from research onto actual product-manufacturing in industry. Copper offers lower resistivity, higher reliability against electromigration, and improved mechanical properties compared to aluminum.[1] It can be easily deposited by chemical vapor deposition (CVD), physical vapor deposition (PVD), electroless deposition, and electroplating. Both CVD and plating techniques have been paid high attention because they provide good step coverage. Plating techniques are especially appealing because of low cost and low processing temperature as well as good via/trench filling capability.

Copper, however, migrates into SiO₂ when subjected to high temperature or high bias.[2] Diffusion in SiO₂ may deteriorate the insulator characteristics and through SiO₂ into silicon may cause device leakage currents. Also copper does not adhere to SiO₂ surfaces because of the inability of Cu in reducing SiO₂ and thus forming strong chemical bonding at the interface.[1] It is thus necessary to employ diffusion barrier layers that prevent copper penetration into the underlying devices and at the same time promote adhesion between SiO₂ and Cu. Several different barrier materials have been researched.[4-8] Most of these have high resistivities (100 – 1000 μΩ-cm range) and have been known to be effective in thickness larger than 10 nm. Future application, in narrower trenches and vias, require barrier with resistivities comparable to copper and very thin (1-3 nm).

Another approach has been considered to barrier formation. The approach uses small amounts of alloying additions to copper leading to self-forming barriers, at the Cu-SiO₂ interface, without significantly affecting the reisistivity of copper.[1] Previous studies have shown that the thin aluminum layer suppressed the migration of copper into the oxide. Bilayers (of Cu/Al) with as low a thickness of aluminum as 5-10 nm have been shown to act as effective adhesion promoter/diffusion barrier.[9] It has also been reported that the Al atoms, on thermal annealing, diffused outward through the Cu to form a passivation layer on copper surface. These results lead to the suggestion that Al is one of the best choices for an alloying additive to copper.[1, 10]

The study of the microstructure of Cu-Al alloy/SiO_2 interfaces, which apparently leads to the bonding and barrier layer at the interface, is necessary in order to generate the understanding of the interfacial mechanisms and obtained reliabilities of Cu(Al) on SiO_2. In this work we have investigated the interfacial reaction of Cu-Al alloy/SiO_2 by TEM analyses, especially focusing on the microstructure evolution of Cu-Al film and metal-oxide interface. This paper reports the results of these investigations. The TEM images show that an interfacial layer is observed in the Cu-Al alloy/SiO_2 interface. This is consistent with the results of our previous XPS study and electrical testing of the MOS capacitors of Cu/Cu-Al alloy/SiO_2/p-Si.[13, 14]

EXPERIMENTAL

P-typed, <100> silicon wafers were used as substrates in all experiments. After RCA cleaning, a high quality of thermal oxide was grown in dry oxygen to a thickness about 100 nm. Thin films of Cu with varying Al contents were deposited by co-sputtering from a Cu and an Al target. The base pressure in the vacuum chamber for all depositions was at the range of 10^{-7} Torr. The Ar pressure during sputtering was 5 mTorr.

XRD experiments were conducted for determining the texture of Cu and Cu-Al alloy films. The grain size and the metal-oxide interface were studied using TEM. The film thickness and Al concentration were measured using RBS. Four-point probe measurements were preformed to determine sheet resistance. All annealings were carried out in vacuum at the range of 10^{-7} Torr.

RESULTS AND DISCUSSION

Figure 1a and 1b show the XRD spectra of as-deposited pure Cu film and of the one that was annealed at 300°C for 3 hours, respectively. The XRD data showed 2 peaks associated with (111) and (200) diffraction. Figure 1c and 1d show the XRD spectra of as-deposited Cu-5 at%Al film and of the one that was annealed at 300°C for 3 hours, respectively. By contrast, the (200) peak is not observable in the XRD spectrum of as-deposited Cu-Al alloy film. Note that for Cu films, (111) and (200) fiber texture components for sputtered samples have been reported. However, pure aluminum film textures consist almost exclusively of (111) fiber and random texture components.[11] It seems that the Al species present in the film during the as-deposited state suppress the (200) texture (and a very strong (111) component). Figure 1d shows that the (200) texture appears, and the integrated intensity of the (111) peak over the (200) peak (3.57) is close to that of the annealed Cu film in Fig 1b (3.21), indicating the aluminum moves out to interface and surface during annealing.

In general, both Cu and Cu-Al film after annealing show an increase in the 2θ value, perhaps indicating a densification. The peak position of (111) peak of Cu-Al alloy film after annealing (43.28) is shifted slightly as compared to that of pure copper (43.36). This perhaps indicates aluminum (in solution in copper) leading to an increase in copper's lattice spacing.

Figure 2 shows the variations of microstructure of Cu-Al films as a function of Al concentration and annealing temperature 300°C or 400°C. In all cases, the as-deposited films exhibited a fine-grained microstructure with grains that are only a few nanometers in diameters.

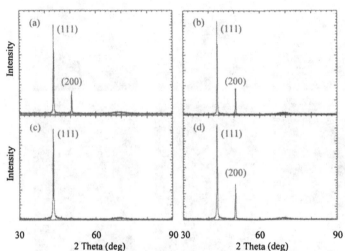

Fig. 1 XRD spectra of pure Cu film (a) as-deposited, (b) annealed at 300°C for 3 hours.
Cu-5 at%Al film (c) as-deposited, and (d) annealed at 300°C for 3 hours.

The pure copper grains grew significantly to a uniform grain size distribution of large grains after annealing. By contrast, the Cu films doped with 1 at%Al shows abnormal grain growth (AGG) leading to a bi-model grain size distribution of small and large grains after annealing at 300°C and AGG was still visible even when film was annealed at a higher annealing temperature 400°C. Twining is observed in the large grains in Cu-1 at%Al film as well as most of the grains in the pure copper films. When the aluminum concentration was 3 at%, the microstructure shows a uniform mono-model grain size distribution of small grains. It is believed that the AGG observed in Cu-1 at%Al film is because of the non-saturated Cu grain boundaries with Al leading to the anisotropic mobility and energy of grain boundaries.[12] Thus AGG can take place when the grain boundaries of a given grain have the growth advantage exclusively over those of the other grains. It is also believed when the aluminum concentration was 3 at.%, the grain boundaries of the Cu were saturated. This saturation of the Cu grain boundaries by the segregation of Al was to minimize the free energy, and the isotropic grain boundary energy led to the uniform distribution of small grains. Similar microstructure evolution in Cu doped with 5 at%Al clearly indicates that the Cu grain boundaries were "stuffed" at this doping level of Al (3 – 5 at%).

Figure 3 shows the variation of resistance of Cu-Al alloy films (of the same thickness) with the Al concentration. As expected, the resistance does increase dramatically with increasing Al content. It is observed that the resistance seems to be consistent with the grain size evolution (in Fig. 2) in the range of Al doping level (1 – 5 at %). It is noted that the resistance of Cu-5 at%Al alloy film after annealing is slightly lower than that of Cu-3 at%Al alloy film. We observed the

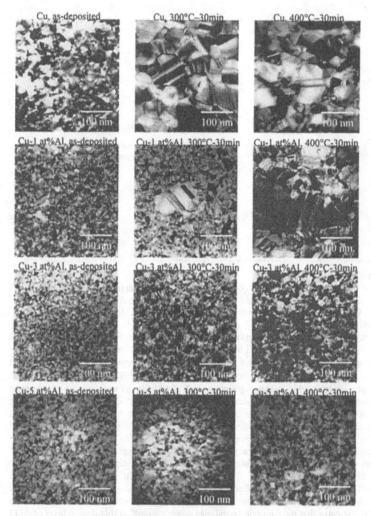

Fig. 2 Plan-view, bright-field, TEM images of Cu and Cu-Al alloy films with various concentration of Al which at different annealed conditions.

second phase, which is possibly $CuAl_2$, starts to appear in the alloy film when the amount of Al content is up to 5 at%Al. It is hypothesized that the nucleation and growth of the Cu-Al phase consume the Al available in Cu, leading to the decrease of the resistance. It is also noted that the resistance of the annealed Cu-5 at%Al alloy film is no more than 3 times of that of the pure Cu film.

Figure 4a shows the cross-sectional view of 200 nm Cu/50 nm Cu-5 at%Al deposited on SiO_2 substrate after annealing at 300°C for 1 hour. An interfacial layer is observed at the metal-oxide interface. This interfacial layer seems to be a multi-layer which is also suggested by our previous XPS study of the buried Cu-Al alloy film/SiO_2 interface.[13] Further study by high resolution TEM is needed to characterize the detail of this interfacial structure. This interfacial layer would explain the good adhesion of the Cu-Al alloy film on SiO_2 substrate and the C-V and I-V characteristics of the MOS of Cu/Cu-Al alloy/SiO_2/p-Si.[14] Figure 4b shows the cross-sectional view of 200 nm Cu-5 at%Al/200 nm Cu on SiO_2 after annealing at 300°C for 1 hour. A very thin and relatively smooth interface is observed in this case. This is expected to be the Al in the upper Cu-Al alloy film moving downwards through the pure Cu to the metal-oxide interface to react with available oxygen in the interface. This result confirms that the Al in the Cu-Al alloy film can segregate to the metal-oxide interface to promote the interfacial reaction at the Cu-Al alloy/SiO_2 interface.

CONCLUSIONS

In this work, the microstructure of the Cu-Al alloy film and the Cu-Al alloy/SiO_2 interface has been studied. The XRD spectra show that the addition of Al into Cu intends to favor the Cu (111) texture. The variation of the resistance of Cu-Al film with Al concentration is consistent with the microstructure evolution in the range of 1 – 5 at%Al. Abnormal grain growth is observed in Cu-1 at%Al film. It is believed that AGG was caused by non-saturated Cu grain boundaries leading to the anisotropic mobility and energy of grain boundaries. Uniform grain size distribution is observed for Cu doped with the amount of Al up to 3 at%. It is believed this is

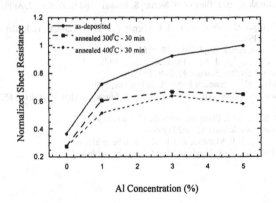

Fig. 3 Normalized sheet resistance of Cu-Al alloy films vs. Al concentration.

Fig. 4 Cross-sectional, bright-field, TEM images of (a) 200 nm Cu/50 nm Cu-5 at%Al deposited on SiO$_2$ substrate and (b) 200 nm Cu-5 at%Al/200 nm Cu deposited on SiO$_2$ substrate after annealing at 300°C for 1 hour.

caused by the saturated (stuffed) Cu grain boundaries with Al. The observed interfacial layer in Cu-5 at%Al/SiO$_2$ explains the good adhesion of Cu-Al alloy film on SiO$_2$ as well as the good C-V and I-V characteristics of its MOS capacitors as reported in our previous work. Furthermore, the interfacial reaction at Cu-5 at%Al/Cu/SiO$_2$ interface confirms that Al can move through the pure Cu into the metal-SiO$_2$ interface. These results indicate that Cu doped with Al appears to act as a proper diffusion barrier/adhesion promoter between Cu and SiO$_2$.

ACKNOWLEDGMENTS

Authors thank Semiconductor Research Corporation for the partial support of this work.

REFERENCES

[1] S. P. Murarka and S. Hymes, Critical Reviews in Solid State and Materials Sciences **20**, 87 (1995).
[2] J. D. McBrayer, R. M. Swanson, and T. W. Sigmon, J. Electrochem. Soc. **133**, 1242 (1986).
[3] Y. Shacham-Diamand, A. Dehia, D. Hoffstetter, and W. G. Oldham, J. Electrochem. Soc. **140**, 2427 (1993).
[4] S. Q. Wang, I. Raaijmakers, B. J. Burrow, S. Suthar, S. Redkar. And K. B. Kim, J. Appl. Phys. **68**, 5176 (1990).
[5] K. Holloway, P. M. Fryer, C. Cabral, Jr., J. M. E. Harper, P. J. Bailey, and, K. H. Kelleher, J. Appl. Phys. **71**, 5433 (1992).
[6] J. Y. Lee and J. W. Park, Jpn. J. Appl. Phys. **35**, 4280 (1996).
[7] M.Takeyama and A. Noya, Jpn. J. Appl. Phys. **36**, 2261 (1997).
[8] M.-A. Nicolet, Applied Surface Science **91**, 269 (1995).
[9] E. J. Kirchnar, PhD Thesis, Rensselaer Polytechnic Institute, 1996.
[10] P. J. Ding, W. Wang, W. A. Lanford, S. Hymes, and S. P. Murarka, Appl. Phys. Lett. **65**, 1778 (1994).
[11] D. P. Tracy and D. B. Knorr, J. Electronic Materials **22**, 611 (1993).
[12] N. M. Hwang, J. Materials Science **33**, 5625 (1998).
[13] Pei-I Wang, G.-R. Yang, S. P. Murarka, and T.-M. Lu, to be published.
[14] Pei-I Wang, S. P. Murarka, G.-R. Yang, and T.-M. Lu, to be published.

Mat. Res. Soc. Symp. Proc. Vol 615 © 2000 Materials Research Society

Observation of Long-Range Orientational Ordering in Metal Films Evaporated at Oblique Incidence onto Glass

David L. Everitt[1], X. D. Zhu*[1], William J. Miller[2], and Nicholas L. Abbott[3]

* to whom correspondence should be made (xdzhu@physics.ucdavis.edu)
[1]Department of Physics, University of California at Davis, Davis, California 95616
[2]Department of Chemical Engineering and Materials Science, University of California at Davis, Davis, California 95616
[3]Department of Chemical Engineering, University of Wisconsin, Madison, Wisconsin 53706

ABSTRACT

We studied long-range orientational ordering in polycrystalline Au films (10 nm - 30 nm) that are evaporated at oblique incidence onto a glass substrate at room temperature. By measuring the averaged optical second-harmonic response from the films over a 6-mm diameter region, we observed a transition from the expected in-plane mirror symmetry at 10 nm to a surprising three-fold in-plane rotational symmetry at 30 nm. X-ray pole figure analysis performed on these films showed the strong <111> fiber texture typical of fcc films, but with a restricted, three-fold symmetric, distribution of crystallite orientations about the fiber axis.

INTRODUCTION

Thin polycrystalline films are used in a wide range of industrial applications, from magnetic storage and read-head media to anchoring and control of functional molecular overlayers in applications such as flat panel displays [1]. Properties of a film depend–among other things–on the film's texture, that is, the distribution of crystalline grain orientations within the film.

Vapor-phase depositions often lead to crystalline grains terminated with Miller-index planes that have the minimum surface energy and that are more or less parallel to the substrate surface. Such films are termed to have out-of-plane textures (fiber texture) [2]. For many metals, the plane parallel to the substrate is the close-packed atomic plane. For example, a {111} plane is usually parallel to the substrate in face-centered cubic (fcc) metals. Note that for consistency of notation we will define this particular {111} plane to be the (111) plane. When the deposition is at an oblique angle or when a thin film is bombarded with ions at an oblique angle during growth, the resultant crystalline grains may also develop a preferred in-plane orientation with respect to the incidence plane of the deposition or bombardment [4,5,6]. Such films are termed to have both out-of-plane and in-plane textures (sheet texture). In-plane textures can be very desirable since they may provide an easy axis of magnetization in the surface plane for magnetic films. They may also cause molecular overlayers to have a preferred orientation with respect to the in-plane texture [1,7]. Recently Gupta and Abbott observed that on an obliquely deposited gold film (on fused silica) covered with an n-alkanethiol self-assembled monolayer (SAM), a liquid crystal overlayer consisting of 5'-pentylcyanobiphenyl (5CB) was aligned either along the in-plane direction of the deposition or perpendicular to it, depending on whether the alkane chain of the n-alkanethiol has an even or odd number of CH_2 groups [8]. Such an odd-even effect may conceivably be related to an in-plane texture if the latter is present in the gold film.

In this work, by measuring the optical second-harmonic response and the X-ray pole figures, we find that an obliquely deposited gold film on fused silica indeed develops an in-plane texture as well as an out-of-plane texture when the thickness increases from 10 nm to 30 nm. Though not discussed in this paper due to space limitations, the development of a preferred in-plane orientation can be attributed to a self-shadowing effect and thus is expected to occur in other metal thin films.

EXPERIMENT AND RESULTS

The gold films are prepared on fused silica substrates at room temperature in an electron beam evaporator. The evaporator chamber has a base pressure of 4×10^{-6} torr. The substrates are cleaned with the same method as reported by Skaife and Abbott [9]. The gold deposition rate is roughly 0.02 nm/sec. To help the gold films to adhere to the fused silica surface, we first deposit a 1-nm titanium (Ti) layer as the buffer. For obliquely deposited gold films, the gold flux is incident at 50° from the substrate normal and along a fixed azimuth. For comparison, we have also made "uniformly deposited" gold films by varying the incidence polar angle from 0° to 70° and the incidence azimuth from 0° to 360° during deposition. Conventional powder X-ray diffraction measurements show that the polycrystalline grains in both obliquely deposited and "uniformly deposited" gold films have <111> fiber texture. We have also measured the X-ray pole figures of these gold films with a Huber Pole Figure Goniometer. The pole figures give the angular distribution of the polycrystalline grain orientation [2]. The results show that (1) the (111) planes for the 30-nm obliquely deposited films are tilted by $\theta \sim 5°$ towards the deposition flux, while the (111) planes for the 30-nm "uniformly deposited" films are parallel to the substrate surface; (2) the polycrystalline grains in the 30-nm obliquely deposited films seem also to have developed a preferred in-plane orientation such that one of the three <110> axes in the (111) plane is perpendicular to the incidence plane of the deposition. Unfortunately the signal-to-noise ratio in the pole figure measurements was not high enough to exhibit the in-plane texture distinctly. The in-plane texture and how it evolves as the film thickness increases are much better exhibited in the optical second-harmonic response of the films.

The measurements of the optical second-harmonic generation (SHG) from the gold films are performed in air. We use 20-picosecond optical pulses at $\lambda_\omega = 532$ nm as the fundamental beams. The optical pulses are obtained by frequency-doubling the output of a Nd:YAG laser with a repetition rate of 10 Hz. In Fig. 1, we show the sketch of the optical set-up.

We choose the substrate surface (also the gold film surface) as the x-y plane and the incidence plane of the optical beams as the x-z plane with the z-axis pointing into the substrate. We employ a co-planar, sum-frequency generation geometry such that one fundamental beam (*p-polarized* with a single pulse energy of 0.1 mJ over an area of 0.3-cm²) is incident at 22° from one side of the substrate normal while the other fundamental beam (also *p-polarized* with a single pulse energy of 0.3 mJ over an area of 0.4-cm²) is incident at 45° from the opposite side. We detect the *p-polarized* optical second harmonics (at $\lambda_{2\omega} = 266$ nm) in reflection direction at 10° from the substrate normal.

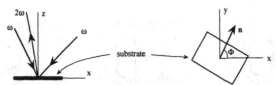

Fig. 1 The geometry of the SHG from an obliquely deposited gold film. The substrate surface coincides with the x-y plane. Two *p-polarized* excitation beams at ω are incident in the x-z plane: one at 22° from one side of the substrate normal and the other at 45° from the opposite side. The reflected SHG at 2ω is emitted at 10° from the substrate normal. The gold flux is deposited onto the substrate at 50° from the normal. The unit vector **n** denotes the projected direction of the deposition in the x-y plane. Φ is the azimuthal angle between **n** and the x-axis.

In this geometry, the measured optical second harmonics is predominantly generated by the x-component of the nonlinear polarization $P_x(2\omega)$ in the gold film. $P_x(2\omega)$ is sensitive to the in-plane structure of the film [10]. We measure the azimuthal dependence by rotating the gold film about the substrate normal over 360° at a step size of 10°. Let **n** be a unit vector in the x-y plane along the projected direction of the deposition. The azimuthal angle Φ is defined as the angle between **n** and the x-axis.

In Fig. 2, we display the SHG signal versus Φ from obliquely deposited gold films of three different thicknesses: 10 nm [Fig. 2(a)], 20 nm [Fig. 2(b)], and 30 nm [Fig. 2(c)]. The SHG signal originates from the electric quadrupole and magnetic dipole responses of the gold film [10]. As a result of the attenuation at the fundamental and the SH frequencies, the SHG signal and its azimuthal dependence comes from the topmost 10 nm of the film and the contribution from the 1-nm Ti buffer layer can be neglected. This is consistent with our observation that the averaged SHG signal is roughly the same at all three film thicknesses. For the 10-nm gold film, the SHG has only a mirror plane coinciding with the incidence plane of the deposition as one would expect from the geometry of deposition. It has a minimum at Φ = 0° and a broad maximum at Φ = 180°. At the film thickness of 20 nm, a small peak emerges at Φ = 0° while the maximum at Φ = 180° shrinks. At the film thickness of 30 nm, the azimuthal dependence has evolved to having a three-fold rotation axis just like the SHG from a single crystalline Au(111) [11]. The peak at Φ = 180° is noticeably smaller than the peaks at Φ = 60° and 300°. It seems that at 20 nm the evolution of the azimuthal dependence is incomplete. For comparison, the SHG from the "uniformly deposited" films is azimuthally isotropic, indicating that the in-plane orientations of the polycrystalline grains in these films are random.

The presence of a 3-fold rotational axis in SHG from 30-nm obliquely deposited gold films, after an average over an area of 0.3 cm², is indicative of a preferred azimuthal orientation of the polycrystalline grains in these films.

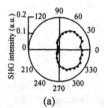 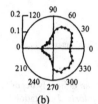 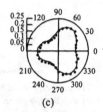

(a)   (b)   (c)

Fig. 2 Azimuthal dependence of the *p-polarized* SHG intensity (in unit of photon counts per pulse) from obliquely deposited gold films vs. the film thickness: (a) 10 nm; (b) 20 nm; (c) 30 nm. The solid line shown in (c) is the best fit of a function $\beta\left(1-\gamma\cos 3\Phi+\eta\cos\Phi\right)^2$ to the data with $\beta = 0.15$, $\gamma = 0.23$ and $\eta = 0.14$.

Our SHG measurement should be compared to the measurement of a non-resonant optical sum-frequency generation from a single crystal Au(111) by Yeganeh and coworkers [11]. Similar to Fig. 2(c), these authors observed a 3-fold rotation axis with the SHG maxima appearing at 60°, 180°, and 300°. As we will show shortly, the asymmetry of the three SHG peaks in Fig. 2(c) can be attributed to a small tilt of the [111] axes of the gold grains towards the deposition flux.

In our case, the optical second harmonics is predominantly generated by the x-component of the nonlinear polarization $P_x(2\omega)$ in the topmost 10-nm layer of the gold films. It is an average of the nonlinear polarization $P_{x,j}(2\omega)$ over many polycrystalline grains. From a single crystalline grain (e.g., the j-th grain), the nonlinear polarization $P_{x,j}(2\omega)$ has an isotropic part $P_{x,j}^{\text{isotropic}}(2\omega)$ and an anisotropic part $P_{x,j}^{\text{anisotropic}}(2\omega)$ [10]. Let θ be the tilt angle of the [111] axis from the substrate normal. Let ϕ_j be the angle between **n** (see Fig. 1) and one of the [211] axes in the terminating facet plane. For a small tilt angle θ, it can be shown that $P_{x,j}^{\text{anisotropic}}(2\omega) \sim$
$\left[-\cos 3\left(\Phi-\phi_j\right)+\left(3/\sqrt{2}\right)\theta\cos\left(\Phi-\phi_j\right)\right]$ and consequently

$$P_x(2\omega)=\sum_j P_{x,j}(2\omega)\sim\sum_j\left[1-\alpha\cos 3\left(\Phi-\phi_j\right)+\alpha\left(3/\sqrt{2}\right)\theta\cos\left(\Phi-\phi_j\right)\right] \tag{1}$$

The total SHG signal as a function of Φ is given by $S(\Phi)\sim\left|P_x(2\omega)\right|^2$. α is a constant determined by the geometry of the experimental set-up. Equation (1) is equivalent to an ensemble average over a distribution of the azimuthal orientations around $\phi = 0°$ for those polycrystalline grains within a macroscopic area of 0.3-cm². As a result, we have

$$S(\Phi)\sim\left|P_x(2\omega)\right|^2\sim\left|1-\alpha\langle\cos 3\phi\rangle\cos 3\Phi+\alpha\left(3/\sqrt{2}\right)\theta\langle\cos\phi\rangle\cos\Phi\right|^2 \tag{2}$$

If the azimuthal orientations of the grains are random (as one would expect of films deposited on amorphous substrates like fused silica) such that ϕ varies from $-60°$ to $+60°$, $\langle\cos 3\phi\rangle$ vanishes and the resultant $S(\Phi)$ only have a weak mirror plane given by the $\alpha\left(3/\sqrt{2}\right)\langle\cos\phi\rangle\cos\Phi$ term. If the tilt angles of the fiber axes are also randomly distributed about $\theta = 0°$, the resultant $S(\Phi)$ becomes isotropic as we found for "uniformly deposited" gold films. If instead there is a preferred orientation along $\phi = 0°$ so that ϕ varies over a significantly narrower range than $120°$, $\langle\cos 3\phi\rangle$ survives the ensemble average and $S(\Phi)$ has a 3-fold rotation axis in addition to a weak asymmetry coming from the tilt angle. This is what we observed for obliquely deposited gold films with thicknesses over 10 nm. The data in Fig. 2(c) is best fit to Equation (2) (solid line) by assuming that *one of the <110> axes in the (111) plane is perpendicular to the incident plane of the deposition* and with $\left(3/\sqrt{2}\right)\langle\cos\phi\rangle/\langle\cos 3\phi\rangle = +0.6$.

The observed in-plane texture and tilt angle θ towards the deposition direction for the polycrystalline grains in an obliquely deposited gold film may have a significant effect on the macroscopic alignment of the terminal methyl groups of an n-alkanethiol self-assembled monolayer (SAM) on such a film [8]. It is known that the methyl groups of a *(2m)*-alkanethiol SAM on Au(111) are more or less oriented along the substrate normal [15]. On an obliquely deposited gold film with the (111) axes tilted towards the deposition direction, the methyl groups of a *(2m)*-alkanethiol SAM on these facets would be also tilted towards the deposition direction. On the other hand, the methyl groups of a *(2m+1)*-alkanethiol SAM on Au(111) are tilted along an in-plane <110> axis [15]. On an obliquely deposited gold film, one of the <110> axes in the (111) plane is perpendicular to the incidence plane of the deposition and the other two <110> axes are at an angle of 60° from the plane of the deposition. On these facets, the orientations of the methyl groups will be determined both by the orientation of the <110> axes and the tilt of the [111] axes. We speculate that short-ranged interactions between the oriented methyl groups and liquid crystal overlayers may provide part of the explanation, at least, for the previously reported effects on liquid crystals of n-alkanethiols with odd and even numbers of CH_2 groups on obliquely deposited gold films [8]. We caution, however, that the orientations of the methyl groups on these substrates need to be determined directly using techniques such as the infrared-visible sum-frequency generation [16].

CONCLUSION

In conclusion, we have shown that a thin gold film obliquely deposited on fused silica (with 1-nm Ti buffer layer) develops an in-plane texture as well as an out-of-plane texture. The out-of-plane texture is such that the (1) the polycrystalline grains have (111) planes parallel to the substrate; and (2) the [111] axes are tilted by a few degrees towards the deposition direction. The in-plane texture is characterized by a preferred azimuthal orientation of the polycrystalline grains with one of the <110] axes in the (111) plane oriented perpendicular to the incidence plane of the deposition. Such in-plane and out-of-plane textures can find desirable applications in magnetic thin films and in anchoring molecular overlayers as soft-hard material interfaces.

ACKNOWLEDGEMENTS

This work was supported by National Science Foundation under NSF-DMR-9808677 and in part under NSF-DMR-9818483. We thank Professor Wenk for performing the X-ray pole figure measurements reported in this work. N.L.A. acknowledges the support by the Office of Naval Research (Presidential Early Career Award for Science and Engineering) and by the Center for Nanostructured Interfaces (NSF-DMR-9632527) at University of Wisconsin-Madison.

REFERENCES

1. J. Cognard, *Alignment of Nematic Liquid Crystals and Their Mixture* (Gordon and Breach, London, 1982).
2. H.J. Bunge, *Texture Analysis in Materials Science* (Butterworths, Toronto, 1982).
3. T.G. Knorr and R.W. Hoffman, Phys. Rev. **113**, 1039 (1959).
4. T.C. Huang, J.-P. Nozieres, V.S. Speriosu, H. Lefakis, and B.A. Gurney, Appl. Phys. Lett. **60**, 1573 (1992).
5. J.M. Alameda, F. Carmona, F.H. Salas, L.M. Alvarez-Prado, R. Morales, G.T. Perez, J. Magn. Magn. Mater. **154**, 249 (1996); K. Itoh, *ibid.* **95**, 237 (1991); K. Okamoto, K. Itoh, and T. Hashimoto, *ibid.* **87**, 379 (1990); Z. Shi and M.A. Player, Vacuum **49**, 257 (1998).
6. J.M.E. Harper, K.P. Rodbell, E.G. Colgan, and R.H. Hammond, Mat. Res. Soc. Symp. Proc. **472**, 27 (1997).
7. M.B. Feller, W. Chen, and Y.R. Shen, Phys. Rev. **A 43**, 6778 (1991).
8. V.K. Gupta and N.L. Abbott, Phys. Rev. **B** *Rapid Commun.*, R4540 (1996).
9. J.J. Skaife and N.L. Abbott, Chem. Mater. **11**, 612 (1999).
10. H.W.K. Tom, T.F. Heinz, and Y.R. Shen, Phys. Rev. Lett. **51**, 1983 (1983); H.W.K. Tom, Ph.D. Thesis, University of California, 1984, p. 144.
11. M.S. Yeganeh, S.M. Dougal, R.S. Polizzotti, and P. Rabinowitz, Phys. Rev. Lett. **74**, 1811 (1995).
12. K.H. Hansen, T. Worren, S. Stempel, E. Lægsgaard, M. Bäumer, H.-J. Freund, F. Besenbacher, and I. Stensgaard, Phys. Rev. Lett. **83**, 4120 (1999).
13. W. Mahoney, S.T. Lin, and R.P. Andres, in *Evolution of Thin Film and Surface Structure and Morphology*, Materials Research Society Symposium Proceedings, Vol. **355**, edited by B.G. 14.O.P. Karpenko, J.C. Bilello, and S.M. Yalisove, J. Appl. Phys. **82**, 1397 (1997).
15. P. Fenter, A. Eberhardt, P. Eisenberger, Science **266**, 1216 (1994); P. Fenter, A. Eberhardt, K.S. Liang, and P. Eisenberger, J. Chem. Phys. **106**, 1600 (1997).
16. X. Wei, X. Zhuang, S.-C Hong, T. Goto, and Y.R. Shen, Phys. Rev. Lett. **82**, 4256 (1999).

Mat. Res. Soc. Symp. Proc. Vol. 612 © 2000 Materials Research Society

Grain Boundary Curvature in Polycrystalline Metallic Thin Films

Alexander H. King[1], Rakesh Mangat[2] and Kwame Owusu-Boahen[3]
[1]School of Materials Engineering
Purdue University, West Lafayette IN 47907-1289.
[2]Massachusetts Institute of Technology
Cambridge, MA, 02139.
[3]Department of Materials Science & Engineering
State University of New York, Stony Brook NY 11794-2275.

ABSTRACT

Well-annealed thin films are typically observed to exhibit mean grain diameters that are approximately equal to the film thickness. The standard explanation of this "sheet thickness effect" is that it results from a balance of grain boundary curvature in two different directions which, in turn, results from pinning at grain boundary grooves. TEM experiments have been performed to assess this model, and it is found that the predicted curvature about axes in the film plane is absent. Alternate explanations of the sheet thickness effect are considered.

INTRODUCTION

Provided that the films are flat, uniform and pore-free, structural impacts on most polycrystalline thin film properties are dominated by the grain size and texture. Our ability to control these aspects of the microstructure is rather limited. The texture can be affected by the substrate preparation [1] or stress in the film [2]. Grain size is, in principle, the simplest microstructural parameter to control, through the temperature of the substrate and the film deposition rate [3]. This control, however, is limited to a regime that is typically bounded by a saturated columnar grain size, with a mean equivalent grain diameter approximately equal to the film thickness. This "specimen thickness effect" was first identified in much thicker sheets than are addressed in this symposium [4], but it appears to operate in the nanometer scale, just as it does in the millimeter scale, and is largely accepted as a fact of life.

The standard explanation for the specimen thickness effect was developed by Mullins [5] and it depends upon pinning of grain boundaries by their surface grooves. As a grain boundary migrates toward its center of curvature, the surface pinning causes the part of the boundary in the center of the specimen to advance ahead of the parts closer to the surface, so the boundary develops a catenoid form, embodying anticlastic curvature: the center of curvature about an axis parallel to the film normal lies on the opposite side of the boundary to the center of curvature about an axis that is parallel to the tangent to the boundary intersection with the film surface, as indicated in Figure 1. When the curvatures about the two orthogonal axes are equal, the net driving force for boundary migration is zero. Mullins shows that this may be expected to occur when the mean equivalent grain diameter is approximately equal to the film thickness. Our purpose in this paper is to examine the nature of the film thickness effect in the regime of film thickness appropriate to current electronic and magnetic state-of-the-art technologies, and to determine whether the Mullins mechanism explains the phenomenon at this length-scale.

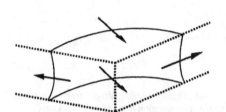

Figure 1: Schematic illustration of the curvature supposed of a grain boundary in a thin film, pinned by grooves on the top and bottom surfaces. The arrows indicate the directions of the driving forces generated by the two curvature components.

Figure 2: TEM bright field image showing the typical structure of a gold film after grain growth saturation. This film is nominally 120nm thick.

EXPERIMENTAL PROCEDURE

Gold thin films were prepared by evaporation from a tungsten filament onto freshly cleaved rock-salt substrates, in a background pressure of 10^{-6} Torr. The substrates were removed by dissolution in water and the films were supported on gold TEM grids. The films were then brought to the point of grain growth saturation by annealing them for 24 hours at 550°C, whereupon they were observed in the transmission electron microscope. Film thicknesses were determined using convergent beam electron diffraction, and grain sizes were determined by measuring the mean grain areas and converting these to equivalent grain diameters.

RESULTS

A typical thin film is shown in Fig. 2. Curvature of some interfaces, about the beam direction, is quite apparent, but there is little evidence of curvature about directions in the film plane. One case in which such curvature is apparent is shown in Fig. 3, but this case appears to represent bulging of the grain boundary, rather than the anticlastic curvature postulated by Mullins. The saturated grain sizes for films of varying thickness are shown in Fig.4, and it is clear that there is a monotonic increase over the thickness range of 20 to 80nm, but the thickest film investigated (120nm) shows a much smaller grain size than anticipated. In order to check this result, this film was re-annealed for a further 7 days, but this resulted in no measurable increase in the grain size.

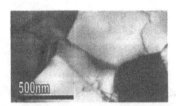

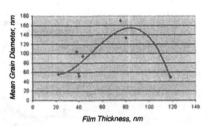

Figure 3: TEM bright field image exhibiting grain boundary curvature through the film thickness.

Figure 4: Mean equivalent grain diameters produced by saturated normal grain growth in gold films of varying thickness. The data show the expected increase with film thickness, except for the thickest film studied. This may indicated a change of mechanism at around 100nm.

DISCUSSION

The extent of the expected grain boundary curvature in the plane of the foil can be deduced following Mullins [5], assuming a simplified groove profile as shown in Fig. 5, and approximating the boundary profile to a circular arc. The projected displacement of the boundary in the film center should reach, but not exceed a value given by

$$d = \frac{h}{2}\left[\frac{1}{\sin\alpha} - \frac{1}{\tan\alpha}\right]$$

where h is the film thickness and α is defined in Fig.5. α is expected to fall in the range of 10° to 15°, corresponding to a ratio of surface energy to grain boundary energy between 3 and 2, and yielding a boundary displacements between 6.5 and 4.5nm, respectively, for a 100nm thick foil. These boundary displacements should be readily observable in the TEM, but we find that it is usually possible to observe grain boundaries "edge-on" without such curvature being apparent. The displacements should be smaller, of course, for thinner specimens, and for smaller surface energies. When the foil thickness is reduced to 10nm, the expected boundary displacement is approximately one unit cell through the film thickness.

An alternative way of estimating the expected boundary projected width is more readily applied to any image. Noting that the boundary segment length is approximately equal to the film thickness, it is simple to deduce that the displacement of any boundary section measured in the plane of the film should be of the same order of magnitude as the displacement through the film thickness, as illustrated in Fig. 6. It is clear that the through-thickness displacements of the boundaries are not comparable to the in-plane displacements, so the observations do not support the Mullins hypothesis concerning the origin of the specimen thickness effect.

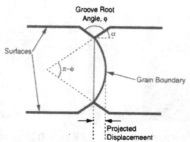

Figure 5: Schematic cross-section of a thin film containing a grain boundary, pinned by surface grooves, to show how the projected displacement of the interface is calculated. The grain boundary can escape from the pinning grooves once it lies perpendicular to the surface at the groove root, as shown here.

We have previously demonstrated that the grain boundary character distribution develops in parallel with grain growth [6]. This appears to be reflected also in a marked preference for symmetric tilt grain boundaries, adjacent to triple junctions [7]. The effect of these components of the grain boundary character is to reduce the overall grain boundary curvature, as discussed in Ref. 8, but they should not have any direct effect upon the through-film curvature of boundary segments that retain curvature about the film normal, such as the one shown in Fig. 6.

Triple junctions do, however, have effects that tend to reduce the through-film curvature from the values estimated by Mullins [5]. Consider the case shown in Fig. 6, where a boundary segment with pronounced curvature is terminated, at both ends, by triple junctions where it meets relatively flat boundary segments. While the triple junction itself may be a curved line, it must contain all three grain boundary planes, at every point. This requirement means that the grain boundaries cannot be arbitrarily curved at the junction, and must adopt some compromise structure that minimizes the overall system energy. The intersection of two planar boundaries defines a straight triple junction line, which seems to be adopted by the curved boundary shown in Fig. 6. This effect may be expected to extend for some distance from the junction itself, and appears to suppress through-film curvature.

Structures of this type are readily observed in films with saturated grain sizes at all thicknesses amenable to transmission electron microscopy. There remains a question, however, concerning the effect of the developing grain boundary character distribution on the types of structure described here. In thicker films grain growth proceeds further, prior to saturation, than in thinner ones that start with the same as-deposited grain size. The saturated microstructure of the thicker film therefore reflects many more grain boundary coalescense events, grain switching events, *etc.*, than does the structure of a thinner film. The thicker microstructure is therefore more mature, and reflects a much greater level of grain boundary character distribution development, that might be expected to show more pronounced triple junction effects. Effects of this type remain to be assessed experimentally, but they may help to explain the non-linear grain-size *vs.* thickness effects indicated in Fig. 3.

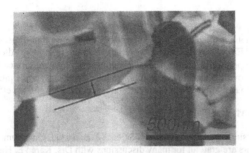

Figure 6: Schematic estimate of the projected image width for a boundary with in-plane curvature matching that postulated by Mullins. If the boundary segment length is equal to the film thickness (as expected) then the displacement of the center of the boundary from the surfaces should be equal to the displacment of the center from the triple junctions, as indicated by the construction shown here. There is clearly no such in-plane curvature, and the analyzed boundary is viewed nearly edge-on at all points.

A final question remains, concerning why grain growth saturation should occur at all, if through-film curvature is absent, while in-plane curvature remains. If the Mullins mechanism does not apply, what other effects can explain the phenomenon? One possibility is the relatively recently discovered phenomenon of triple-junction drag [9], which is expected to retard grain growth [10] and may be sufficient, under certain circumstances, to arrest the migration of curved boundaries such as those observed in this work. It is interesting to note that triple-junction drag is temperature sensitive, and is less pronounced at higher temperatures [9]. If this hypothesis is correct, then grain growth stagnation, induced by triple junction drag, should be relieved by increasing the temperature. Recent work on silver thin films, reveals that stagnated grain sizes are, in fact dependent upon the annealing temperature, as anticipated [11].

Alternate explanations of grain growth stagnation include the effect of twinning, as observed very profusely in our specimens. When twins span a grain, the growth of the grain may require the growth of the twins, causing an increase in the twin boundary area. Because of the energy of the twin boundaries, their presence reduces the driving force available for a grain to grow, but increases that available to drive the shrinkage of a small grain. The net effect is thus the difference of the contributions to the shrinkage of small grains and the growth of large ones. The low value of the twin boundary energy necessarily makes this a small contribution to the overall stagnation effect.

Finally, we note that Sanchez and Arzt have provided an analysis of surface pinning [12] that differs from that of Mullins, and predicts stagnation that is independent of the in-plane curvature, though it is still directly related to the film thickness – a result which may be questionable in the light of the data shown in Fig.4.

CONCLUSIONS

The conventional explanation of grain growth stagnation in thin films, due to Mullins [5], is not supported by TEM observations of polycrystalline structure in gold. Alternative explanations, based upon triple junction drag, twin growth or alternate surface pinning morphologies may provide plausible explanations of the effect. Only triple junction drag has been shown, so far, to explain the temperature dependence of the stagnation effect.

ACKNOWLEDGEMENTS

This work is supported by the National Science Foundation, grant number DMR 0096147. The authors are grateful for many discussions with Drs. Karen E. Harris, Varun V. Singh, and Rand Dannenberg.

REFERENCES

1.   D.B. Knorr and K.P. Rodbell, *J. Appl. Phys.*, **79**, 2407 (1996).
2.   C.V. Thompson and R. Carel, *J. Mech. Phys. Sol.*, **44**, 657 (1996).
3.   C.R.M. Grovenor, H.T.G. Hentzell and D.A. Smith, *Acta Metall.*, **32**, 773 (1984).
4.   P.A. Beck, J.C. Kremer, L.J. Demer and M.L. Holzworth, *Met. Trans.*, **175**, 372 (1948).
5.   W.W. Mullins, *Acta Metall.*, **6**, 414 (1958).
6.   V. Singh, G. Dixit and A.H. King, *J. Elec. Mater.*, **26**, 987 (1997).
7.   V. Singh and A.H. King, Scripta Mater., **34**, 1723 (1996).
8.   A.H. King in *Grain Growth in Polycrystalline Materials III*, Eds. H. Weiland, B.L. Adams and A.D. Rollett, TMS, Warrendale, PA (1998) p. 333.
9.   G. Gottstein, V. Sursaeva and L.S. Shvindlerman, *Interface Science*, **7**, 273 (1999).
10.   G. Gottstein, A.H. King and L.S. Shvindlerman, *Acta Mater.*, **48**, 397 (2000).
11.   M. Adamik, P.B. Barna and I. Tomov, *Thin Solid Films*, **359**, 33 (2000).
12.   J.E. Sanchez and E. Arzt, *Scripta Met. et Mat.*, **26**, 1325 (1992).

Mat. Res. Soc. Symp. Proc. Vol. 612 © 2000 Materials Research Society

INTERFACIAL TIAL3 GROWTH : IN-SITU TEM OBSERVATIONS AND NUMERICAL SIMULATIONS.

X.FEDERSPIEL, M.IGNAT, L.GUETTAZ[1], C.BERGMAN[2], J.PHILLIBERT[3], A.MACK[4], H.FUJIMOTO[4], T.MARIEB[4].
INPG/LTPCM, Saint Martin d'Hères, France.
[1]CEA/CENG, Grenoble, France.
[2]LMMP Marseille, France.
[3]LMP Université d'Orsay, France.
[4]Intel Components Research Santa Clara, CA.

ABSTRACT

It is well known that solid state reactions forming intermetallics in interconnect systems, will degrade their electrical reliability and produce mechanical stresses which in certain circumstances will induce damage.

For example islands of TiAl₃ can be nucleated at the interfaces of stacks of Al and Ti thin films under the effect of heating, producing the above mentioned awkward effects.

A solution to avoid these effects is to control the growth of a thin layer of the compound, in conditions that will maintain it stable, during any further thermomechanical solicitation.

From in-situ TEM observations, and DSC results obtained previously, we develop a numerical model for TiAl₃ growth. The model is based on interfacial diffusion mechanisms.

INTRODUCTION

Integrated circuits (IC's) present complex architectures, established from stacks of thin layers of different sort of materials. These materials provide the principal functions of the device (conduction, isolation, protection). Because of their structural arrangement, the interfaces and the singularities of IC's are potential sites were mechanisms can be activated, degrading the device performances. Typical examples are the internal residual stresses outcoming from strain incompatibilities which can produce cracks in the layers or at the interfaces; and also interfacial reactions. As a matter of fact, when current flows producing heating, depending on the chemical affinity of thin materials which are in contact, the nucleation and growth of a new phase can be activated. This reactive process may also induce mechanical stresses (raised up at singularities) simultaneously with loss of conductivity. The understanding of the process which drive a device to failure, is essential to improve it's reliability. Information can be obtained trough experiments and analysis (TEM, DSC, SIMS…) which allow to establish realistic hypothesis to model the failure process.

During the last decade, emphasis has been devoted to study interfacial reactions. The Al/Ti system represents a particular interest, because the constitutive materials correspond to those widely used for conducting (Al) and diffusion barrier (Ti) functions [1,2]. Eventhough nowadays new materials and new architectures for IC's are developed, a lack of information on material reactions in thin films and their dependence on constitutive parameters subsists. Consequently the Al/Ti system remains an excellent model to improve the knowledge on interfacial reactions and their evolution. From TEM in-situ experiments and previous results obtained by differential scanning calorimetry (DSC), a simulation of an interfacial growth process is presented.

MICROSTRUCTURAL OBSERVATIONS : IN-SITU TEM

Our initial purpose was to observe in-situ, the main features of $TiAl_3$ formation and growth from an Al/Ti interface. The samples consisted in a 2800Å thick Al layer and a 1000Å thick Ti layer. The Al/Ti films are deposited by magnetron sputtering process on a <100> silicon substrate with a 500 Å thick SiO_2 layer. Each Al film contains 0.5 wt% of copper.

The TEM thin samples were obtained using a tripod thinning device, producing a thin wedge with the interfaces perpendicular to the surface. The thin samples were fixed to a copper ring by a high temperature resistant glue.

The microstructural in-situ observations were performed in a 300KV JEOL TEM, having a heating sample holder. This device allowed to reach temperatures as high as 800K in our samples. The sample was heated continuously up to 720K (450°C) without detecting any microstructural changes. However from 720K (450°C) to 800K (530°C), we observed a microstructural evolution of several grains. It was characterized by a migration of grain boundaries into the Al rich zone. In some cases the migration occurred sequentially for a stack of grains (see Figure 1). Taking into account these observations, which were video recorded, we assumed that the moving boundaries corresponds to the $TiAl_3$ growth.

When increasing the temperature above 800K (530°C), we did not observe any further microstructural changes. After cooling the sample, nanodiffractions were performed. They showed the evidence that the $TiAl_3$ formation was completed (see Figure 2).

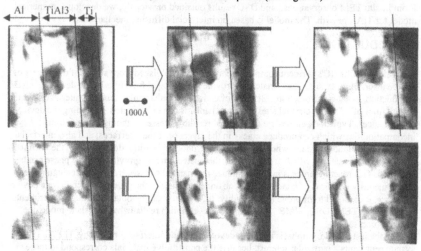

Figure 1. *From top left to bottom right, the pictures show selected steps (0, 10, 68, 80, 89 and 107 sec.) during the in-situ observation. In the first micrographs, the full line corresponds to the $TiAl_3$/Ti interface. The dotted lines represent the $TiAl_3$/Al interface of 2 $TiAl_3$ grains.*

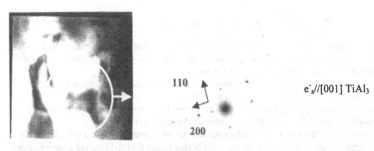

Figure 2. TiAl₃ grain and corresponding diffraction pattern.

NUMERICAL SIMULATION : INTERFACIAL TIAL3 GROWTH

Usually, the growth of an intermetallic is modeled by a unidirectional diffusion flux across the grown intermetallic layer. This layer is assumed to be continuous and to have a homogeneous thickness. These types of models are characterized by a parabolic evolution of the thickness in time. Although these models are validated through the second Fick equation, they do not take into account the topology of the nucleated grains. Moreover, our TEM and DSC experiments suggest the existence of different growing steps that would correspond to different regimes.

In previous works [3,4], we presented a simulation of the TiAl₃ growth including a nucleation controlled regime, followed by a diffusion limited regime. This was an attempt to develop a phenomenological model characterized by two steps.

Our objective here is to simulate the growth of individual grains on the interface, by taking into account the interdiffusion of Al and Ti at their boundaries. As the Al and Ti diffusionnal fluxes are non-correlated, the growth towards the Ti and Al regions can be solved separately. The growth would be controlled by the Al diffusion along a Ti/TiAl₃ phase boundary and by the Ti diffusion along a Al/TiAl₃ boundary. In this case, the diffusion in volume is neglected.

On the Figure 3, the arrows denote the Ti fluxes along the Al/TiAl₃ phase boundary and the consumption flux perpendicular to the phase boundary. After lateral contact, the TiAl₃/TiAl₃ grain boundaries act in series with the Al/TiAl₃ phase boundaries.

Along the TiAl₃ boundary, the consumption of matter modifies the second Fick law. Then along the interface, the growth rate of the intermetallic, is only proportional to the Ti concentration (considering that the Al concentration is not a limiting factor). The following equation corresponds to the modified second Fick law, taking into account the matter consumption :

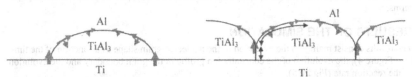

Figure 3. Schematic representation of the Titanium fluxes along the phase boundaries before (left) and after (right) lateral contact of the TiAl₃ growing grains.

$$\frac{\partial C}{\partial t} - (D\frac{\partial^2 C}{\partial x^2} - K.C - K.C.Vs \times dl/l) = 0 \qquad (1)$$

C is the Ti concentration, D the diffusion coefficient of the Ti at the phase boundary, Vs the specific volume, K the coefficient of transfer which corresponds to the Ti flux consumed by the TiAl$_3$ formation and dl/l is the local mesh elongation.

Because we cannot solve analytically this problem for a non stationary regime, we solve it numerically using a finite differences procedure.

The diffusion occurs along the grain boundary surface; but by considering an axial symmetry, then it is possible to solve the concentration distribution using a 1D mesh.

Relation (1) can be discretized to express the evolution of the concentration at a node i of a 1D mesh as a function of the concentration of the neighboring nodes (i+1, i-1) during a time step Δt:

$$\frac{C_i^{t+\Delta t} - C_i^t}{\Delta t} - (D\frac{C_{i+1}^{t+\Delta t} + C_{i-1}^{t+\Delta t} - 2.C_i^{t+\Delta t}}{\Delta x^2} - K.C_i^{t+\Delta t} - K.Vs.C_i^{t+\Delta t} \times dl/l) = 0 \qquad (2)$$

Here, K.C$_i$ is the flux consumed to supply the intermetallic growth, the term K.Vs.C$_i$.dl/l represents the apparent dilution due to the interface elongation.

This procedure provides us an approximate solution for the concentration distribution. As a consequence relation (1) is not equal to 0. Then the remaining r$_i$ which corresponds to the approximation can be expressed as :

$$D\frac{C_{i+1}^{t+\Delta t} + C_{i-1}^{t+\Delta t}}{\Delta x^2} + \frac{C_i^t}{\Delta t} - C_i^{t+\Delta t}(\frac{1}{\Delta t} + \frac{2D}{\Delta x^2} + K - K.Vs \times dl/l) = r_i \qquad (3)$$

To converge to the solution we have to minimize r$_i$. For this we used the relaxed Gauss-Seidel method [6]. This method consist in applying a ΔC_i to obtain r$_i$ = 0 , on each node of the mesh.

$$\Delta C_i = -\frac{r_i}{\frac{1}{\Delta t} + \frac{2D}{\Delta x^2} + K - K.Vs \times dl/l} \qquad (4)$$

The value C$_i$+ΔC$_i$ is called the implicit solution C$^*_i$. To ensure convergence, we used a mixed solution :

$$C_i^{n+1} = \rho.C_i^* + (1 - \rho).C_i^n \qquad (5)$$

C$_i^n$ is the result of the previous iteration and ρ is the relaxation factor. The calculation is repeated on the C$_i$ matrix until the C$_i$ values stabilize.

Since the Titanium distribution is known for a time step, the local growth rate is calculated. Then the nodes of the interface are moved perpendicularly with respect to the local surface orientation. This algorithm can simulate the growth of a nucleated grain until it reaches contact with the neighboring grains. When a grain boundary is created, the transport of matter can no longer be solved with the equation (1), but simplifies to the standard Fick equation without consumption terms.

RESULTS OF THE SIMULATION

The results consist mainly in the evolution of the nucleated grain shape as a function of the time (see Figure 4), the evolution of the concentration profile in the grain boundary and the evolution of the reaction rate (Figure 5).

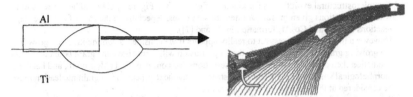

Figure 4. *Simulation of the nucleated grains shape evolution. The scheme on the left corresponds to the region where the diffusion growth is simulated. On the right scheme, the arrows indicate the growing direction.*

It is also possible to simulate the same parameters during isothermal and non-isothermal reactions. The control of the grain boundary transport on this reaction reveals two original behaviors : a non isotropic growth and an evolution of the kinetic regime.

Because of the competition between diffusion along the boundary and the local matter consumption, a sharp chemical potential sets up. Then the local $TiAl_3$ formation rate decreases continuously from the edge to the top of the grain. This leads to high ratio of length over height when the contact with the neighboring grains occurs.

Concerning the simulated growth regime, it is noticeable that when the $TiAl_3/TiAl_3$ grain boundaries get into contact, the reaction rate decreases. This simulated behavior can be related to the inflexion obtained during the DSC experiments. When the intergranular contact is completed, only one source of Ti feeds the $TiAl_3$ growth through the grain boundary, while before,two distinct sources were acting.

DISCUSSION AND CONCLUSION

We shall remind first, that our comments, are in agreement with previous DSC experiments. These experiments confirmed the existence of two steps during the reaction [3]. Actually, the simulation produces a microstructural evolution compatible with the TEM observations, and the DSC results. Also the simulation describes the roughness of the $TiAl_3/Al$ interface during the growth, producing shapes in agreement with the corresponding observations in TEM.

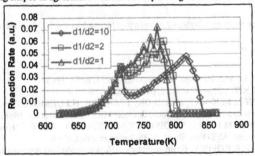

Figure 5. *Effect of the ratio between the diffusion coefficient of the phase boundary D1 and the diffusion coefficient of the grain boundary D2 on the evolution of the global reaction rate with the temperature. The global reaction rate takes into account the total amount of intermetallic grown all along the interface.*

The microstructural evolution as the kinetics of growth (see Figure 5) obtained here, seems to fit with interpretations given in similar interdiffusion cases, especially in the case of multiple step reactions (NbAl$_3$ and Co$_2$Al$_5$ formation in [5] and [7]).

However a limitation with respect to reality appears if we consider a continuous grain growth: the growing grains will become flat , then a protrusion will form above the grain boundaries. The simulation then diverts from our TEM observations. To continue the TiAl$_3$ growth, and keep a morphological similarity to the real microstructure, the next generation of grain nucleation must be considered at the triple points (see Figure 6) .

As a matter of fact, the free energy variation associated to the nucleation is expressed by the sum of the surface energy variation corresponding to the working forces balance at triple points (see Figure 6) and the free energy of the TiAl$_3$ formation :

$$\Delta G = \int_{\alpha_{eq}}^{\alpha_c} (\gamma_{gb} - 2.\gamma_{ip}.\sin(\alpha).(L + \tan^2\alpha).d\alpha - \frac{r_n^2}{2}\left[\Omega + \frac{\sin^2\Omega}{\tan\Pi/3} - \cos\Omega\sin\Omega\right]\Delta G_{TiAl3} \qquad (6)$$

r_n is the nucleation radius, γ_i correspond to surface energies, the other terms are defined by the geometry of the system (see Figure 6).

The two critical parameters r_n and α_c can not be found with only equation (6). We may consider the first derivatives with respect to α and r as a second equation to find the critical parameters. The nucleation of the second generation grains will change again the diffusion path and will also change the interfacial roughness. We are convinced that the thermodynamic condition based on surface tensions must be considered. Finally, both the interfacial roughness and the grain size will depend on the initial nucleation points distribution and on the "second generation of grains" conditions to nucleate.

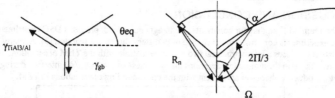

Figure 6. *Surface tensions γ (left) and geometry of the nucleus (right), R_n is the nucleus radius .*

REFERENCES

[1] D.PRAMANIK , Mat. Res.Soc. Bulletin,vol.20, n.11, pp57, 1995.
[2] J.TARDY, K.N.TU, Phys.Rev. B 32, p2070-2080, 1985.
[3] X.FEDERSPIEL, M.IGNAT, F.VOIRON, T.MARIEB, H.FUJIMOTO, "Interfacial reactions in multilayers intended for microelectronic devices, MRS Spring meeting, 1998.
[4] X.FEDERSPIEL, M.IGNAT, F.VOIRON, T.MARIEB, H.FUJIMOTO, "Kinetics of the TiAl$_3$ Formation from Al/Ti thin films", MRS Spring meeting 1999.
[5] J.BARMAK, K.R.KOFFEY, Grain boundary diffusion controlled precipitation as a model for thin film reactions, MRS Spring meeting 1993.
[6] M.RAPPAZ, M.BELLET, M.DEVILLE, "Modélisation numérique en science et génie des matériaux", vol. 10, pp184-194, Presses Polttechniques et universitaires Romandes, 1998.
[7] E.EMERIC, P. GAS, G. CLUGNET and C. BERGMAN, Microelectronic Engineering 50, pp285-290, 2000.

Joint Session:
Process Integration and
Manufacturability

Mat. Res. Soc. Symp. Proc. Vol. 612 © 2000 Materials Research Society

Technique of surface control with the Electrolyzed D.I.water for post CMP cleaning

Mitsuhiko Shirakashi*, Kenya Itoh*, Ichiro Katakabe*, Masayuki Kamezawa*,
Sachiko Kihara*, Manabu Tsujimura*, Takayuki Saitoh**, Kaoru Yamada**,
Naoto Miyashita***, Masako Kodera***, Yoshitaka Matsui***

* Precision Machinery Group, Ebara Corporation
4-2-1,Honfujisawa, Fujisawa-shi 251-8502, Kanagawa, Japan
** Center of Technology Development, Ebara Research CO., Ltd.
4-2-1,Honfujisawa, Fujisawa-shi 251-8502, Kanagawa, Japan
*** Manufacturing Engineering Center, Toshiba Corporation Semiconductor Company
8,Shinsugita-cho, Isogo-ku, Yokohama 235-8522, Kanagawa, Japan

ABSTRACT

Chemical mechanical planarization (CMP) has been widely used for planarization of ILD, STI, plug and wiring processes. In post metal CMP cleaning processes, there are still many problems to be solved. There are several surfaces of materials, such as wiring materials, barrier materials, dielectric materials etc., on the wafer that must be cleaned at the same time,. It is also important to clean these different surfaces without any chemical or mechanical damage. We have confirmed that the Electrolyzed D.I.water is effective in post CMP cleaning for controlling the surface condition during cleaning and leaving a robust surface after CMP. We describe the Electrolyzed D.I.water system and present some results on the cleaning capability and control of the metal surface for application to cleaning after a metal CMP process.

INTRODUCTION

Wet processes such as CMP and electrochemical deposition (ECD) have recently received a tremendous amount of attention in the semiconductor industry in accordance with the progressive down sizing of the technical node on semiconductor devices. The integration of these processes into the semiconductor industry is primarily for the superior planarity (CMP) or gap filling (ECD) capabilities compared to the existing dry processes. The main reason why such wet processes have not been adopted into semiconductor processing in the past has been because wet processing has been considered "dirty" both from a process perspective and from a tool design perspective. The recent acceptance of wet processes and tools into semiconductor processing facilities is due to the recognition that such processes can be done at lower temperatures (<100°C) and the introduction of the "dry-in dry-out" concept allows for cleaning at these low temperatures prior to future high temperature dry processes. This "dry-in dry-out" concept, which is now accepted as the industry standard, uses built-in cleaning and drying technologies that are critical to these wet processes.

The RCA cleaning approach which uses hydrochloric acid, ammonia, sulfuric acid, hydrofluoric acid etc. has been widely adopted so far. However, as the industry moves to larger wafer sizes (200 mm and 300 mm), it is very important to minimize the cost of chemical consumption and waste treatment.. The electrolyzed D.I. water system and processes were developed in accordance with this cost and environmental requirement. This report focuses on control of the oxidization level of the copper metal wiring surface after CMP using Electrolyzed D.I.water. Although the main purpose of these cleaning experiments is to remove foreign materials, control of the surface condition after cleaning is also important for the next process step. The superiority of electrolyzed water in semiconductor device cleaning over water containing dissolved gasses is demonstrated

EXPERIMENTAL

The cleaning mechanisms for the following experiments are defined as follows:

(1) Physical removal of foreign materials from surface.
(2) Control of the electrical potential and pH.
(3) Chemical etching of wafer surfaces.
(4) Removal of foreign materials in controlled chemical environment.

The principle of the Electrolyzed D.I.water is shown on Figure 1.

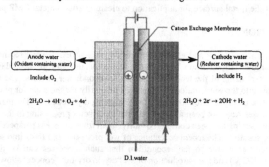

Figure 1 : Principle of generating the Electrolyzed D.I.water

The electrolyzing cell utilizes a cation exchange membrane between the anode and cathode separating the electrolyzed D.I.water in each compartment. Reactions occurring in the anode and cathode compartments are as follows:

Anode side: $2H_2O \rightarrow 4H^+ + O_2 + 4e^-$
Cathode side: $2H_2O + 2\ e^- \rightarrow 2O\ H^- + H_2$

In this reaction, oxidant containing solution (named anode water) is generated and reductant containing solution (named cathode water) is generated in the cathode compartment.

Figure 2 shows a schematic of the flow in the electrolyzed D.I.water supply system.

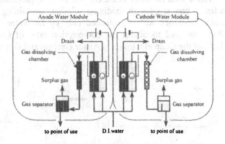

Figure 2 : Flow of the Electrolyzed D.I.water supply system

The system has two electrolyzing cells to allow the user to generate Anode and Cathode water simultaneously or independently. Gas dissolving chambers are mounted in sequence with the cells in order to enhance thorough dissolution of the generated gasses into solution. Surplus gasses are extracted from the anode and cathode water using gas separators before, each is supplied to the point of use.

Figure 3 shows a picture of the electrolyzed D.I.water supply system used in this study.

Figure 3 :
Appearance of
the Electrolyzed D.I.water supply unit

Figure 4 shows a typical experimental procedure. 200 mm diameter blanket wafers with electroplated copper were used. The wafer surface is initialized by rinsing with D.I.water followed by light etching of the oxidized surface with DHF. The cleaning steps were done using an ultrasonic nozzle with DI water as the control, or the appropriate solution. The effect of anode water is tested compared with oxidizing solutions containing dissolved gases (O_3 water and O_2 water). Reference wafer is cleaned only by DI water. X-ray Photoelectron Spectroscopy (XPS) was used to obtain the composition ratio of Cu, Cu oxide and Cu hydroxide by analysis of the Cu2p electron, Cu Auger electron and O1s electron.

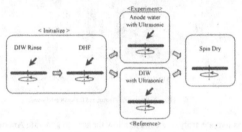

Figure 4 : Procedure of experiment

RESULTS AND DISCUSSION

In order to confirm the surface control by Anode water, the level of copper oxidization was analyzed by XPS. After four kinds of cleaning treatments, (1)D.I.water (2)O_3 water (3)O_2 water (4)Anode water, XPS results were obtained and compared on wafers 24 hours and 2 weeks after cleaning.

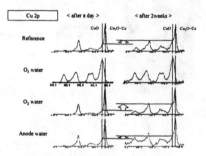

Figure 5 : Results of Cu2p spectrum analyses by XPS

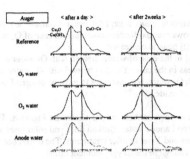

Figure 6 : Results of Cu Auger spectrum analyses by XPS

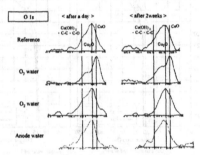

Figure 7 : Results of O1s spectrum analyses by XPS

The results of Cu2p spectrum analyses by XPS are shown on Figure-5. The results of Cu Auger spectrum analyses are shown on Figure-6. The results of O1s spectrum analyses are shown on Figure-7.

<u>24 hours after cleaning treatments</u>:

CuO peak of O_3 water treatment is biggest than peaks of other three cases in Figures-5, -6, -7. CuO peak of O_2 water is higher than other treatments in Figure-6. CuO peak of DI water is higher than O_2 water and Anode water cases in Figure-7. The data shows that copper oxidization levels were in the following order:

O_3 water > DI water > O_2 water > anode water.

with O_3 water exhibiting the most oxidation and anode water showing the least amount of oxidation.

<u>2 weeks after cleaning treatments</u>:

In case of O_3 water, oxidization level is still remained high even after 2 weeks.

CuO levels of all treatment cases except O_3 water treatment are almost same as shown on Figure-5. Figure-7 shows that oxidization continued in case of O_2 water because CuO peak is increased and $Cu(OH)_2$ peak is decreased.

The data suggests that in the case of wafers treated with O_3 water oxidation is the highest and remains stable, whereas in the case of wafers treated with O_2 water oxidation is not as high as on wafers treated with O_3 water, but continues to grow as oxygen from the environment is able to penetrate the surface oxide.

Depth analysis of the oxygen content ratio is shown on Figure-8. The data compares wafers treated with O_2 water and anode water. Broken lines and solid lines show respectively oxygen ratio data both 24 hours and 2 weeks after treatments. The difference between broken lines and solid lines shows the change in the oxygen content ratio.

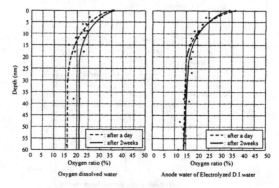

Figure 8 : Oxygen content in Cu

Oxygen ratio for wafers treated with O_2 water has clearly increased up to a depth of 60 nm, but the ratio for wafers treated with anode water is stable. It is postulated that the nature of the thin oxide film that is created by anode water treatment prevents oxygen penetration and prevents the progression of further oxidation.

CONCLUSION

The treatment of copper wafers with anode water has been demonstrated as a viable method to prevent post CMP oxidation of copper lines in semiconductor devices. Oxidization levels on copper surfaces after four kinds of treatments, (1) D.I.water, (2) O_3 water, (3)O_2 water, and (4)Anode water, were compared.

(1) CuO ratio on wafers treated with anode water was stable and low compared with the treatments by DI water, O_2 water and O_3 water even 2 weeks after treatment.
(2) Oxidization on wafers treated with anode water was shown to be stable and low even 2 weeks after treatment using depth analysis.

Based on the results obtained in this report, further studies on actual devices will be conducted.

REFERENCES

[1] N.Miyashita, Y.Mase, J.Takayasu, Y.Minami, M.Kodera, M.Abe, T.Izumi: "Mechanism of a New Post CMP Cleaning for Trench Isolation Process", Materials Research Society 1999 Spring Meeting, San Francisco, CA, April 1999.

[2] D.Briggs, M.P.Seah: "Practical Surface Analysis by Auger and X-ray Photoelectron Spectroscopy", John Wiley & Sons Ltd.

[3] The surface science society of Japan: "Method of X-ray Photoelectron Spectroscopy", Maruzen

Mat. Res. Soc. Symp. Proc. Vol. 612 © 2000 Materials Research Society

A NEW POLY-Si CMP PROCESS WITH SMALL EROSION FOR ADVANCED TRENCH ISOLATION PROCESS.

Naoto Miyashita*, Shin-ichiro Uekusa**, Takeshi Nishioka*** and Satoko Iwami***
* Dept of Electrical and Electronic Engineering, Meiji Univ., Toshiba Co, Semiconductor
Company 8, Shinsugita-cho, Isogo-ku, Yokohama 235-8522, Yokohama Japan
** Dept of Electrical and Electronic Engineering, Meiji Univ., Kawasaki Japan
*** Mechanical Systems Laboratory, Corporate R&D Center, Toshiba Corporation, 1,
Komukai-Toshiba-cho, Saiwai-ku, Kawasaki Japan

ABSTRACT

Chemical-Mechanical Polishing has been revealed as an attractive technique for poly-Si of
trench planalization. Major issue of the process integration is pattern erosion after over polishing.
A new process with silica slurry containing organic surfactant is reported in this paper. A
patterned wafer after conventional CMP process is eroded by over polishing, however, the new
process conducts small erosion for wide trenches. The organic surfactant is well known as a
inhibitor for the protection of poly-Si from alkaline, and the new slurry shows a large pH
dependency of the viscosity. The experimental work has been focused on the viscosity, and the
mechanism of the small erosion is discussed. This new process should be useful for recessing
poly-Si by CMP, because it keeps the erosion level very low.

INTRODUCTION

Fine device patterns are formed using CMP techniques in the production of semiconductor
devices. Advanced trench isolation technology has been developed and applied to high-speed
Bi-CMOS LSI production. Poly-Si CMP technique has made much improvement on the deep
trench planarizing process. [1] Major issue of the process integration has been the erosion
problem.

In this process, dishing has been caused by over polishing in poly-Si CMP process.
During poly-Si CMP process, over-polish is necessary to remove all the residual poly-Si globally
for the achievement of high degree of the planarity. Due to the high selectivity of poly-Si
removal rate to Oxide by typical poly-Si CMP slurry, over-polish inevitably introduces poly-Si
dishing in trench structures. The poly-Si dishing in turn causes oxide erosion and active area's
Si3N4 film in the neighboring oxide, attributable to the concentrated cap oxide stress. These
stresses introduce many crystal defects. Trench layer should be planarized without dishing on the
local scale (distance < 700nm) and the global scale(distances > 1000nm) in poly-Si CMP
process. However it is difficult to reduce the dishing by the conventional poly-Si CMP technique.
Therefore we studied a dishing less new CMP method using high viscosity slurry.

EXPERIMENTAL

Figure1 shows a cross sectional image of the trench isolation pattern. The patterned wafers used in this experiment the test structure especially designed for Poly-Si CMP process development. Trench structure with 4500-5000nm of depth is defined and patterned on Si wafer. Before patterning, 100 nm thick thermal oxide and 70 nm thick Si3N4 layers were grown on Si wafers.

LP CVD poly-Si was used for the trench filling. In this trench process, Si3N4 film was used as a stopper for poly-Si CMP and LOCOS musk. Two kinds of films, that is SiO2 and Si3N4 appeared on the wafer surface after poly-Si CMP. Over polish inevitably introduces poly-Si dishing in trench filling poly-Si and Si3N4 erosion on the surface. In this experiment, we paid attention to the viscosity of slurry in order to improve the Si3N4 erosion and poly-Si dishing. Step heights of trench were measured using the probe type apparatus after CMP.

The schematic diagram of the polisher, EBRA EPO-112, used in this study is shown in Fig2. The polisher consists of polishing unit and wafer cleaning unit. The polishing unit is composed of wafer carrier and turntable. The wafer cleaning unit is composed of PVA brush, cloth brush and spin-dry stations [2]. The wafer is held on the wafer carrier and is rotated for CMP. Rodel IC 1000/ SUBA IV stacked polishing pad were applied in this study.

RESULTS AND DISCUSSION

1. Conventional method

Trench filling poly-Si dishing and local Si3N4 erosion is serious issue in poly-Si CMP process. The post-CMP structure is shown schematically in Fig.3. Field Oxide loss is defined as the amount of oxide loss in region with no poly pattern or wide trench pattern. This loss is small typically. Poly-Si dishing and local SiN erosion are then defined as the amount of recess of patterned trench or SiN, relative to filled oxide surface. It is clear that local Si_3N_4 erosion are 1-50nm in our conventional poly-Si CMP process. Poly-Si dishing and local Si_3N_4 erosion associate with stress after trench cap oxidation. In advanced trench process, the wafer after Poly-Si CMP is oxidized as next process step. Therefore, it is necessary to keep the erosion very low. In order to optimize the process, two kinds of slurry mixing CMP process was investigated. The corresponding experiment was produced with two poly-Si CMP slurry. We have measured the profile of total erosion of trench after poly-Si CMP. It is appear that 90-300 nm-erosion is occurred. The viscosity of slurry A is 10-12mPa-s and the pH value of slurry is 10.2-10.5 at the use point of poly-Si CMP.

2. New CMP method

Erosion of trench is serious issue in poly-Si CMP. In order to optimize the process, high viscosity CMP process was investigated. We focused on the viscosity of slurry and increased it in poly-Si CMP process. We have been revealed as an attractive technique for poly-Si of trench planalizing process using two slurry including organic surfactants.

Slurry A is conventional colloidal silica slurry having a 1-5wt% concentration of alkaline abrasive silica content. Slurry B is prepared as the cleaning slurry for post CMP cleaning process.

[4] In the new CMP method, the wafer is polished in two-steps. The first step is the main poly-Si removal step. The second step is dishing less CMP process. In the first polishing step, the slurry A is dropped on to the rotating table. The second step, the slurry B dropped on to the rotating table after optically measured end point signal. At this time the transmittance of slurry were investigated by spectrophotometer. The characteristics of each process step are shown in table 1. Fig.4 shows the relationship of the viscosity of the mixed slurry and pH to the ammonia solvent concentration. The viscosity of the mixed slurry increased rapidly from 0.5% as the alkaline concentration in slurry decreased in our experiment. The pH of mixed slurry decreased rapidly from pH 10.4. In this process step, organic surfactants of slurry B condensed with silica abrasives as core on the table. Fig.5 shows the relationship between pH of slurry B and DI water dilution ratio. The pH value dramatically decreased by using the DI water dilution. In other experiment, the viscosity of slurry B did not increase, because it is not contain silica abrasive.

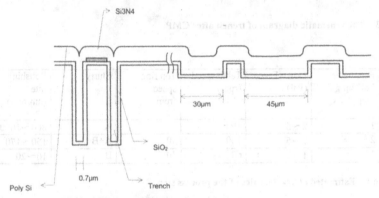

Fig. 1   A cross sectional image of the trench test pattern

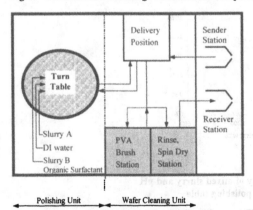

Fig. 2   Schematic diagram of the CMP equipment (EPO-112)

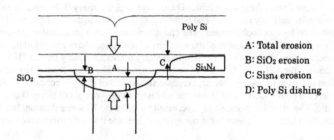

A: Total erosion
B: SiO₂ erosion
C: Si₃N₄ erosion
D: Poly Si dishing

Fig.3 The schematic diagram of trench after CMP

Parameter Polishing Step	Down Force (Psi)	Table speed (rpm)	Top ring speed (rpm)	Slurry	Polishing rate (nm/min)
1	2～5	100	40	A	800～1000
2	2～5	100	40	A+B	150～170
3	1～2	60	40	B	10～20

Table.1 Estimated characteristics of the process step

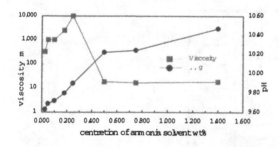

**Fig.4 the relation of viscosity of mixed slurry and pH
of anemone solvent on the polishing table**

ACKNOWLEDGMENT

The authors would like to thank Mr. Takayasu and Mr. Minami who has made a great effect to do these experiments. And authors gratefully acknowledge the personnel in the Silicon Facility at TOSHIBA Micro electronics center for processing the wafers. Special thanks go to S.Kikuchi and M.Terasaki for the lithography, K.Doi for the trench fill, Y.Otani for the trench etching, and K.Iwade for the line process.

REFERENCES

[1] S. A. Abbas : "Silicon on poly silicon with deep dielectric isolation", Proc. of IBM' Technical Disclosure Bulletin Vol. No.7 Dec. 1997 P.2754

[2] D. L. Hetherington : "The effects of double-sided scrubbing on removal of particles and metal contamination from chemical-mechanical-polished wafers", Proc. of DUMIC Conference 1995 ISMIC-101D/95/0156

[3] N. Miyashita, Y. Minami, I. Katakabe, J. Takayasu, M. Abe, T.Izumi: "Characterization of new post CMP cleaning method for trench isolation process" Proc. of Proceedings of 14th International Vacuum Congress. (1999)P.71

[4] N.Miyashita, Y.mase, J.Takayasu, Y.Minami, M.Koderà, M.Abe, and T.Izumi: "Mechanism of a new post CMP cleaning for trench isolation process." Proc. of Proceedings of MRS.Vol.566 (1999)P. 253

[5] T.Nishioka, K.Sekine and Y.Tateyama: "Modeling on hydrodynamic effects of pad surface roughness in CMP process" Proc. IEEE 1999 Int. Interconnect Tech. Conf. (1999) p.89

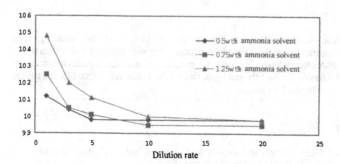

Fig.5 The relation of slurry B pH to DI water dilution ratio.

By using this new CMP method, the erosion of trench on wafer is below 5nm (7nm width) trench structure and below 50nm(30-100 nm width) recess. Two possible mechanisms are considered for the excellent planarity. One is that the condensed surfactants make the pad harder, and the other is the increase of the hydrodynamic pressure of the slurry. Nishioka, et.al. proposed a model on the hydrodynamic effects of pad surface roughness, and reported the effective hydrodynamic pressure occurs with more than 100 mPa-s viscosity of slurry [5]. Farther investigation should be necessary to make the mechanism clear.

CONCLUSIONS

From these results, we conclude that:

(1) The poly-Si of trench after poly-Si CMP process were eroded by over in the case of conventional method.

(2) This new CMP method with the organic surfactants of slurry could reduce the erosion of the e poly-Si of trench on the wafer below 5nm and below 50nm(30-100 nm width) recess. It is applicable to CMP planarization process for VLSI.

(3) The mechanisms of the excellent planarity are considered as the effect of the condensation of the surfactants and the increase of the viscosity of the slurry.

Mat. Res. Soc. Symp. Vol. 612 © 2000 Materials Research Society

REMOVAL RATE, UNIFORMITY AND DEFECTIVITY STUDIES OF CHEMICAL MECHANICAL POLISHING OF BPSG FILMS

Benjamin A. Bonner, Boris Fishkin, Jeffrey David, Chad Garretson, and Thomas H. Osterheld
CMP Division, Applied Materials
3111 Coronado Drive, Santa Clara, CA 95054

ABSTRACT

Wafers where deposited with BPSG films having phosphorus concentration varying from 3.65 to 6.25% and boron concentration varying from 4 to 5.7%. These wafers were polished using CMP and the rates were found to depend on dopant concentrations. A fit to the data indicated that removal rates were more than 3 times as sensitive to boron concentration compared to phosphorus concentration. For a constant phosphorus concentration of 5%, each percent increase in boron increases CMP removal rate by 340 Å/min. For a constant boron concentration of 5%, each percent increase in phosphorus increases CMP removal rate by 96 Å/min.

INTRODUCTION

Borophosphorosilicate glass (BPSG) is currently a film of choice as pre-metal dielectric [1-4]. The addition of phosphorous to silicate films may lower the migration of alkali ions, while boron addition lowers the glass transition temperature of the film allowing it to flow at lower temperatures to give better local planarization [1,2,5]. The move toward sub-0.25 micron line width requires global planarization to achieve good depth of focus. This global planarization can be achieved by chemical mechanical polishing (CMP).

EXPERIMENTAL DETAILS

Polishing experiments were performed on an Applied Materials Mirra® CMP system equipped with the Oxide Plus hardware package. All experiments used SS-12 slurry from Cabot Corporation and IC1010 pads from Rodel.

Oxide thickness was measured on a Thermawave Optiprobe 3260. Oxide thickness was measured using a standard 49-point contour map having a 5 mm edge exclusion. Removal rates were calculated from average thickness using the 49-point contour map. Within-wafer-nonuniformity (WIWNU) was expressed as a percentage and calculated using standard deviation of removal divided by average removal.

BPSG wafers were deposited in an Applied Materials Giga-Fill™ SACVD chamber. Boron concentration was varied between 4 and 5.7% while phosphorus concentration was varied between 3.65 and 6.25%. Wafers were annealed by standard methods after deposition.

RESULTS AND DISCUSSION

Doping an oxide film with boron or phosphorus increases oxide removal rate significantly [2,5-7]. Under the baseline BPSG polish conditions used for these experiments, thermal oxide

Mat. Res. Soc. Symp. Vol. 612 © 2000 Materials Research Society

(undoped) has a removal rate of 1590 Å/min (standard ILD polish conditions yield a much higher removal rate). In contrast, all of the oxide films doped with boron and phosphorus had removal rates of more than 4000 Å /min. To better understand this behavior, a series of experiments were performed on wafers deposited with varying amounts of boron and phosphorus.

For the first series of experiments, wafers were prepared with 5 different phosphorus concentrations by varying the flow rate of the phosphorus precursor during deposition. The boron precursor flow-rate was held constant during the deposition of conditions 1-5. The weight percent of each dopant was measured after deposition using X-ray fluorescence and the resulting dopant concentrations are summarized in Table I. Table I shows that phosphorus concentration is varied from 3.65% by weight to 6.25% by weight. Some variation in boron concentration is also observed even though the precursor is held at a constant flow rate.

Table I. Boron and Phosphorus concentrations for Conditions 1-5

Condition	B wt.%	P wt.%
1	4.80	3.65
2	4.90	4.40
3	5.00	5.10
4	5.25	5.65
5	5.25	6.25

For the second series of experiments, wafers were prepared with 5 different boron concentrations by varying the flow rate of the boron precursor during deposition. The phosphorus precursor flow-rate was held constant during the deposition of conditions 6-10. The weight percent of each dopant was measured after deposition using X-ray fluorescence and the resulting dopant concentrations are summarized in Table II. Table II shows that boron concentration is varied from 4.0% by weight to 5.7% by weight. Some variation in phosphorus concentration is also observed even though the precursor is held at a constant flow rate.

Table II. Boron and Phosphorus concentrations for Conditions 6-10

Condition	B wt.%	P wt.%
6	4.00	5.60
7	4.50	5.35
8	5.00	5.10
9	5.35	4.95
10	5.70	4.80

Two wafers were polished for each dopant condition using a standard BPSG polishing process. Average removal rate and WIWNU for the polish process were obtained using pre and post-polish measurements. Results of these experiments are summarized in Table III.

Inspection of conditions 1 through 5 in Table III clearly shows that removal rate increases as phosphorus concentration increases. For example, condition 1 with 3.65% phosphorus has a rate of 4154 Å/min while condition 5 with 6.25% phosphorus has a higher removal rate of 4599 Å/min. Similarly, inspection of conditions 6-10 shows that removal rate increases as boron concentration increases. For example, condition 6 with 4.0% boron has a rate of 4202 Å/min while condition 10 with 5.7% boron has a higher removal rate of 4680 Å/min.

Table III. Removal rate and WIWNU for polished BPSG wafers

Condition	B (wt%)	P (wt%)	Removal Rate (Å/min)	WIWNU (%)
1	4.80	3.65	4154	3.44
2	4.90	4.40	4303	3.32
3	5.00	5.10	4325	2.83
4	5.25	5.65	4460	2.99
5	5.25	6.25	4599	3.59
6	4.00	5.60	4202	2.68
7	4.50	5.35	4255	3.97
8	5.00	5.10	4325	2.83
9	5.35	4.95	4581	3.36
10	5.70	4.80	4680	2.70

Removal rate dependence on varying boron and phosphorus concentration is plotted in Figure 1. The best-fit lines for both sets of conditions are also provided in the graph because the dependence on P or B concentration appears reasonably linear in this range. The slope for the fit to varying phosphorus concentration is 160 (Å/min)/(% P) with an R^2 of 0.95 while the slope for the fit to the varying boron concentration is 290 (Å/min)/(% B) with an R^2 of 0.885.

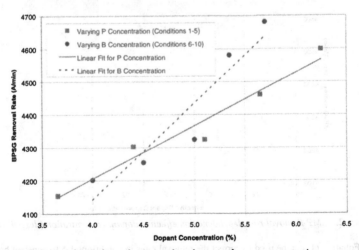

Figure 1. *CMP Removal rate dependence on phosphorus or boron concentration*

Mat. Res. Soc. Symp. Vol. 612 © 2000 Materials Research Society

Inspection of Tables I and II shows that boron concentration was not constant for conditions 1-5 (designed to only vary phosphorus concentration) while phosphorus concentration was not constant for conditions 6-10 (designed to only vary boron concentration). At the same time, Figure 1 shows that removal rate depends on both phosphorus and boron concentration. It would be valuable to isolate the contribution of phosphorus and boron to removal rate.

Because the slopes are so different for rate dependence on boron vs. phosphorus concentration, one would not expect a dependence on total dopant concentration (sum of phosphorus and boron concentration). This conclusion is borne out by a plot of rate vs. total dopant concentration where conditions 1-5 are not collinear with conditions 6-10. Instead, a multiple regression analysis was performed to obtain the dependence on boron and phosphorus concentrations. This analysis resulted in Equation (1) having an R^2 of 0.83.

$$\text{Rate (Å/min)} = 335 \text{ (Å/min/\%B)} \times (\text{B Conc}) + 100 \text{ (Å/min/\%P)} \times (\text{P Conc}) + 2235 (\text{Å/min}) \quad (1)$$

The ratio of slopes from this equation is 335/100=3.35. The ratio of slopes indicates that doping with boron is 3.35 times as effective as phosphorus in increasing removal rate. An effective dopant concentration can be calculated by adding 3.35 times the boron concentration to the phosphorus concentration. A plot of removal rate dependence on this effective dopant concentration is provided in Figure 2. The best-fit line for all of the data (conditions 1-5 and conditions 6-10) is also provided. This fit has an R^2 of 0.87 indicating a reasonably valid correlation.

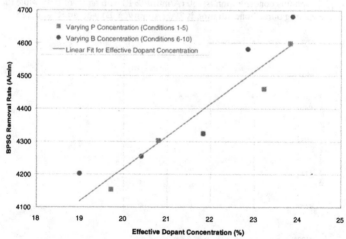

Figure 2. *CMP Removal rate dependence on effective dopant concentration (described in text)*

Equation (1) can be used to correct the removal rates in conditions 1-5 to constant boron concentration and the rates in conditions 6-12 to constant phosphorus concentration. The mid-point of 5% was selected in both cases. Table IV provides the results of this calculation.

Mat. Res. Soc. Symp. Vol. 612 © 2000 Materials Research Society

Corrected removal rate dependence on varying boron and phosphorus concentration is plotted in Figure 3. The best-fit lines for both sets of conditions are also provided in the graph. The slope for the fit to varying phosphorus concentration is 96 (Å/min)/(% P) with an R^2 of 0.86 while the slope for the fit to the varying boron concentration is 340 (Å/min)/(% B) with an R^2 of 0.915. For a constant boron concentration of 5%, each percent increase of phosphorus increases the film removal rate by 96 Å/min. For a constant phosphorus concentration of 5%, each percent increase in boron concentration increases the film removal rate by 340 Å/min.

Table IV. Removal rate corrected to constant B or P concentration for polished BPSG wafers

Condition	B (wt%)	P (wt%)	Corrected B (wt%)	Corrected P (wt%)	Removal Rate (Å/min)	Corrected Removal Rate (Å/min)
1	4.80	3.65	5.00		4154	4221
2	4.90	4.40	5.00		4303	4337
3	5.00	5.10	5.00		4325	4325
4	5.25	5.65	5.00		4460	4376
5	5.25	6.25	5.00		4599	4515
6	4.00	5.60		5.00	4202	4142
7	4.50	5.35		5.00	4255	4220
8	5.00	5.10		5.00	4325	4315
9	5.35	4.95		5.00	4581	4586
10	5.70	4.80		5.00	4680	4700

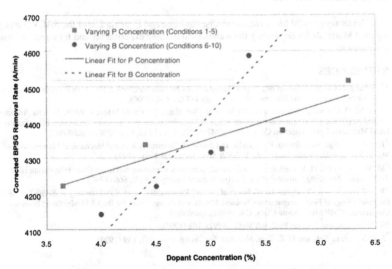

Figure 3. *Corrected CMP Removal rate dependence on phosphorus or boron concentration*

Several previous studies have investigated removal rate dependence on phosphorus concentration (no boron dopant) [1,2,5,6]. We are also aware of a study that tested sensitivity to boron concentration (no phosphorus dopant) [5]. Only one study showed non-normalized data and found approximately 150 Å/min increase in rate for each percent increase in phosphorus concentration [6]. This compares reasonably to 96 Å/min in this study. One would not necessarily expect good aggreement given that different process conditions and consumables were used in the two studies.

CONCLUSIONS

CMP of BPSG films showed that removal rates depend linearly on dopant concentrations. A fit to the data indicated that removal rates were more than 3 times as sensitive to boron concentration compared to phosphorus concentration. Coefficients from least-squared fits to the data were used to calculate a linear dependence on both boron and phosphorus concentration with an R^2 value of 0.87. These calculations were also used to isolate the removal rate dependence on both boron and phosphorus. For a constant phosphorus concentration of 5%, each percent increase in boron increases CMP removal rate by 340 Å/min. For a constant boron concentration of 5%, each percent increase in phosphorus increases CMP removal rate by 96 Å/min.

ACKNOWLEDGMENTS

The authors would like to acknowledge Paul Gee and Eugenia Liu of the SACVD group at Applied Materials for depositing the wafers used in these experiments and for numerous useful discussions.

REFERENCES

[1] S. J. Fang, S. Garza, H. Guo et al., "Optimization of the Chemical Mechanical Polishing Process for Premetal Dielectrices," Journal of the Electrochemical Society 147 (2), 682 (2000).

[2] C.-T. Ni, H. C. Chen, D. Huang et al., "A Study of CMP Slurry Chemistry Effect on BPSG Film for Advanced DRAM Application," presented at the Proceedings of Third International Chemical-Mechanical Planarization for the ULSI Multilevel Interconnection Conference (CMP-MIC), Santa Clara, CA, 1998 (unpublished).

[3] Joseph M. Steigerwald, Shyam P. Murarka, and Ronald J. Gutmann, Chemical Mechanical Planarization of Microelectronic Materials, 1 ed. (John Wiley & Sons, Inc., New York, 1997).

[4] M. Yoshimaru and H. Wakamatsu, "Microcrystal Growth on Borophosphosilicate Glass Film during High-Temperature Annealing," Journal of the Electrochemical Society 143 (2), 666 (1996).

[5] W.J. Schaffer, J. W. Westphal, H. W. Fry et al., "CMP Removal Rate and Nonuniformity of BPSG," presented at the Proceedings of First International Chemical-Mechanical Planarization for the ULSI Multilevel Interconnection Conference (CMP-MIC), Santa Clara, CA, 1996 (unpublished).

[6] S. Pennington and S. Luce, Proc. 9th VMIC 9 (92), 168 (1992).

[7] S. C. Sun, F. L. Yeh, and H. Z. Tien, Mat. Res. Soc. Symp. Proc. 337, 139 (1994).

Mat. Res. Soc. Symp. Vol. 612 © 2000 Materials Research Society

Using Wafer-Scale Patterns for CMP Analysis

Brian Lee[1], Terence Gan[1], Duane S. Boning[1], Jeffrey David[2], Benjamin A. Bonner[2],
Peter McKeever[2], and Thomas H. Osterheld[2]
[1]Massachusetts Institute of Technology, Cambridge MA
[2]Applied Materials, Santa Clara, CA

ABSTRACT

A new set of wafer-scale patterns has been designed for analysis and modeling of key CMP effects. In particular, the goal of this work is to develop methods to characterize the planarization capability of a CMP process using simple measurements on wafer scale patterns. We examine means to pattern large trenches (e.g. 1 to 15 mm wide and 15 mm tall) or circles across 4" and 8" wafers, and present oxide polish results using both stacked and solo pads in conventional polish processes. We find that large separation (15 mm) between trenches enables cleaner measurement and analysis. Examination of oxide removal in the center of the trench as a function of trench width shows a saturation at a length comparable to the planarization length extracted from earlier studies of small-scale oxide patterns. Increase in polish pressure is observed to decrease this saturation point. Such wafer scale patterns may provide information on pad flexing limits in addition to planarization length, and promise to be useful in both patterned wafer CMP modeling and studies of wafer scale CMP dependencies such as nanotopography.

INTRODUCTION

Current techniques for characterizing CMP typically involve patterning test dies onto a wafer, running polish experiments, and analyzing measurements to obtain characterization parameters for the process [1]. A key parameter known as planarization length is typically used to describe the length scale over which feature-induced pattern density on the wafer affects the polishing at a particular point on the die. By adding feature-scale step-height considerations to planarization length-based density evaluation, accurate models of post-CMP oxide thickness (with ~100 Angstrom error) have been demonstrated for conventional stacked pads and processes [2].

An alternative approach using wafer-scale patterns has previously shown promise as a tool to study CMP pattern dependencies [3]. With an increased interest in harder polishing pads (to increase planarization length and reduce within-die variation), the planarization length is approaching the size of the typical die. In addition, harder pads may induce a "pad flexing limit" in which the contact of the pad in large "low" regions of the die is decreased. For these reasons, as well as interest in simplified measurement and analysis of planarization length, wafer-scale patterns for detailed CMP planarization characterization are explored further in this work.

In the next section, we describe new mask designs for wafer-scale patterns. Key issues include the range of trench (or circle) sizes that should be included, as well as the separation between them on the wafer. In addition, we describe two patterning methods used here, including traditional mask plates and acetate-based masks. The following section then summarizes the sets of wafer fabrication and polishing experiments conducted, followed by analysis and discussion of the experimental results. The relationship between the observed results and previous pattern density or contact wear models is briefly considered. Finally, we offer conclusions and suggestions for further work.

MASK DESIGN

Three sets of wafer-scale patterns are described here. Two of the designs are used for patterning 4-inch wafers, and the third design is used to pattern 8-inch wafers. The guiding principle behind the design of the masks is to fabricate structures of various sizes separated by relatively large distances to reduce interaction between these structures. The use of wafer scale patterns enables much larger structures and separations to be fabricated than is usually possible with conventional die patterns.

The layouts of the three patterns are summarized in Figure 1, where the width or diameter and relative positions of the trenches or circles are shown. The first pattern (Pattern A) is implemented as a standard quartz mask for use in a Karl Seuss contact aligner. The total pattern size is 70 mm x 70 mm, consisting of rectangular trenches 8 mm in height, ranging in width from 20 μm to 8 mm. The second pattern (Pattern B) is implemented using alternative transparency masks on 4" wafers. There are 30 circular trench structures in the layout, with 17 distinct widths ranging from 2 mm to 8 mm, and replicates of several of the structure sizes. These structures are separated from each other and from the edge of the wafer by at least 10 mm to reduce edge effects and interaction among structures. The structures are arranged in three concentric rings on the wafer. The final pattern (Pattern C) is implemented as a standard quartz mask for use in a contact aligner, with a total pattern size of 140 mm x 140 mm. It consists of 25 rectangular trench structures, each of which is 15 mm in height. There are 22 distinct widths, ranging in size from 20 µm to 15 mm, with replicates of the 1mm, 5mm, and 10mm trenches. These structures are separated from each other and from the edge of the wafer by at least 15 mm to further reduce edge effects and interaction among structures.

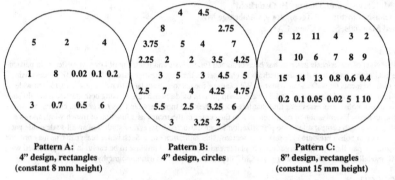

Pattern A:	Pattern B:	Pattern C:
4" design, rectangles	4" design, circles	8" design, rectangles
(constant 8 mm height)		(constant 15 mm height)

Figure 1. Mask floor plan for wafer-scale masks (dimensions in mm).
Positions of numbers indicate relative location of structures on the wafer.

In addition to consideration of structure size and separation issues, in this work we investigate an alternative means of patterning wafer-scale structures. Patterns A and C are implemented using traditional mask fabrication facilities. Pattern B, on the other hand, is implemented as a lithographic print on a translucent sheet of acetate (i.e., transparency), which is then used in a Karl Seuss contact aligner to pattern wafers. The lithographic print method is chosen to insure sufficient opaqueness of the pattern. This method results in poor structure edge resolution, and restricts the minimum structure size on the mask to 2 mm. For the study of large scale polishing behavior (large multi-mm structures), these limitations are not critical. The benefit of this method is that it enables inexpensive and rapid production of distinct wafer-scale patterns for characterization work.

EXPERIMENTAL DETAILS

Pattern A and Pattern B wafers are polished on a Strausbaugh 6EC CMP tool using a standard pad stack (IC1000/SubaIV) and slurry (Cabot SS-12). Pattern C wafers are polished on an Applied Materials Mirra™ tool using standard and harder pad stacks and standard slurry.

Pattern A wafers have 1.5 µm of CVD oxide, and are patterned and etched to produce 0.6 µm trenches using a wet etch process. The wafers are then polished under the following process conditions: 25 rpm table speed, 55 rpm quill speed, 3 psi down force, 1 psi back pressure, modifying the polish time to yield splits of three different thickness removals: 0.4 µm, 0.6 µm, and 0.8 µm.

Pattern B wafers have an initial CVD oxide of 1.5 µm, and are patterned (using the alternative mask method), and then etched to produce 0.7 µm trenches using a wet etch process. The wafers are then polished under the following process conditions: 25 rpm table speed, 15 rpm quill speed, 2.5 psi down force, 1 psi back pressure, modifying the polish time to yield splits of three different thickness removals: 0.3 µm, 0.5 µm, and 0.7 µm.

Pattern C wafers have trenches etched in silicon to an etch depth of 0.82 µm, with 1.5 µm of oxide then deposited across the wafer. Wafers are polished under three different process conditions (constant speed and varying pressures), but all splits are targeted towards the same amount of oxide removed (0.5 µm).

Thickness measurements are taken across the trenches using standard profilometry scans (in Pattern A), optical measurements (in Pattern B), or high-resolution optical measurements (in Pattern C).

EXPERIMENTAL RESULTS

One proposed method of using wafer-scale post-polish measurement data to characterize the CMP process is to analyze the amount of material removed in the center of trench structures as a function of the structure size. Wider trenches should result in more material removed, since the CMP pad can deform into the trenches to a greater degree. In this section, we examine the polish data from the trench structures of different sizes. Results indicate that wafer scale uniformity, as well as structure separation, are important considerations for wafer scale patterns. Given sufficient separation, trends in trench removal vs. structure size are clearly discernible.

Polish Depth and Structure Separation Guidelines

As expected, the amount of trench oxide removal increases as the etched structure size increases, as shown in Figure 4 for Patterns A and B. We see that the curve for Pattern B (which uses a non-traditional patterning step and circular trenches) exhibits the same general signature as the curve for Pattern A (which uses a traditional patterning step and rectangular trenches). Figure 4a shows that the general signature of the process does not change with the amount of material removed (approximately scaling with amount removed), provided that one does not polish past the depth of the trench itself. The curve for the 0.4 μm and 0.6 μm target removals show similar signatures, while the curve for the 0.8 μm removal exhibits a much noisier signature. We conjecture that wafer scale polish nonuniformities are exacerbated or that trench structures interact more strongly for large material removals, suggesting that moderate amounts of polish are best for wafer scale pattern studies.

Figure 4b shows the trend of trench area removal vs. structure size for the Pattern B wafers, where structures are replicated in three concentric rings on the wafer. We see that the general range of values and curve trends do not change depending on ring position. However, the structures nearer to the edge seem to exhibit a much noisier signal than the structures in the center of the wafer. Here again the 10 mm separation distance between structures and from the edge of the wafer may not be sufficient to block against structure interaction. The potential for wafer scale polish uniformity to affect structure polish also suggests that replication of trench structures is important in order to separate polish uniformity from trench size dependencies.

Pad and Process Impact on Trench Removal vs. Trench Width

Pattern C wafers, consisting of structures with larger 15 mm heights and separations, are polished using two different pads and three different polish processes; the trench removal vs. trench width plots are shown in Figure 5. The left side of Figure 5 shows results using a standard stacked pad, while the right side plots the results with a hard polishing pad. Three polishing processes in which only the head pressure varies are shown from top to bottom, where increasing pressure is from curves (a) to (c) and from curves (d) to (f).

Considering first the standard standard stacked pad and process of Figure 5(a), we see that the trench removal reaches a saturation point for trench sizes in the 5-6 mm range, at which point the trench removal is nearly constant (but less than in the surrounding unetched regions) for increasing trench widths. We also see that the trench removal appears to linearly decrease toward zero for smaller trench widths. As the polish pressure increases from Figure 5(a) to (c), we see that the saturation point becomes more pronounced and appears to occur at smaller trench widths. The difference between the trench and non-trench oxide removal for the largest trench sizes also decreases for the larger pressure processes.

Comparing these stacked pad results with the hard pad results shown in Figure 5(d) to (f), we see a dramatic difference in the saturation point for the harder pads. Indeed, for the lower pressure the saturation point is not reached for the largest 15 mm structure examined here. In the case of the highest pressure process, we see what may be a fairly sharp saturation at a trench width of 15 mm, compared to 4-5 mm for the stacked pad at the same pressure. For study of emerging hard pad CMP processes, these results suggest that wafer scale patterns with even larger structures and separation distances should be considered in the future.

DISCUSSION AND MODELS

In this section, we first consider the relationship between the observed data and a pattern density-based CMP model, and then briefly consider the above data from the perspective of a contact wear model.

Planarization Length and Pattern Density-Based Model

The "planarization length" is used to describe the ability of a CMP process to remove variation on a die [1]. The saturation point on the curve for the standard pad is in the 4-6 mm range, which is comparable to planarization lengths previously extracted for this process using pattern density test masks [4]. An idealized pattern density-based CMP model is considered here to show that this saturation point is the same as the planarization length parameter.

Figure 2 illustrates the trench polish problem using an idealized analysis, in which a simple square averaging window of size equal to the planarization length is used to calculate the "effective density" of raised topography around and within a trench. The effective density as calculated as the ratio of raised material to total area within some region defined by the planarization length. The effective density-based polish model assumes that raised or "up" areas of regions will polish as the blanket rate divided by the effective density of that region, and once the up structures are removed the region polishes at the blanket polish rate [1].

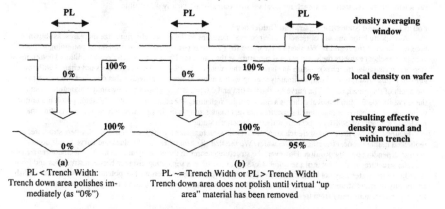

(a)
PL < Trench Width:
Trench down area polishes immediately (as "0%")

(b)
PL ~= Trench Width or PL > Trench Width
Trench down area does not polish until virtual "up area" material has been removed

Figure 2. The trench polish problem using a planarization length (PL) and effective density analysis.

We see in Figure 2 that trenches with widths larger than the planarization length of the process have a trench central region that polishes as a "0% effective density area" surrounded by areas of monotonically increasing density. Regions more than half a planarization length away from the trench edge polish as 100% effective density regions (or at the blanket polish rate). In the "0% effective density area" the wafer should polish as the blanket rate. For trenches with lengths equal to or less than the planarization length, the trench central region will evaluate to a non-zero effective density, so that no down area (i.e. trench center) polish occurs until the up area material is removed. For the trench case where we have non-zero density points inside the trench, this would refer to virtual "up area" since points inside the trench have no real up area material. The hypothetical trench removal vs. trench width plot resulting from this effective density polish model is summarized in Figure 3.

While the effective density models can relate the saturation point in the plots of Figure 5 to the notion of planarization length, other aspects of the data are not well explained using such a model. First, as discussed earlier, at saturation the trench removal amount is less than the corresponding amount removed from outside the trench, while the density model suggests they should both polish at the blanket rate (the dashed line in Figure 3(a)). Second, the data for both the standard and harder pads indicate that trench down area removal is zero only at very small trench sizes, while the density model predicts a substantial range of zero polish. Down-area polish before complete removal of local step height has been modeled [2; however, the step height at which this begins to occur (e.g. 2000 Angstroms) is less than the final trench height in this polish experiment. The polish results presented here may indicate that such step "contact heights" may be very large for such large structures.

Contact Wear Model

Since the effective density/planarization length approach was originally formulated consider polishing on the feature scale, an alternative method of analysis may be more suited for approaching the macroscopic wafer-scale polishing problem – that of considering pad/wafer contact mechanics [5,6]. This approach considers the physical interaction of the contact between an elastic pad and the wafer, and forms a relationship between the displacement of the pad and the pressure on the wafer. We implement a contact mechanics model similar to that in [5,6] and apply the model to the trench polish problem resulting in a hypothetical trench removal vs. trench width curve as shown in Figure 3(b). Using a contact mechanics argument, the pressure on the pad at the bottom of a trench is less than the pressure on the raised area, which directly translates to less material removed in the trench center than on the area outside the trench, thus explaining that characteristic of the observed data. The contact mechanics approach can also result in zero down polish for non-zero trench widths (i.e. there may be a width at which the pad does not contact the wafer). This aspect of the data observed in Figure 4 needs further exploration.

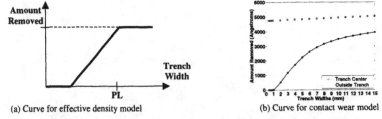

(a) Curve for effective density model   (b) Curve for contact wear model

Figure 3. Hypothetical amount removed vs. trench width plots, for pattern density-based model (a), and contact wear model (b).

CONCLUSIONS

We have examined wafer-scale patterns as an alternative means for characterizing and modeling CMP processes. Polishing results suggest that such wafer scale patterns benefit from very large separations of structures from each other and the wafer edge to avoid interactions, particularly for harder pad or longer planarization length processes. Trench removal vs. trench width plots provide useful insight into the polish process, and can indicate a planarization length parameter. Such plots also reveal dependencies that merit further exploration (the difference between large trench and outside trench polish, the width at which trenches begin to polish) using contact mechanics and other modeling approaches. Future models of CMP pattern evolution may well require integration of both macroscopic consideration and feature scale behavior.

Wafer-scale pattern experiments and modeling may prove particularly useful in the study of "nanotopography" or "nanotopology" related to the nanometer-scale surface variations (occurring over mm length scales across the wafer) that may be present on bare silicon wafers [7]. It has been proposed that natural "random" nanotopography that occurs on an unpatterned raw silicon wafer can be approximated by using a fixed grid of randomly sized cylindrical posts on a wafer-scale pattern [8]. Polishing of such patterned films could lead to insights on how to model the CMP of natural nanotopography on wafers.

ACKNOWLEDGMENTS

The authors acknowledge the MIT Microsystems Technology Laboratories technicians for assistance in some of these experiments. We thank Peter Burke for discussions about his wafer-scale patterns, and Alvaro Maury from Lucent for early discussion of this approach. We also thank Michael Oliver from Rodel Inc. and Dale Hetherington at Sandia National Laboratories for discussions on this topic. This work has been supported in part by a DARPA subcontract with PDF Solutions.

REFERENCES

1. D. Ouma, *et al.*, "An Integrated Characterization and Modeling Methodology for CMP Dielectric Planarization," *International Interconnect Technology Conference*, San Francisco CA, June 1998.
2. T. Smith, *et al.*, "A CMP Model Combining Density and Time Dependencies," *CMP-MIC Conference*, Santa Clara, CA, Feb. 1999.
3. P. Burke, *et al.*, MRS, Oct. 1996.
4. R. Jin, *et al.*, "A Production-Proven Shallow Trench Isolation (STI) Solution Using Novel CMP Concepts," *CMP-MIC Conference*, Santa Clara, CA, Feb. 1999.
5. O.G. Chekina, *et al.*, "Wear-Contact Problems and Modeling of Chemical-Mechanical Polishing," *J. Elec. Soc.*, Vol 145, No. 6. June 1998.
6. T. Yoshida, "Three-Dimensional Chemical-Mechanical Polishing Process Model by BEM," *ECS*, Oct. 1999.
7. K.V. Ravi, "Wafer Flatness Requirements for Future Technologies," Future Fab International, Issue 7, pp. 207.
8. N. Poduje, *et al.*, "Nanotopology Effects in Chemical Mechanical Polishing," *SEMI-AWG Nanotopology Workshop*, Tokyo, Japan, Nov. 1999.

Mat. Res. Soc. Symp. Vol. 612 © 2000 Materials Research Society

Planarization of Copper Damascene Interconnects by Spin-Etch Process: A Chemical Approach

Shyama P. Mukherjee, Joseph A. Levert
Honeywell Electronic Materials, 1349 Moffett Park Dr., Sunnyvale, CA, USA
Donald S. DeBear,
SEZ America Inc, 4824 South 40th St., Phoenix, AZ 85040, USA

ABSTRACT

The present work describes the process principles of "Spin-Etch Planarization" (SEP), an emerging method of planarization of dual damascene copper interconnects. The process involves a uniform removal of copper and the planarization of surface topography of copper interconnects by dispensing abrasive free etchants to a rotating wafer. The primary process parameters comprise of (a) Physics and chemistry of etchants, and (b) Nature of fluid flow on a spinning wafer. It is evident, that unlike conventional chemical-mechanical planarization, which has a large number of variables due to the presence of pads, normal load, and abrasives, SEP has a smaller number of process parameters and most of them are primary in nature. Based on our preliminary works, we have presented the basic technical parameters that contribute to the process and satisfy the basic requirements of planarization such as (a) Uniformity of removal (b) Removal rate (c) Degree of Planarization (d) Selectivity. The anticipated advantages and some inherent limitations are discussed in the context of process principles. We believe that when fully developed, SEP will be a simple, predictable and controllable process.

INTRODUCTION

Copper dual damascene architectures are increasingly being used for interconnects in integrated circuits. The copper is electrodeposited with processes that generate a nonplanar surface [1]. The current method for the planarization of copper dual damascene features and their underlying barrier metals is Chemical-Mechanical Planarization (CMP) process [2]. In CMP processes, the mechanical forces play a dominant role. The present work describes an emerging method of planarization called "Spin-Etch Planarization" (SEP). This is a chemical approach involving no mechanical force. The process involves the uniform removal as well as planarization of copper surface topography by dispensing abrasive free etchants onto a rotating wafer using a commercial spin-etch tool (SEZ 203) [3,4]. In this context, it is worth referring another chemical approach of planarization called Electrochemical Planarization based on the principle of anodic leveling or electropolishing [5]. This process also does not have applied mechanical forces. We believe that the advent of copper damascene interconnects and the polymeric and nanoporous low-k dielectric are the driving force for chemical approaches.

The removal process, of metal as used in SEP, can be described as "controlled" wet chemical dissolution of a metal surface in electrolytes on a rotating wafer. The term "etch" here should be expressed as a controlled wet "chemical polishing" rather than "etching" which does not fulfill the requirement of smooth surface finish. The wet chemical etching of metals in electrolytes is an electrochemical process. A large number of galvanic cells are created on the metal surfaces after immersion in electrolytes. Hence, the surface of a metal may be regarded as a complex multi-electrode system. The metal surface finish after wet chemical removal is controlled by the relative potential difference between the anodic and cathodic regions at a particular time, viscosity and local current density [6]. Another aspect of the chemical polishing is the formation of a viscous layer on the metal surface whose thickness in cavities is greater than on projections and edges and as a result, the dissolution rate of projections is higher and leads to leveling or planarization. The nature and the dimension of the viscous boundary layer developed during polishing play important roles in the SEP process.

OBJECTIVES

We have already reported some preliminary experimental results of SEP of copper interconnect [3,4]. The objectives of this paper are:

1) To identify the key basic technical principles and parameters that play a role in satisfying the primary requirements of planarization such as (a) the uniform removal of copper (b) the removal rate (c) the selectivity of copper removal (d) the degree of planarization, and

2) To evaluate qualitatively the processing advantages and limitations in the context of processing principles and characteristic features of the tool.

SPIN-ETCH PLANARIZATION TOOL

The tool (SEZ Spin Processor 203) is a commercially available single-wafer etch system. Spin etch tools are currently used for silicon wafer backside post-grind etch and for removing metal contamination from the wafers [7]. In this work we will emphasize the key elements of the tool that play important roles in the SEP process. The details have been described elsewhere [3, 4]. The key features of the tool providing simplicity and flexibility are as follows:

1) Capability of spinning 200 mm wafers on a contamination free nitrogen cushion at a wide range of rotation rates.

2) A programmable chemistry dispense arm that can deliver fluids at a selectable flow-rate and over any desired dispense profile sweeping radially across the wafer's surface, allowing for flexibility in controlling removal rate non-uniformity or for correcting for as-deposited copper thickness non-uniformities.

3) Three separate chemistry sources and dedicated process chambers for different etchants used during subsequent phases of the planarization process, and an additional chamber for cleaning/rinsing and drying the wafer.

An additional benefit of the SEP process and the Spin Processing system is its inherent low cost of ownership. The requirement of consumable materials is less; and because the SEZ 203 has integrated wafer cleaning and drying, no additional post-process cleaning equipment are required as part of the process flow. Fast etch rates and in-situ cleaning and drying will yield short process times, maintaining a high wafer throughput. Costs of ownership factors include waste disposal which is also lessened because all effluent is aqueous and contains no suspended particles. Relative ease of neutralization and the potential for recovery and re-use may further reduce costs.

SPIN ETCH PLANARIZATION PROCESS

A proposed flow of process steps for accomplishing copper dual damascene planarization is as follows:

Phase 1. A uniform removal and planarization of excess copper to the barrier layer interface with a single etchant that removes copper selectively without any dissolution of barrier layer such as Ta/TaN. Our present etchant system developed for copper planarization has no reactivity with either Ta or TaN.

Phase 2. The phase 2 involves a selective passivation of copper and a subsequent removal of the barrier layer, with a second step, by a second etchant that has a high selectivity to the barrier material.

EXPERIMENTAL RESULTS

The following typical experimental data is presented to help illustrate the basic principles as well as the connection between these principles and the controlling SEP process parameters. Pattern and blanket electroplated 200 mm copper wafers were used to obtain this data using Phase 1 SEP (described above).

The average blanket copper wafer removal rate was 14,000. Å/min with a 3σ non-uniformity of 9.17%. The copper removal rate was typically a function of the etchant chemistry and the spin speed while the uniformity was primarily a function of the etchant dispenses profile. The surface texture of plated copper before and after SEP was observed using a scanning electron microscopy (SEM). The SEM photomicrographs of copper surfaces before and after polishing are shown in Figure 1. The post-SEP surface was visually shiny and was smoother when measured by stylus profilometry (KLA-Tencor HRP-220). The post-SEP surface was 81. Å. (root mean square roughness) smoother than the as deposited copper. The post-SEP surface roughness was typically a function of the etchant chemistry as well as the spin speed.

We have evaluated the planarization performance by monitoring the change of feature recesses or protrusions near the end of Phase 1. Planarity is illustrated by the Degree of Planarization (DoP) by the following expression:

$$DoP\ (\%) = 100\ (1\text{-}R_f/Ri\ )$$

Where: R_i = initial recess before etching, (Å), and
                 R_f = final recess after SEP (Å).

Where 100% indicates no copper feature recess/dishing after SEP, while 0% indicates that all of

| Copper Surface Before Polish | Copper Surface After Polish |
| 40,000x Magnification | 40,000x Magnification |

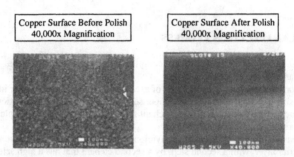

Figure 1. *SEM photomicrographs of Copper surfaces: (left) before Spin-Etch Planarization, (right) after Spin-Etch Planarization.*

the copper was removed from the bottom of the etched feature. The typical planarization for small features was a primarily a function of the etchant selection is given in the table below after 85% of the as deposited copper was removed.

Feature Width, μm	Initial Recess (Protrusion), Å	DoP, %
1.50	2,740.	81.
2.50	3,262.	78.

Erosion, of the dielectric spaces between dense small features, was evaluated using stylus profilometry (KLA-Tencor HRP-220) of a pattern electroplated copper wafer. The copper was removed from the field with a minor over-etch of the copper features (<500 Å recess/dishing) prior to profilometry. The resulting trace in Figure 2 shows no erosion. Erosion was only a function of the etchant chemistry.

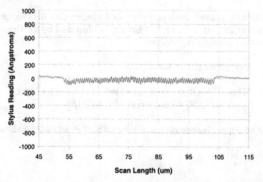

Figure 2. *Profilometry trace showing the absence of dielectric erosion after Spin Etch Planarization of dense featured patterned copper wafer surface.*

DISCUSSION

In the light of initial experimental results and from information from the existing literature, we anticipate that the following two key technical parameters play major roles in controlling SEP.

1) Physics and Chemistry of Etchants:

The composition and chemistry of the etchants determines the oxidizing power and dissolution behavior of the etchant and the selectivity of dissolution to different layers (such as copper layer, barrier layer and dielectric layer). The etchant used for this work was selective to copper only. The elimination of dielectric erosion is important for the selection of etchant for the barrier removal. The viscosity and the chemistry of the etchants controls the thickness and the nature of the diffusion boundary layer produced during the etching process as a function of spin speed.

2) Nature of fluid flow on a spinning wafer with non-planar topography:

This fluid flow phenomenon, with a particular etchant is controlled by the following key process parameters: (a) Spin Speed, (b) Flow Rate, and (c) Etchant Dispense Profile. The selected etchant contributes to the kinematic viscosity and chemical nature of interfacial layer developed during the etching process.

Based on the above information and from our experimental results it is evident that the chemistry and physics of the etchant was the dominant parameter for the copper removal rate, surface finish, and DoP. Spin speed was a secondary parameter, which affected the copper removal rate, surface finish, and DoP. Our experimental results on the surface roughness show that the surface roughness increases with spin speed. This is an indication that the spin speed might affect the viscous boundary layer, which consequently controls roughness and planarization. However, with a particular etchant ,the uniformity of removal is controlled primarily by the dispense profile which influences the distribution of the etchant across the wafer.

We anticipate that a copper diffusion boundary develops during the spin-etch process. The following relationship has been found to hold for the diffusion boundary layer representing mass transport on a rotating disk electrode process of metal removal:

$$\delta_d = D^{1/3} \upsilon^{1/6} \omega^{-1/2}$$

Where: δ_d = Diffusion boundary layer thickness,
D = Diffusion coefficient of Cu^{+2} ions,
υ = Kinematic viscosity,
ω = Rotation rate.

We believe that this relationship may be qualitatively applicable to the boundary layer thickness developed in SEP. The chemical nature and thickness of the diffusion boundary layer play key roles in controlling the SEP.

SEP performance was most affected by three process parameters: Etchant Chemistry, Spin speed, and Dispense profile. These three parameters are analogous to Slurry chemistry,

Platen speed, and Polishing head motion profiles of mechanically based polishing processes. However, mechanically based polishing is also affected by additional critical process parameters such as Normal load, and Pad material properties which consequently have complex secondary variables such as contact pressure, and pad conditioning. Although the SEP process has not yet been developed to the maturity of conventional planarization processes , SEP does have an advantage of a smaller process parameter set requiring control. The small SEP process parameter set is governed by two basic technical factors: the chemistry of etchants, and nature of fluid flow on a patterned wafer, which in turn are governed by well-known principles of chemistry and physics.

The absence of mechanical contact as well as the selectivity of abrasive free etchants allows SEP to eliminate the problems of dielectric erosion and surface scratching and abrasive particle contamination. The elimination erosion is an important requirement for the electrical performance of future high-density devices having polymeric or porous low-k dielectrics.

CONCLUSIONS

The SEP process has an intrinsic ability to impart high surface finish (Figure 1) and planarize copper surfaces. SEP is a controlled wet chemical removal process of copper involving an electrochemical dissolution process. Dissolution reactions are modified by the diffusion of chemically reactive species within a diffusion boundary layer modified by etchant flow on the rotating wafer. The relatively small number of SEP process parameters has the prospect of yielding a simpler process based on previously codified principles of chemistry and physics.

ACKNOWLEDGMENTS

The authors want to thank Lynn Forester and Michael Thomas of Honeywell Electronic Materials STAR Center and Michael West of SEZ for their support.

REFERENCES

1. P. C. Andricacos, C. Uzoh, J. O. Dukovic, J. Horkins, and H. Deligianni, IBM J. Res. Development 42 (5) 567-574 (1998).
2. P. Wrschka, J. Hernandez, G. S. Oehrlein and J. King, J. Electrochemical Soc. 147 (2) 206-712 (2000).
3. J. A. Levert, S. P. Mukherjee, D. S. DeBear, Semicon Japan 99: SEMI Technology Symposium, 9 4-73 to 4-81 (1999).
4. D. S. DeBear, J. A. Levert, S. P. Mukherjee, Solid State Technology, March, 53-58.(2000).
5. R. J. Contolini, S. T. Mayer, R. T. Graf, L. Tarte, A. F. Bernhardt, Solid State Technology, June 155-158 (1997).
6. J. W. Bloor, J. Metals of Australia 4, 276-282 (1972).
7. P. S. Lysaght, and M. West, Solid State Technology, November 63-70 (1999).

AUTHOR INDEX

SUBJECT INDEX